C0-ATX-001

ALLE ZEIT WACH
1842

Gerhard Eisenbeis · Wilfried Wichard

Atlas on the Biology of Soil Arthropods

With a Foreword by Friedrich Schaller

With 1133 Scanning Electron Microscope
Photographs in 192 Plates
and 219 Figures in the Text

Springer-Verlag
Berlin Heidelberg New York
London Paris Tokyo

Dr. GERHARD EISENBEIS
Department of Biology
Institute of Zoology
Saarstraße 21
D-6500 Mainz, FRG

Dr. WILFRIED WICHARD
Siebengebirgsstraße 221
D-5300 Bonn 3, FRG

Translator:
ELIZABETH A. MOLE
Nördliche Hauptstraße 66
D-6940 Weinheim, FRG

Original edition:
Eisenbeis/Wichard, Atlas zur Biologie der Bodenarthropoden

ISBN 3-540-17764-7 Springer-Verlag Berlin Heidelberg New York
ISBN 0-387-17764-7 Springer-Verlag New York Berlin Heidelberg

Library of Congress Cataloging-in-Publication Data. Eisenbeis, Gerhard, 1943– Atlas on the biology of soil arthropods. Translation of: Atlas zur Biologie der Bodenarthropoden. Bibliography: p. 411 Includes index. 1. Soil fauna. 2. Arthropoda. I. Wichard, Wilfried, 1944– . II. Title. QL110.E3413 1987 595'.2'09148 87-4948

Printed in Germany

Typesetting: K+V Fotosatz GmbH, Beerfelden
Offsetprinting: Beltz, Offsetdruck, Hemsbach/Bergstr.
Bookbinding: J. Schäffer OHG, Grünstadt
2131/3130-543210

Foreword

Towards the end of 1941 my studies of zoology in Vienna had reached a point at which I was able to look for a subject for my doctorate. As every "proper" zoologist is primarily a visual person, his choice of research object is often also motivated by aesthetics. Thus, I approached Prof. Kühnelt with the wish to carry out research on dragonflies, as they are strikingly beautiful and remarkably mobile. But as a simple student I had not considered that such a young lecturer as Kühnelt naturally also harboured his own ideas about his desired future research project. At that time he and the soil scientist W. Kubiena had already begun to study soil biology, the zoological aspects of which he was the first to present so vividly several years later (W. Kühnelt, *Bodenbiologie mit besonderer Berücksichtigung der Tierwelt.* Herold, Wien 1951). To cut a long story short, our conversation was brief and the outcome was that I undertook a thesis on soil biology, namely the Collembola fauna of the soil in the Vienna Woods.

Naturally, even in those days a student at my stage had to know that so-called soil animals, and among them Collembola (Springtails), exist and I had certainly already been shown some of them, but I had hardly really looked at them. After all, apart from the mole and the earthworm, they are small and insignificant.

In the subsequent 40 years I was able to revise this prejudiced opinion thoroughly and permanently. Anyone who has glimpsed a sample of soil animals, especially the soil arthropods, through the binocular microscope will never forget the fascination of their abundant shapes and structures. And this abundance of forms immediately provokes innumerable questions:

What are these animals under our feet up to? How do they live? What do they need to live? And what significance do they have for the soil, for the plants and thus, in the long run for us?

In "my" day the only way to view this "underworld" was through the binocular and the light microscope. In the meantime, electron microscopy has added another dimension which multiplies the initial fascination. When I leaf through this "atlas" my attention is repeatedly caught by the numerous pictures. If something like this had existed in those days I would certainly not have enquired about dragonflies first.

In the meantime the richness of form and function of the soil arthropods has proved to be an inexhaustible source of biological interest. The authors of this book demonstrate this in a praiseworthy manner. Specialists will certainly miss certain details, but those who seek a first insight, an overview and a stimulating selection will be rewarded. Even the specialist will not be disappointed. The authors have included the latest literature and they also characterize specific physiological capabilities of the soil animals using selected examples.

But, as I have already mentioned, this atlas will be valued especially for its visual qualities, and not only by soil biologists. It is to be hoped that it will also extend general ecological and biological knowledge and thus increase environmental consciousness of the biotope "soil". Our soils and their biotic communities are and will certainly remain the essential biologically productive basis of our present and future existence. Therefore soil biology, and particularly the "biology of the soil arthropods", are especially relevant research and teaching subjects.

If soil biology has not always taken its rightful place in general ecology up to now, this was largely due to the fact that an adequate overview of its abundance of form and function was almost impossible to obtain within a reasonable period of time. This atlas offers every teacher the opportunity of rapidly obtaining an overview and thus of compiling the most vivid teaching material. As the authors have made every effort to give a short characterization of the fundamentals and concepts of soil science as well as a brief morphological and biological description of each animal group, the non-zoologist will also benefit from looking at this atlas. Finally, the index offers easy access to each individual aspect of the subject to those who are motivated to study further.

I am very pleased that this book originated in the "Mainz School of Zoology" where I was also able to make my contribution to extending our knowledge of the "underworld of the animal kingdom."

Prof. Dr. FRIEDRICH SCHALLER
Vienna, September 1984

Preface

This book was inspired by the fascinating variety of soil arthropods with their diverse forms of adaptation to the differentiated life in the soil. Underlying this diversity of form, which we have presented in the light of the electron microscope by means of selected examples rather than exhaustively, are physiological mechanisms for the ecological adaptation of the animals. Therefore, we have tried, wherever possible, to fit the functional-morphological discussion of the diversity of forms as expressed in this atlas into the wider fields of ecology and soil biology like pieces of a mosaic.

We were also inspired to write this book by our concern for the biotope which we are constantly treading on. It is not only intended for scientists and students of biology, ecology, forestry and agriculture, but should also interest naturalists and environmentally conscious readers and show them the bizarre soil arthropods which, as consumers, participate in the decomposition of plant litter.

It would have been impossible to produce this book in its present form without the numerous scientific research publications on soil arthropods which have, fortunately, appeared with increasing frequency in the last decade and which have made important new contributions to our knowledge of their biology, morphology, physiology and ecology. We hope that we are not only presenting an atlas on the biology of soil arthropods, but that we will also stimulate further research in this field by demonstrating numerous unknown structures.

During our work many of our colleagues contributed to the success of this project by providing us with valuable advice and practical help. We would especially like to thank Dr. B. Baehr (München), Prof. Dr. L. Beck (Karlsruhe), Prof. Dr. J. Bitsch (Toulouse), Prof. Dr. A. Brauns (Braunschweig), Mr. W. Brück (Mainz), Prof. Dr. N. Caspers (Köln), Mr. H. Diehlmann (Karlsruhe), Dr. M. Geisthardt (Wiesbaden), Dr. H. Günther (Ingelheim), Dr. U. Hoheisel (Mainz), Dr. K. Honomichl (Mainz), Prof. Dr. C. Jura (Krakow), Prof. Dr. O. Larink (Braunschweig), Prof. Dr. J. Martens (Mainz), Dr. E. Meyer (Innsbruck), Dr. H.-W. Mittmann (Karlsruhe), Dr. K. Renner (Bielefeld), Mr. O. Rehage (Münster, Heiliges Meer), Prof. Dr. R. Rupprecht (Mainz), Mr. R. Schreiber (Mainz), Dr. H. Späh (Bielefeld), Prof. Dr. W. Topp (Köln), Mr. W. Verhaagh (Karlsruhe). Mrs. K. Rehbinder and Mrs. M. Ullmann kindly helped us with the drawings and the photographs. Our very special thanks are due to Prof. em. Dr. H. Risler for his continual and benevolent promotion of our research in Mainz.

October 1984 GERHARD EISENBEIS · WILFRIED WICHARD

Contents

1 General Introduction

2 Systematic Chapters

1 General Introduction

The soil arthropods illustrated in this atlas represent a large number of soil-inhabiting arthropods of the classes: Crustacea, Arachnida, Myriapoda and Insecta. Their biotope is the soil, formed from the Earth's outer rock layer by weathering and the action of organisms. According to the degree of weathering and activity of the organisms, differently structured horizons are formed which are visible in the soil profile. The effective physical, chemical, climatic and biological factors are, at the same time, the abiotic and biotic, ecological factors which influence the soil arthropods. Certain forms of life have developed among the soil arthropods as adaptations to these ecological factors and to life in the soil, these being euedaphic, epedaphic and hemiedaphic life forms. Thus, the soil arthropods with their specific life forms take part in the biocenosis of the entire soil organisms, known as the edaphon. The position of the soil arthropods within the biocenosis corresponds to the trophic level in the food chain; as consumers they lie between the producers and the reducers, and participate in the consumer food web by degrading plant debris.

This short overview and the following summarized introduction (Part 1: General Introduction) serve as a guide to the ecological characterization of the soil arthropods presented in this atlas (Part 2: Systematic Chapters). Soil biology and the biology of the soil arthropods has been clearly presented by Kühnelt (1950, 1961), Kevan (1962), Schaller (1962), Palissa (1964), Ghilarov (1964), Müller (1965), Burges and Raw (1967), Brauns (1968), Trolldenier (1971), Wallwork (1970, 1976), Dunger (1983), Brown (1978) and Topp (1981).

1.1 The Soil as a Biotope

1.1.1 Soil Profiles

A vertical section of a soil profile exhibits several horizons which reflect the successive steps during weathering of the original rock and biological decomposition of the deposited vegetation in the upper organic layer. Although soils are differently structured according to their types, the layer between the original rock and the organic litter can be divided into four successive horizons (Kubiena, 1953; Scheffer and Schachtschabel, 1979):

O-horizon: organic upper layer of plant debris on the mineral soil.
A-horizon: upper, fine mineral soil interspersed by permeated organic material.
B-horizon: weathered, rough mineral soil coloured brown by small deposits of humus.
C-horizon: original, unweathered rock under the soil.

The vast majority of soil arthropods live in the organic upper layer (O-horizon). The thickness of this layer depends on the amount of vegetation deposited annually and the intensity of its degradation by soil organisms. As decomposition in woodland soil progresses over a period of several years, layers of variable thickness depending on their year of origin and their degree of decomposition are formed within the organic upper layer. These layers correspond to the subhorizons of the O-horizon (Scheffer and Schachtschabel, 1979):

O_l: litter layer; not yet decomposed deciduous and coniferous plant debris.
O_f: fermentation layer; partly decomposed deciduous and coniferous plant debris with macroscopic and microscopic vegetation fragments.
O_h: humus layer; lacking plant residues recognizable in the light microscope.
The O-horizon merges into:
A_h: a humus-enriched, dark upper mineral layer.

According to Zachariae (1965), the litter of the beech woodland soil, which serves as an example of a biotope for soil arthropods here (Fig. 1), extends through the L-layer (O_l), the F-layer (O_f) and the mineralized H-layer, the latter being called the A_h layer (Beck and Mittmann, 1982) as the O_h layer in this woodland soil is already mixed with the upper stratum of mineral soil.

1.1.2 Ecological Factors

The ecological factors affecting the soil, to which the soil arthropods mainly adapt, include the abiotic factors: pore volume, soil moisture, soil ventilation and soil temperature.

Like a labyrinth, the soil is riddled with pores („Porosphäre“; Vannier, 1983). In the loose litter layer (L-layer) of a beech woodland soil there are large hollow spaces between the fallen leaves. As decomposition of plant debris proceeds, the pore diameter decreases by about 3 – 0.3 mm from the F-layer downwards, thus providing a suitable biotope for the euedaphic mesofauna. In the A-horizon the size of the pores is primarily dependent on the size of the grains of the weathered mineral soil, but the hollow spaces are mostly filled by permeated humus material so that only small pore volumes remain unoccupied in the A_h-layer.

The size of the pores corresponds to the size of the soil arthropods of the mesofauna and macrofauna (van der Drift, 1951 and Dunger, 1983; Fig. 2). Starting with the F-layer, the pore system of deeper soil layers is colonized by air-breathing, photophobic, euedaphic mesofauna, while the epedaphic macrofauna live on the surface of the soil or in the hollow spaces of the loose litter. Also included in the macrofauna are those soil arthropods which, independent of the pore system of the soil, burrow deep into the soil and live partly hemiedaphically in their self-made tunnels.

The pore system of the soil is often partly filled with water which is either precipitated as rain or rises as capillary groundwater. The remainder of the pore system contains water vapour saturated air, the relative humidity of which decreases towards the layer of air just above the soil. There is a steep gradient in the relative moisture especially at the transition zone between the F- and L-subhorizons, if the air layer just above the soil has

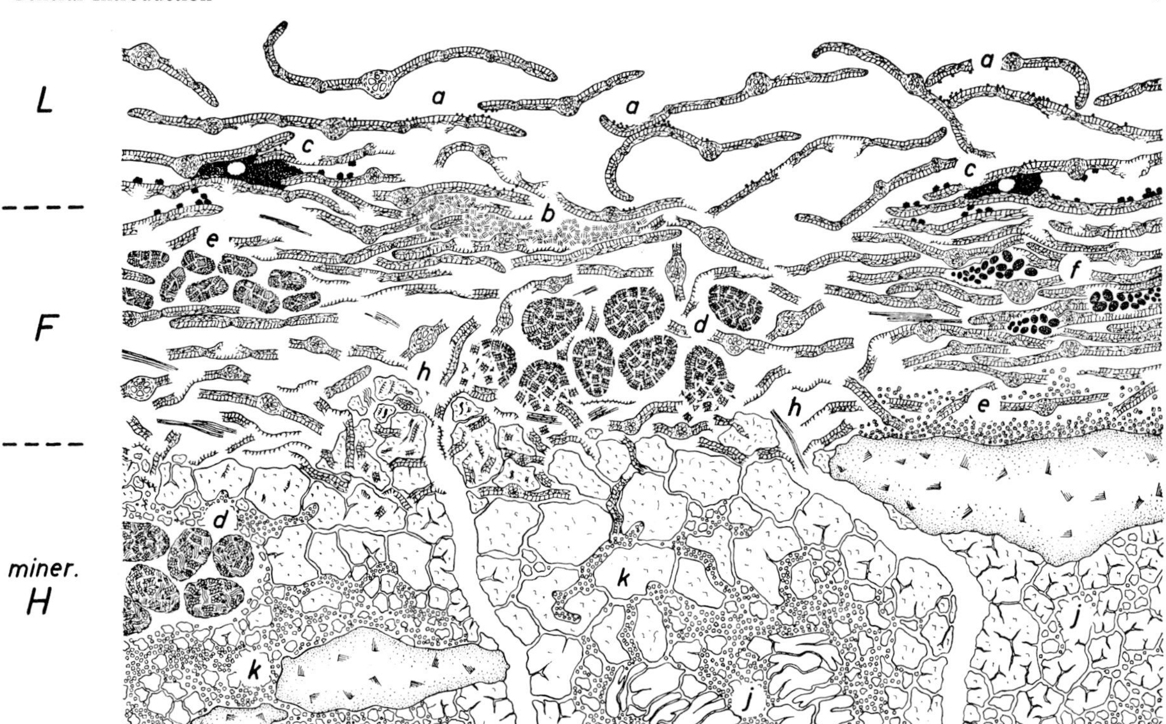

Fig. 1. Degradation of litter in woodland soil by animal decomposer groups. *L*, *F*, *miner. H* denote the subhorizons: litter layer (*L*), fermentation layer (*F*) and the partly mineralized humus horizon (*miner. H*) (Zachariae, 1965). **a** L-leaves with droppings of microphytophagous and macrophytophagous Collembola and Oribatidae. **b** Faeces of small, macrophytophagous and saprophytophagous (L/F-leaves) Diptera larvae. **c** Droppings and excrement tunnels of microphytophagous and saprophytophagous *Dendrobaena* and Enchytraeidae (Oligochaeta). **d** (*Centre*) Faecal pellets of saprophytophagous (F-leaves) Diplopoda. **d** (*Left*) Droppings of saprophytophagous (leaf remains of F-layer) Tipulidae larvae deposited under the surface of the mineralized humus layer. **e** (*Left*) Droppings of coprophagous (arthropod faeces) and saprophytophagous (leaf remains of F-layer) *Dendrobaena* (Oligochaeta). **e** (*Right*) Droppings of coprophagous (arthropod faeces) Enchytraeidae, concentrated on top of a stone. **f** Pile of leaves with faeces of macrophytophagous Phthiracaridae (Oribatei). **h** Entrances to tunnels of *Lumbricus terrestris*; fresh earthworm excrement lies at the left entrance. **j** (*Centre*) Loose mass of excrement of *Allolobophora* (Oligochaeta). **j** (*Right*) Compressed mass of excrement of *Allolobophora*, concentrated and with desiccation fissures. **k** (*Centre*) Concentrated earthworm droppings with tunnels bored by coprophagous Enchytraeidae. **k** (*Left*) Concentrated droppings of Enchytraeidae

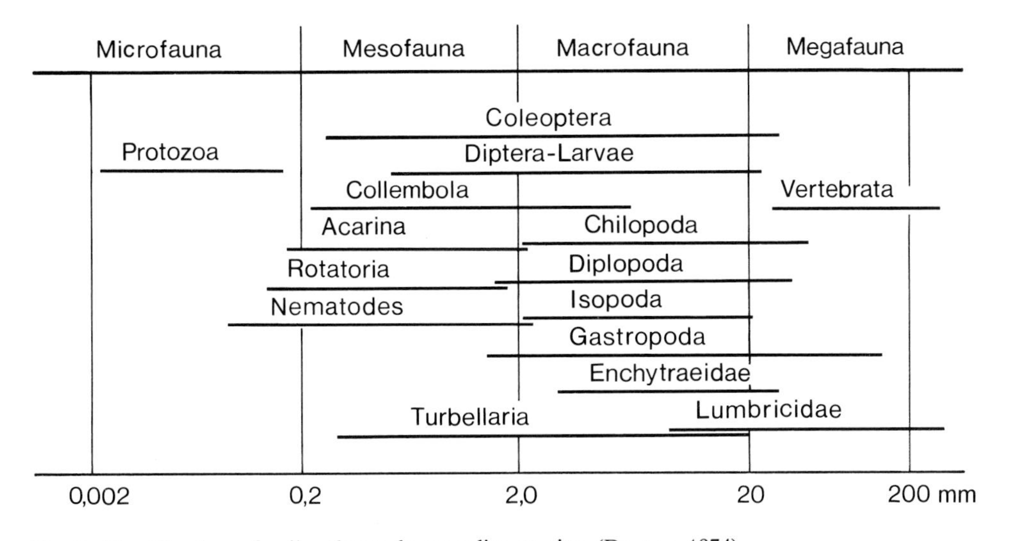

Fig. 2. Classification of soil arthropods according to size. (Dunger, 1974)

Table 1. Average rates of transpiration (in percent) measured as the change in water mass (Δm %/h) at various relative humidities (% r. h. at 22 °C) for the slow, linear component of transpiration (basic, cuticular component) and average surface constants (k_s) to calculate the mean, macroscopic body surface area of various soil arthropods[a]

	% r. h.					k_s	Moisture type[b]
	100	98	76	33	0		
	Δm %/h						
Pseudoscorpiones							
Neobisiidae							
Neobisium sp.	–	0.10	0.40	0.69	0.82	10.11	m
Isopoda							
Ligiidae							
Ligidium hypnorum	–	4.7	22.5	26.0	34.5	12.97* (15.49)**	h
Trichoniscidae							
Trichoniscus pusillus	–	10.0	22.6	50.5	74.1	10.66*	eh
Oniscidae							
Oniscus asellus (without oostegites)	–	1.1	6.0	10.0	13.7	12.52* (15.39)**	h
Oniscus asellus (with oostegites)	–	0.8	4.7	7.7	9.9	–	h/m
Porcellionidae							
Cylisticus convexus	–	1.2	5.4	10.4	15.5	–	h
Porcellio scaber	–	0.5	1.6	3.2	4.6	–	h/m
Chilopoda							
Lithobiidae							
Lithobius sp.	–	2.2	10.9	21.9	28.1	11.85	h
Diplopoda							
Glomeridae							
Glomeris marginata (rolled up)	–	0.13	0.18	0.34	0.42	3.75	m
Glomeris marginata (unrolled)	–	0.69	2.43	–	–	11.18	
Polyxenidae							
Polyxenus lagurus	–	0.08	0.28	0.36	0.61	12.81	m
Symphyla							
Scutigerellidae							
Scutigerella sp.	1.7	5.1	19.2	38.3	78.4	12.49	eh
Diplura							
Campodeidae							
Campodea sp.	2.4	5.3	23.0	44.3	77.4	11.44	eh
Collembola							
Onychiuridae							
Onychiurus sp.	7.2	17.2	94.5	253	325	8.84	eh
Tetrodontophora bielanensis	1.1	2.5	9.6	21.8	31.9	8.78	h
Entomobryidae							
Orchesella villosa	0.27	0.6	6.0	14.5	24.9	10.70	h
Tomoceridae							
Tomocerus flavescens	0.28	1.6	9.0	16.8	33.9	10.53	h
Sminthuridae							
Sminthurides aquaticus	–	22.1	93.7	171	276	8.24	eh
Allacma fusca	–	0.46	0.8	1.4	2.1	7.46	h/m
Archaeognatha							
Machilidae							
Trigoniophthalmus alternatus	–	0.15	0.5	0.8	1.0	11.13	m
Zygentoma							
Lepismatidae							
Lepisma saccharina	–	0.1	0.16	0.27	0.37	12.85	m

[a] The rates Δm %/h are a measure of the change in basic water mass m_0 of normally hydrated animals. k_s serves to determine the mean, macroscopic body surface area S from Meeh's formula: $S = k_s \cdot w_0^{2/3}$ when the body mass of normal hydrated animals (w_0) is known

[b] Classification of moisture types (Eisenbeis, 1983 b) based on transpiration rates in 0% r. h./22 °C as Δm %/h (m = mass of water; h = hours):

a high saturation deficit because of aridity and high temperature. The soil arthropods are adapted to soil moisture (Edney, 1977; Eisenbeis, 1983); their transpiration rate at various relative air humidities serves as an indirect measure of their moisture requirement (Table 1). Euedaphic soil arthropods, e.g. the highly adapted *Onychiurus* species (Collenbola), have a considerable transpiration rate even in air saturated with water vapour so that they are dependent on a constant supply of water.

When the soil becomes desiccated, especially during dry summer months, vertical and horizontal migrations take place to satisfy moisture preferences. Euedaphic soil arthropods of the mesofauna follow the pore system vertically downwards and remain temporarily in deeper layers. Similarly the hemiedaphic soil arthropods of the macrofauna burrow into the soil to reach tolerable or optimal moisture conditions. Epedaphic soil arthropods of the macrofauna, which cannot penetrate the soil, wander horizontally to fulfil their moisture requirements in more suitable habitats.

If the pore system becomes waterlogged due to rainfall, rising groundwater or flooding in marshy on fen woodland, the soil becomes an "aquatic system" (Vannier, 1983) and most of the epedaphic and hemiedaphic soil arthropods seek sheltered habitats in the horizontal direction. For the euedaphic soil arthropods, with their restricted mobility within the pore system, movement to another habitat is hardly possible. Instead they are often adapted to flood conditions. The bodies of many euedaphic arthropods are covered by hydrophobic cuticular structures like trichomic hairs, which trap a layer of air, known as the plastron, between the body surface and the surrounding water. It has two physiological effects.

1. The plastron prevents the osmotic influx of water through permeable surfaces into the body. Unlike aquatic insects, which have adapted by developing effective osmoregulation and renal excretion for water which has entered the body by osmosis, soil arthropods do not normally exhibit a highly developed osmotic regulation. Influx of water into their bodies would cause the animals to swell, leading to their death (Potts and Parry, 1964).

2. The plastron allows air-breathing arthropods to breathe underwater. The boundary between the plastron and the surrounding water functions as a respiratory surface through which oxygen can diffuse from the water towards the O_2-consuming body. As in the case of some air-breathing aquatic insects with gills, euedaphic arthropods are able to extract a continuous supply of oxygen from water and, for a certain time, obtain more oxygen than was originally present in the plastron. There is a danger of oxygen deficiency during long-lasting flooding, however, as consumption by soil bacteria reduces the oxygen content of the pore water (Rahn and Paganelli, 1968).

The composition of the air in the loose litter layer hardly differs from that of the atmosphere, but the lower pore spaces frequently contain a higher concentration of carbon dioxide than atmospheric air. As the carbon dioxide content increases there is an antagonistic reduction of the oxygen concentration (Verdier, 1975; Vannier, 1983). This is due to the metabolic activity of plant roots and soil organisms which actively participate in the decomposition of plant debris. For the euedaphic soil arthropods the concentration changes of CO_2 and O_2 in the air they breathe necessitate a higher resistance to CO_2 and a differentiated adaptation to a low O_2 partial pressure. According to Zinkler (1966, 1983) small euedaphic Collembola are not particularly sensitive to oxygen deficiency. At 3% oxygen concentration in its respiratory air, *Onychiurus armatus* has an average oxygen intake of 93% of its normal consumption, and also has a distinctly higher CO_2 resistance than epedaphic Collembola.

---

eh: extremely hygric $>50\%/h$
h: hygric $>10\%/h$ $<50\%/h$
h/m: hygric/mesic (transition form) $>1\%/h$ $<10\%/h$
m: mesic $>0{,}1\%/h$ $<1\%/h$
x: xeric $<0{,}1\%/h$
* without pleopods (pleopods are integrated into the ventral side)
** with pleopods

Soil temperature influences the poikilothermic arthropods, as their temperature does not generally differ from the ambient temperature. Many soil arthropods react to seasonal temperature changes with a diapause and vertical migrations towards their temperature optimum which can lie between 5°–10 °C for animals active in winter and between 10°–18 °C for those active in summer. The respective vertical migrations thus occur in opposite directions. Heating effects can occur in the litter layer within short intervals. Thus, a temperature of 15 °C measured near the surface of shaded beech woodland litter rose to about 30 °C within a few minutes as a result of direct insolation. The magnitude of the resulting temperature differences and the consequent vertical migrations depend on the type and depth of soil. Temperature amplitudes in dry, south-facing, sandy soils are greater than in damp, overgrown, clay soils. As soil depth increases, temperature variation rapidly decreases so that euedaphic soil arthropods in moist, overgrown clay soils are independent of variations in the surface temperature and avoid vertical migrations. Epedaphic soil arthropods are more easily exposed to temperature variation and retreat into the litter or burrow into the soil to survive the cold winter period as hemiedaphic animals. In contrast, extremely cold stenothermic soil arthropods often reach their temperature and activity optimum at low temperatures of −5° to +5 °C like some species of the genus *Isotoma* in the Collembola or the mecopteran *Boreus* and the dipteran *Chionea* which live on perpetual snow and ice. Altogether tolerance of low temperatures is an important physiological adaptation (Joosse, 1983; Block, 1982; Sømme, 1982; Vannier, 1983).

1.2 Forms of Life in the Soil

The distribution of animals in the soil is correlated to the ecological factors affecting it. As the soil is not homogeneous, but stratified, the collective ecological factors influence soil fauna vertically in the successive horizons. Soil arthropods can be grouped into different life forms which have adapted to the combination of these prevailing ecological factors.

Gisin (1943) was first to clearly recognize different life forms in the edaphon and to differentiate between euedaphon, hemiedaphon and atmobios. The inhabitants of the lowest soil layers belong to the euedaphon. The hemiedaphon comprises the inhabitants of the upper soil and litter layers, while the fauna of the soil surface, including the upper strata (herbaceous, shrub and tree layers), are part of the atmobios. Although it has been modified several times, Gisin's classification of life forms (1943) has been generally accepted. However, Gisin's (1943) priority was to arrange the Collembola into different life forms. Therefore, inconsistencies often become apparent when this scheme of classification is extended to include the whole edaphon. His definitions are, in many cases, too narrow to be generally applicable. For instance, they do not take many arthropods with their developmental stages into account. As Dunger (1974) justifiably remarked, there are difficulties in integrating the micro-, macro- and megafauna into this system.

In order to overcome these difficulties we propose a modified system which contributes, at least in this book, to a clearer classification of all soil arthropods. In accordance with the intentions of this atlas and the definition of life forms, the prime criteria in the differentiation of life forms are the mechanisms of ecological adaptation. Only in the light of these ecological criteria does the vertical distribution of life-form types become understandable.

Three general life forms are to be found in the biotope of the edaphon: euedaphon, epedaphon, hemiedaphon.

1.2.1 Euedaphon

The pore system and its effective ecological factors give the euedaphic life form its particular character. The biotope of the euedaphon is mostly confined to the natural pore system, but insect larvae extend this living space by digging and burrowing as they increase in size. Euedaphic soil arthropods are small and distinguished by their round or worm-like body form. Their diameter corresponds to that of the pore system. With the limitations imposed by this way of life the extremities are frequently reduced. They can hardly escape predators, instead many euedaphic animals possess defensive or toxic glands. Their sensory perception has also undergone adaptation to subterranean life. The animals, most of which are photophobic and lack pigmentation, either do not possess eyes or have eyes which have degenerated according to the degree of their adaptation to euedaphic life. To the same degree mechano- and chemosensitive organs have developed to compensate for the absence of functional eyes. Their heightened resistance to CO_2, their restriction to moist habitats due to their high transpiration rates (hygrophilic animals) as well as their ability to form a plastron as a respiratory and osmoregulative protection when the pore system is flooded all belong to the adaptations evolved by this group.

1.2.2 Epedaphon

Soil arthropods living on the surface and in the litter layer have an epedaphic life form. They are not adapted to the combined factors affecting the pore system. Epedaphic soil arthropods are often restricted to the soil surface and larger hollow cavities in the litter (L-layer) because of their size. They exhibit great diversity of body forms, are strongly pigmented and often dorso-ventrally flattened with well-developed extremities. Eyes and sensory organs, often concentrated on long filamentous antennae, enable the animals to perceive their environment. Further characteristic features of the epedaphic soil arthropods are their great mobility and their diurnal activity rhythms.

The epedaphon colonize the soil surface and the shaded litter layer, while the euedaphon inhabit the dark pore system in layers beneath the L-layer. Euedaphon and epedaphon complement each other in the soil ecosystem and are mutually exclusive because the mechanisms of ecological adaptation are not interchangeable, though the life forms of certain soil arthropods show a fluent transition between the two.

1.2.3 Hemiedaphon

The hemiedaphon does not occupy an intermediate position between the euedaphon and epedaphon. It is usually a temporary form of life, adopted by epedaphic and atmobiotic arthropods, enabling them to occupy existing or more often self-made burrows in the soil. Ecological adaptations to this life form include the ability to excavate using mouthparts or fossorial legs, or locomotory power with which soil arthropods can enlarge existing cracks and pores to burrow deeper into the soil.

Epedaphic soil arthropods and atmobiotic arthropods from higher strata (herbaceous, shrub and tree layers) adopt the hemiedaphic life form for various reasons. Some soil arthropods dig channels and craters, then lie in wait for prey which lives on the surface [tube-web constructing spiders, larvae of tiger beetles (Cicindelidae) and larvae of the ant lion (Myrmeleonidae)] or burrow through the soil hunting small, epedaphic arthropods. Others enter the soil temporarily to escape unfavourable ecological factors, e.g. to avoid desiccation and cold. Many arthropods regularly overwinter in self-made or pre-existing burrows, the rest of the year they live on the surface or as part of the atmobios. Furthermore, reproductive cycles and care of their young bind many arthropods temporarily to the soil. They deposit their eggs in the soil and build simple or complex subterranean nests and brood chambers (e.g. several Hymenoptera or Coleoptera). After hatching larvae burrow in the soil as they increase in size while passing through the various larval stages. In the course of their development they undergo fluent transitions between the hemiedaphic, epedaphic and euedaphic forms of life, especially those which grow in a brood chamber and are not fully adapted to edaphic conditions.

1.3 Soil Communities

1.3.1 Decomposition of Plant Debris

Soil arthropods are part of the biocenosis of the edaphon, which comprises the flora and fauna of the soil. They are consumers and as such participate in the decomposition of plant debris as primary and secondary decomposers.

Decomposition of the plant debris, documented in Figs. 1 and 3–23, progresses steadily in the upper organic layer of the soil (O-horizon; L-layer, F-layer and H- or A_h-layer resp.) and continues for several years in a beech woodland soil. After leaf-fall in autumn and if weather conditions are damp, the process of decay due to enzymatic activity of microorganisms begins in the lower stratum of the litter (L-layer). At the same time earthworms, snails and soil arthropods begin to feed on the leaves. Phytophagous animals do not eat indiscriminately, they prefer soft leaves with a high water content (elder, ash, alder and lime) or hard leaves, several years old, which have been thoroughly soaked and have already undergone initial microbial degradation. Under these saprophytophagous conditions even oak leaves, which are least palatable to herbivores when fresh, become acceptable to them. First the leaves are nibbled, then perforated and they are gradually broken down until only their vascular bundles remain. This decomposition by microbial and animal primary decomposers provides the nutritional basis for secondary decomposers. The decomposition process is accelerated in the F-layer where the mechanically fragmented parts of plants and excrement pellets of primary decomposers are colonized by microorganisms and ingested again by the secondary decomposers. These are the saprophytophages, coprophages and microphytophages in the decomposer food web. Moreover, zoophagous and necrophagous consumers also participate indirectly in decomposition. These secondary decomposition processes are repeated till vegetation in the H- or A_h-layer is no longer recognizable as such in the light microscope (Zachariae, 1965; Brauns, 1968; Funke, 1971; Dunger, 1958, 1974; Dickinson and Pugh, 1974; Herlitzius and Herlitzius, 1977; Herlitzius, 1982).

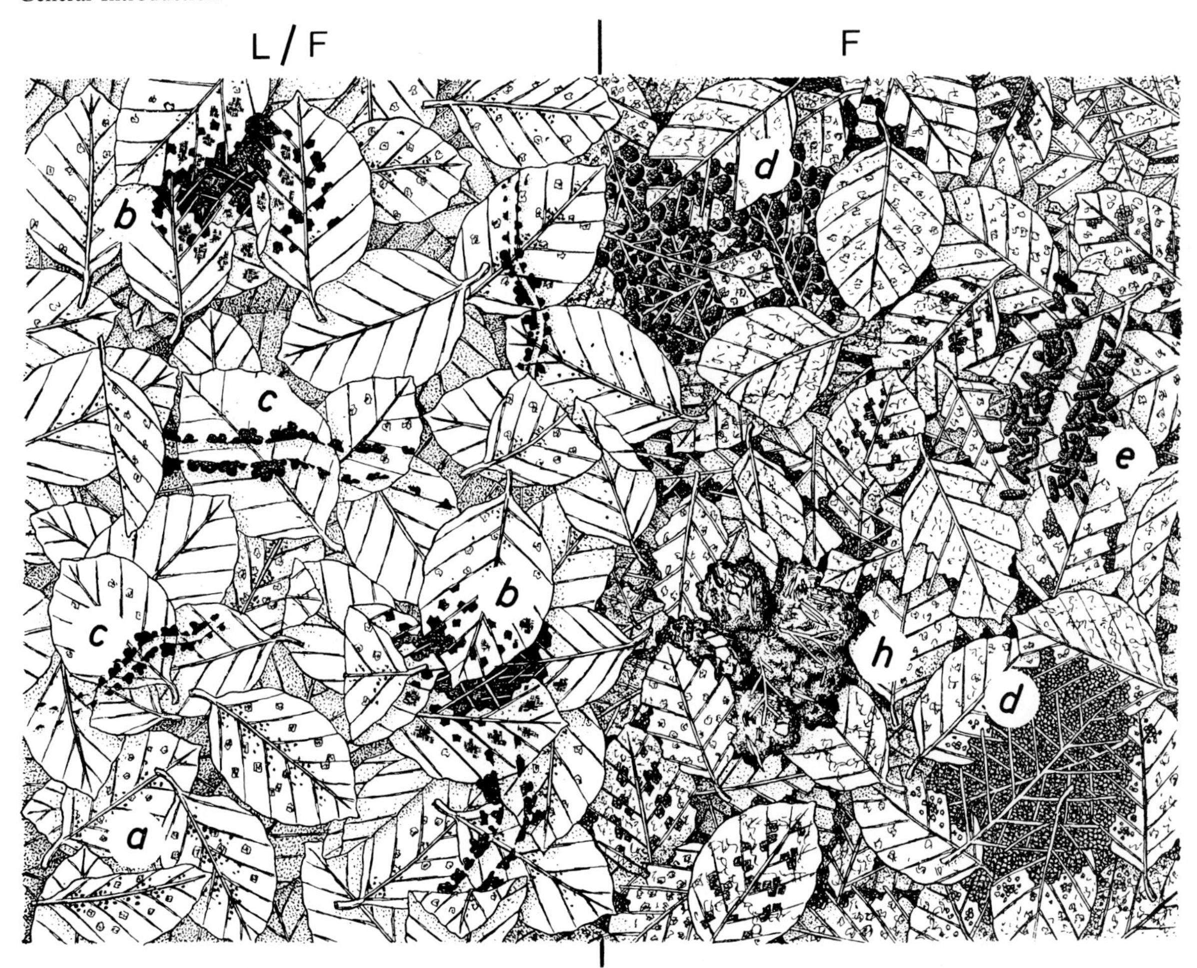

Fig. 3. Combined horizontal and vertical mosaic of the decomposition of leaf litter. The *left* half of the picture shows the transition zone between the L- and F-layers (the upper layer of litter has been removed); on the *right* sections of the F-layer are shown (Zachariae, 1965).

L/F Transition zone. **a** Traces of feeding (beginning of perforation and skeletonization) and excrement of macro- and microphytophagous Collembola and Oribatidae. **b** Excrement and traces of feeding left by small macro- and saprophytophagous Diptera larvae. **c** Faecal tracks and opened excrement tunnels of *Dendrobaena* (*larger*) and Enchytraeidae (*smaller*) after micro- and saprophytophagous feeding activities.

F-Horizon. **d** (*Top*) Traces of feeding with faecal pellets of larger saprophytophagous Diplopoda. **d** (*Bottom*) Traces of feeding left by saprophytophagous Bibionidae larvae (Diptera). **e** Droppings of *Dendrobaena* (Oligochaeta) after coprophagous feeding on arthropod faeces. **h** Fresh excrement of *Lumbricus terrestris* which extends from the H-layer into the F-layer.

The remaining small traces of feeding and faecal pellets can be attributed to the feeding activities of micro- and saprophytophagous Oribatidae, Collembola and other arthropods of the meso- or macrofauna

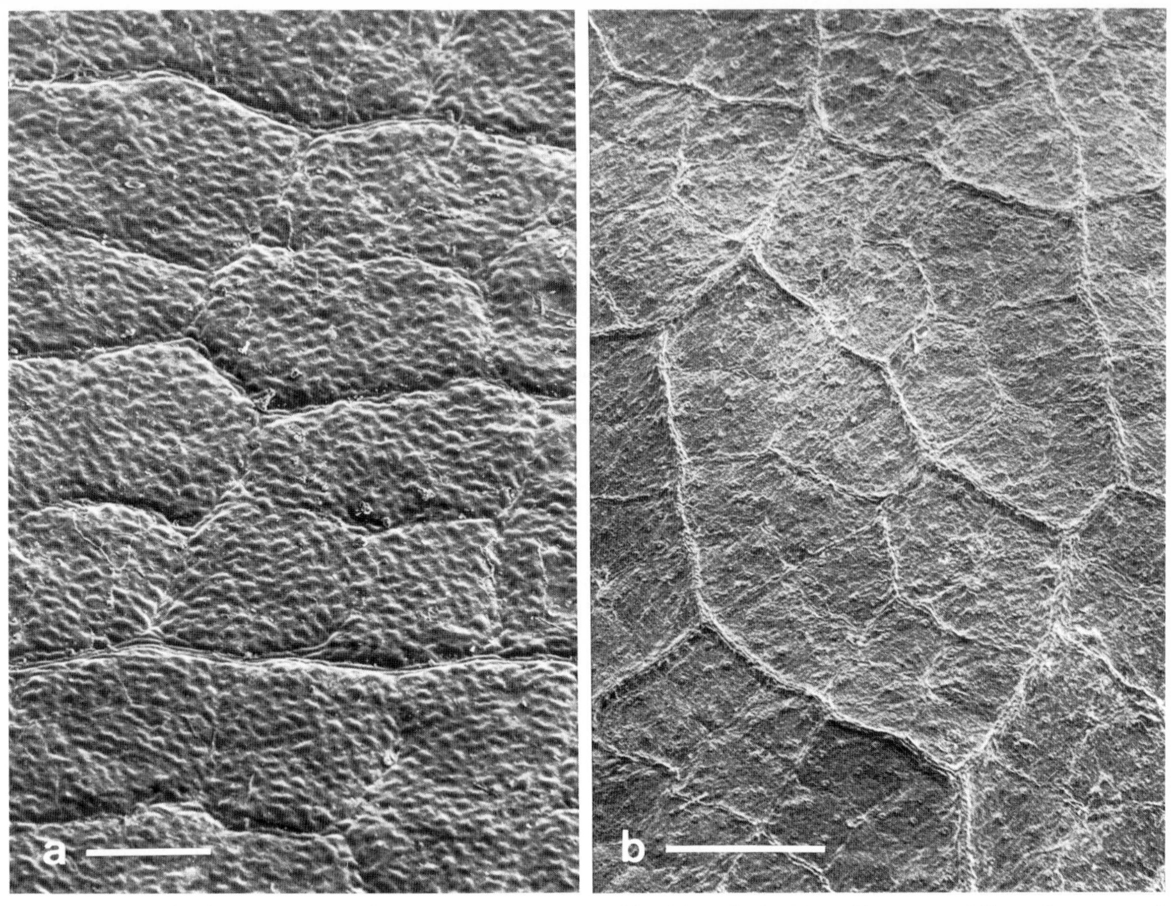

Fig. 4. Surface of a fresh green beech leaf in October. **a** Upper side with cuticula. The cell structure follows the course of the vascular bundles. 0.5 mm. **b** Lower side. The cell structure follows the course of the vascular bundles. 200 μm

Fig. 5. Beech leaf from the litter surface (litter from the previous year, dry). **a** Upper side with fungal hyphae. 100 μm. **b** Lower side with fungal hyphae. 80 μm

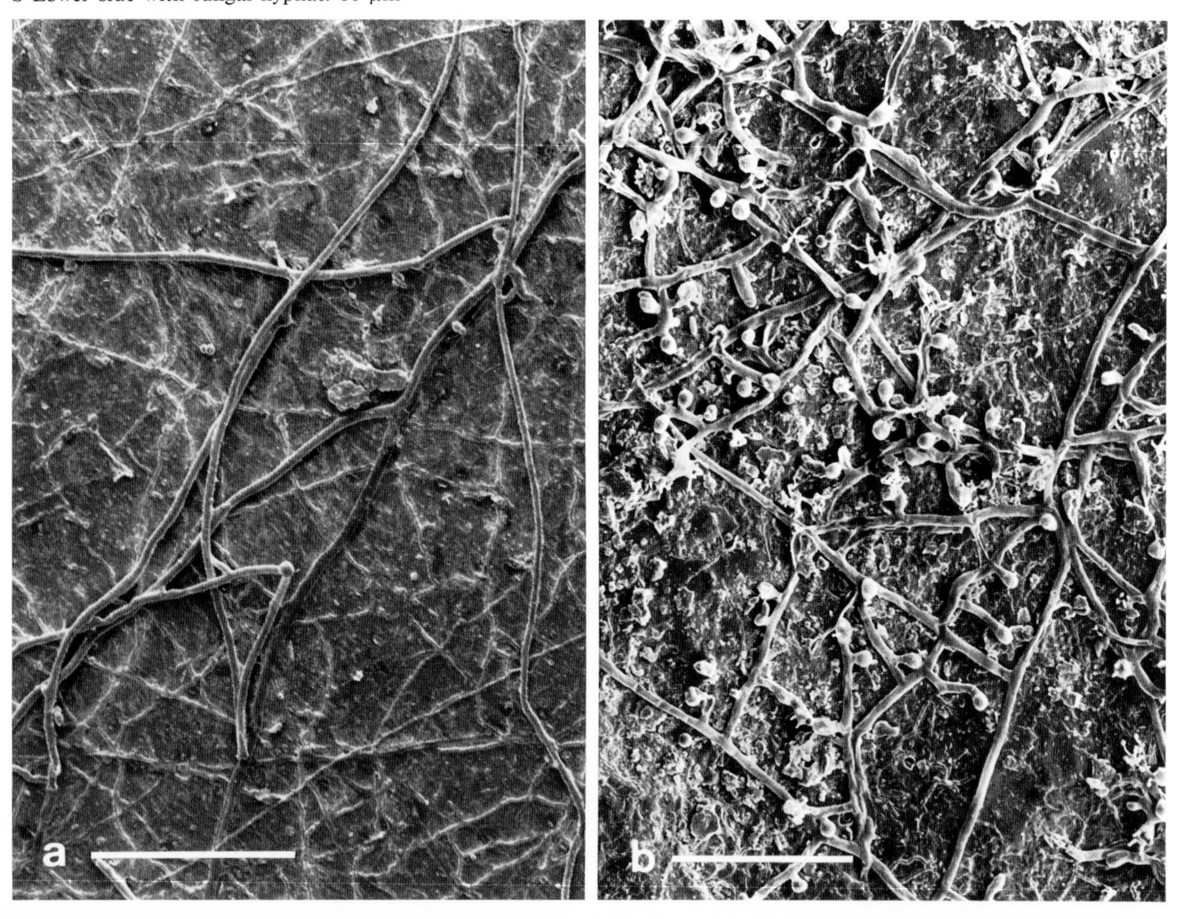

Fig. 6. Surface of a beech leaf from a deeper litter level (several years old); lower side of the leaf shows deep fissures and initial traces of feeding. 400 μm

Fig. 7. Surface of a beech leaf from a deeper litter level (several years old); upper side with traces of feeding. The fungal hyphae penetrate the leaf tissue (palisade parenchyma). 100 μm

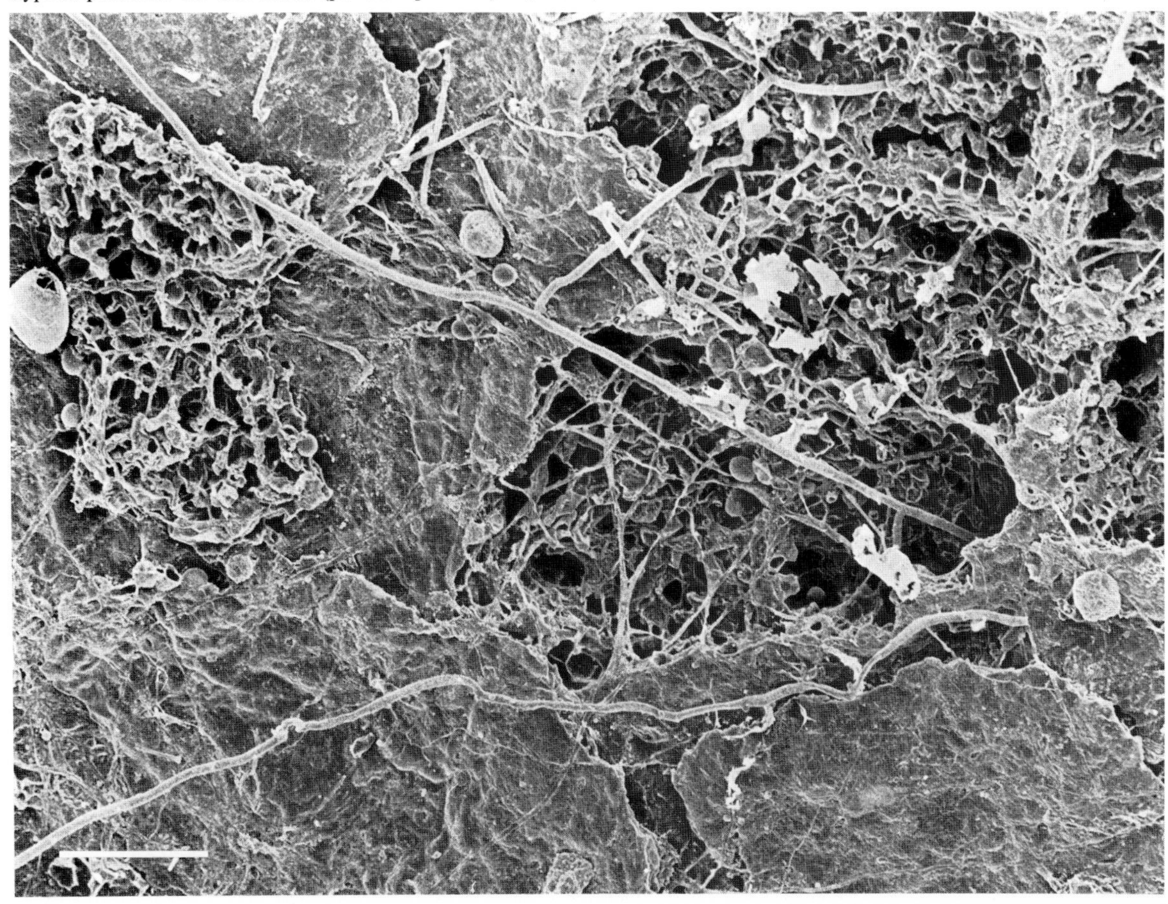

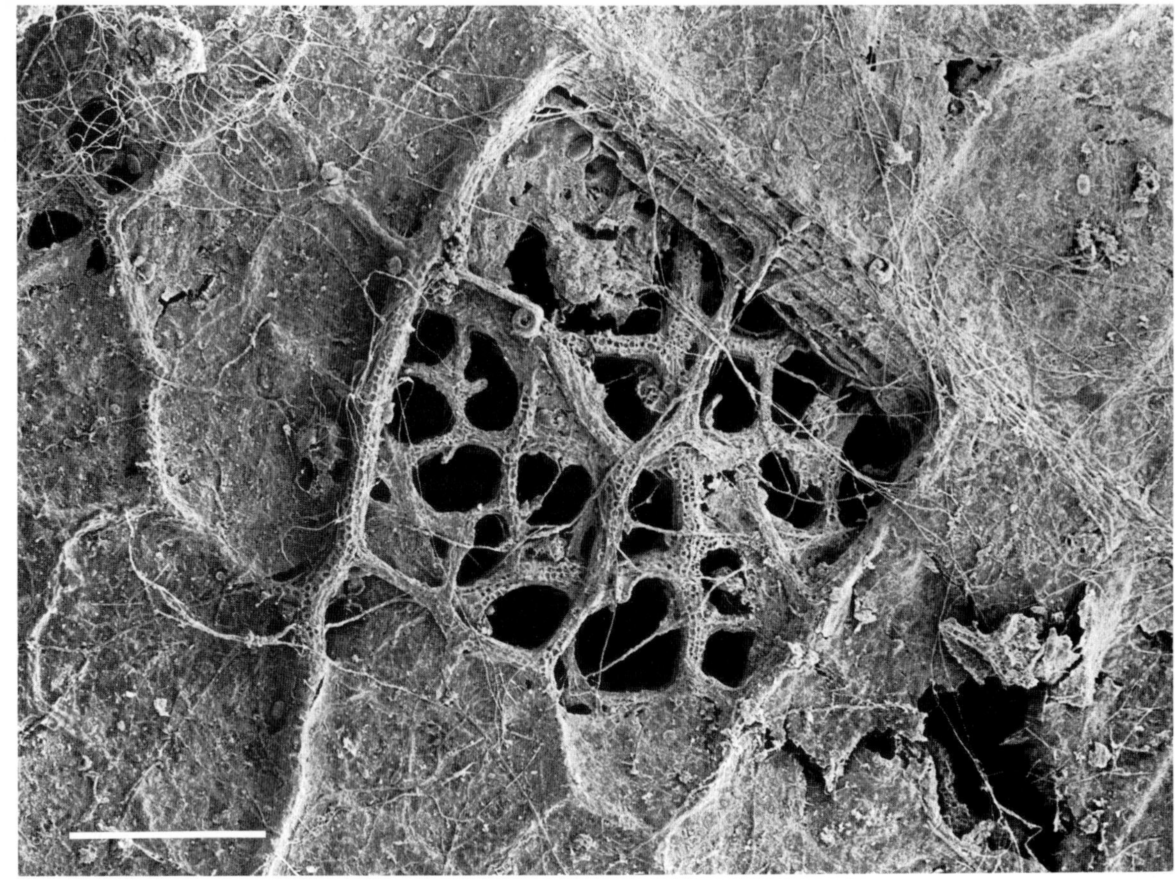

Fig. 8. Beech leaf from the fermentation layer; lower side with window-like perforations due to feeding activities. 400 μm

Fig. 9. Beech leaf from the fermentation layer; lower side. Most of the leaf tissue has been eaten away, leaving part of the more durable upper side and the vascular bundles. 400 μm

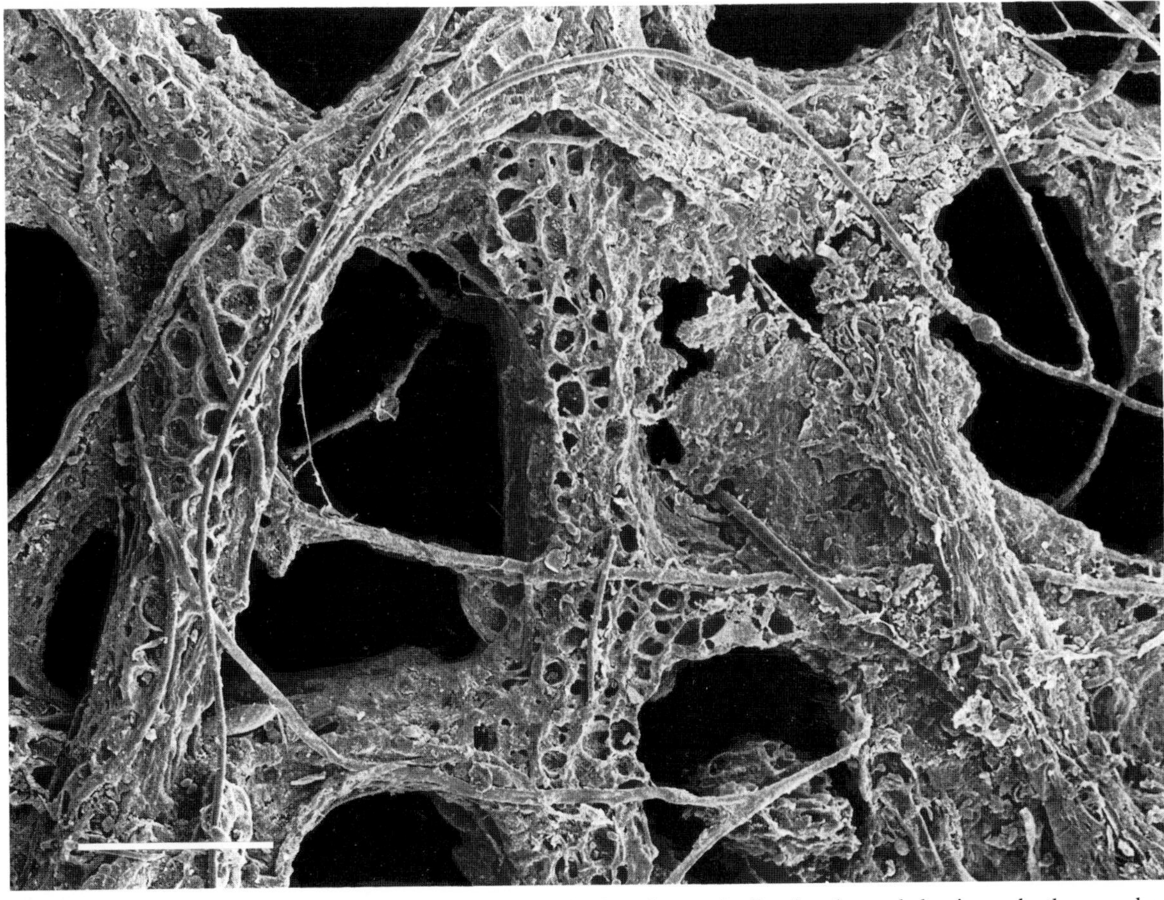

Fig. 10. Beech leaf from the fermentation layer. Skeletonization due to feeding is advanced, leaving only the vascular bundles. 80 μm

Fig. 11. Beech leaf from the fermentation layer in a state of advanced decay. **a** Upper side. 200 μm. **b** Lower side. 400 μm

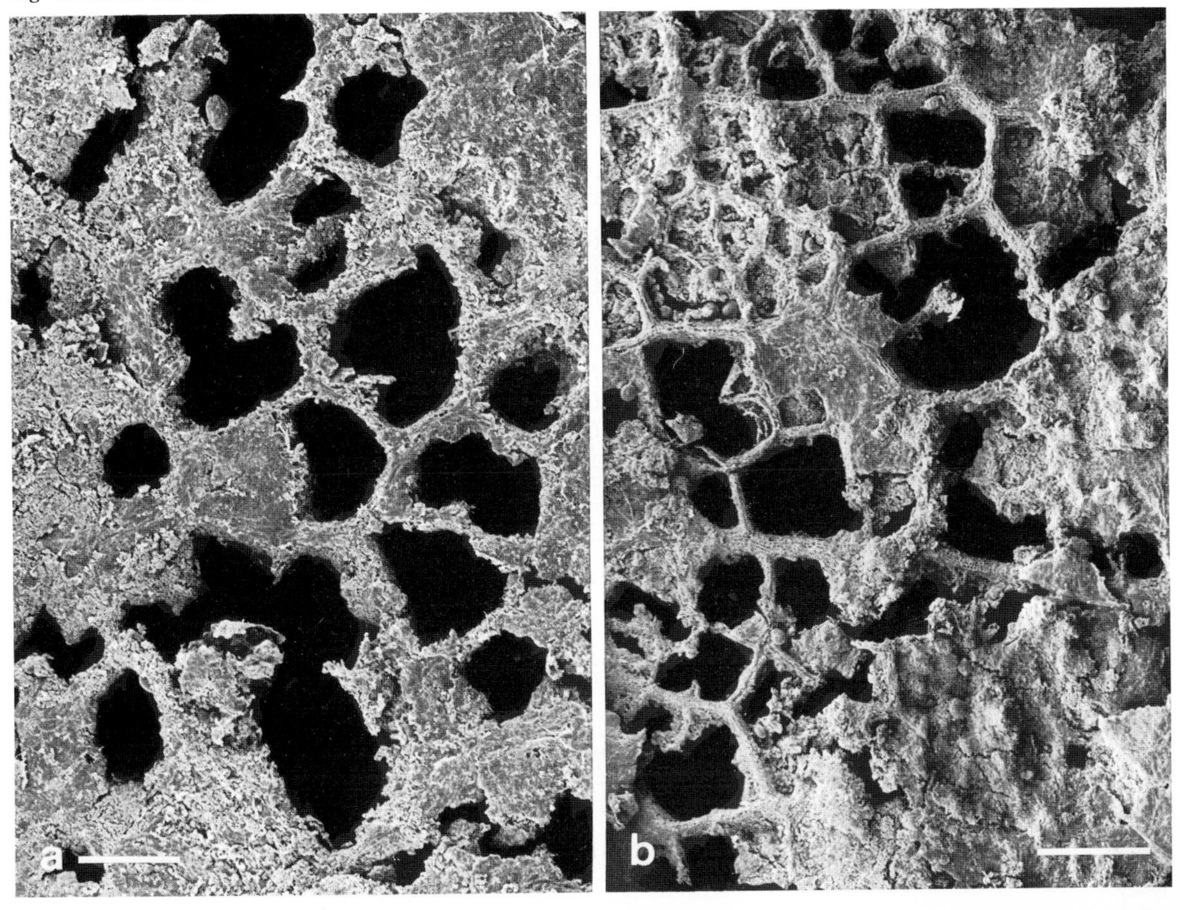

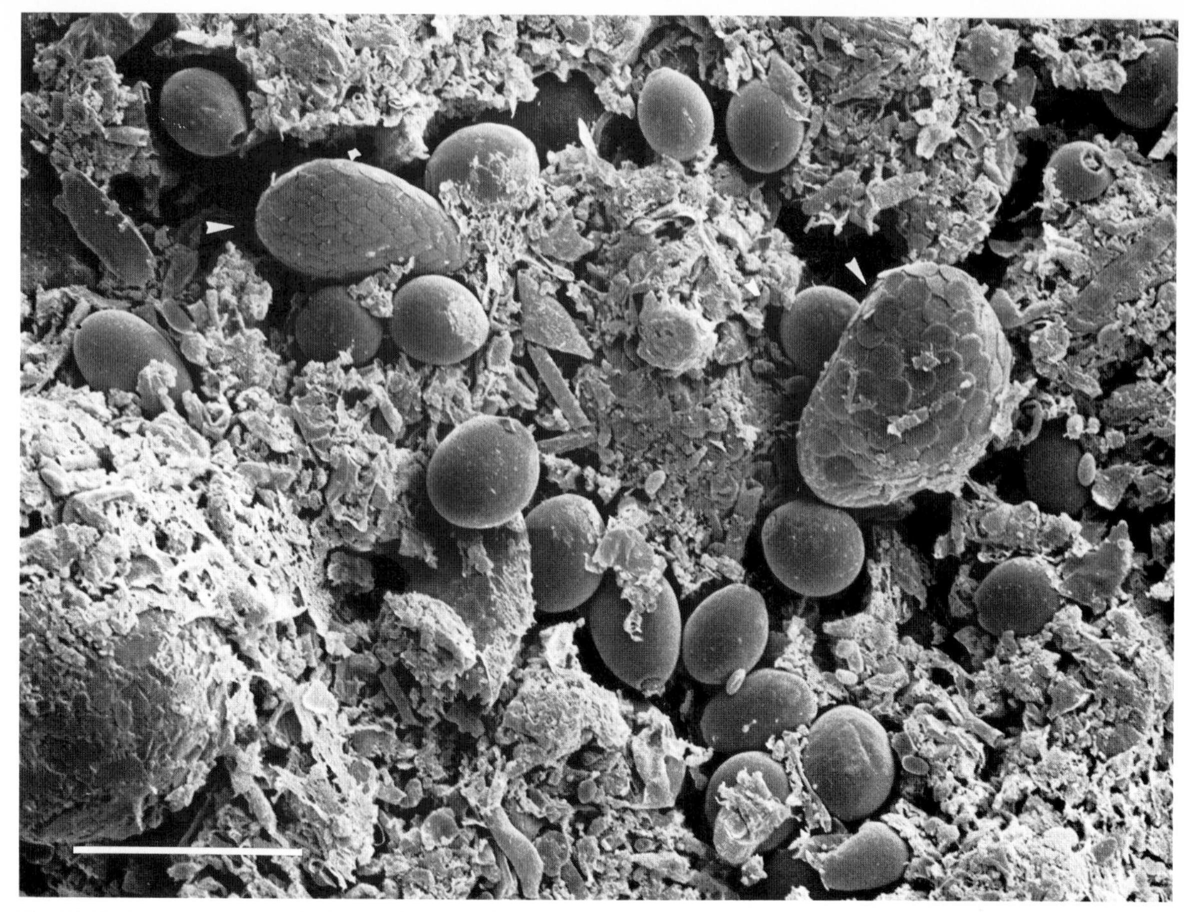

Fig. 12. Surface of an extremely decayed beech leaf from the fermentation layer with eggs and Testacea (shelled rhizopods) (*arrows*). 40 μm

Fig. 13. Surface of a beech leaf from the fermentation layer with the Testacea *Nebela collaris* which lives in acid woodland soils. 20 μm

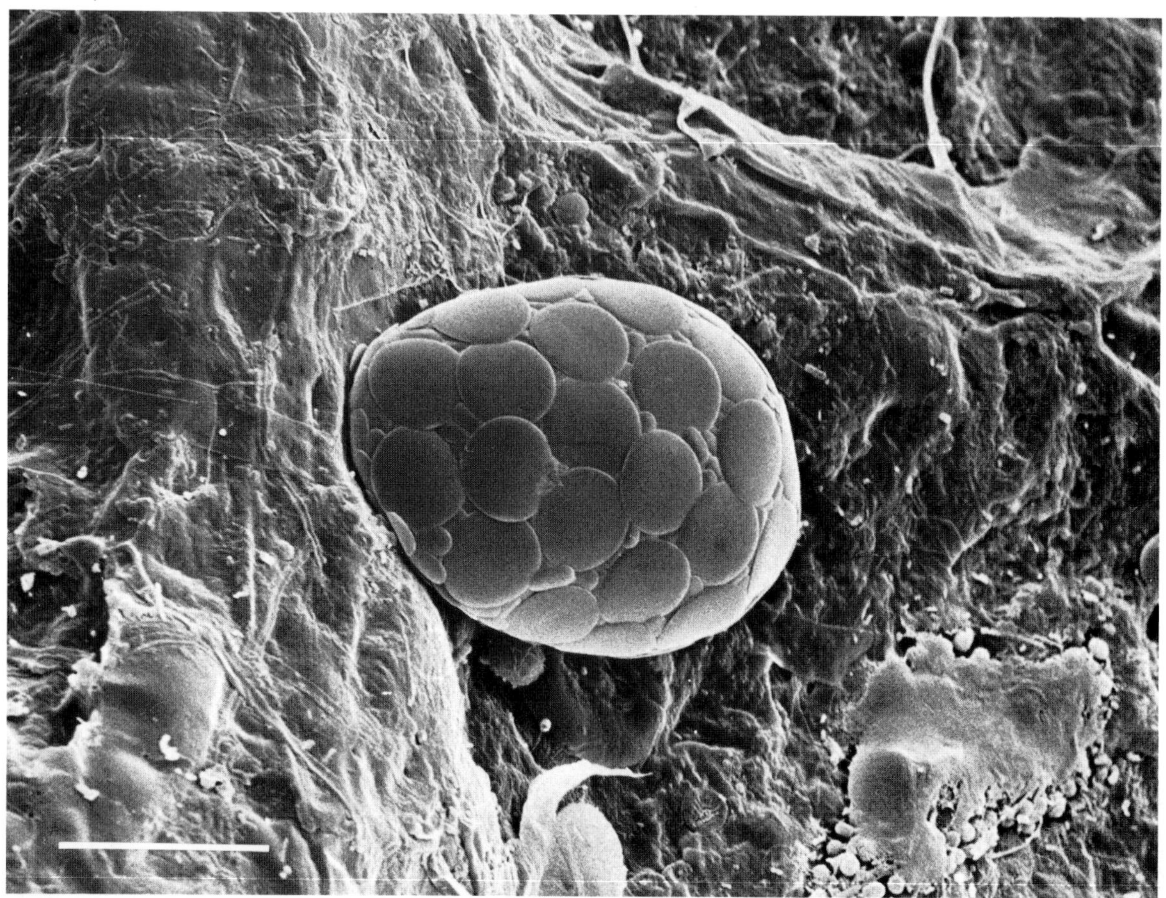

Fig. 14. Beech leaves from the fermentation layer with a pile of faecal pellets (animal droppings). 0.5 mm

Fig. 15. Surface of a faecal pellet. **a** Faecal pellet with the peritrophic membrane opened at one end. 100 µm. **b** *Inset* from **a** with faecal pellet material consisting of fragmented but not decayed plant remains. The *arrow* shows pit structures of vascular vessels. 50 µm

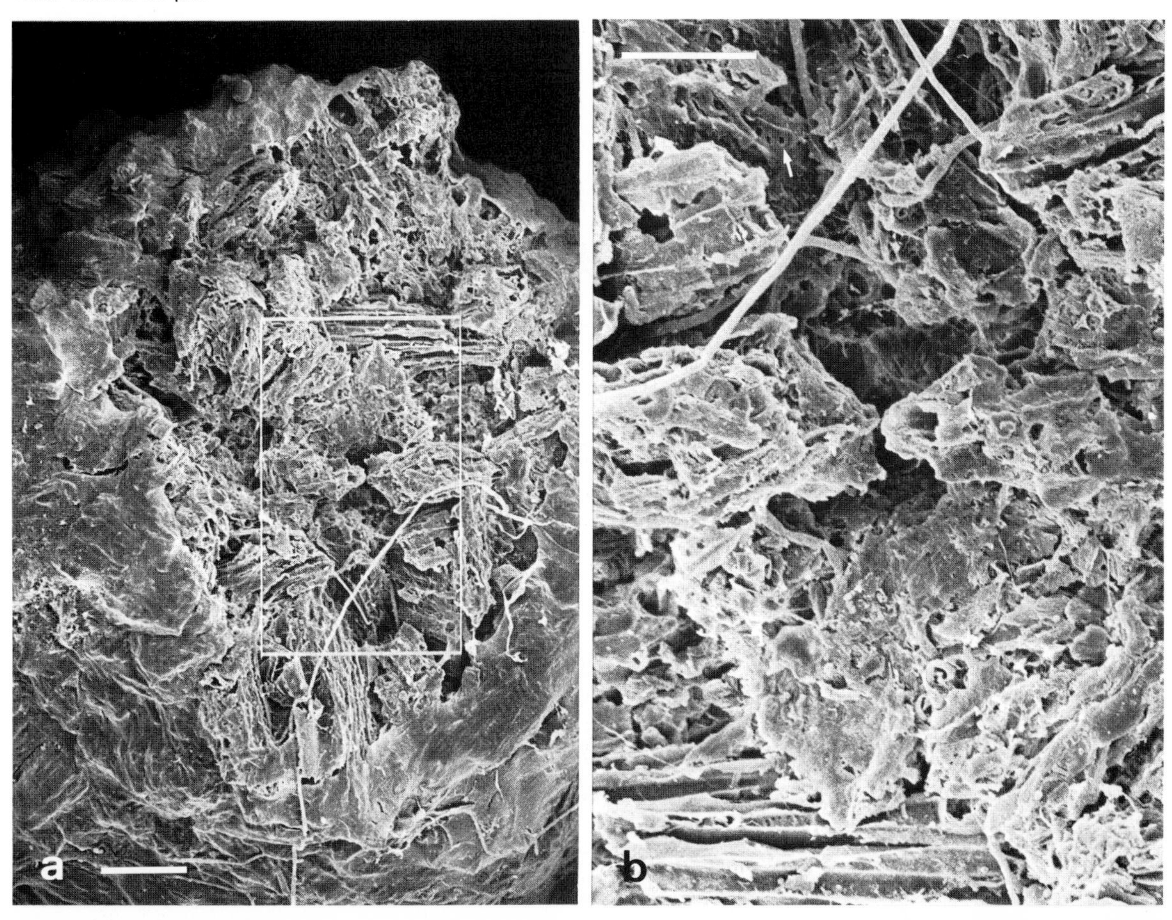

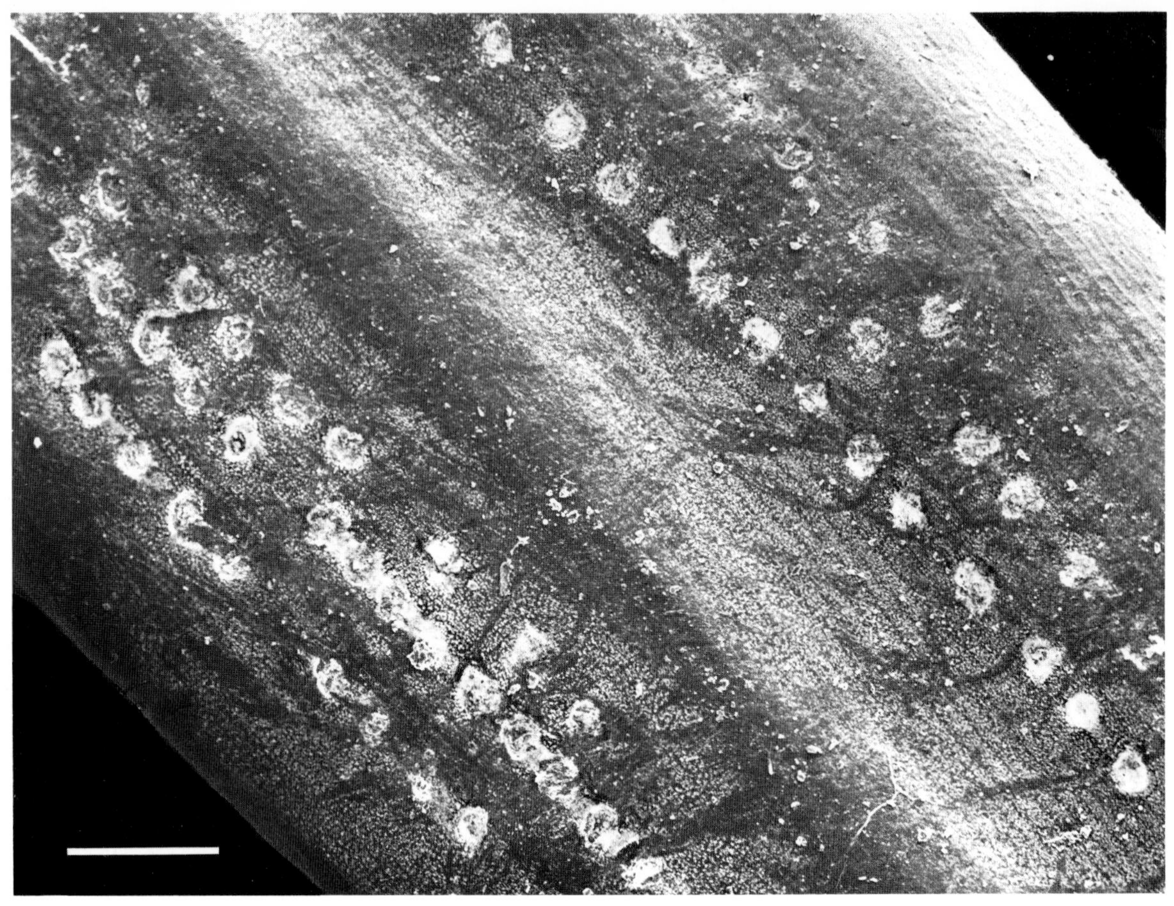

Fig. 16. Surface of an approximately 2-year-old spruce needle with cuticular wax layer. The *white spots* are waxy secretions on top of the stomata (respiratory pores). 200 μm

Fig. 17. Surface of a 2-year-old spruce needle. **a, b** Stomata with deposited, porous wax cap which acts as a filter und protection against transpiration. 40 μm, 20 μm

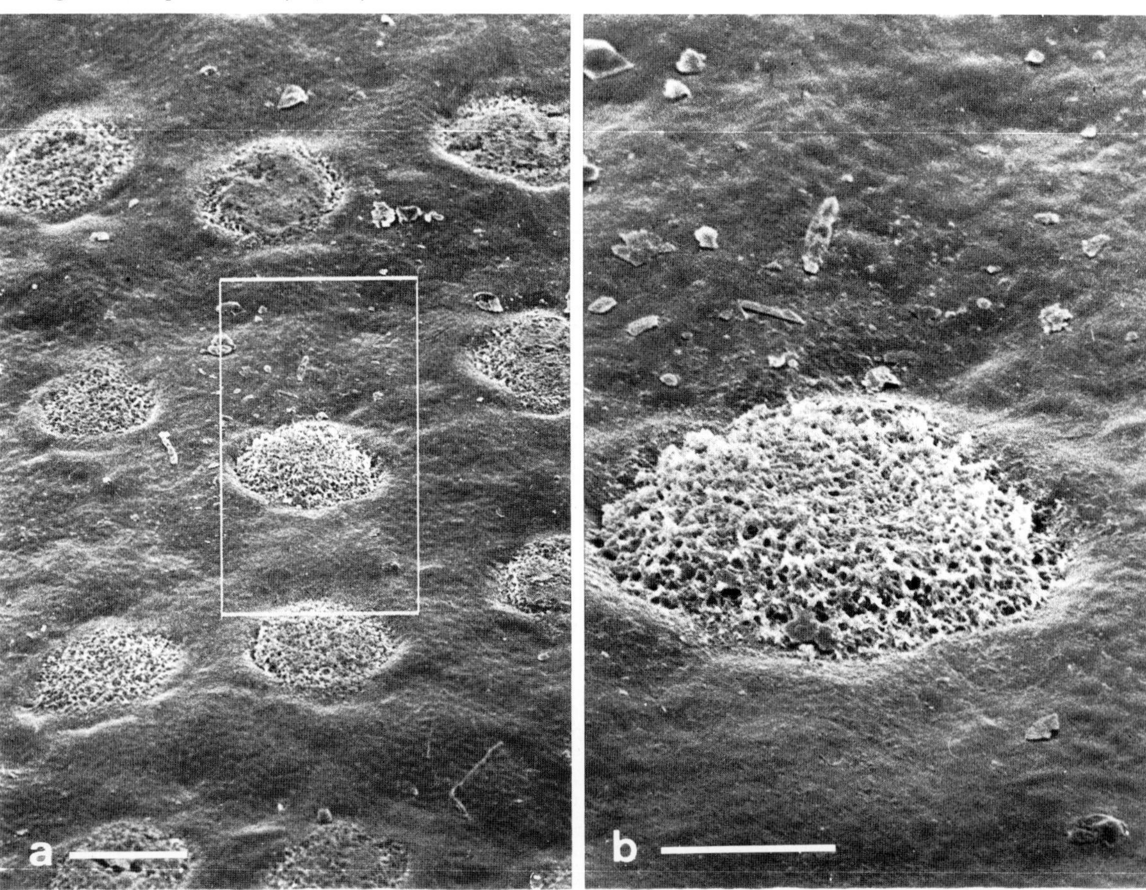

Fig. 18. Surface of a spruce needle from a deeper litter level. The respiratory pores (stomata) are free. Shrinkage and cracks are visible. 200 μm

Fig. 19. Surface of a spruce needle from a deeper litter level with free respiratory pores (stomata). Beside them pollen grains. 40 μm

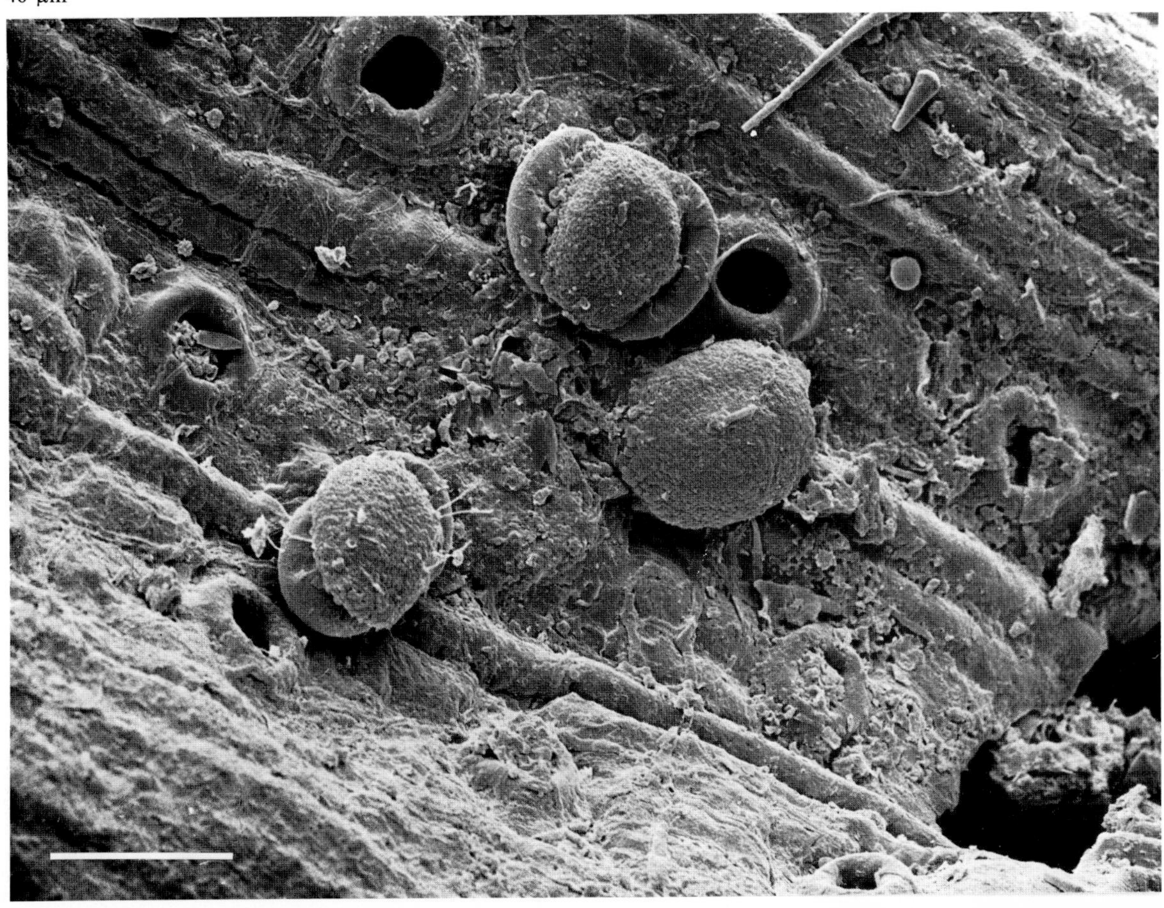

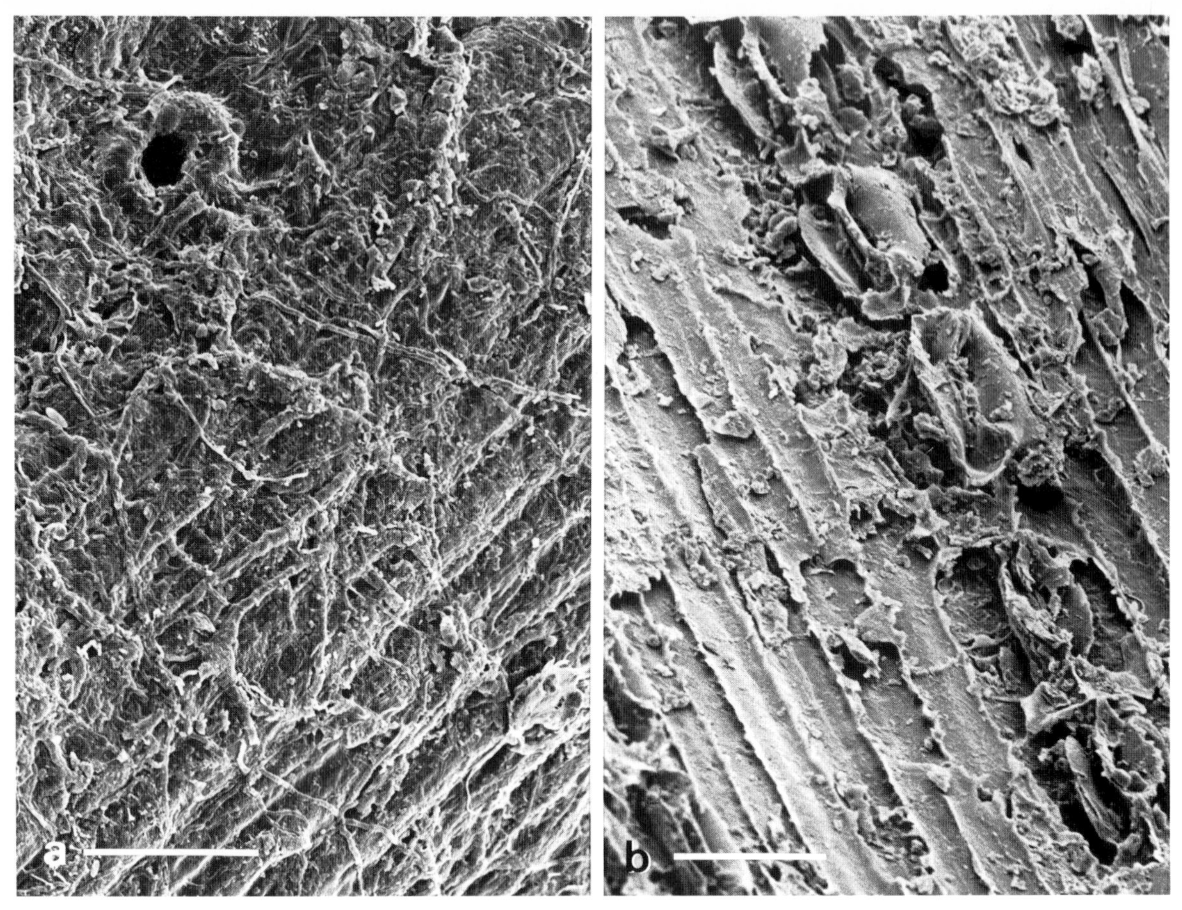

Fig. 20. Spruce needle from a deeper litter level. **a** Surface extensively attacked by fungal hyphae. 40 μm. **b** Inner surface of an extensively decayed needle showing a row of stomata. 50 μm

Fig. 21. Remains of spruce needles from raw humus. **a** Only thin, transparent and perforated remnants of the needles remain. 100 μm. **b** Extensively decayed needle sheath with enlarged respiratory holes. 100 μm.

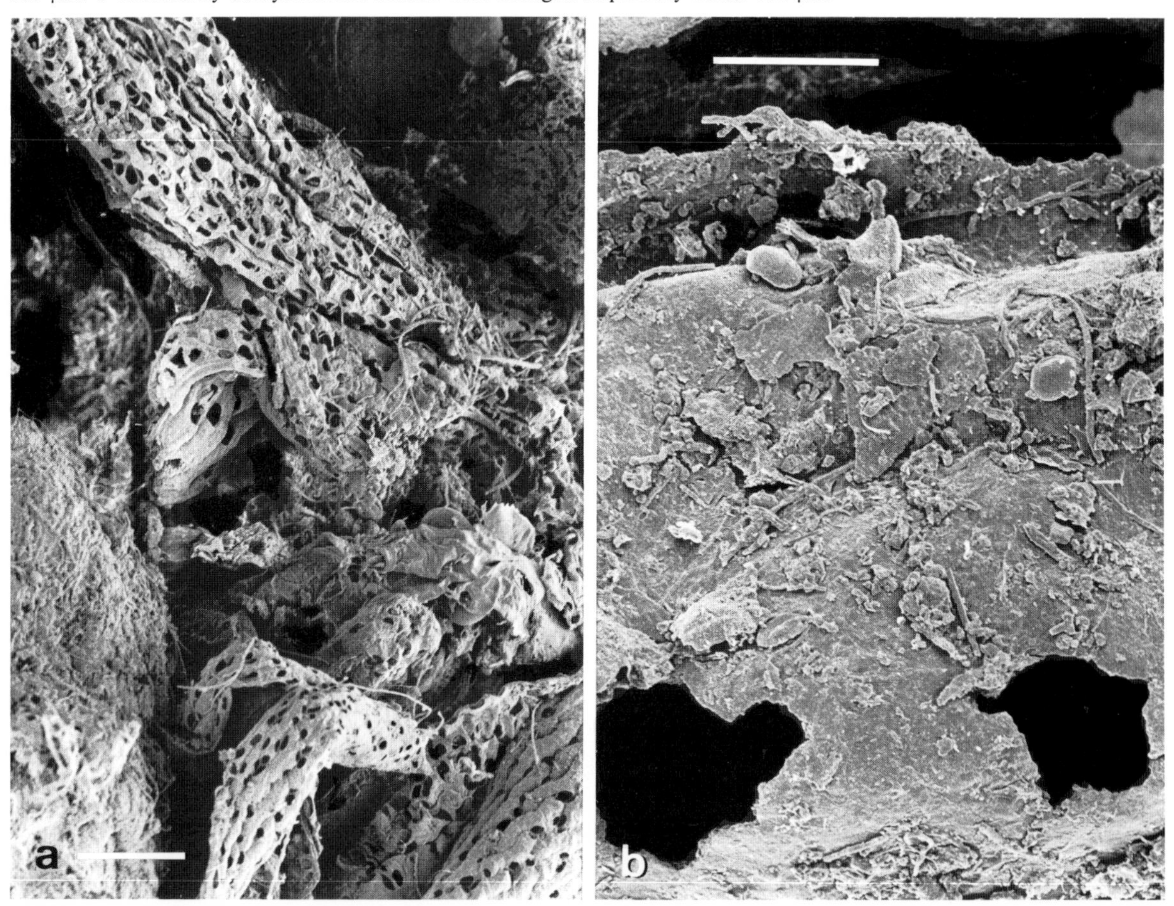

Fig. 22. Section of spruce needle raw humus. Root hairs and fungal hyphae are closely interwoven with the residues of the needles. 400 μm

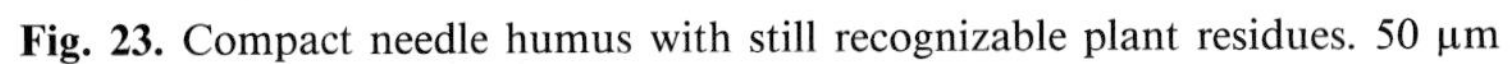

Fig. 23. Compact needle humus with still recognizable plant residues. 50 μm

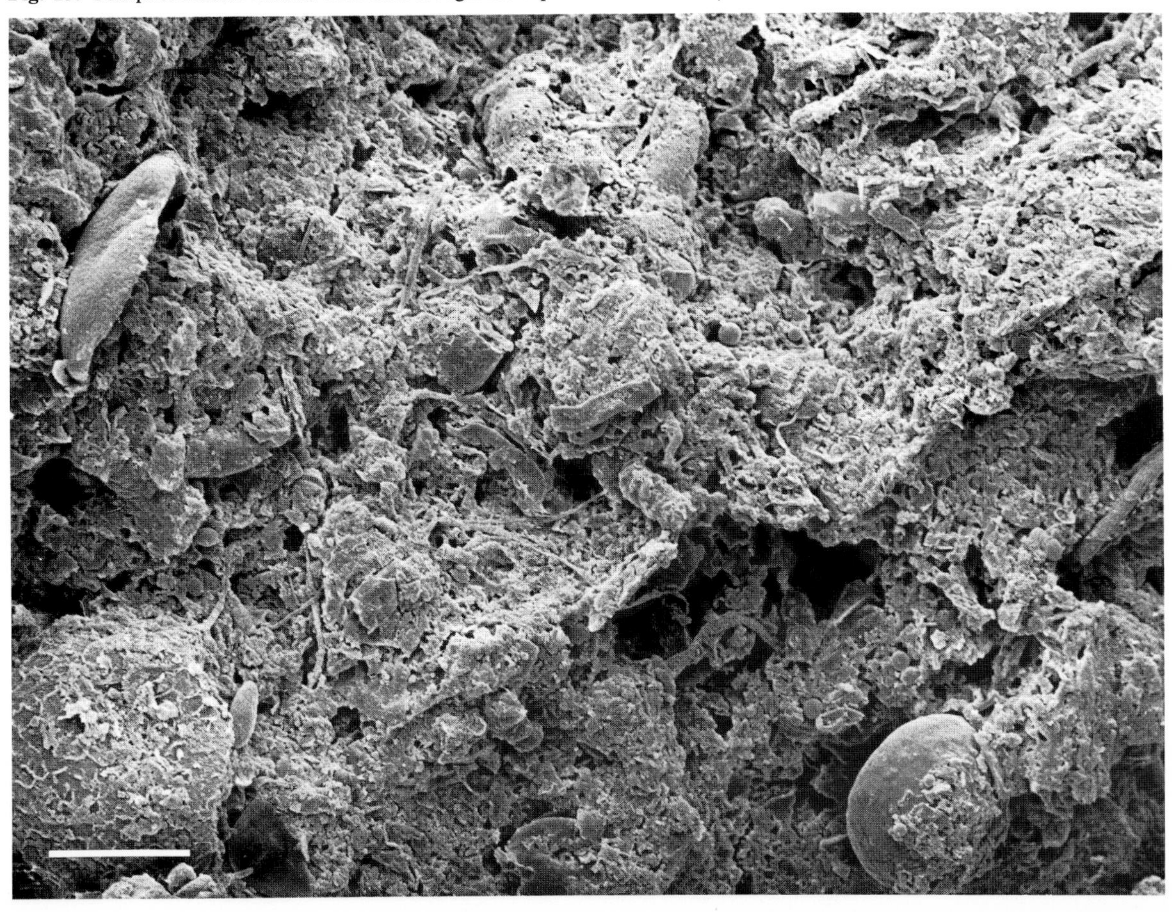

1.3.2 Decomposer Food Web

Producers, consumers and reducers form the functional unit of a food chain in the ecosystem. Considered systematically, the biotic components of this functional unit are the species of plants, animals and microorganisms which interact in a life community (biocenosis). Interspecific competition and coexistence as well as predators and food belong to the biotic ecological factors influencing this biocenosis. These factors give rise to a trophic interrelationship (known as the food web) among the consumers (Pimm, 1982; Petersen and Luxton, 1982; Schaefer, 1982; Beck, 1983; Fitter et al., 1985). The arthropods, belonging to the food web and engaged in the decomposition of plant debris in the soil, are food specialists which are classified according to their feeding habits: macrophytophagous, microphytophagous, saprophytophagous and zoophagous, coprophagous and necrophagous soil arthropods (Fig. 24):

Macrophytophages: consumers which feed on living plants including litter leaves, not yet attacked by microorganisms.

Microphytophages: consumers feeding on pollen, algae, fungal spores and hyphae, and microorganisms which participate in the decay of plant and animal debris.

Saprophytophages: consumers which feed on dead vegetation, i.e. material attacked and degraded by microbes (macerated leaves, raw humus).

Zoophages: consumers which prey on living animals (predators).

Necrophages: consumers which feed on dead animal substance (carrion).

Coprophages: consumers which feed on animal excrement.

Some soil arthropods are polyphagous. Consumers of plant material in the litter layer are often macrophytophagous and thus primary decomposers. At the same time they are saprophytophagous, often preferring plant material which has been softened and predigested by microorganisms.

It cannot be ruled out that saprophages, i.e. saprophytophagous, coprophagous and necrophagous consumers actually prefer the microorganisms which can constitute a considerable part of their food. In this case they must be regarded as microphytophagous consumers.

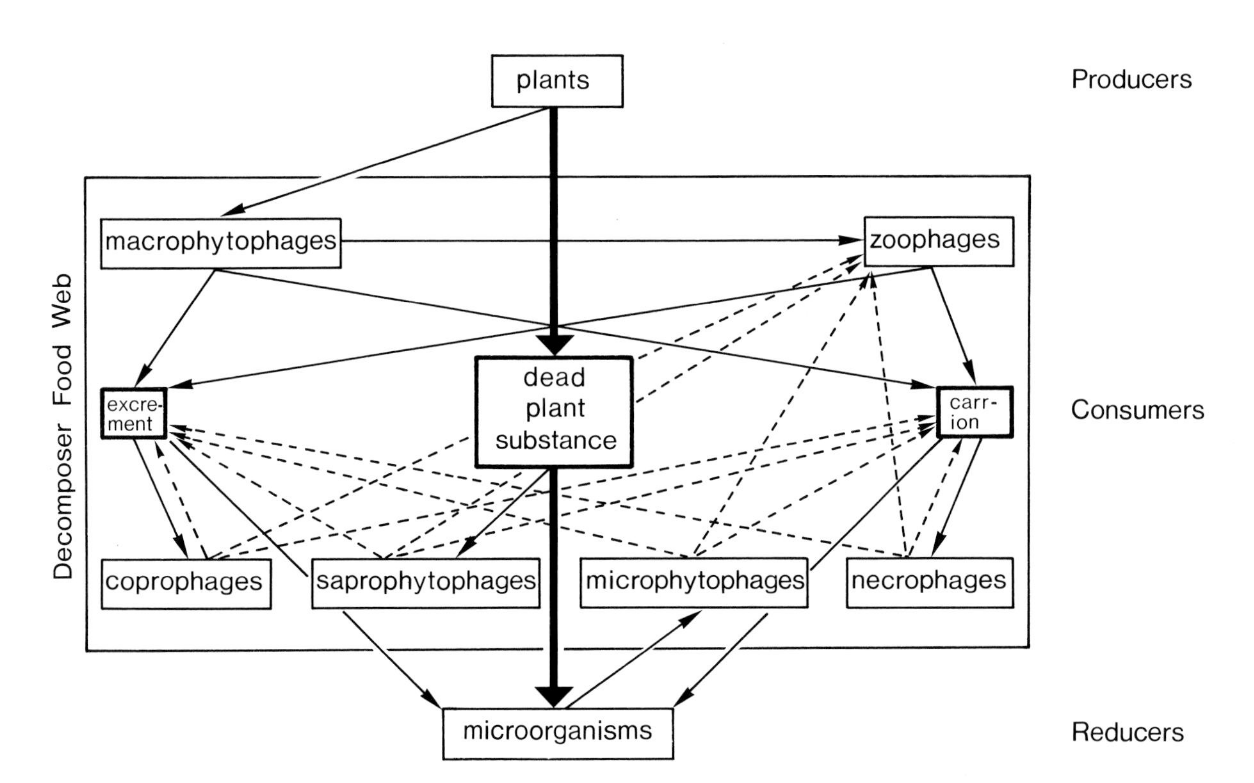

Fig. 24. The decomposer food web in the soil, combined according to various authors

2 Systematic Chapters

2.1 Order: Araneae – Spiders (Arachnida)

General Literature: Bristowe, 1958; Kullmann and Stern, 1975; Foelix, 1979; Jones, 1983; Nentwig, 1986; Roberts, 1985, 1986

2.1.1 Soil-Dwelling Spiders

There is still considerable controversy about the significance of spiders in soil biology. This is hardly surprising as the ecology of spiders is only in its initial stages. For the time being Duffey (1966) offers a useful, vertical classification of the biotopes of spiders into four zones:

1. Soil zone:	leaf litter, stones	0 – 15 cm
2. Herbaceous zone:	low vegetation	15 – 180 cm
3. Shrub zone:	bushes, trees	1.8 – 4.5 m
4. Woodland zone:	trees	over 4.5 m

A considerable proportion of spiders live in the soil zone, their population density reaching 50 to 150 animals/m^2 in woodland soil (Dunger, 1983), and 85% of all indigenous spiders overwinter in the litter layer, sheltered from extreme temperatures and aridity (Edgar and Loenen, 1974; Schaefer, 1976; Foelix, 1979; Albert, 1977; Dumpert and Platen, 1985).

Those spiders, which spin tubular webs in the earth or directly on the ground as dwellings and for capturing prey, belong without doubt to the soil arthropods, e.g. purse-web spiders (Atypidae), sack spiders (Clubionidae), tube spiders (Eresidae), funnel-web spiders (Agelenidae), soil spiders (Hahniidae), etc. Their webs often consist of a tubular retreat in the soil and a trapping web of crimped or viscid threads above ground. Migrant and hunting spiders, e.g. jumping spiders (Salticidae) and many wolf spiders (Lycosidae) also play a regulatory role in soil life. As predators they help to maintain a natural balance in the biocenosis.

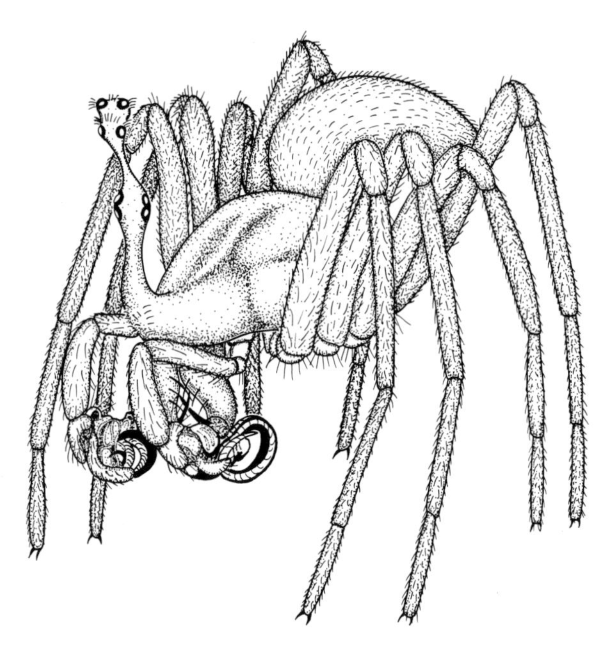

Fig. 25. Male of *Walckenaera acuminata* (Micryphantidae) with eye stalk and pedipalps which have been modified to form copulatory organs

Plate 1. *Walckenaera acuminata* (Micryphantidae), male.
a Overview, lateral. 1 mm.
b Overview, caudal. 1 mm.
c Overview, dorsal. 0.5 mm.
d Frontal view with eye stalk, pedipalps and legs. 0.5 mm.
e Eye stalk and pedipalps, dorsal. 250 µm.
f Overview, ventral. 1 mm

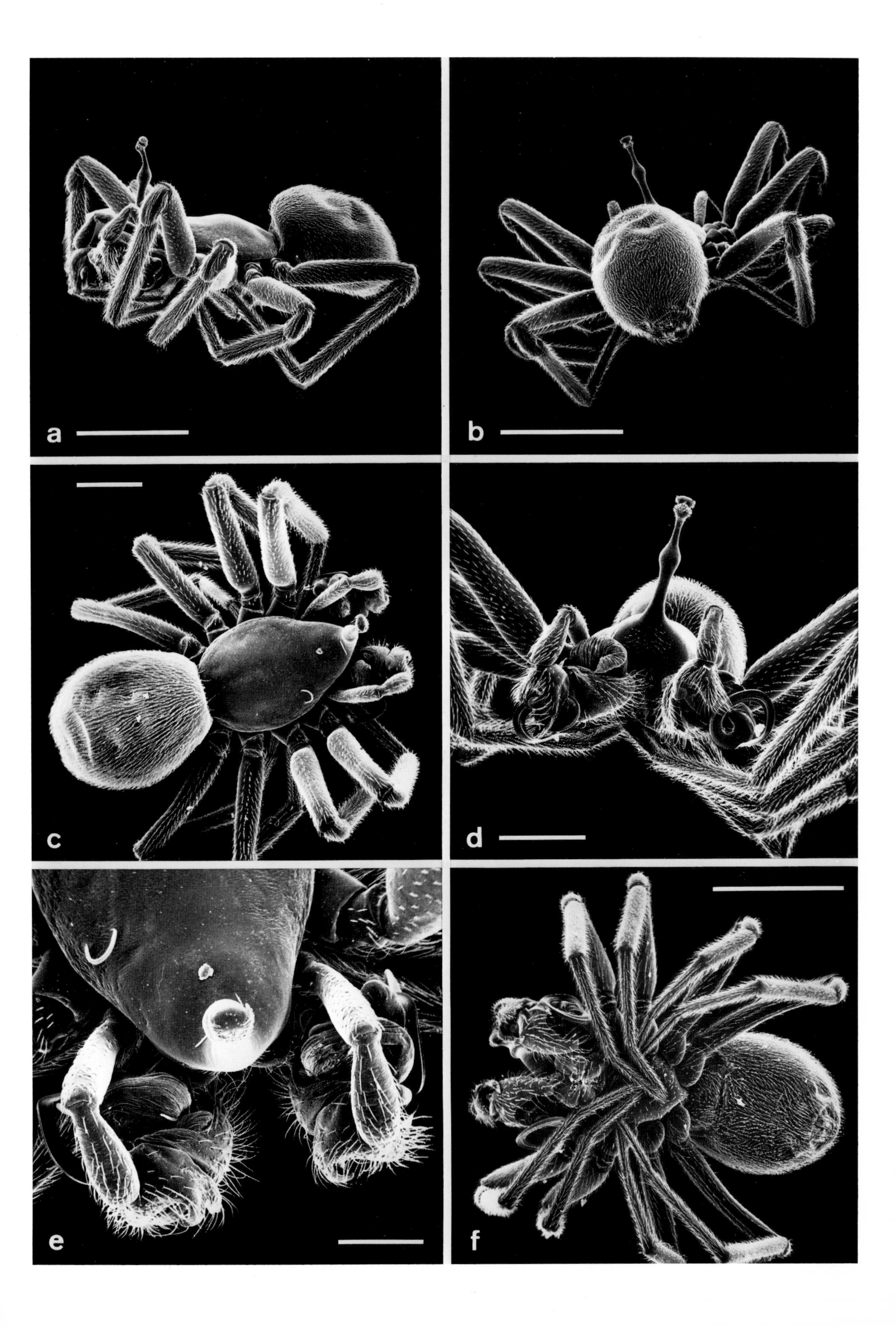
a
b
c
d
e
f

2.1.2 Structural Features of the Body of Spiders

The body of a spider consists of two main parts. The anterior part (prosoma) is connected to the abdomen (opisthosoma) by means of a constricted waist (petiolus) (Fig. 26). The lung slits, genital furrow, tracheal spiracle, anal cone and spinnerets are recognizable on the exterior of the opisthosoma. Besides four pairs of legs, the prosoma bears the pedipalps and chelicerae. On the carapace (dorsal shield) of the prosoma most spiders have eight eyes, which were investigated by Homann (1971).

The 1–2-mm-long spiders belonging to the Micryphantidae family are strictly limited to the epedaphic biotope. Wiehle (1960) presented 144 dwarf, micryphantid spiders in a comprehensive monograph. *Walckenaera acuminata* with its extraordinary modification of the prosoma deserves special attention. Above the carapace rises an eye stalk, bearing the four lateral eyes on a bulge about halfway up its length. Its head-like distal end is also broadened and carries comb-like sensory hairs as well as the four medial eyes so that *Walckenaera* has almost complete allround vision (Plate 2; Figs. 25 and 27).

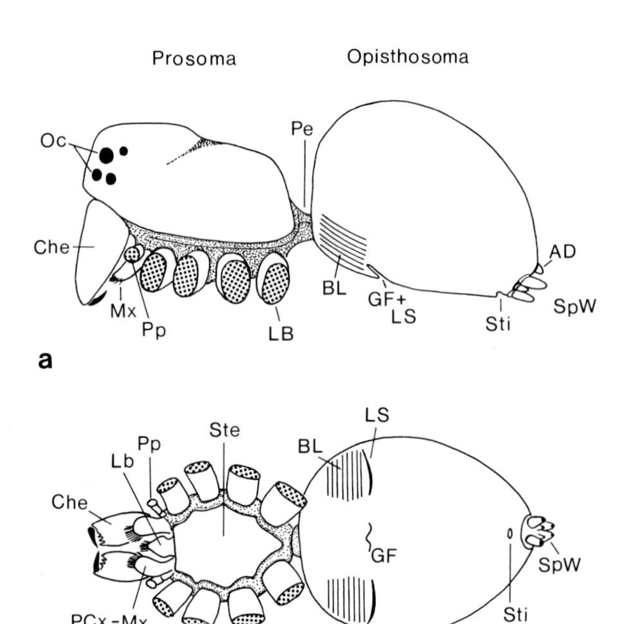

Fig. 26. Basic arrangement of the body of a spider (After Foelix, 1979)
a Lateral view; **b** ventral view. *AD* Anal tubercle; *BL* book lung; *Che* chelicerae; *GF* genital furrow; *LB* coxae of legs; *Lb* labium; *LS* lung slits; *Mx* maxilla; *Oc* ocelli; *Pe* petiolus; *PCx* pedipalp coxa; *Pp* pedipalp; *SpW* spinnerets; *Ste* sternum; *Sti* tracheal spiracle. Lung slits and genital furrow can be united to form the epigastric fold

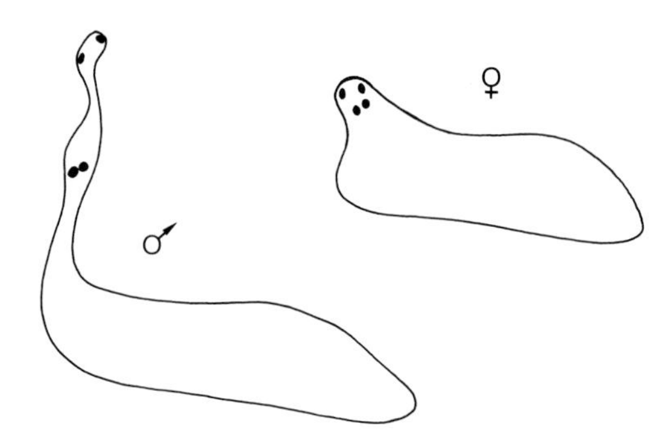

Fig. 27. *Walckenaera acuminata* (Micryphantidae). Male and female trunk in profile. (After Locket and Millidge, 1968)

Plate 2. *Walckenaera acuminata* (Micryphantidae), male.
a Eye stalk, caudal. 250 µm.
b Eye stalk, oblique frontal. 250 µm.
c "Head" of the eye stalk. The *arrow* shows the position of the frontal median eyes. 50 µm.
d Hair structures on the eye stalk. 5 µm.
e Bulge on the eye stalk with two lateral eyes. 50 µm.
f Eye stalk, distal with the caudal median eyes (*arrows*). 50 µm

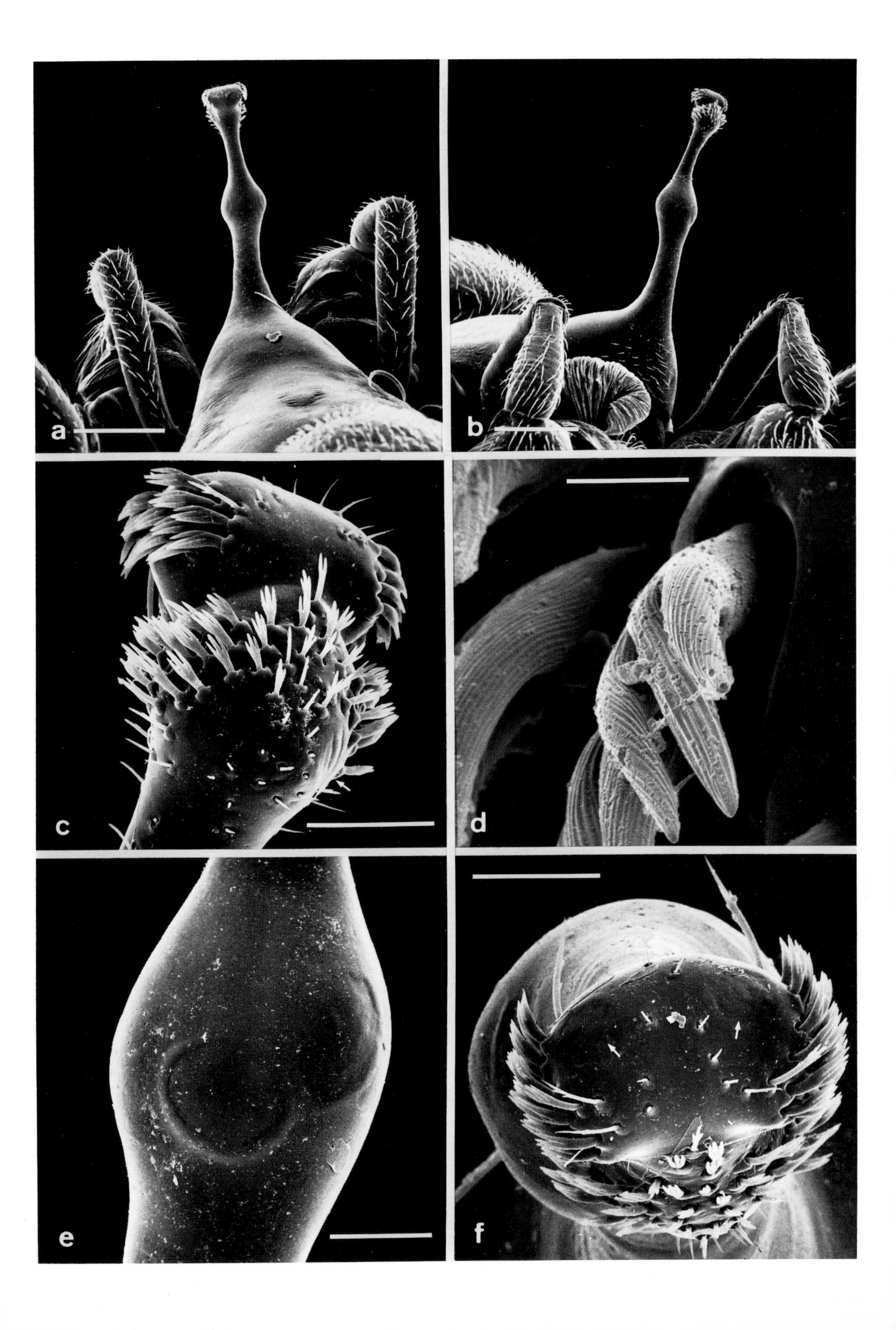
a
b
c
d
e
f

2.1.3 The Pedipalps of the Male Spider

The maxillary palps or pedipalps of the male spider have been modified to form copulatory organs which serve to transfer sperm (Wiehle, 1961). In its simplest form the palpal tarsus bears a bulb which is drawn out to a point. The sperm duct (spermophore) with its opening at the tip (embolus) extends spirally into the bulb where it terminates in a blind ending (Fig. 28). Before courting a female, the male spider fills the sperm duct by spinning a small sperm web, pressing a drop of sperm from his genital opening by vigorous movements of his abdomen and depositing it on the sperm web, and then taking the drop of sperm into the sperm duct, probably by means of capillary suction (Harm, 1931).

Male copulatory organs show various degrees of differentiation and in micryphantid spiders they are remarkably complicated (Plates 1 and 3). These functional-morphological developments consist of modification of the bulb, change in the shape of the embolus, which often becomes long and rolled up like a spiral hose, as well as the formation of various characteristic bulb sclerites (e.g. conductor). The differentiation of the male copulatory organs is a reflection of the equivalent differences in the female genitalia, especially in the epigyne. And they are, to a large extent, characteristic features of each species.

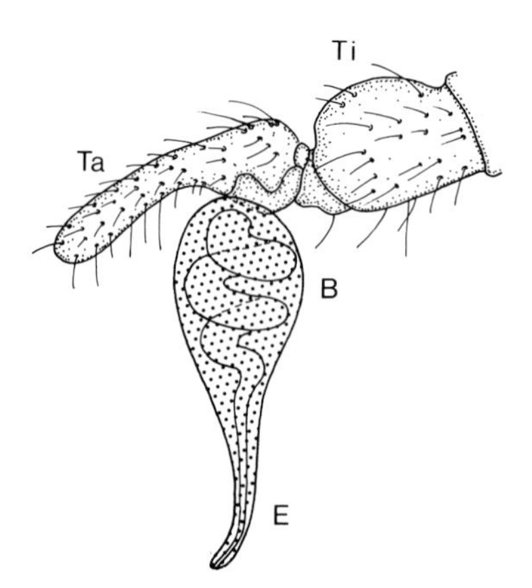

Fig. 28. *Segestria florentina* (Dysderidae). Simply structured palp of the six-eyed spider. (After Harm, 1931, and Foelix, 1979). An inflatable bulb (*B*) is inserted in the tarsus (*Ta*). It contains the sperm duct and merges into a copulatory tip, the embolus (*E*), at the distal end. More highly developed spiders have a more complicated arrangement because of the formation of additional sclerites and transformation of the embolus into a highly differentiated sclerotized appendage. The tarsus is then called the cymbium; *Ti* Tibia

Plate 3. *Walckenaera acuminata* (Micryphantidae), male.
a Prosoma, lateral, glandular area. 10 μm.
b Prosoma, lateral, normal cuticle. 10 μm.
c Opisthosoma, caudal, furrowed cuticle with hair base. 3 μm.
d Male pedipalp with femur, patella, tibia, tarsus (cymbium, *arrow*) and embolus (*arrowhead*). 250 μm.
e Part of embolus. 50 μm.
f Male pedipalp with femur, patella, tibia, tarsus, bulb and embolus. 250 μm.
g Surface of bulb. 10 μm

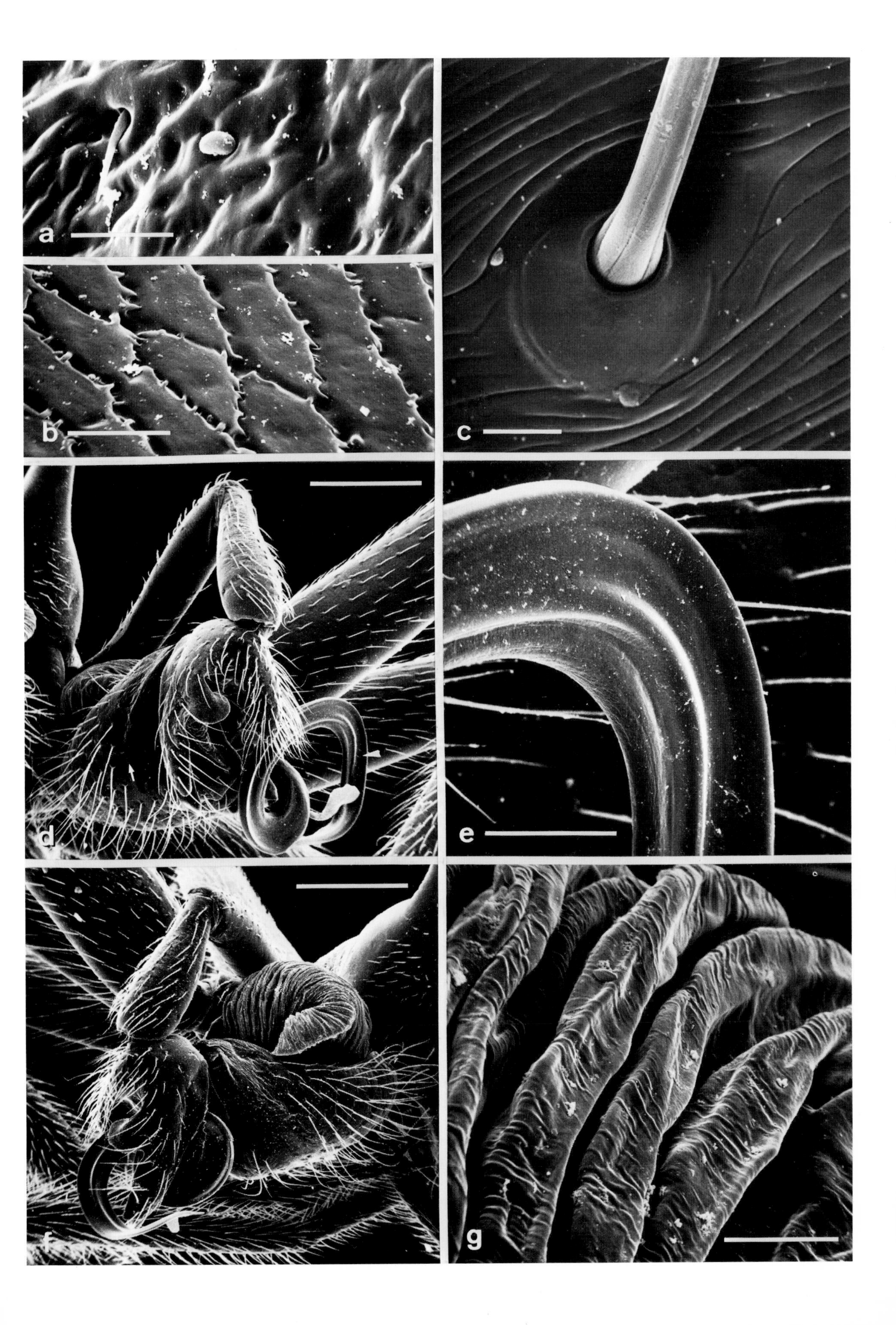
a
b
c
d
e
f
g

2.1.4 The Chelicerae

The foremost extremities on the prosoma in front of the mouth are the chelicerae. They are divided into two segments, a powerful base (paturon) and a mobile fang (unguis). The paturon and the fang are so articulated that they can grasp prey like pincers. In the resting position the fang is folded into a furrow in the paturon like the blade of a penknife. Many spiders have a row of teeth along each edge of this furrow. These are used together with the fangs to cut up their prey (Plate 4; Fig. 30). Spiders lacking these rows of teeth are only able to suck out their prey (Foelix, 1979). The outlets of the paired poison glands, which take up considerable space in the prosoma (Fig. 29), are situated at the tip of the conically curved, pointed fangs (Plate 4). Poison is injected to kill prey, but it also seems to speed up digestion. The glands of the spitting spider, *Scytodes thoracica*, are particularly differentiated. The posterior part of the gland produces glue, while the anterior part produces poison. The spider catches its prey by suddenly spitting glue onto the animal, wrapping it in sticky threads and then killing it with an injection of poison (Dabelow, 1958).

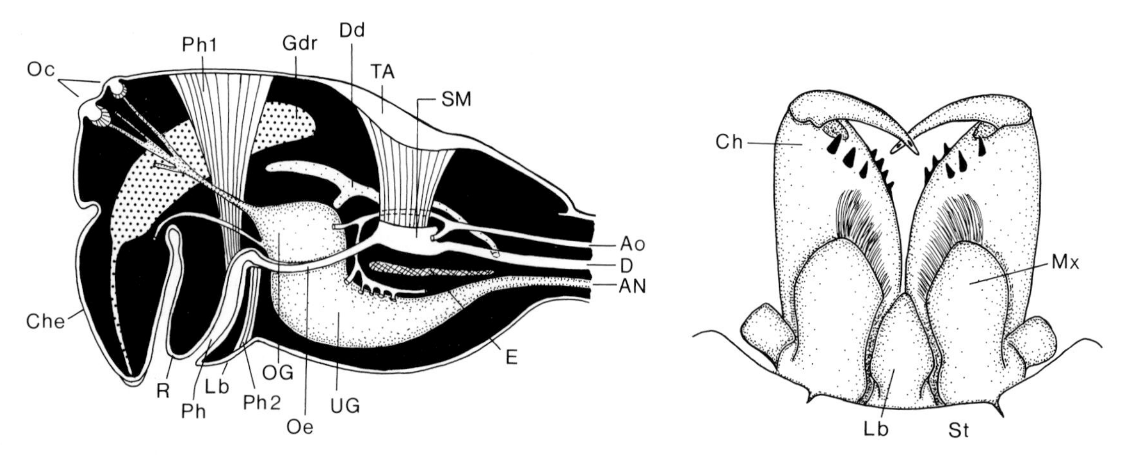

Fig. 29. Schematic drawing of a section through the prosoma of a spider. (After Foelix 1979). *AN* Abdominal nerve; *Ao* aorta; *Che* chelicera; *D* gut; *Dd* gut diverticulum; *E* endosternite; *Gdr* poison gland; *Lb* labium; *Oc* eyes; *Oe* oesophagus; *OG, UG* upper and lower pharyngeal ganglia; *Ph* pharynx; *Ph 1, 2* pharyngeal elevator, pharyngeal depressor; *R* rostrum; *SM* pumping stomach; *TA* tergal apodem

Fig. 30. Mouthparts of a spider (After Foelix, 1979). Arrangement of the mouthparts of a wolf spider (*Lycosa*), ventral view. *Ch* Chelicera with fang; *Mx* maxilla (Pedipalp coxa); *Lb* labium (lower lip); *St* sternum (ventral shield)

Plate 4. *Meta menardi* – Araneidae (orb weavers) cave spider.

a Prosoma with grouped eyes, pedipalps and chelicerae, dorsal 0.5 mm.

b Pedipalp, chelicerae, maxillae and labium, lateral. 1 mm.

c Chelicerae, maxillae, labium and coxae of the forelegs, ventral. 1 mm.

d Chelicerae with rows of teeth and fang. The outlet of the poison gland is situated at the distal tip of the fang (*arrow*). 250 µm.

e Filter apparatus in the mouth region with bundles of setae. 200 µm.

f Pedipalp, distal, with claw. 100 µm

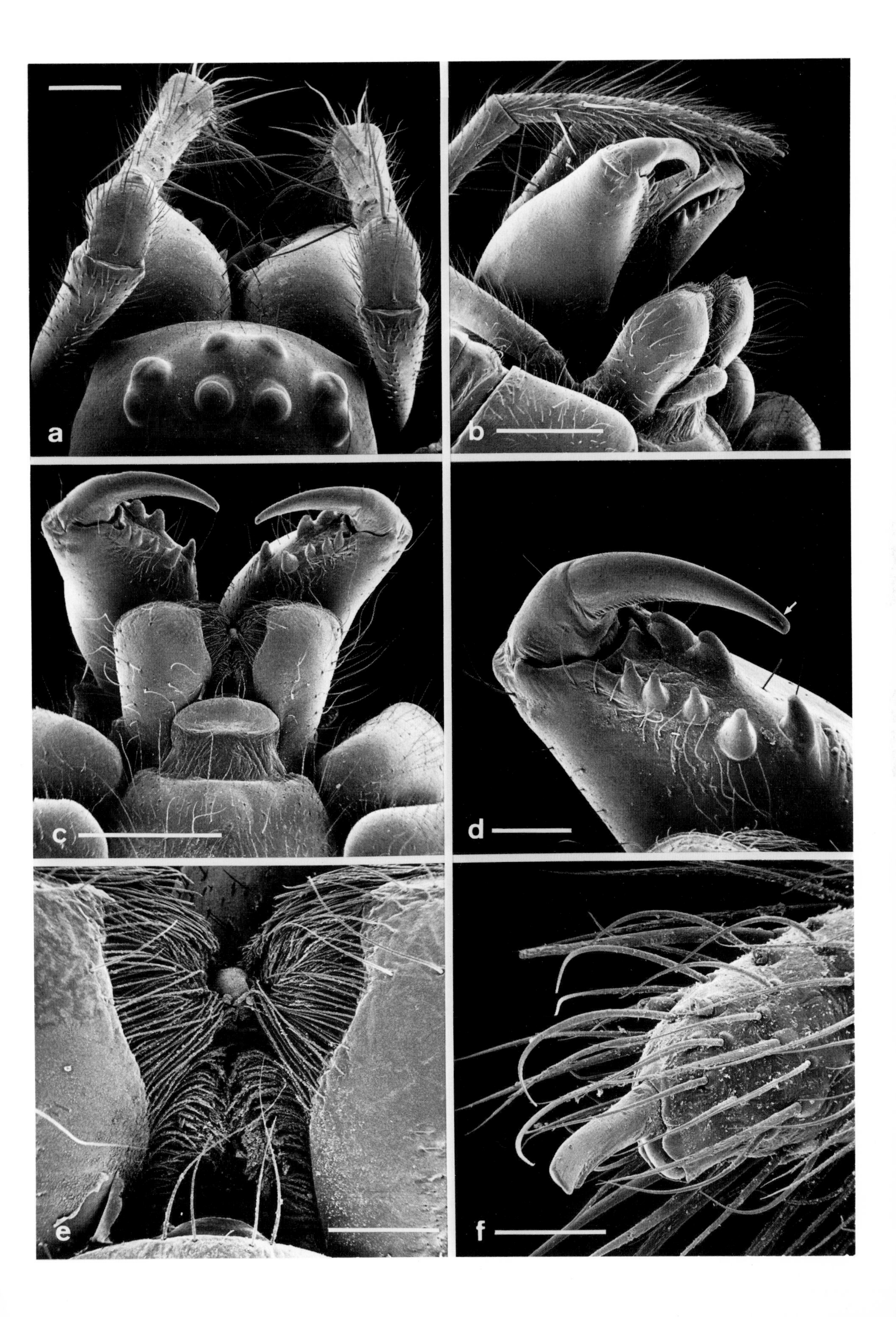
a
b
c
d
e
f

2.1.5 Sheet Web Constructing Spiders

Funnel-web spiders or Agelenidae are well known in human habitations through the house spider *Tegenaria atrica* (Collatz and Mommsen, 1974). Numerous funnel-web spiders live near the ground. They spin horizontal sheet webs on the grass, in the shrub layer and near the ground. At one end of these webs the threads converge to build a funnel-shaped tube where the spider lies in wait for insects to fall into the web.

In the same way baldachin spiders (Linyphiidae) build inverted hammock-like webs supported by vertical silk threads above ground. In contrast to funnel-web spiders the baldachin spider hangs beneath its slightly concave web (Fig. 31). When an insect falls onto the web the spider bites through the threads and pulls its prey downwards.

The predatory life of these spiders has only a limited importance for the soil as their nutrition largely consists of flying insects. Only their metabolic products and food leftovers, which fall to the ground, contribute to the soil contents (Wiehle, 1949).

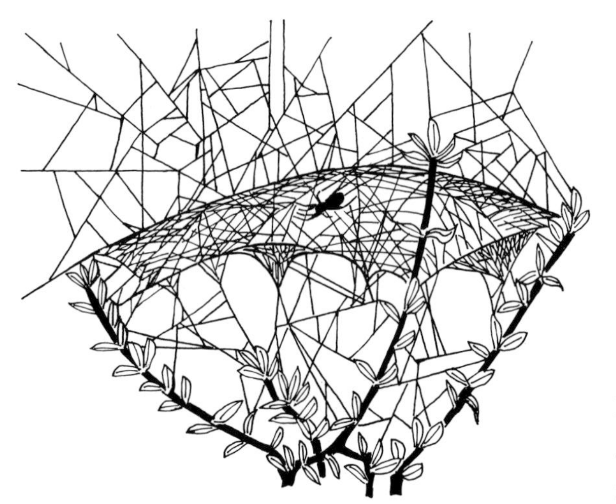

Fig. 31. Sheet web of the baldachin spider, *Linyphia triangularis* (Linyphiidae). The spider hangs upside down beneath the concave part of its hammock-like web. (After Foelix, 1979)

Plate 5. *Macrargus rufus* (Linyphiidae).
a Overview, dorsal. 1 mm.
b Frontal view of the prosoma, chelicerae and pedipalps. 0.5 mm.
c Overview, ventral. 1 mm.
d Prosoma with two rows of eyes. Parts of the left foreleg are shown: femur, patella (*arrow*) and tibia. 250 μm.
e Epigyne at the edge of the epigastric fold (*arrow*), the visible part of the female genital apparatus. 100 μm.
f Pair of eyes (revolved 180°) combined from the front and back row of eyes. 50 μm

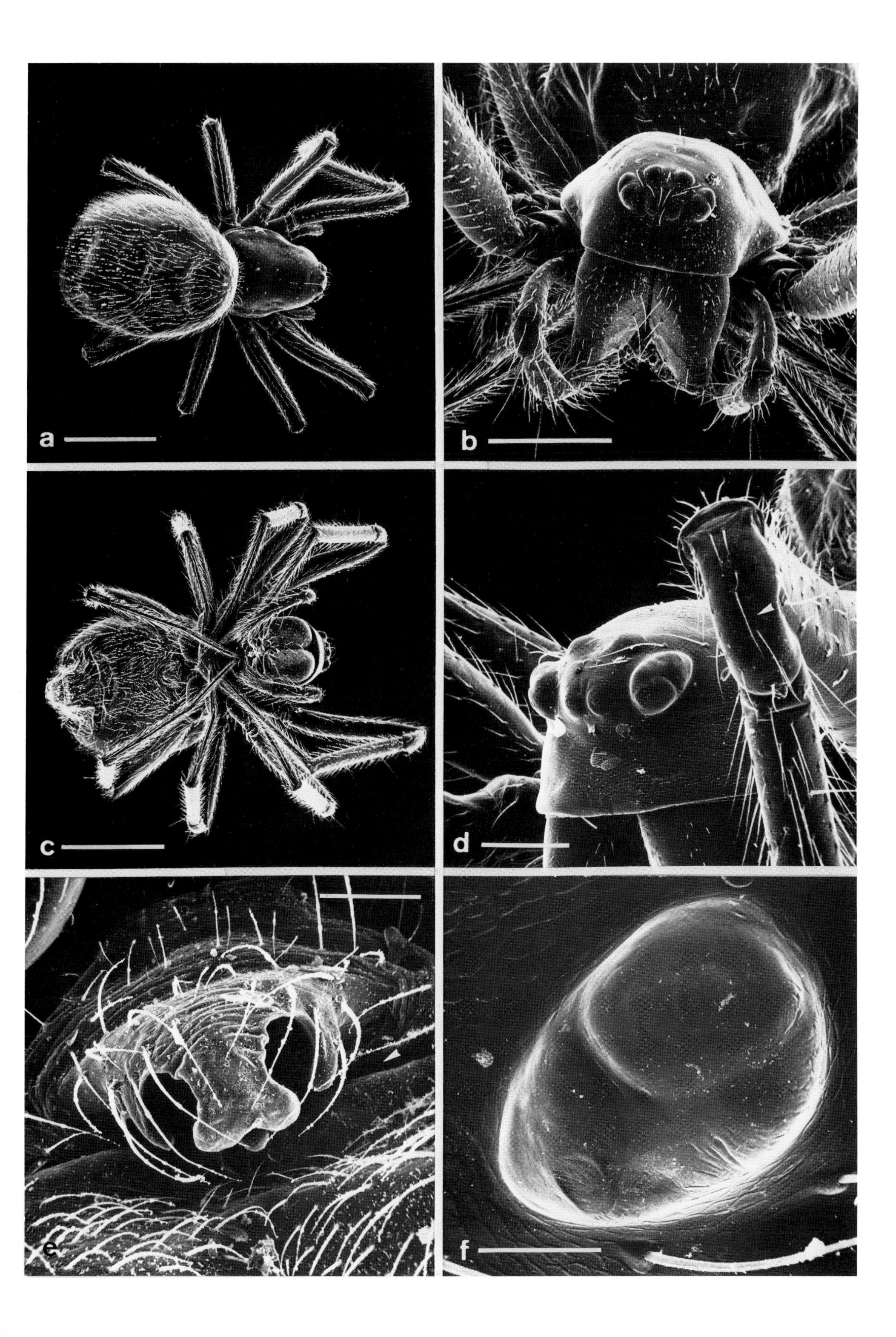
a
b
c
d
e
f

2.1.6 The Spinning Apparatus

All spiders possess silk glands which are situated in the opisthosoma. Different kinds of silk are produced in several (4–8) types of silk glands (Peters, 1965; Richter, 1970; Glatz, 1972, 1973). Their secretory ducts lead to spinning tubes (spools or fusulae, or spigots) situated on the spinning field of spinnerets (Figs. 32–34; Plate 6). The spinnerets are singly or poly-articulated, movable, paired appendages on the opisthosoma. The spinning apparatus usually consists of three pairs of spinnerets. Their phylogenetic origin is thought to be from abdominal limbs, of which there were originally four pairs.

The shape of the spinning tubes standing beside each other on the spinneret tips already indicates the consistency of the silk thread they produce. Tubes which rise from phial-like sockets secrete a solid silk thread. Other, shorter tubes supply threads which only solidify when they reach the air. These threads facilitate better attachment onto the web and the periphery. The thickness of these threads amounts to about 1 μm. Only the fine threads of the trapping wool produced by about 40,000 silk glands on the cribellum of the cribellate spiders are much thinner (about 100–200 Å; Foelix, 1979).

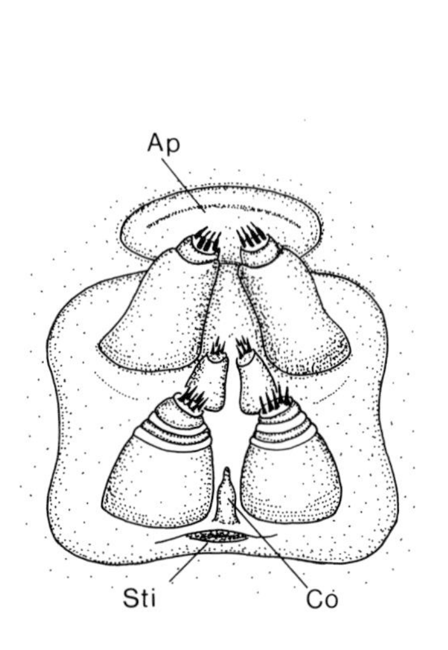

Fig. 32. Arrangement of the spinnerets of *Segestria senoculata* (Dysderidae) (six-eyed spiders). In front of the spiracle (*Sti*) there is a scale-like appendage, the colulus (*Co*), presumably a non-functional relic of a frontal, median spinneret pair in ecribellate spiders. *Ap* Anal plate. (After Glatz, 1972, and Foelix, 1979)

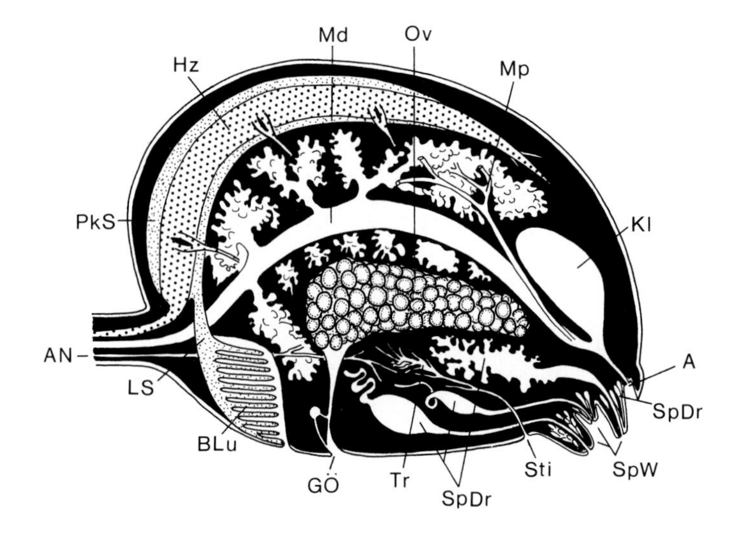

Fig. 33. Schematic diagram of a section through the opisthosoma of a female spider. (After Glatz, 1972, and Foelix, 1979). *A* Anus with anal operculum; *AN* abdominal nerve; *BLu* book lung; *GÖ* genital aperture with receptaculum seminis; *Hz* heart; *Kl* cloaca; *LS* lung sinus; *Md* midgut with diverticuli; *Mp* Malpighian tubules; *Ov* ovary; *PkS* pericardial sinus; *SpDr* silk glands (several different types); *SpW* spinnerets; *Sti* tracheal spiracle; *Tr* tracheal branches

Plate 6. *Amaurobius fenestralis* (Amaurobiidae).
a Spinnerets with anal operculum and cribellum. 200 μm.
b Single anterior spinneret with spinning tubes and setae. 50 μm.
c Spinning field with two types of spinning tubes (numerous small spools and two bigger spigots). 10 μm.
d Spinning tubes (spools or fusulae) of an anterior spinneret. 5 μm.
e Feather hair at the edge of a spinning field. 5 μm.
f Spools distal, with remains of emerging silk. 3 μm

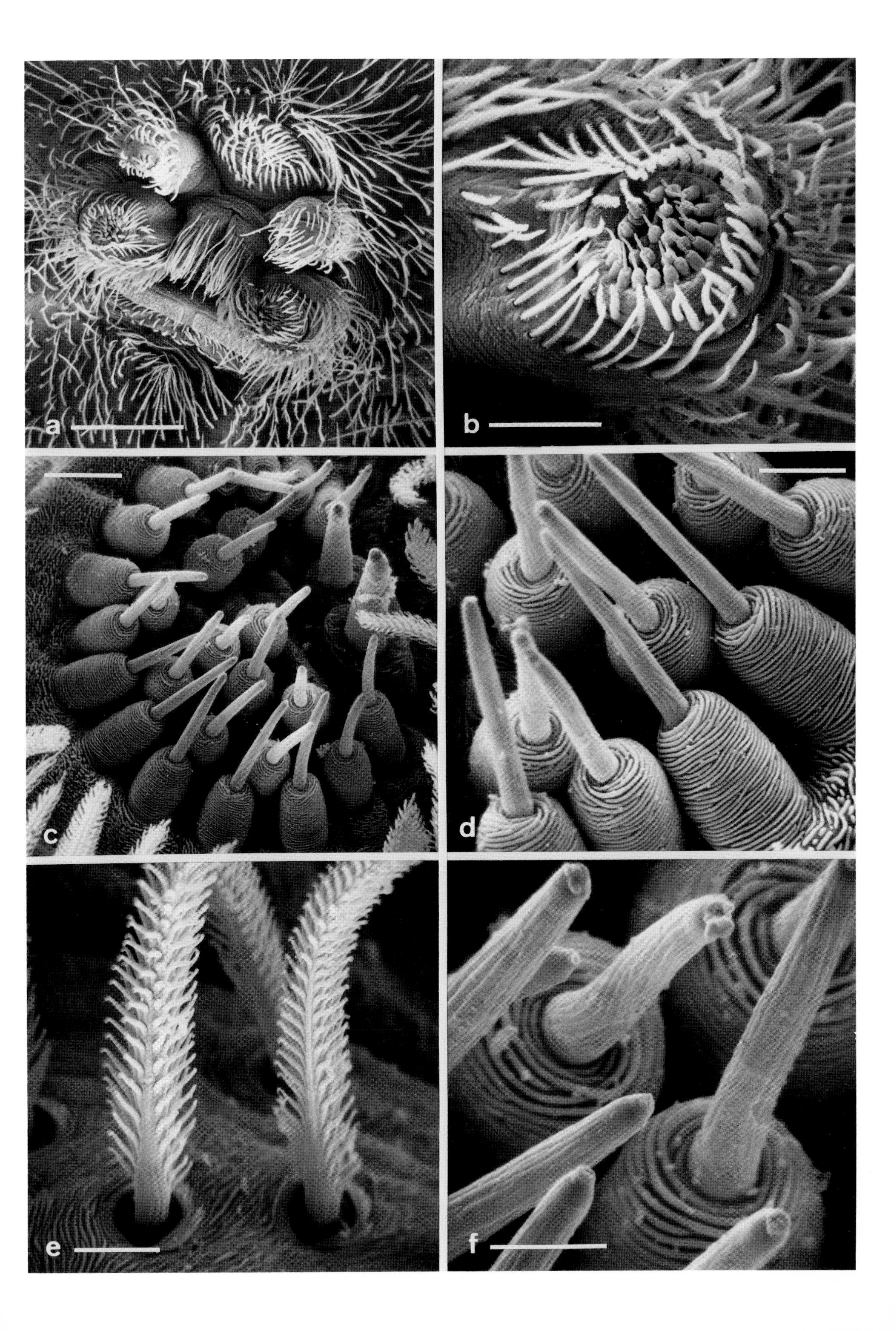

2.1.7 Cribellum and Calamistrum of *Amaurobius*

The Amaurobiidae belong to the cribellate spiders which possess a cribellum as well as three pairs of spinnerets. In *Amaurobius*, as in most cribellate spiders, the cribellum is paired and is inserted in front of the spinnerets in its resting position (Fig. 34; Plate 7). The cribellum is a glandular area with thousands of densely crowded, small spools, known as the tubuli textori. These spools secrete elementary threads with a thickness of only 10–20 nm. The result is a fine wool which is wrapped in one to two axial threads to make trapping threads. The wool is combed out of the glandular area by a bristle comb, the calamistrum, on the metatarsus of the fourth pair of legs (Fig. 35; Plate 8) and laid on the simultaneously spun, axial thread. Cribellate thread owes its effectiveness to the crimped wool in which the extremities of the prey become entangled. Ecribellate spiders produce trapping threads coated with drops of glue in which small animals also become stuck (Lehmensick and Kullmann, 1956; Kullmann, 1969; Friedrich and Langer, 1969).

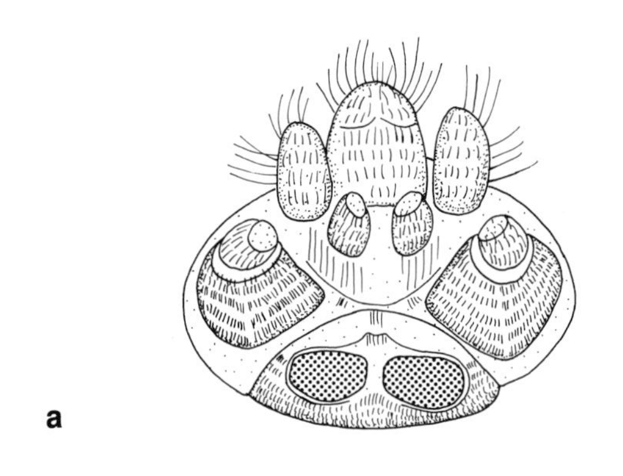

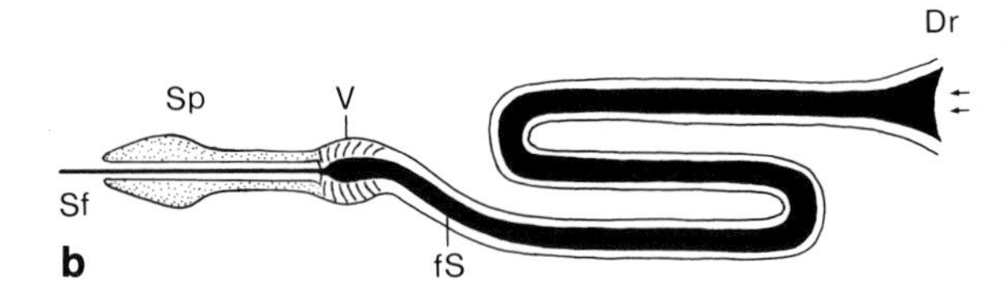

Fig. 34. Arrangement of the spinnerets in a cribellate spider and construction of a spinning tube.
a Cribellum (*dotted area*), spinnerets, and anal operculum of *Eresus niger* (Eresidae). (After Wiehle, 1953). **b** Spinning tube (*Sp*) with glandular duct (*Dr*) in an araneid spider. The thickness of the silk thread (*Sf*) is regulated by a pinch valve (*V*). Liquid silk (*fS*) solidifies when it reaches the air, i.e. already at the base of the nozzle inside the valve, when the valve is narrowed. (After Foelix, 1979)

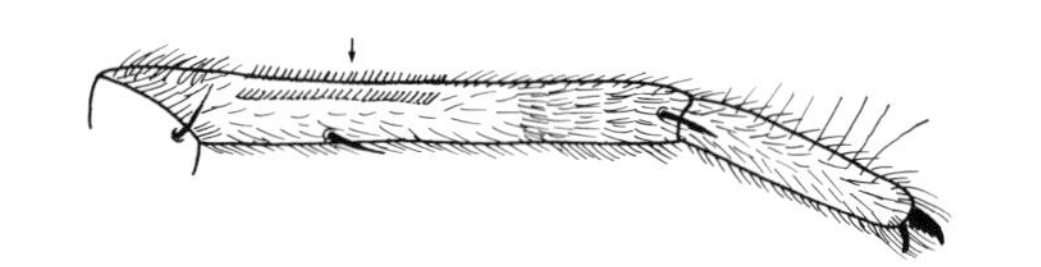

Fig. 35. Calamistrum (double row of bristles) on the metatarsus of the fourth pair of legs of *Amaurobius ferox*. The calamistrum is used to comb cribbellum secretion. (After Wiehle, 1953)

Plate 7. *Amaurobius fenestralis* (Amaurobiidae).
a Spinnerets with cribellum. 200 μm.
b Epigastric fold with lateral entrances to the book lungs. 250 μm.
c Cribellum. 100 μm.
d Book lung entrance. 100 μm.
e Cribellum area, lateral, with fine glandular spools. 10 μm.
f Muscle attachment (*arrow*) and glandular opening in the area anterior to the book lung entrance. Slit (*arrowhead*) possibly belongs to a slit sensory organ. 50 μm

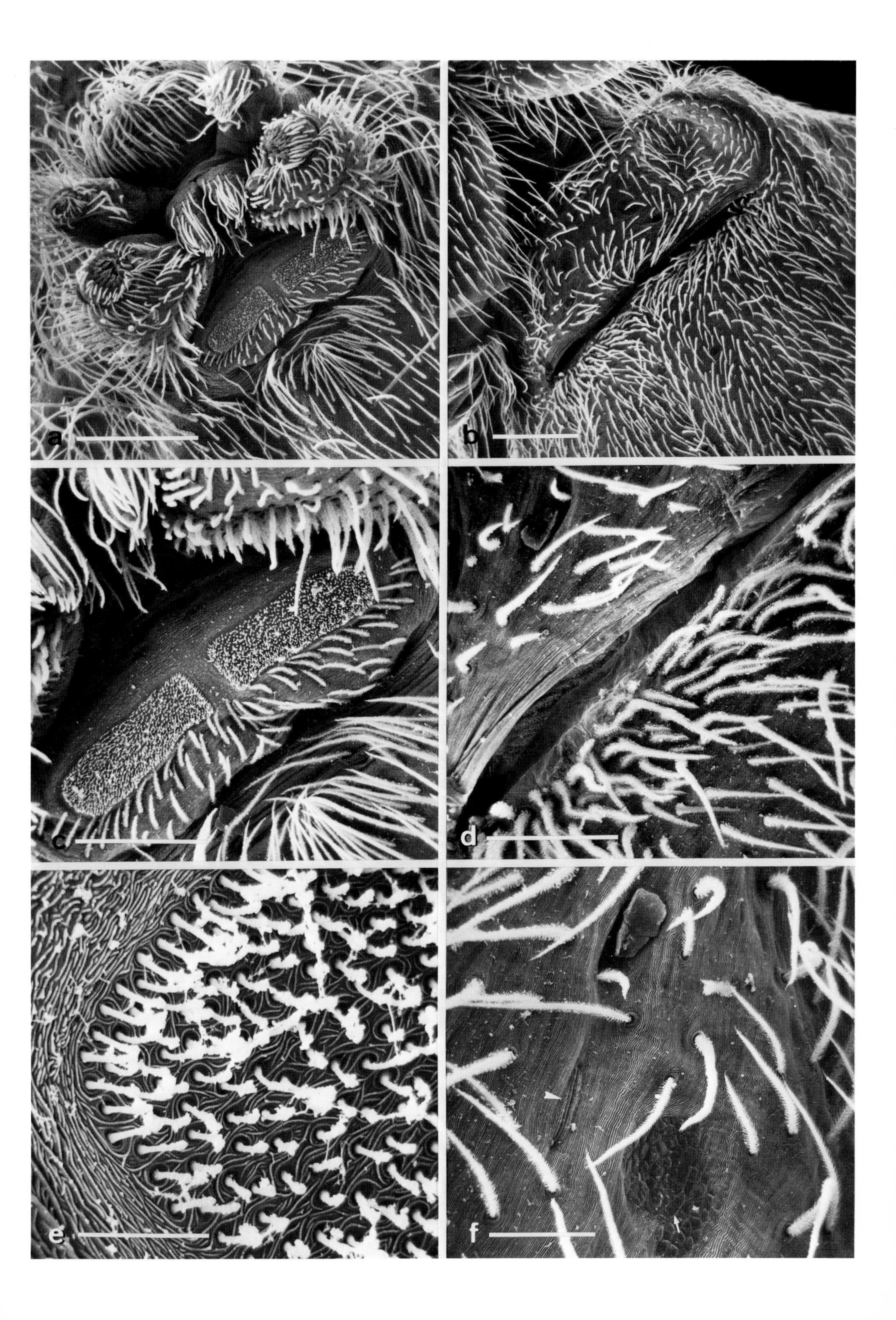
a
b
c
d
e
f

2.1.8 Spiders which Spin Tubular Webs

Many epedaphic and hemiedaphic spiders spin tubular webs which serve as a dwelling and also partly as a trap for prey. The purse-web spider *Atypus piceus* (Atypidae) constructs a deep silk tube which extends several centimetres vertically into the earth and a few centimetres along the ground (Fig. 36). This is the spider's permanent dwelling. It feeds on insects which walk over the horizontal part above ground. These insects are ambushed by the spider, bitten from below and pulled into the tunnel.

As primitive web builders the Amaurobiidae and Eresidae also make tubular webs as dwellings on the ground with several trapping threads radiating from the entrance. They retreat to their tube to lie in wait for small insects which become stuck in the threads. Others, for instance, *Eresus niger*, build a funnel-shaped web over the tunnel which functions both as a protection and as a trap (Fig. 37).

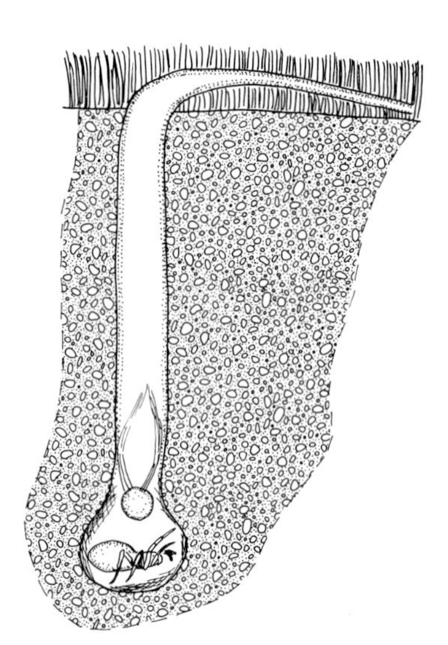

Fig. 36. Tubular cell of an atypid spider, *Atypus piceus*. The web is fastened to the tunnel wall with silk threads. (After Dunger, 1983)

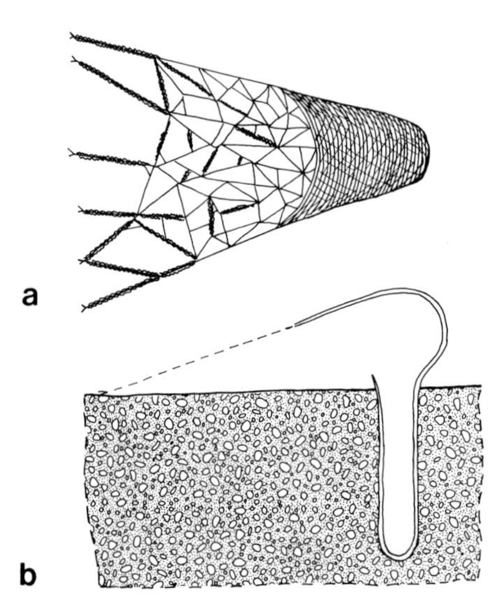

Fig. 37. Funnel web of *Eresus niger* (Eresidae). (After Wiehle, 1953).
a Dorsal view of the web with the tightly woven protective part and loosely woven trapping part. The threads of the trapping web are reinforced by cribellum secretion (trapping wool) in which prey becomes entangled. **b** Schematic section through the tunnel with its funnel-shaped entrance and the web with protective and trapping part

---

Plate 8. *Amaurobius fenestralis* (Amaurobiidae).

- **a** Calamistrum (*arrow*) on the metatarsus of the last pair of legs. 250 μm.
- **b** Calamistrum from the side. 100 μm.
- **c** Calamistrum from above. 250 μm.
- **d–f** Fine structure of the calamistrum bristles. 100 μm, 50 μm, 25 μm

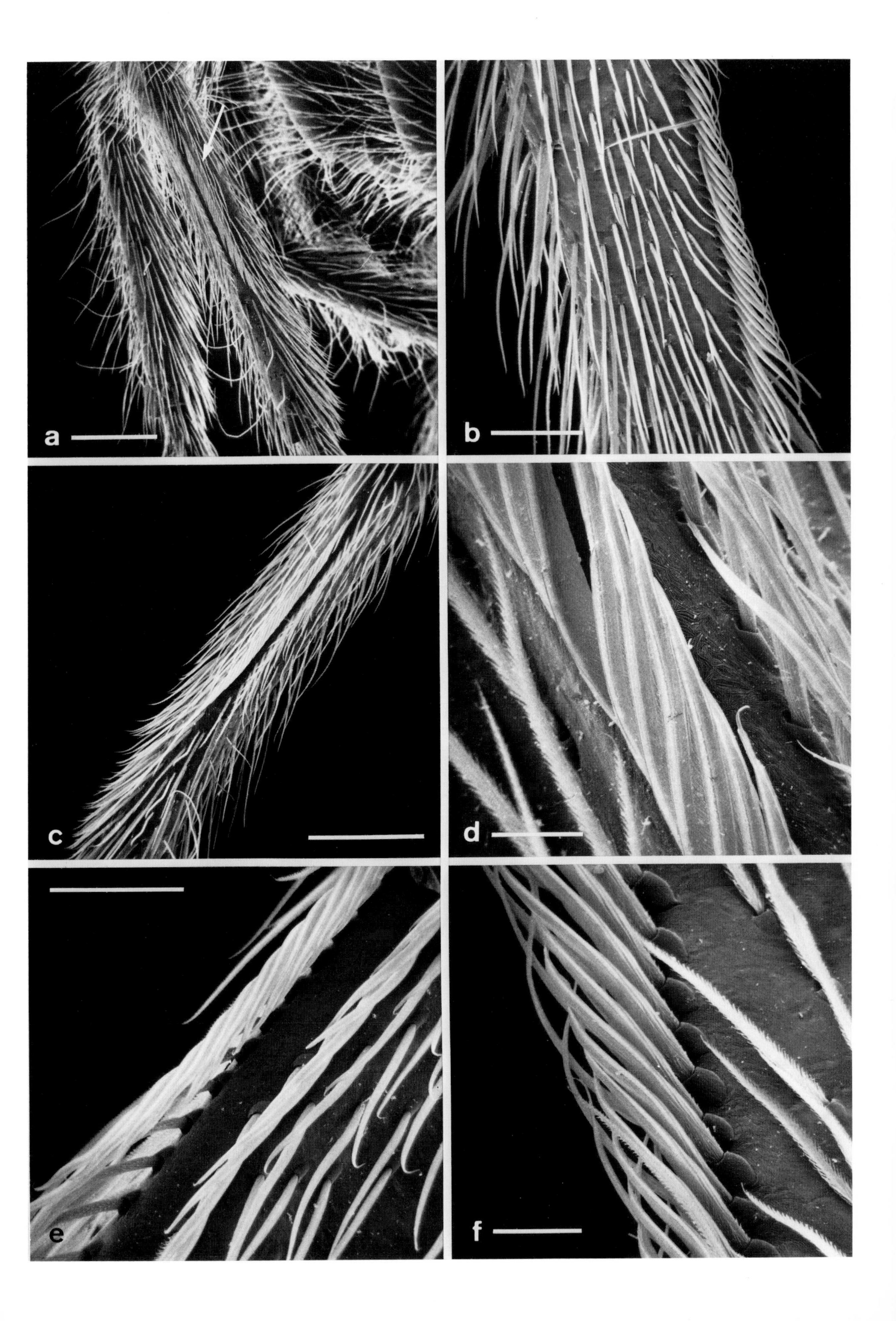
a
b
c
d
e
f

2.1.9 Free Hunting Spiders

The majority of the epedaphic spiders spend the day in their silk-lined cell dwellings, under stones, in leaf litter or moss and become active at night when they go hunting. Lycosid wolf spiders (Lycosidae) (Schaefer, 1972, 1974), clubionid (Clubionidae) and drassodid (Drassodidae) spiders belong to this group. *Callilepis nocturna* (Drassodidae) preys exclusively on ants. The spider searches out ant nests and attacks the ants in front of the nest entrance. It quickly bites its prey at the base of the antenna, then withdraws and waits approximately one minute till the poison has taken effect before devouring the immobilized ant (Heller, 1976).

The small (3–10 mm long), attractive jumping spiders (Salticidae), whose most striking characteristic is their front row of four eyes on the brow area of the prosoma, are very sensitive to optical stimuli (Fig. 38; Plate 9). They can detect their prey from a distance of 10 cm, creep towards it and pounce on it from a short distance away. They always attach a safety thread to the ground before they jump in case their leap carries them into empty space. The structure and function of the eyes of jumping spiders were investigated by Land (1969a, b, 1972), Eakin and Brandenburger (1971) and Homann (1981).

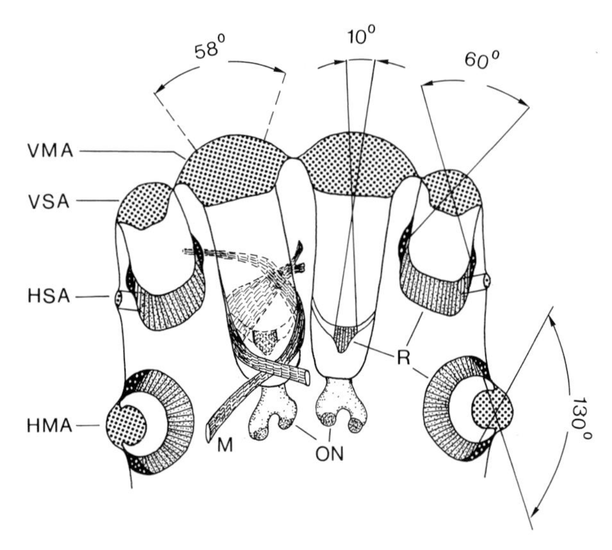

Fig. 38. The horizontal field of vision of a jumping spider (Salticidae) (modified after Foelix, 1979). The anterior eyes (*VMA, VSA*) resemble car headlights. The posterior lateral eyes (*HSA*) have been reduced. The anterior median eyes (*VMA*) have an extremely limited angle of vision of 10°. This is increased to a value of 58° by their mobility, for which the retina muscles (*M*) are responsible.
HMA, HSA Posterior median and posterior lateral eyes; *M* retina muscles; *ON* optic nerve; *R* retina; *VMA, VSA* anterior median and anterior lateral eyes.

Plate 9. *Euophrys* sp. (Salticidae), jumping spiders.
a Side view. 1 mm.
b Frontal view with "headlight" eyes. 0.5 mm.
c Prosoma with large, posterior, median eye. The posterior lateral eye (*arrow*) is reduced. 250 µm.
d Prosoma with "headlight" eyes. 200 µm.
e Anterior median eye. 50 µm.
f Anterior median eye, cuticle of the cornea. 10 µm

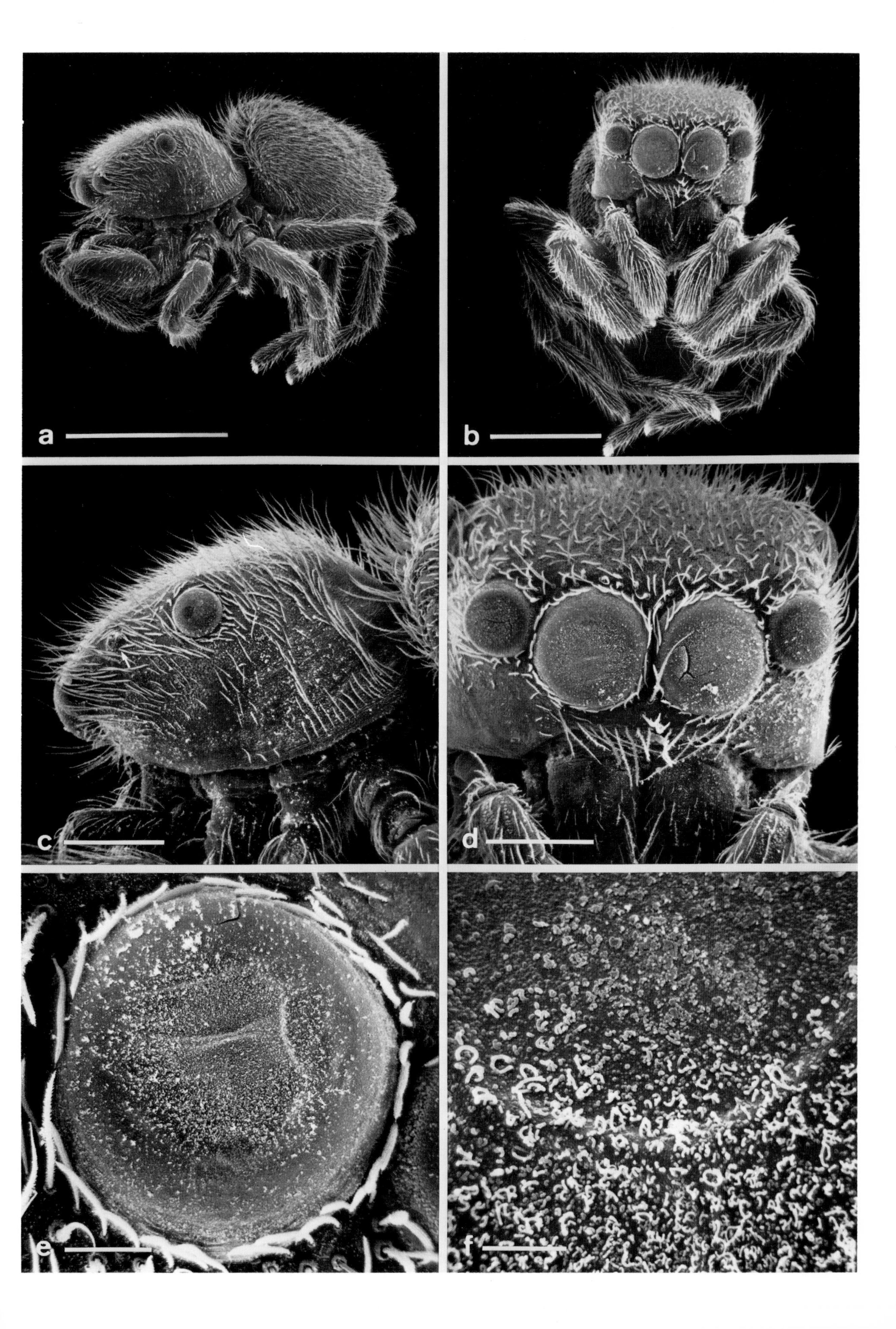
a
b
c
d
e
f

2.1.10 Biotope of Edaphic Spiders

The biotope of edaphic spiders is concentrated in the soil zone as characterized by Duffey (1966). Tubular web building and hunting spiders thus lead hemiedaphic and epedaphic lives. Within these life forms species-characteristic differentiations have taken place as adaptations to the various prevailing microclimatic and biological factors. Thus, species can be assigned to an ecological niche (Baehr, 1983; Baehr and Eisenbeis, 1985).

Nørgaard (1951, 1952) and Gettmann (1976) observed small, soil-dwelling wolf spiders (Lycosidae), which live in their tubular webs during the day, but are nocturnal hunters. *Lycosa pullata* and *Pirata piraticus* both occur in *Sphagnum* moss; *L. pullata* lives on the moss cushion and is exposed to large fluctuations in temperature, while *P. piraticus* prefers the sheltered interior of the moss cushion where temperature variation is less extreme (Fig. 39). In this case, temperature proves to be one of the many ecological factors which characterizes a specific ecological niche.

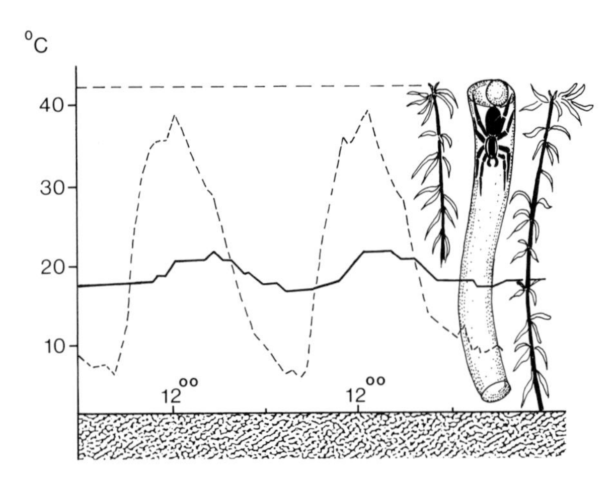

Fig. 39. Microclimate in the biotope of the wolf spider *Pirata piraticus* (Lycosidae). On the surface of the peat moss cushion the circadian temperature variation is approx. +8° to +38 °C (*dotted curve*). In the interior of the moss cushion, where the tubular web is situated, the temperature fluctuates only slightly around 20 °C (continuous curve). When it is sunny females expose their egg sacs to the sunshine at the entrance of their tubular web. (After Nørgaard, 1951, and Foelix, 1979)

Plate 10. *Pirata uliginosus* (Lycosidae), wolf spiders.
a Front view. 1 mm.
b Prosoma and part of the opisthosoma, dorsal. 1 mm.
c Opisthosoma, terminal, with spinnerets and anal tubercle. 250 µm.
d Epigastric fold with epigyne (*arrow*). The *arrowheads* mark areas with actively secreting, glandular hairs. 250 µm.
e Spinnerets and anal tubercle. 200 µm.
f Slender spinning tubes together with long setae of a posterior spinneret. 50 µm

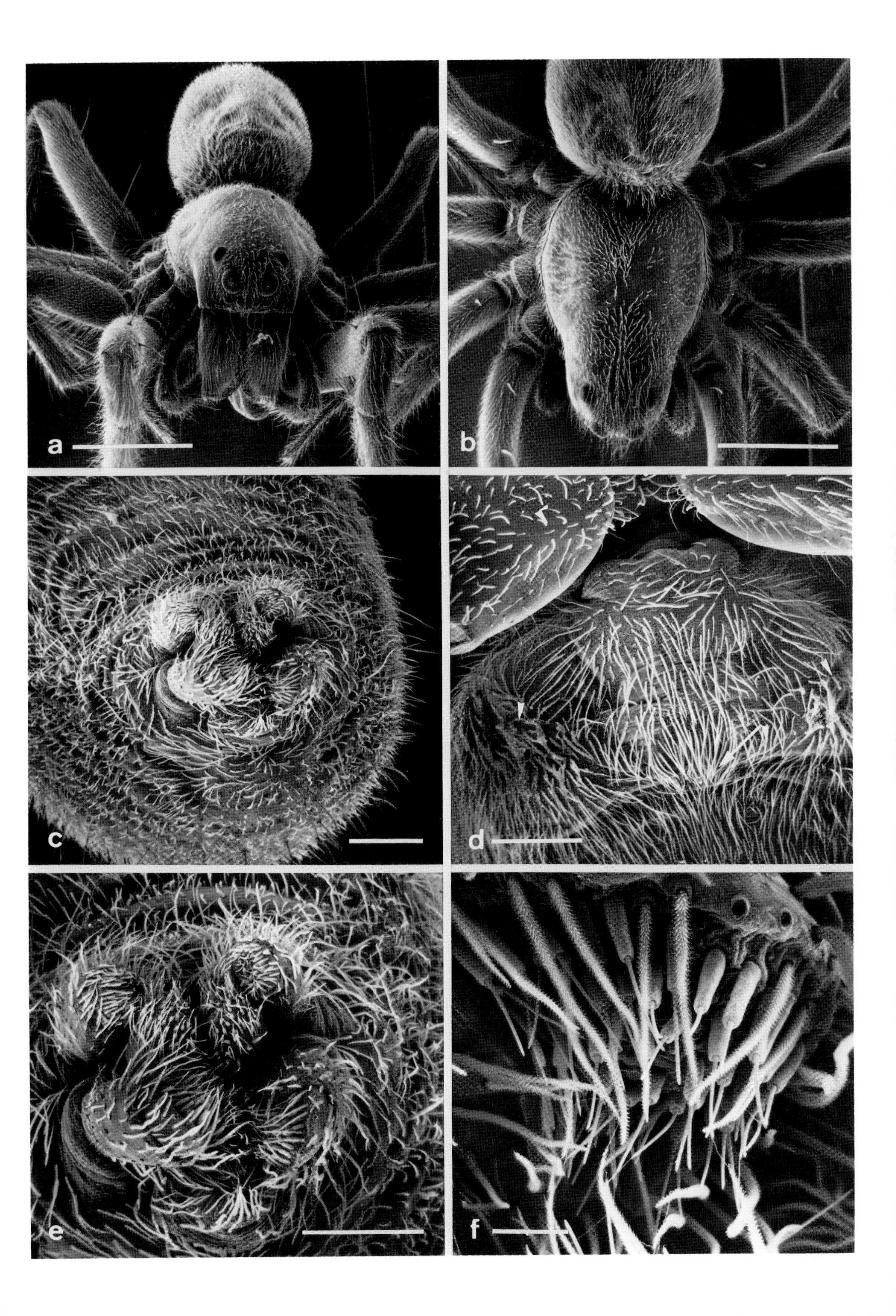
a
b
c
d
e
f

2.1.11 Care of the Young – The Egg Sac

Before laying her eggs the female wolf spider begins to make a cocoon, usually in the following sequence. First, a basic web is woven which is reinforced towards a central basal plate. This plate is formed into a bowl-shaped egg container by adding a ring wall. During the subsequent oviposition the eggs, which have been fertilized in the uterus by sperm stored in the receptaculum seminis, swell out of the genital opening. Then the female spins a cover for the egg container. Finally, the cocoon is enclosed in a mesh of threads to form a papery sheath. This cocoon remains attached to the spinnerets till the young spiderlings emerge (Crome, 1956; Kullmann, 1961; Pötzsch, 1963) (Plates 11 and 12).

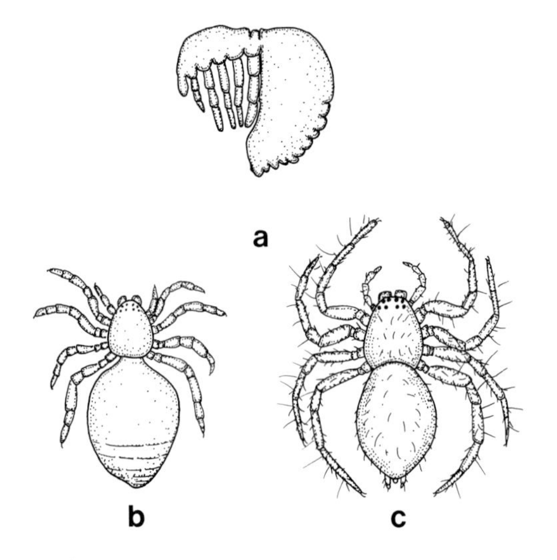

Fig. 40. Juvenile stages of a spider. (After Foelix, 1979, and Vachon, 1957).
a Prelarva with incomplete extremities. It has reached this stage at the end of embryonic development and usually emerges as such from the egg. It subsists on yolk reserves. There are one to two prelarval stages. **b** Larva with fully segmented legs, but hardly any microstructures, such as hairs and spines, on the cuticle. It still lives from yolk reserves, is helpless and hardly mobile. The chelicerae still lack a poison duct and the spinnerets have no spinning tubes. There are one to three larval stages. **c** Nymphal stage with fully developed external structures such as chelicerae, spinnerets and sensory hairs. The genitalia are non-functional. There are five to ten nymphal stages, then moulting to an adult (sexually mature)

Plate 11. *Pirata uliginosus* (Lycosidae), wolf spiders.
a Cocoon with pore zone. 0.5 mm.
b Silk mesh of the outer wall. 20 μm.
c Pore zone. 100 μm.
d Cocoon mesh, from exterior. 5 μm.
e Pore in the cocoon wall. 10 μm.
f Cocoon mesh, from exterior. 3 μm

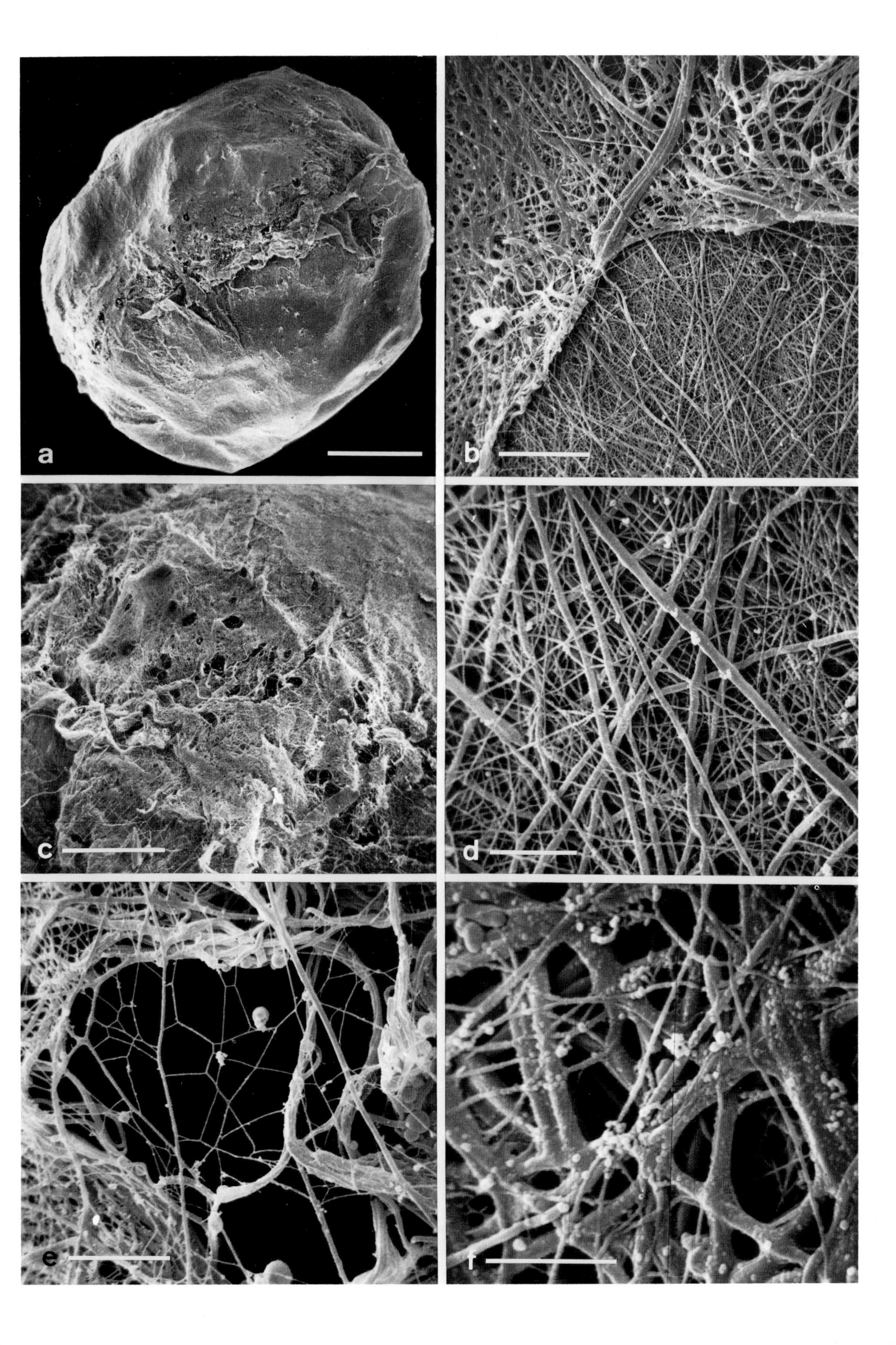
a
b
c
d
e
f

2.1.12 Care of the Young – Brood Care

Wolf spider females help their young to emerge from the egg sac by cutting an opening at the side of the solid cocoon with their chelicerae. Once free the young spiderlings climb on to their mother's back where they cling to hairs. Up to 100 spiderlings, closely packed and on top of each other, have been counted on their mother's back. Thus protected they are carried around for several days until their yolk reserves are exhausted (Engelhardt, 1964; Higashi and Rovner, 1975; Foelix, 1979).

Many spider mothers not only make and guard the egg sac, they also tend their young after they have left the cocoon. In addition to protecting them, they have also been observed to feed the spiderlings either with liquid nourishment regurgitated from their gut (Fig. 41) or immediately with prey they have caught. This behaviour has been observed in the soil-dwelling, funnel-web spider *Coelotes terrestris* (Agelenidae). The spiderlings remain for about a month in the mother's funnel web. They use their palps and front legs to stroke the chelicerae of their mother, who then drops the prey in front of them. One of the spiderlings carries its prey to a safe place to suck it out in peace.

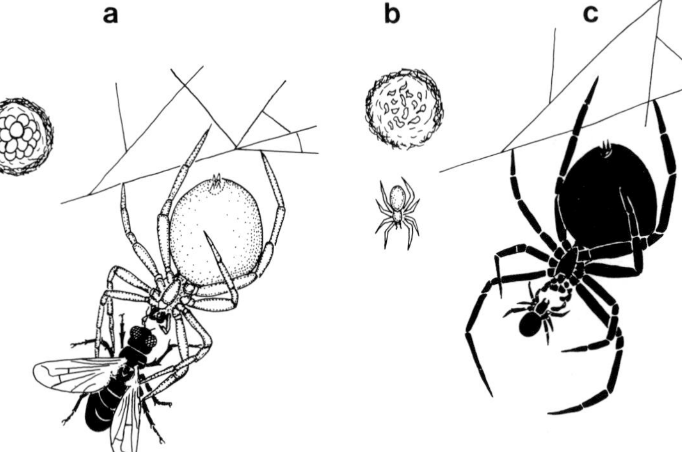

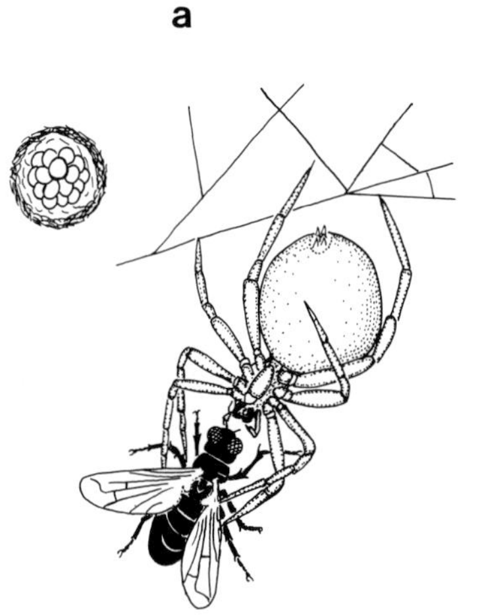

Fig. 41. Evidence of mouth-to-mouth feeding of a spiderling (with regurgitated food) by its mother. (After Foelix, 1979, and Kullmann and Kloft, 1969). **a** Spider mother ingests the liquified body content of a radioactively labelled fly. **b** A young, unlabelled spider hatches from the cocoon. **c** The mother feeds the juvenile spider. The spiderling becomes radioactive after feeding

Plate 12. *Pirata hygrophilus* (Lycosidae), wolf spiders.
a Cocoon attached to the anterior spinnerets. 200 μm.
b Young larvae emerging. The cocoon wall has been cut open. 1 mm.
c Young larva, lateral. 250 μm.
d Opithosoma of a young larva with embryonic segmentation. 200 μm.
e Spinnerets and anal tubercle of a young larva. As yet the spinnerets are non-functional as they lack spinning tubes. 50 μm.
f Cuticle of the opisthosoma. No real hairs have yet been formed. 20 μm

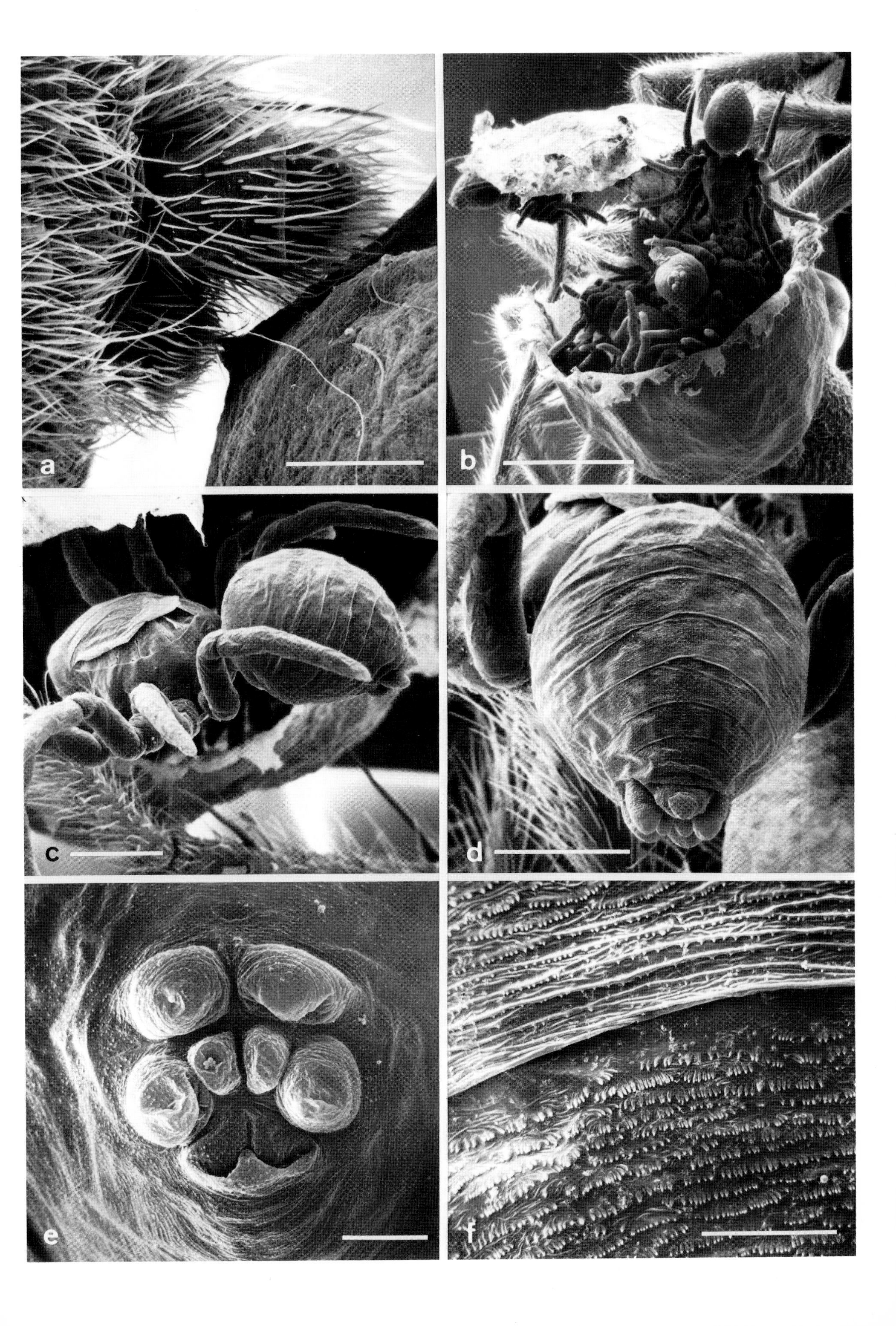
a
b
c
d
e
f

2.1.13 Surface Structures of a Spider

The surface of a spider's body is covered by numerous hairs. *Pirata uliginosus* has feather-like hairs which are mechanoreceptors and as such perceive tactile stimuli. Besides these simple hair sensilla, trichobothria are often distributed over the body and the appendages.

The fine structure of the cuticular surface of *P. uliginosus* consists of strikingly regular lamellae due to fine folding of the cuticle. This surface structure occurs in many spiders but has only been documented so far in the water spider, *Argyroneta aquatica* (Agelenidae) (Braun, 1931; Kullmann and Stern, 1975).

The layer of air which surrounds the opisthosoma and sternum of the prosoma of the water spider is retained by feathery setae, also called pillar hairs for that reason, not by the cuticular structure. This layer of air allows the water spider to breathe when submerged and its function corresponds to that of the compressible gas gills of aquatic insects.

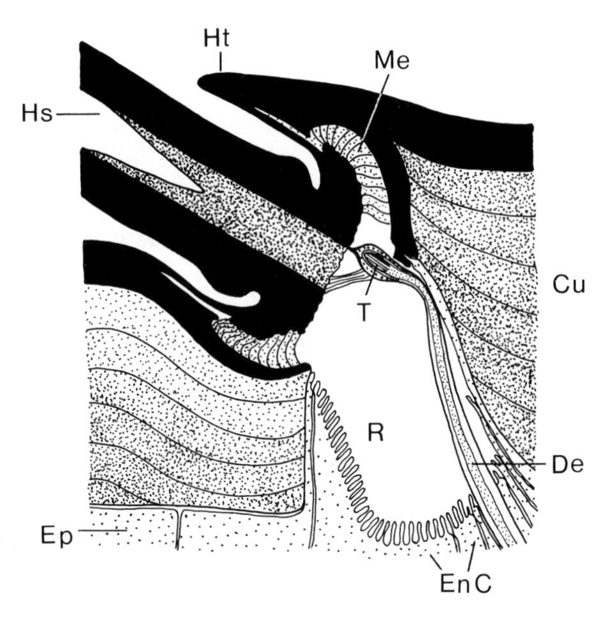

Fig. 42. Diagram of the base of a tactile hair (mechanoreceptor). The dendrite (*De*) extends from the sense cell (not shown) to the hair shaft (*Hs*) within an insulating sheath and terminates in the tubular body (*T*). The tubular body is regarded as an important structure in the transformation of the mechanical stimulus into electrical impulses, as it is displaced (compressed or extended) by movement of the hair. The receptor lymph cavity (*R*) is bounded by envelope cells (*EnC*) which, as well as their secretory function, also maintain a certain ion milieu in the lymph cavity.
Cu Cuticle; *Ep* epidermis; *Ht* hair socket; *Me* suspending membrane, flexible. (After Foelix, 1979)

Plate 13. *Pirata uliginosus* (Lycosidae), wolf spiders.

a Opisthosoma, dorsal, with hair sensilla and glandular area. 20 μm.

b Opisthosoma, complex glandular area. 20 μm.

c Opisthosoma, basic structure of the cuticle. This pattern is typical for many spiders. 3 μm.

d Opisthosoma, cuticle with hair base and glandular area divided into four parts. 10 μm.

e Opisthosoma, base of hair shaft, with hair socket. 5 μm.

f Opisthosoma, glandular area divided into two parts. 10 μm

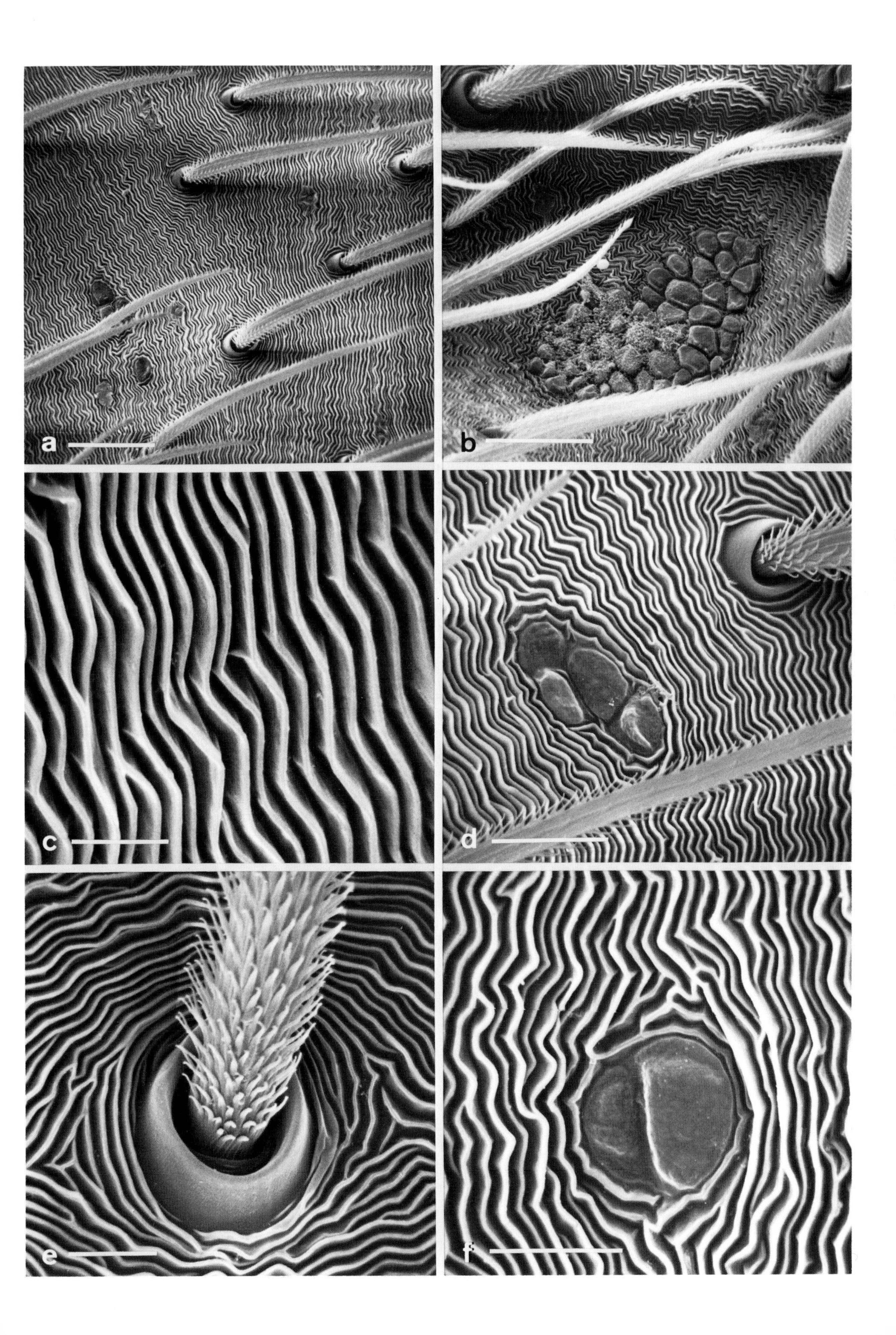
a
b
c
d
e
f

2.2 Order: Pseudoscorpiones – Pseudoscorpions (Arachnida)

General Literature: Kaestner, 1927; Vachon, 1949; Beier, 1963; Weygoldt, 1966a; 1969

2.2.1 Characteristics of Pseudoscorpions

The animals which constitute the order of pseudoscorpions are of attractive appearance. They are distributed among three taxonomic groups, the suborders of the Chthoniinea, Neobisiinea and Cheliferinea (Beier, 1963). Though only 2–4-mm-long, the animals are conspicuous for their large pedipalps which form powerful pincers at the distal ends (Plate 14). Together with the chelicerae these belong to the external mouthparts. Pseudoscorpions participate as carnivores in the biocenosis of the soil, preying on the mesofauna; their diet consists partly of Collembola, mites and nematodes. Not all species of these arachnids, which are predominantly distributed in tropical and subtropical regions, inhabit forests, the bark of trees and the leaf litter layer, but of the 22 species occurring in central Europe almost half exclusively prefer the soil as their biotope (Beier, 1950; Ressl and Beier, 1958; Gabbutt and Vachon, 1965; Goddard, 1976; Wäger, 1982; Braun and Beck, 1986).

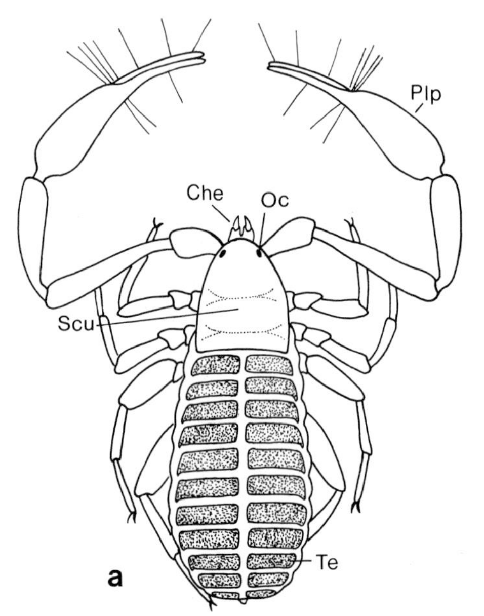

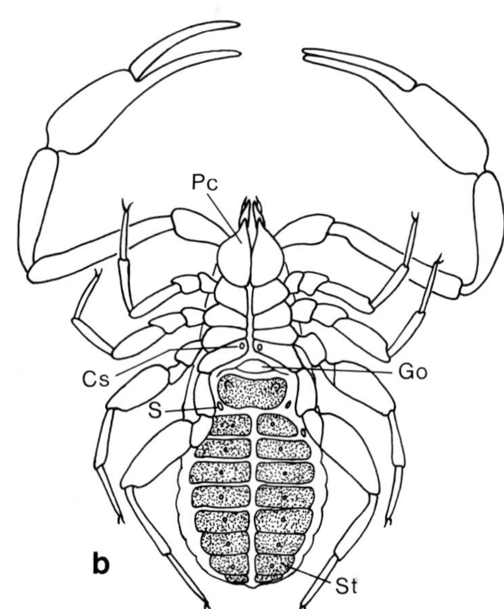

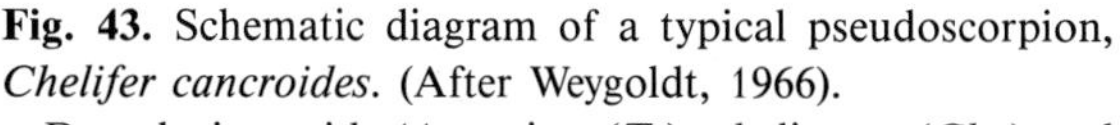

Fig. 43. Schematic diagram of a typical pseudoscorpion, *Chelifer cancroides*. (After Weygoldt, 1966).
a Dorsal view with 11 tergites (*Te*), chelicerae (*Che*) and pedipalps (*Plp*). *Oc* Eye; *Scu* scutum.
b Ventral view with nine sternites (*St*). The mouth region is bordered by the enlarged basal segments (coxae) of the pedipalps (*Pc*). *Cs* Coxal sac; *S* tracheal spiracle; *Go* genital operculum

Plate 14. *Neobisium* sp. (Neobisiidae).
a Dorsal view with pedipalps, chelicerae, scutum and opisthosoma as well as four pairs of walking legs. The tergites of the opisthosoma are not divided. 0.5 mm.
b Ventral view. The basal segments of the legs and pedipalps are arranged in one plane. 1 mm.
c Frontal view showing the powerfully developed pincers on the pedipalps. 1 mm

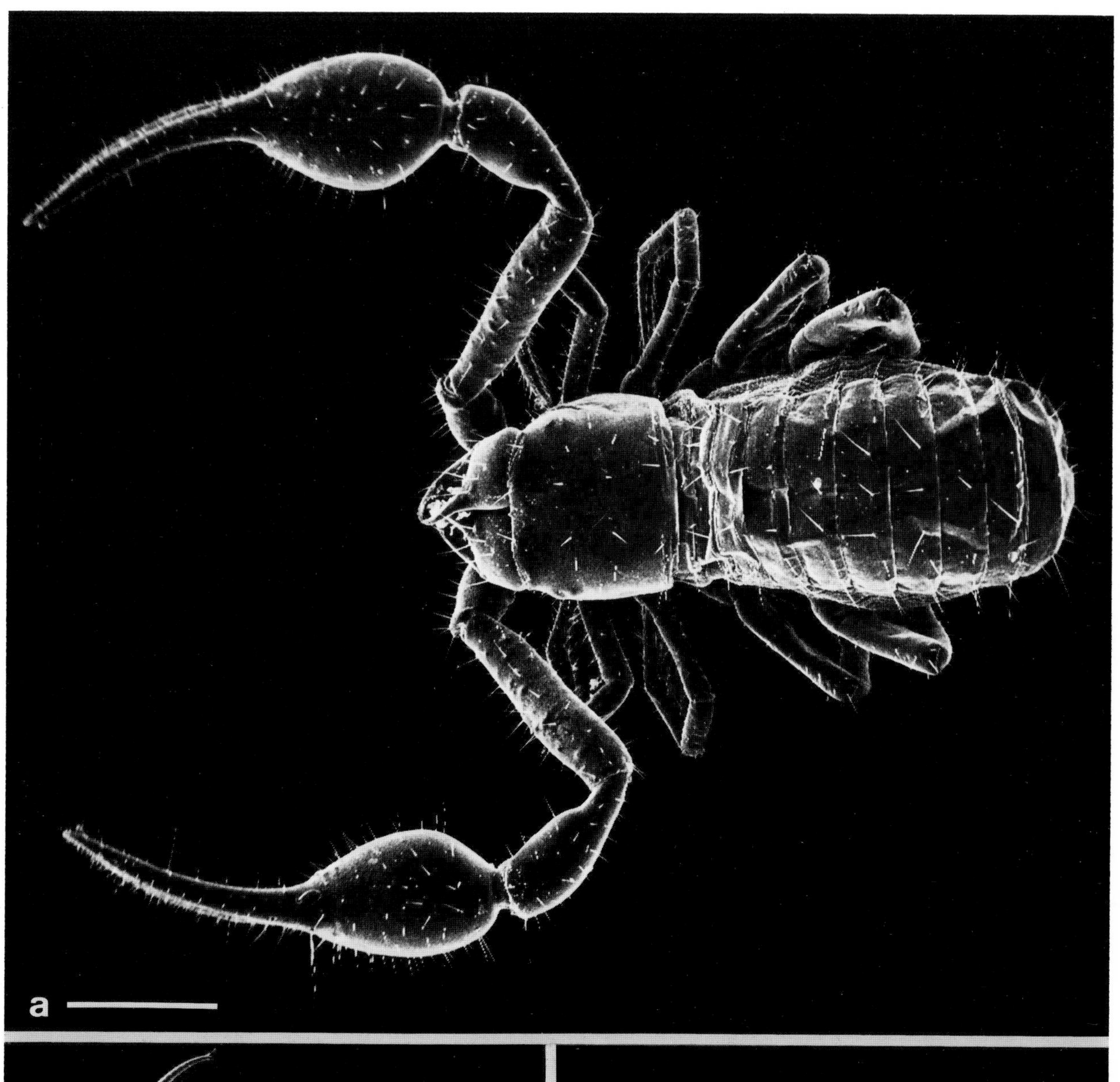
a

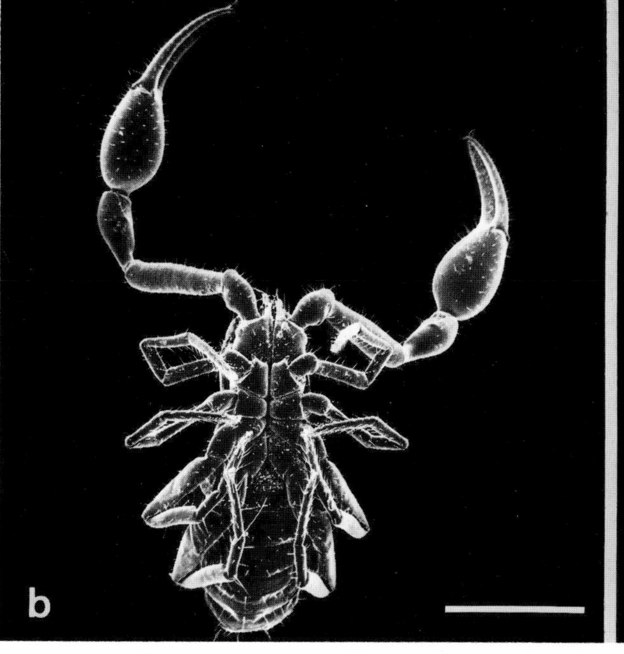
b

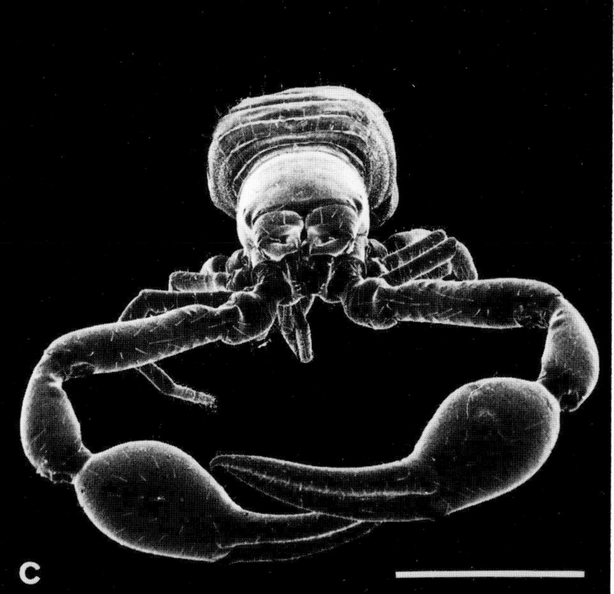
c

2.2.2 Predatory Behaviour in Pseudoscorpions

Most pseudoscorpions actively seek their food. They detect their prey with the help of trichobothria on the large pincers of the pedipalps (Plate 15), rapidly seize it with the teeth of the pincer fingers and convey it – sometimes still struggling – to their chelicerae. The pseudoscorpions obviously possess poison glands (Fig. 44) to immobilize larger prey. These glands are variously arranged with ducts opening on the inwards curving tip of the pincers. Pseudoscorpions thrust or bite into the body of their prey with their chelicerae and inject it with digestive fluid. The liquid food is subsequently sucked out of the body until only an empty shell remains (Fig. 45) (Vachon, 1949; Ressl and Beier, 1958).

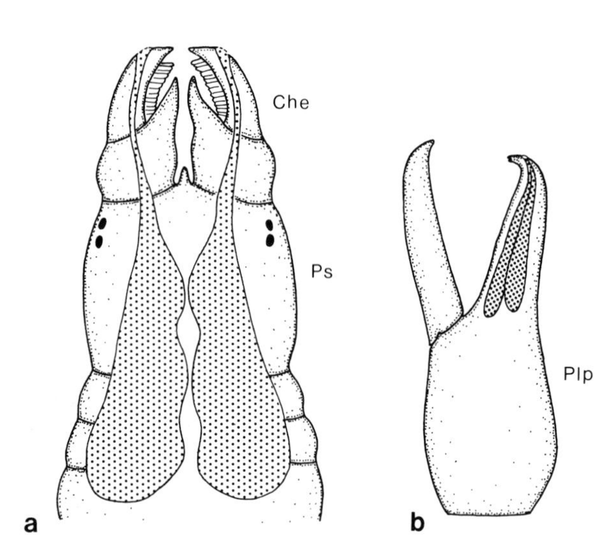

Fig. 44. Arrangement of the mouthparts of pseudoscorpions. (After Vachon, 1949).
a Position of silk glands which extend from the opisthosoma through the prosoma (*Ps*) to the opening in the movable cheliceral fingers (*Che*) of *Neobisium simoni*. **b** Poison glands in the immovable finger of the palp of *Neobisium flexifemoratum*

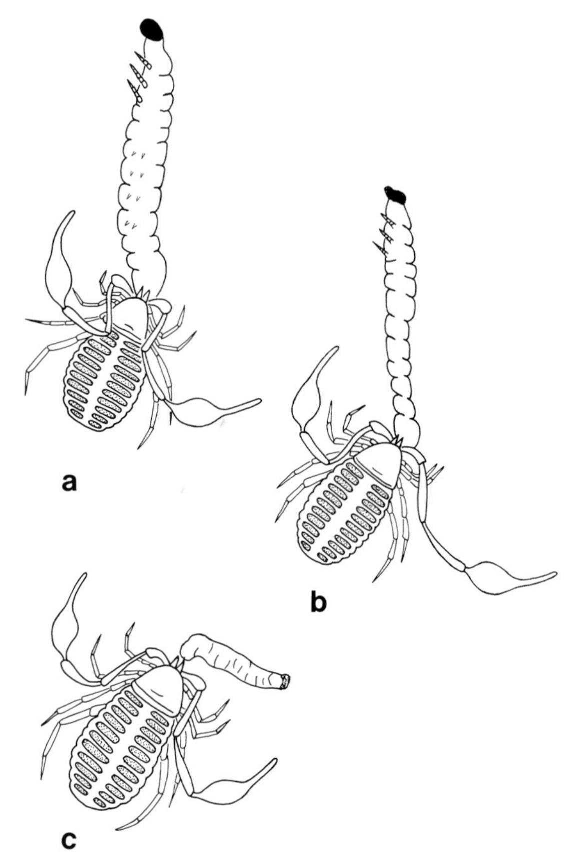

Fig. 45. *Chelifer cancroides* (Cheliferidae), three phases of ingestion of a flour moth larva. (After Roewer, 1940).
a Extraintestinal digestive phase. A slightly alkaline digestive fluid from the glandular intestine is injected through the mouth into the larva. Its body swells. **b** Ingestion: the liquified "chyme" is sucked out of the larva. **c** The shrivelled-up body of the larva remains

Plate 15. *Neobisium* sp. (Neobisiidae).
a Hand of the pedipalp. 0.5 mm.
b Pedipalp, rows of teeth on the pincer fingers. 10 µm.
c Pedipalp hand, distal, with movable and immovable pincer finger. The long, slender sensory hairs, the trichobothria, which are extremely sensitive to air vibrations are characteristic (see enlarged holes in the cuticular surface). 200 µm.
d Pedipalp, rows of teeth on the pincer fingers. 5 µm.
e Pedipalp, view of a median part of the pincer fingers with rows of teeth and trichobothria. 50 µm.
f Teeth on pincer fingers, with microridges on the teeth surface. 3 µm

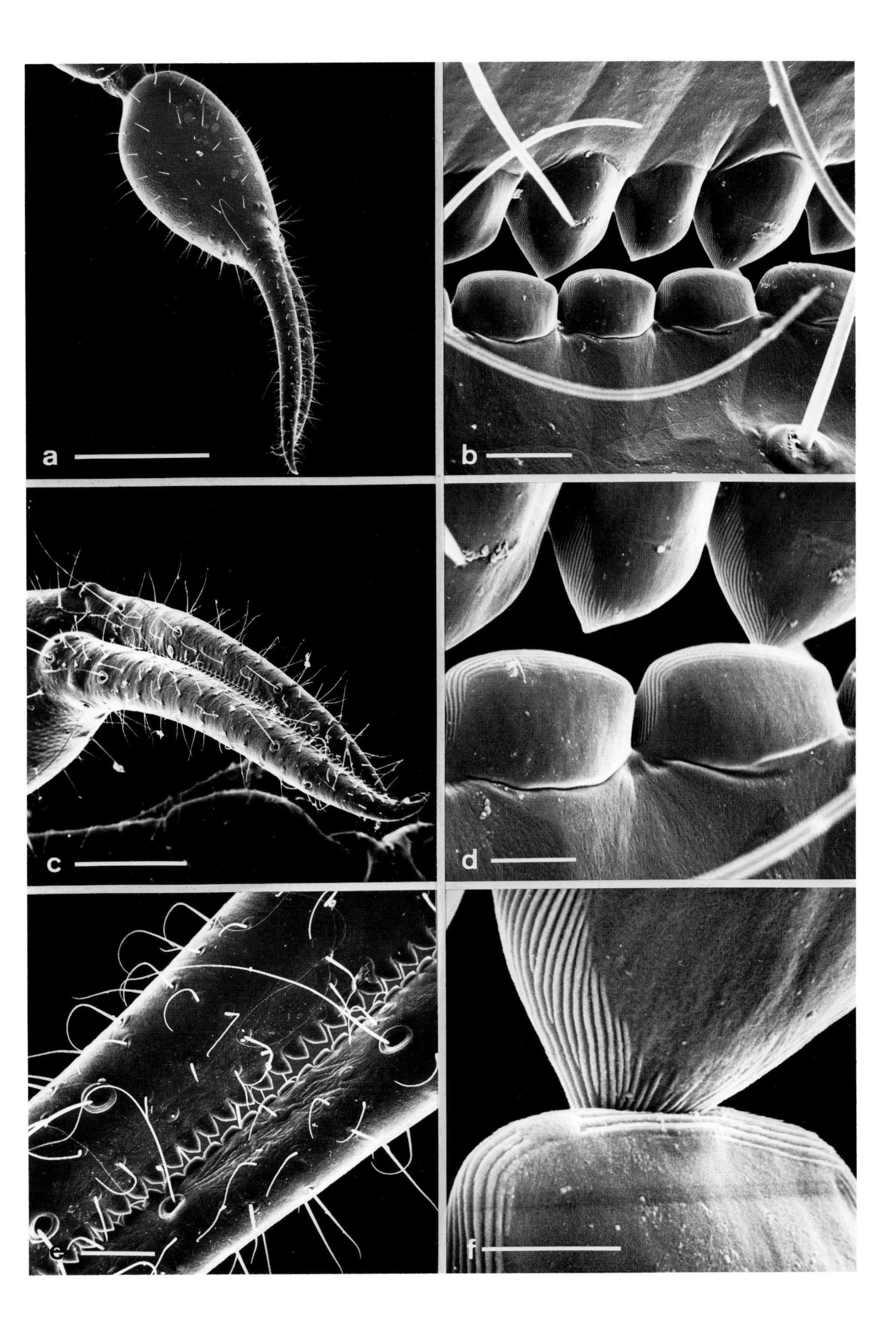
a
b
c
d
e
f

2.2.3 Food Preferences

The predatory pseudoscorpions feed on small soil animals, especially the numerous Collembola. Though epedaphic Collembola of the upper soil layers are part of their diet, not all Collembola species are consumed. The euedaphic Onychiuridae are rejected. As their furca is either reduced or completely lost due to adaptation to life in the air-filled pore system, they cannot spring to escape from predators. Instead they have numerous pseudocelli distributed over their body surface which emit a caustic liquid to repulse their enemies (Karg, 1962 from Weygoldt, 1966a).

Other small soil animals, such as mites and the larvae of various soil arthropods, also provide nutrition for the pseudoscorpions.

Thus, they probably play a regulatory role within the soil community as a result of their predatory way of life (Fig. 46).

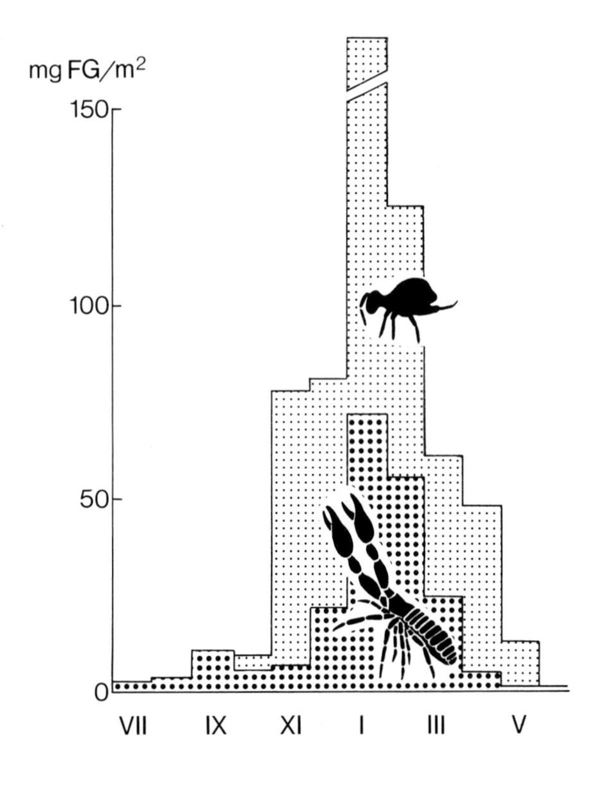

Fig. 46. Activity density of predator and prey populations from soil litter of an acid-humus beech wood for the pseudoscorpion *Neobisium muscorum* and the collembolan *Dicyrtoma ornata* (Beck, 1983). *mg FG/m*^{2} milligram fresh body mass per square metre

Plate 16. *Neobisium* sp. (Neobisiidae).

a Prosoma, lateral, with scutum, chelicerae and leg bases. On the front lateral area of the scutum two lateral eyes like ocelli are visible. 0.5 mm.

b Prosoma, ventral 0.5 mm.

c Chelicerae and basal segments of the pedipalps (coxognathids), lateral. 100 µm.

d View of the mouth region with chelicerae and basal segments of the pedipalps (coxognathids). 100 µm.

e Pincer fingers of the chelicerae with movable and immovable fingers. 100 µm.

f Cleaning combs on the movable fingers. 50 µm

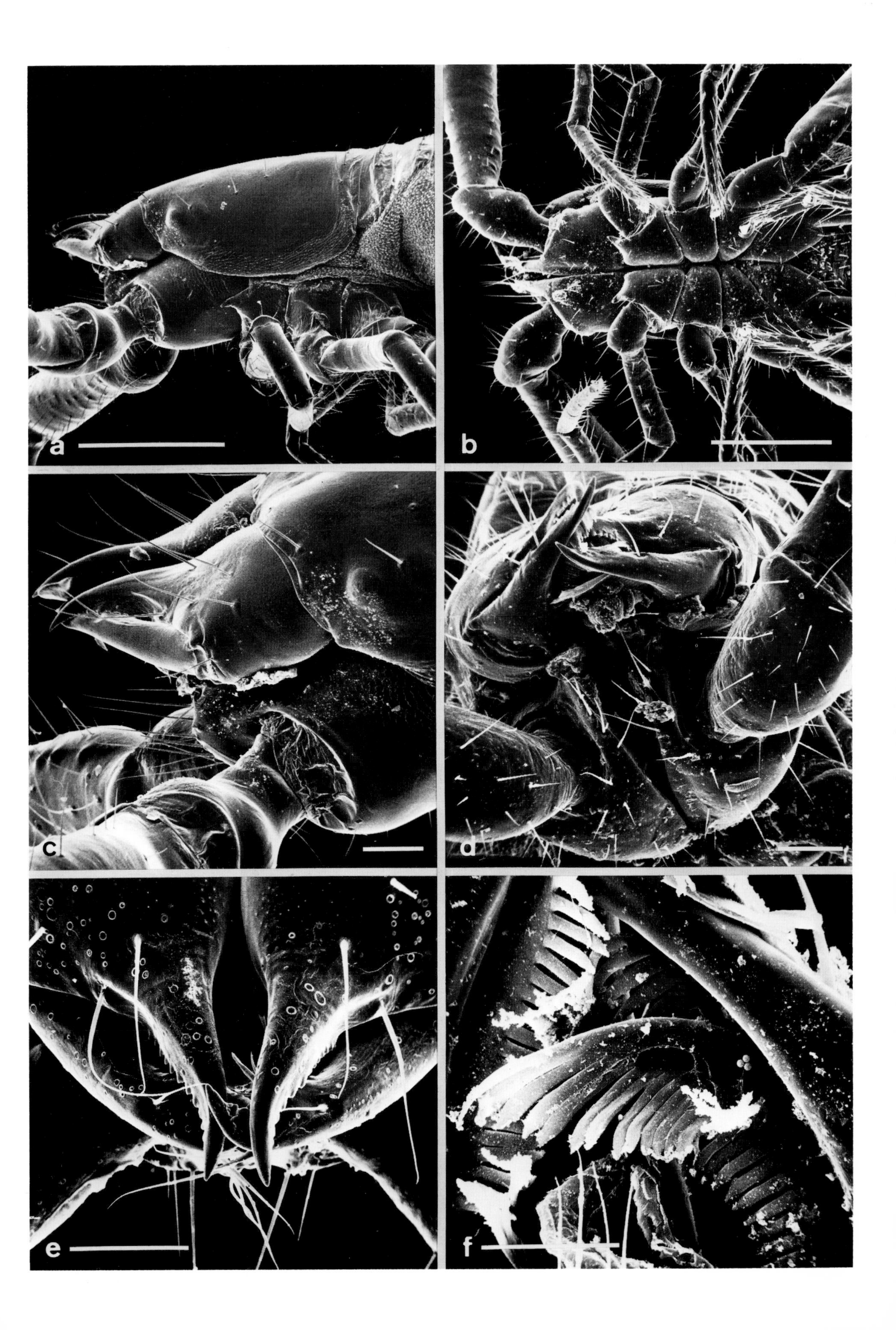

a
b
c
d
e
f

2.2.4 Mating and Brood Biology

Pseudoscorpions belong to those soil arthropods in which transfer of sperm from the male to the female is achieved indirectly, by means of a spermatophore. Apparently unexpectedly and without any sign of courtship (Chthoniidae, Neobisiidae) or after a courtship dance (Chernitidae, Cheliferidae) the male presses his genital aperture onto the ground and rises slowly up again. A thread with a drop of sperm on top emerges from the genital aperture. Sperm transfer without courtship necessitates production of a large number of spermatophores at the appropriate humidity, so that the female is attracted by the scent of the sperm and takes them into her genital aperture. In contrast, transfer of sperm from the spermatophore to the female is achieved during the courtship dance by the male's skilful guidance.

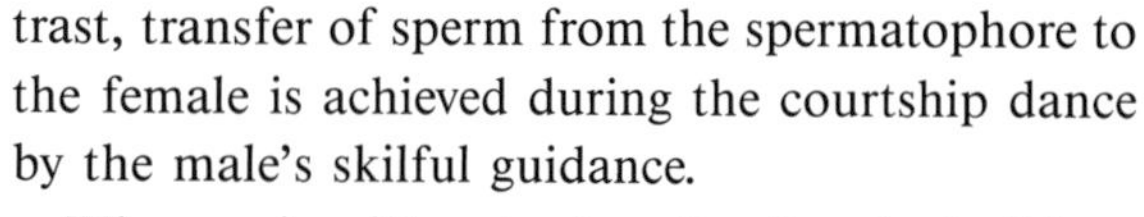

When oviposition is due the female builds a brood nest of grains of sand and tiny fragments of plants which are grasped in the pedipalps, placed in the appropriate position on the ground and glued together with the secretion from the silk gland on the chelicerae to form a small circular wall. The female then retreats into the completed nest. In most species she carries 10–40 eggs in a small brood sac under her genital opening (Fig. 47). The brood sac is made of a secretion from the glands in the genital region. This secretion forms a lattice which hardens in the air (Janetschek, 1948; Schaller, 1962; Weygoldt, 1965; 1966 a, b).

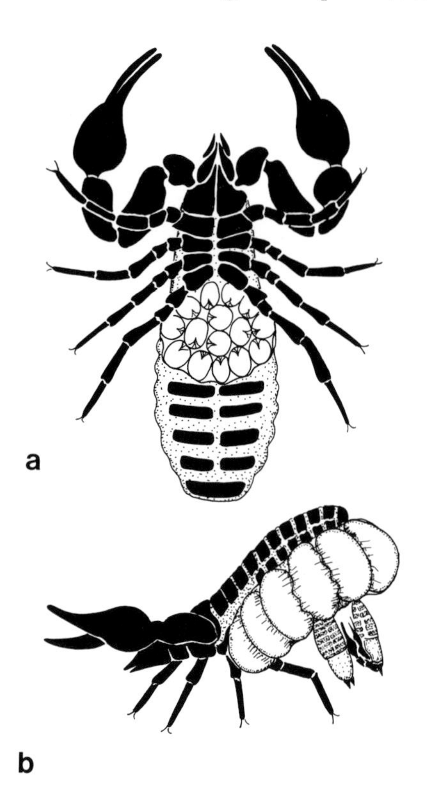

Fig. 47. Brood care in *Pselaphochernes.* (Weygoldt, 1966). **a** The brood sac is still small and is attached to the ventral side of the opisthosoma. **b** The brood sac occupies the whole of the ventral side of the abdomen. Embryonal development is complete, the young protonymphs are in the process of emerging

---

Plate 17. *Neobisium* sp. (Neobisiidae).
a Opisthosoma, lateral, with tergites, sternites and granular skin on the pleura. 0.5 mm.
b Opisthosoma, obliquely caudal, with anal segment. 100 μm.
c Opisthosoma, tergal and pleural cuticle. 100 μm.
d Anal segment with anal papilla. 100 μm.
e Opisthosoma, dorsal, with bristle hairs along the tergite fold. 50 μm.
f Tergite, dorsal, base of a bristle and porous cuticle. 5 μm.
g Tarsus of a leg with two claws (ungues) and a tarsal sac (arolium) which improves the adhesive function of the tarsus. 25 μm

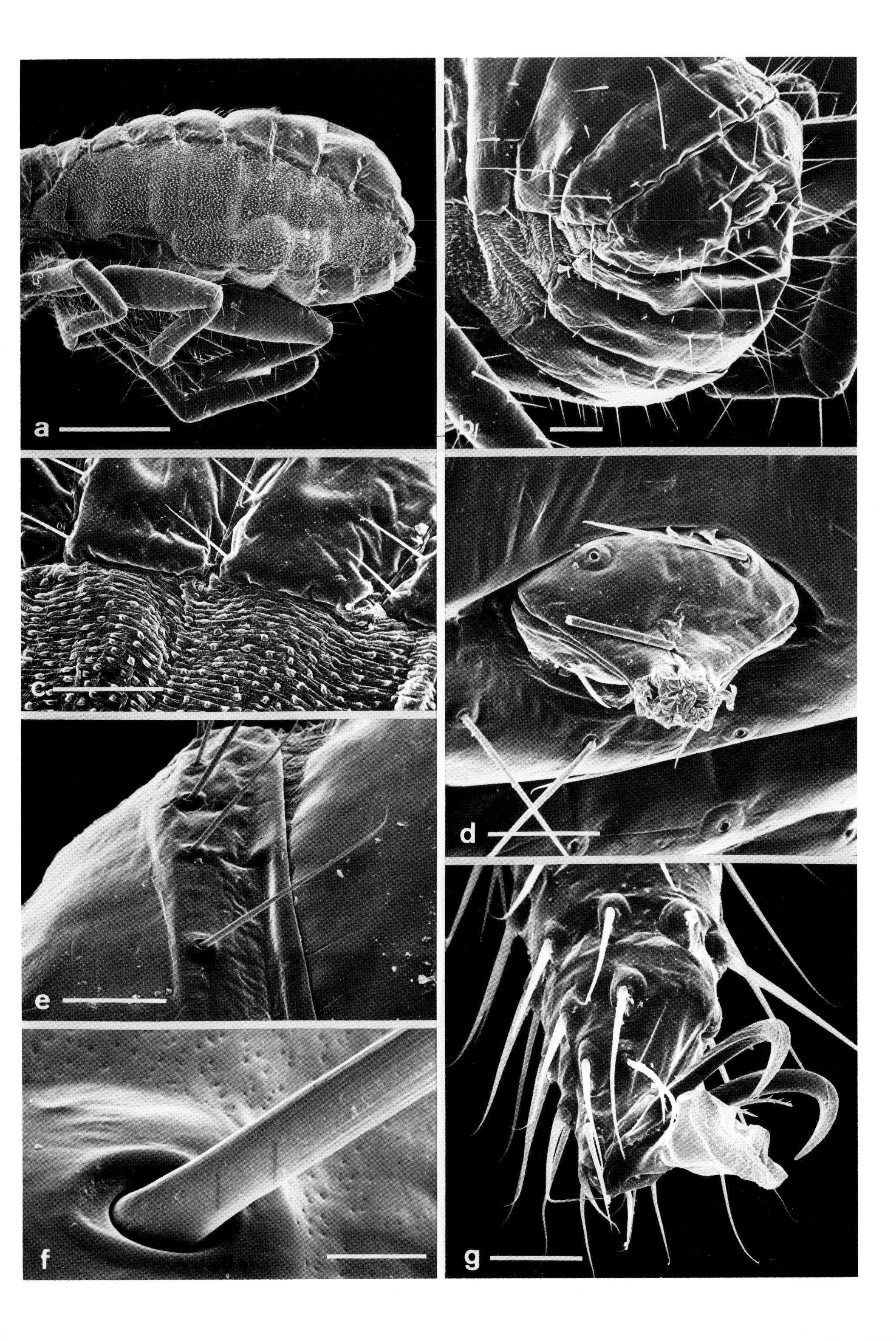
a
b
c
d
e
f
g

2.3 Order: Opiliones – Harvestmen (Arachnida)

General Literature: Kaestner, 1928, 1935–1937; Martens, 1978, 1986

2.3.1 Anatomy of Harvestmen

Harvestmen are arachnids, conspicuous when rapid movements of their long stilt-like legs carry the swinging egg-shaped or round body over the ground. The prosoma (= cephalothorax) and the segmented opisthosoma (= abdomen) are united over a broad area (Fig. 48), giving their body its compact shape. The extremities are situated on the prosoma: the toothed chelicerae, divided into three segments, often bear secondary male sexual characters (apophyses and glands) (Martens, 1967, 1973; Martens and Schawaller, 1977), the pedipalps, which look like walking legs and are equipped with sensory hairs (Rimsky-Korsakow, 1924; Wachmann, 1970), and serve as tactile organs or raptorial feet for obtaining food as well as clinging organs for climbing (Martens, 1978), and finally the walking legs (Barth and Stagl, 1976; Gnatzy, 1982) (Plate 24) possessing slit sensory organs and with poly-segmented tarsi which are thrown like a lasso round blades of grass and thin branches to support the agile harvestmen.

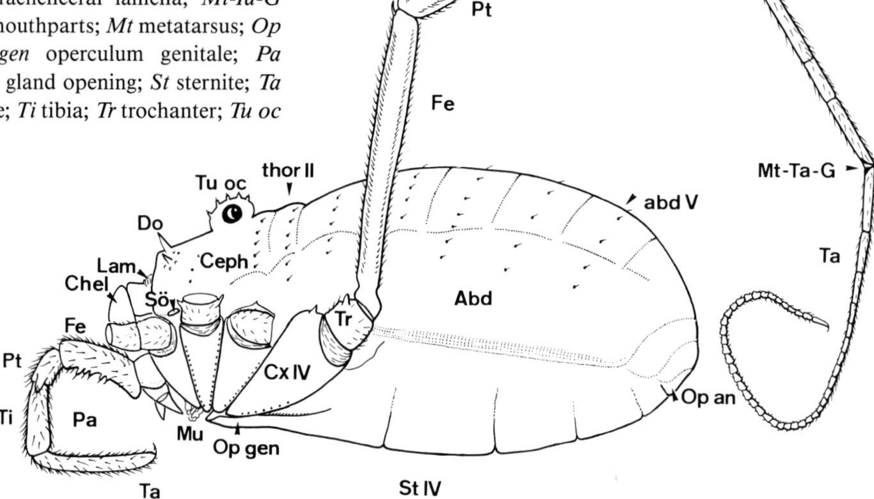

Fig. 48. Lateral view showing the arrangement of a harvestman's body using the Phalangiidae as an example. (Martens, 1978). *Abd* Abdomen (opisthosoma); *Ceph* cephalothorax (prosoma); *Chel* chelicerae; *Cx* coxa; *Do* frontal group of spines; *Fe* femur; *Lam* supracheliceral lamella; *Mt-Ta-G* metatarsus-tarsus joint; *Mu* mouthparts; *Mt* metatarsus; *Op an* operculum anale; *Op gen* operculum genitale; *Pa* pedipalp; *Pt* patella; *Sö* stink gland opening; *St* sternite; *Ta* tarsus; *Thor II* thoracic tergite; *Ti* tibia; *Tr* trochanter; *Tu oc* tuber oculorum

Plate 18. *Paranemastoma quadripunctatum* (Nemastomidae).

a Overview, frontal. *Arrows* mark the outlets of the cheliceral glands. 0.5 mm.
b Dorsal process on the opisthosoma with glandular outlets. 25 μm.
c Opisthosoma, cuticle from the pleural region. 50 μm.
d Tergal cuticle of the opisthosoma with tubercles. 20 μm.
e Pincers of the chelicerae, dorsal. 50 μm.
f Eye tubercle (tuber oculorum), lateral. 100 μm.
g Comb on the chelicerae. 10 μm

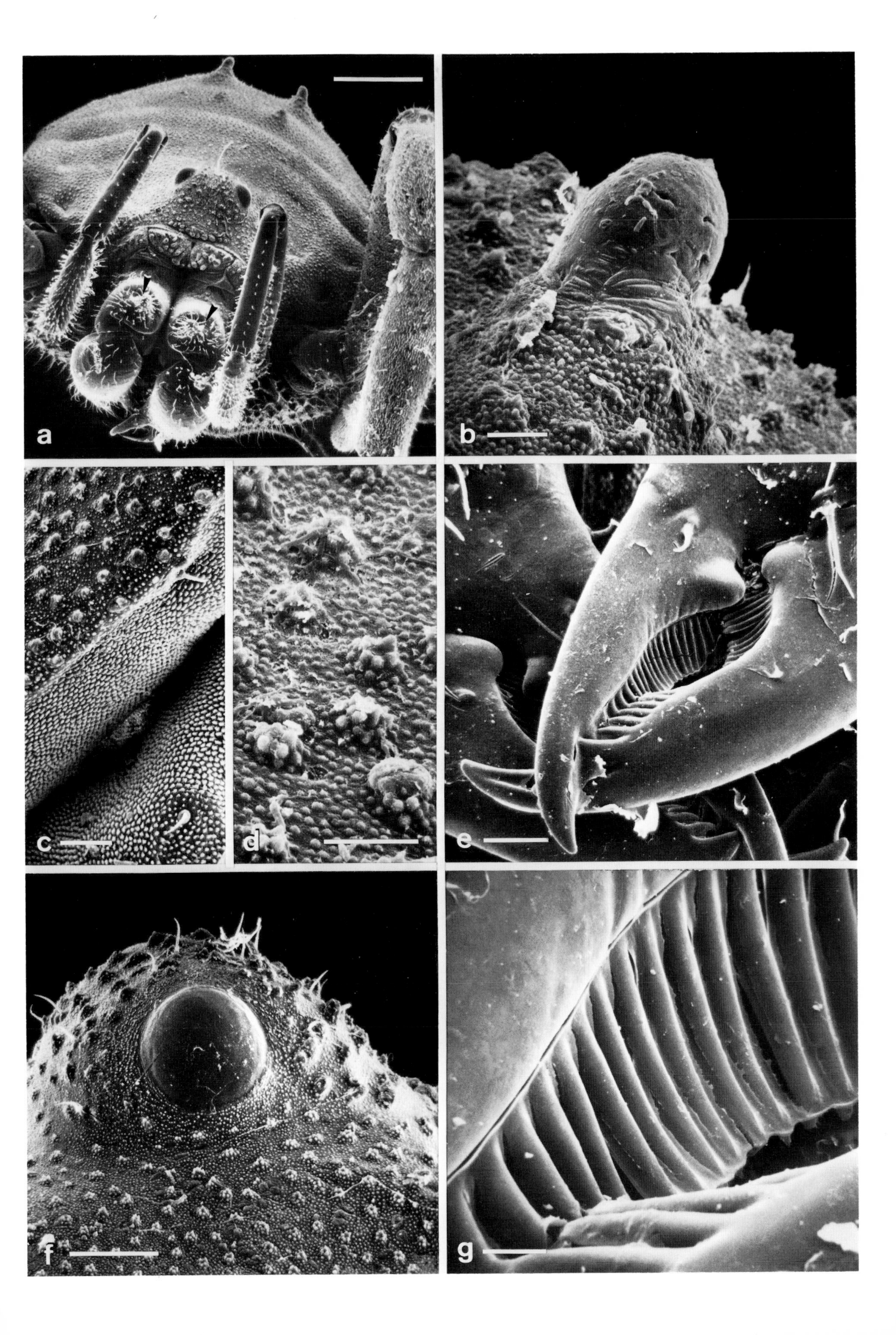
a
b
c
d
e
f
g

2.3.2 The Biotope of Harvestmen

With regard to the vertical distribution of harvestmen in their biotopes, Pfeifer (1956) suggested a classification into four groups similar to that suggested by Duffey (1966) for spiders:

1. Soil dwellers
2. Inhabitants of the herbaceous layer
3. Dwellers in shrubs and bushes
4. Tree dwellers

Many harvestmen are terrestrial, belonging to the soil community, but they prefer the epedaphic way of life in the upper soil layers and only burrow deeper into the soil in the hemiedaphic manner in exceptional cases (*Siro*). The soil-dwelling harvestmen are found especially among the Sironidae, Nemastomatidae, Ischyropsalididae, Trogulidae and members of Phalangiidae (Table 2). These terrestrial forms all have high moisture requirements at moderate temperatures. However, they have adopted very different forms as a result of adaptation to the diverse ecological niches (Franke, 1985).

For instance, the Nemastomatids, to which *Paranemastoma quadripunctatum* belongs, live in moist-soil, woodland communities, preferring deciduous and mixed woodland, and are usually to be found in moist leaf litter, detritus, on rotting wood and in damp, stony rubble (Immel, 1954; Gruber and Martens, 1968; Martens, 1978).

Table 2. Classification of harvestmen according to their vertical distribution in biotopes. (After Pfeifer, 1956)

Family/Species[a]	Soil	Herb	Shrub	Tree
Sironidae	×			
Nemastomatidae	×			
Trogulidae	×			
Ischyropsalididae	×			
Phalangiidae:				
Oligolophus tridens	×	×		
Mitopus morio	(×)	×		×
Lacinius horridus	×	×		(×)
Lacinius ephippiatus	×	×	(×)	(×)
Lophopilio palpinalis	×			
Phalangium opilio	×	(×)		×
Rilaena triangularis		×	×	
Platybunus pinetorum	×	×	×	
Leiobunum rotundum		×	×	
Leiobunum blackwalli		×	×	

[a] Genus and species names according to Martens (1978)

Plate 19. *Paranemastoma quadripunctatum* (Nemastomatidae).
a Overview, oblique caudal. 1 mm.
b Overview, lateral. 1 mm.
c Opisthosoma, caudo-ventral. 0.5 mm.
d Genital operculum, flanked by the coxae of the legs, coxognathids of the pedipalps (*arrow*) and chelicerae. 250 µm.
e Anal region. 250 µm.
f Coxae of the legs with trabecular structures. 250 µm

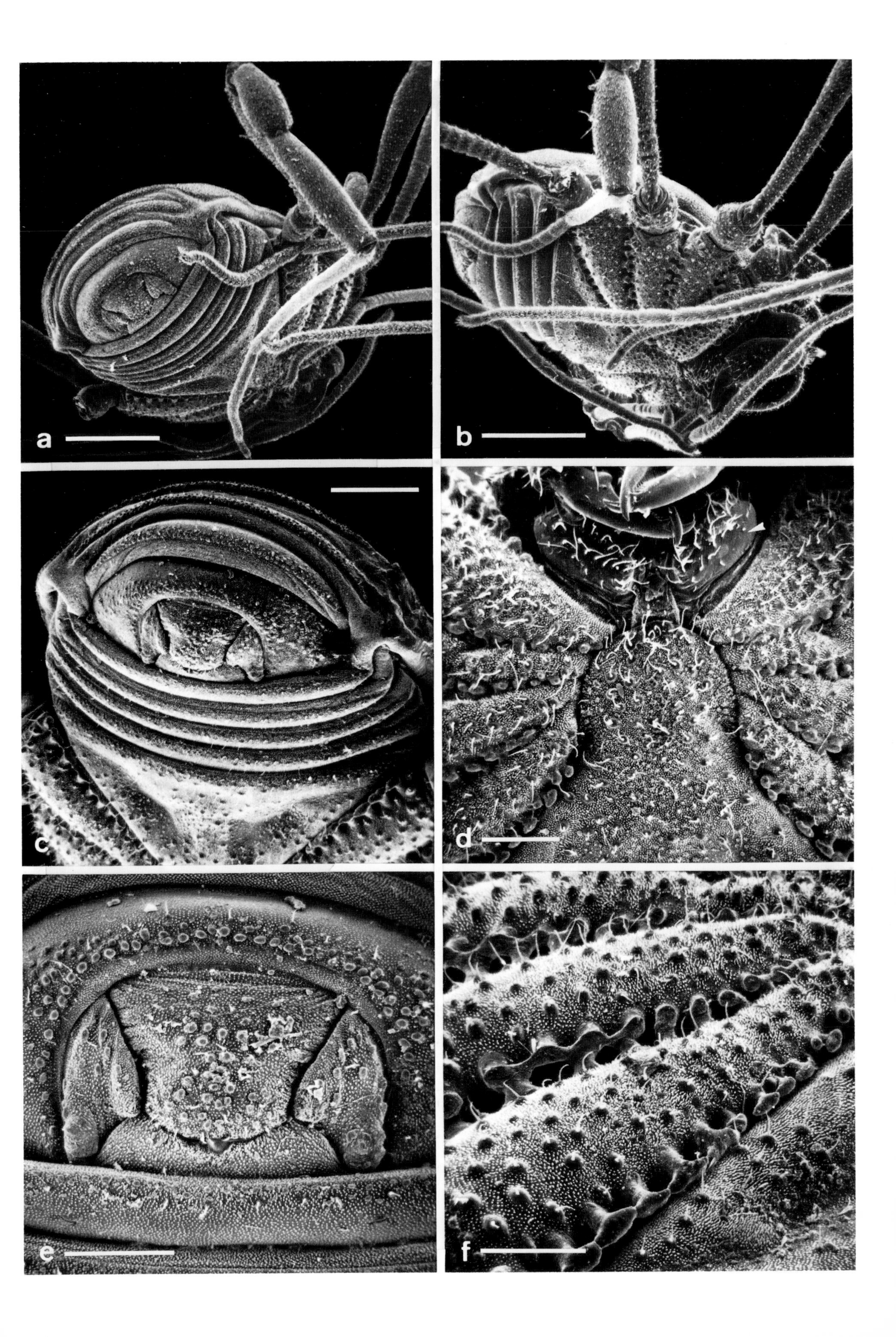
a
b
c
d
e
f

2.3.3 The Diet of Harvestmen

Harvestmen are predators which seize their prey with the chelicerae, pedipalps (raptorial legs) and walking legs (second pair). Afterwards the prey is shredded using the chelicerae, kneaded and conveyed in small balls to the entrance of the mouth and then to the pharynx.

They feed on arthropods and snails. The snail-eating harvestmen belong to the Trogulidae and Ischyropsalididae families. All other harvestmen catch mites, Collembola and larvae on the ground. In addition to their predatory feeding habits, the Phalangiidae are reported to be necrophagous, occasionally feeding on fresh carrion (crushed snails) (Martens, 1978). The phalangiid harvestman *Mitopus glacialis*, which occurs in alpine and nival regions, sometimes walks over the icy surface of glaciers in search of stranded flying insects, and has also been observed to catch glacial Collembola, the so-called glacier fleas (Collembola) (Steinböck, 1939).

Plate 20. *Trogulus nepaeformis* (Trogulidae).
a Overview, dorsal. 1 mm.
b Overview, oblique ventral. 1 mm.
c Prosoma with hood, dorsal. On the posterior edge of the hood lie two ocelli. 250 µm.
d Hood, oblique frontal. *Arrows* mark an ocellus and the outlet of a stink gland. 200 µm.
e Frontal view with chelicerae beneath the hood. 250 µm.
f Lower side of the hood. 100 µm

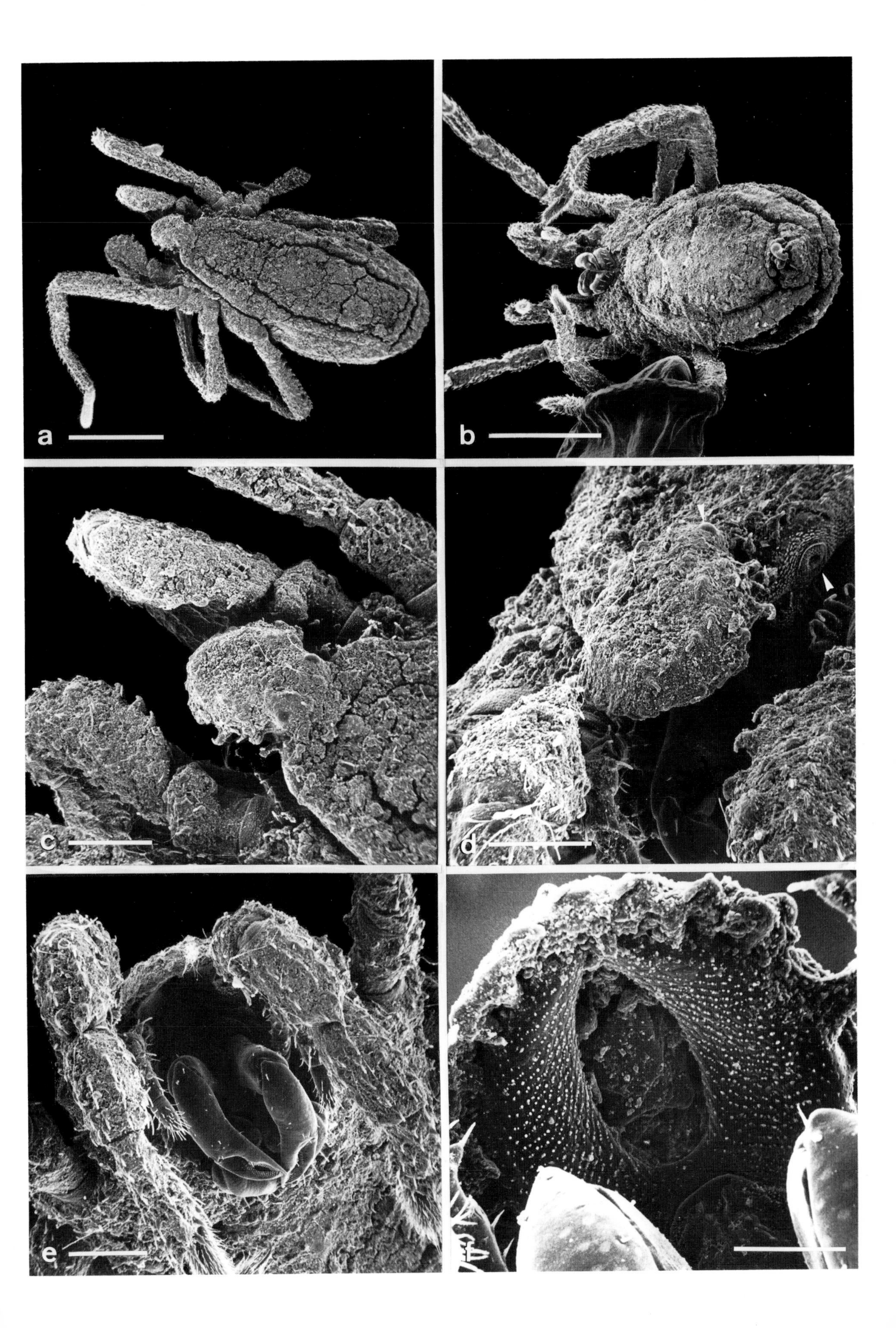
a
b
c
d
e
f

2.3.4 Snail-Eating Harvestmen

Trogulus nepaeformis, which is encrusted with earth particles, belongs to the trogulid species which live in the litter of woodland soils characterized by constant moisture and a high calcium content. Their preference for calcareous soil is obviously related to the occurrence of the snails on which the Trogulidae and Ischyropsalididae species feed (Pabst, 1953; Martens, 1965, 1969a, 1975b).

Trogulus nepaeformis holds a captured snail so that the opening of the snail shell is freely accessible. The soft body of the snail is cut off piece by piece with the chelicerae. As feeding progresses the harvestmen gradually pushes its way into the front portion of the shell and extracts all the soft parts without destroying the snail shell. The powerfully developed chelicerae of *Ischyropsalis hellwigi*, which are longer than the body in some Ischyropsalididae species, are used to break off the edges of large shells to reach the soft parts of the snail, which has withdrawn into the shell (Fig. 49).

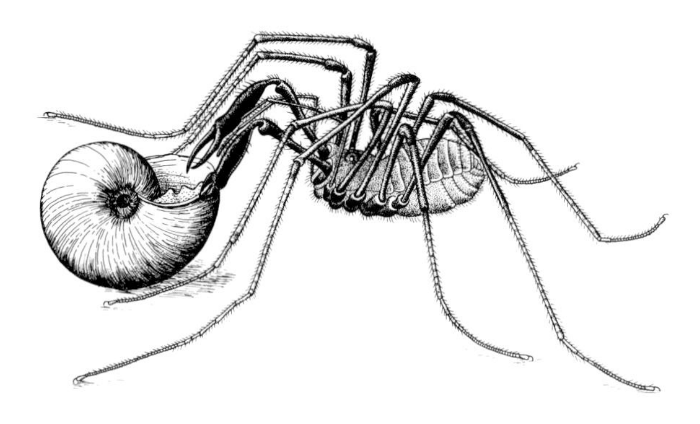

Fig. 49. A snail-eating harvestman (*Ischyropsalis*) in front of its prey, the snail *Aegopinella nitens*. (After Engel, 1961)

Plate 21. *Trogulus nepaeformis* (Trogulidae).
a Chelicerae and gnathites. The latter are formed by the coxae of the pedipalps. 200 µm.
b Comb on the chelicerae. 10 µm.
c Cleaning scales on the gnathites. 10 µm.
d Teeth of the comb on the chelicerae. 3 µm.
e Walking leg, distal, with tarsus. 100 µm.
f Claw segment, distal, with hook. 25 µm

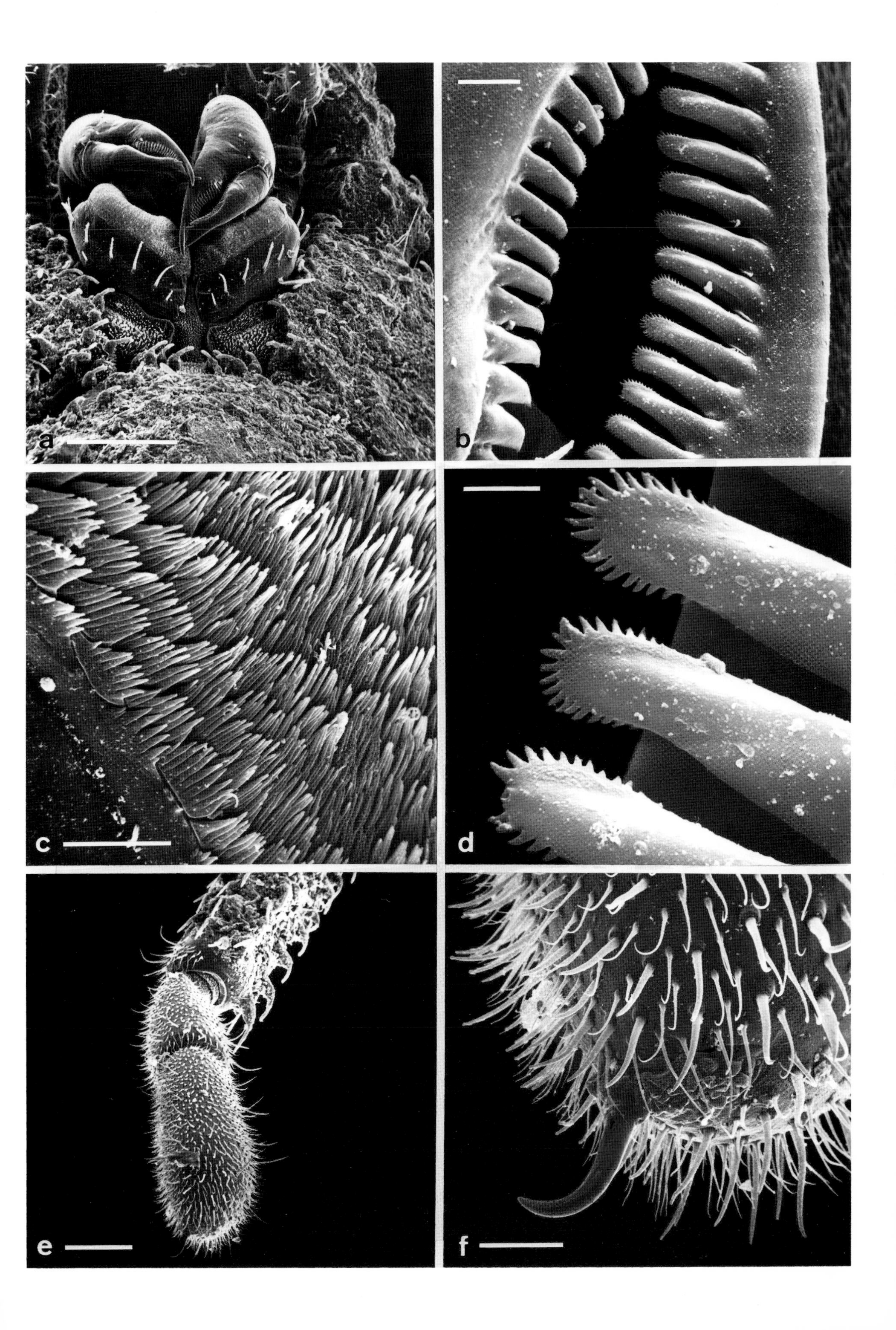
a
b
c
d
e
f

2.3.5 The Mating Behaviour of Harvestmen

Most harvestmen begin mating without preliminary courtship. The partners stand facing each other, the male guides his penis between the chelicerae into the female's genital aperture and transfers the sperm. This simple mating behaviour is extended and complicated by a gustatory courtship, at least in the *Ischyropsalis* species and *Ischyropsalis hellwigi.* On the basal segment of their chelicerae the males have a glandular area which secretes a pheromone during courtship. If the female is willing to mate, she raises her prosoma and chelicerae so that her mouthparts are exposed. Then the male approaches her mouthparts with the basal segment of his chelicerae extended, to present her with the pheromone which subsequently triggers off copulation (Fig. 50) (Martens 1967, 1969b, 1975a).

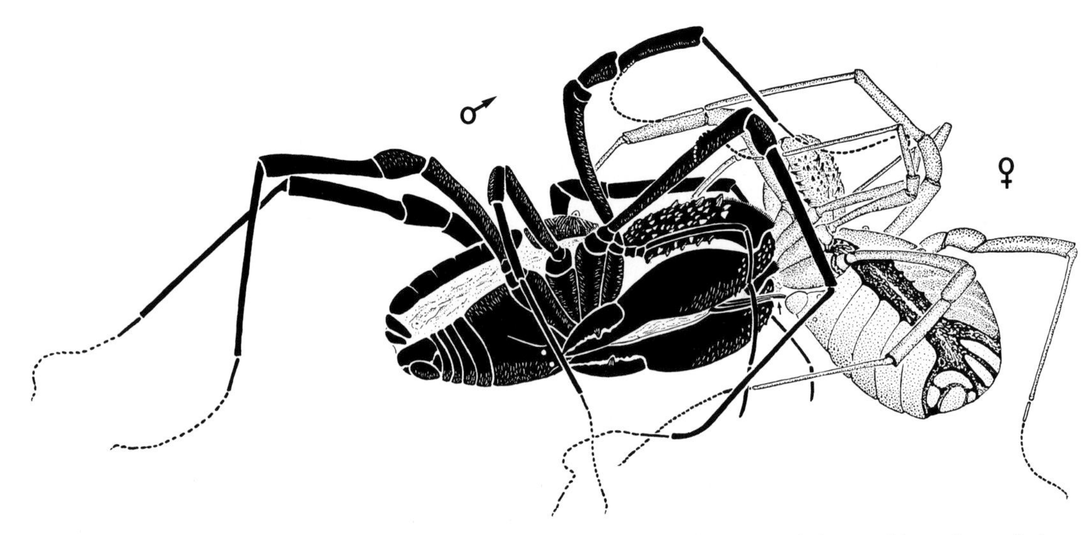

Fig. 50. *Ischyropsalis hellwigi* (Ischyropsalididae), copulating pair. The penis (*arrow*) is inserted into the genital aperture of the female. The partners encircle and touch one other with their extremities. (After Martens, 1969b)

Plate 22. *Ischyropsalis luteipes* (Ischyropsalididae).
a Overview, oblique lateral. The second walking leg has been removed. 1 mm.
b Overview, lateral. Note the powerful chelicerae with their spiny basal segments. 1 mm.
c Frontal view. 0.5 mm.
d Chelicerae, distal. 250 μm.
e Tarsus, distal. 50 μm.
f Spiny surface of the chelicerae. 25 μm

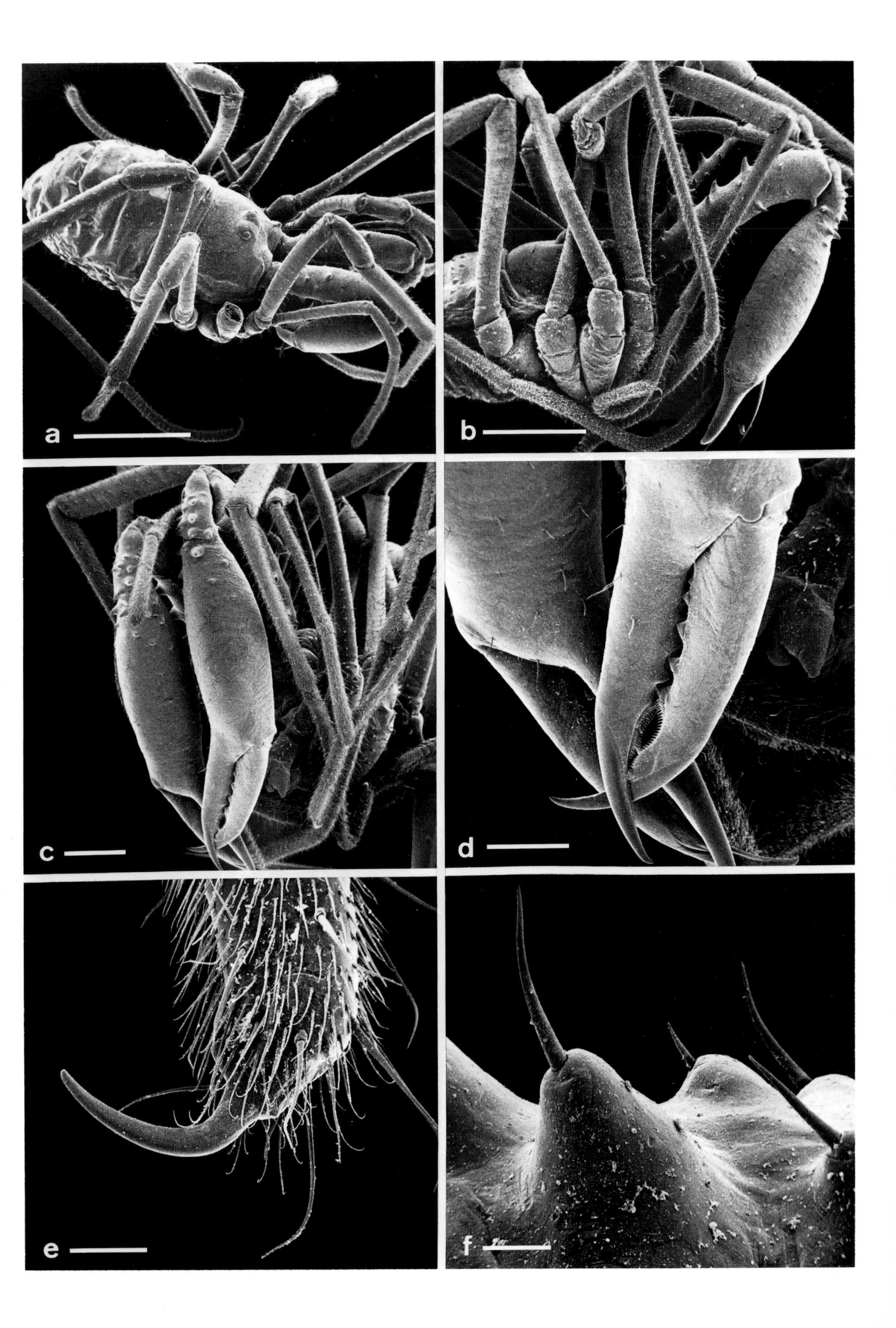
a
b
c
d
e
f

2.3.6 On the Phenology of the Phalangiidae

Phalangium opilio (Phalangiidae) has a Palaearctic distribution and is one of the most frequently occurring harvestmen. Its biotope requirements differ from those of the stenotopic soil-dwelling harvestmen. It is eurypotent with regard to air humidity although it avoids dry soil and prefers the high humidity of open deciduous or fen woodland. At the same time *Phalangium opilio* shuns shady places and seeks sunny areas in clearings or at the edge of woodlands. During the day this thermophilic harvestmen lives on the leaves in the herbaceous and shrub layers and often shelters on the ground only at night (Pfeifer, 1956).

The phalangiid harvestman, *Oligolophus tridens*, is widely distributed in central Europe, mainly in varied, open woodlands. It generally colonizes the edge of woods, small copses and bushes, preferring semi-shady places. Like *Ph. opilio, O. tridens* also occurs in open, cultivated land such as orchards and overgrown gardens, parks and even in open countryside, provided the soil is sufficiently moist. The temperature preference of the two species (27.6 °C for *Ph. opilio*, 10.3 °C for *O. tridens*) reflects their different living requirements. *O. tridens* is more restricted to the soil, usually occurring in litter, under stones and wood as well as in turf. When population densities are high, the adult animals invade the herbaceous and shrub layer up to a height of 1 m to rest on leaves and branches during the day (Martens, 1978).

Plate 23. *Oligolophus tridens* (Phalangiidae).
a Prosoma, lateral, with eye tubercle. 0.5 mm.
b Prosoma, dorsal, with eye tubercle and outlets of the stink glands (*arrows*). 0.5 mm.
c Chelicerae with mouth area, frontal. 250 µm.
d Opisthosoma, lateral. The *arrow* marks a glandular pit (cf. Plate 24f). 1 mm.
e Bristle humps on the eye tubercle. 10 µm.
f Opisthosoma, anal area. 250 µm

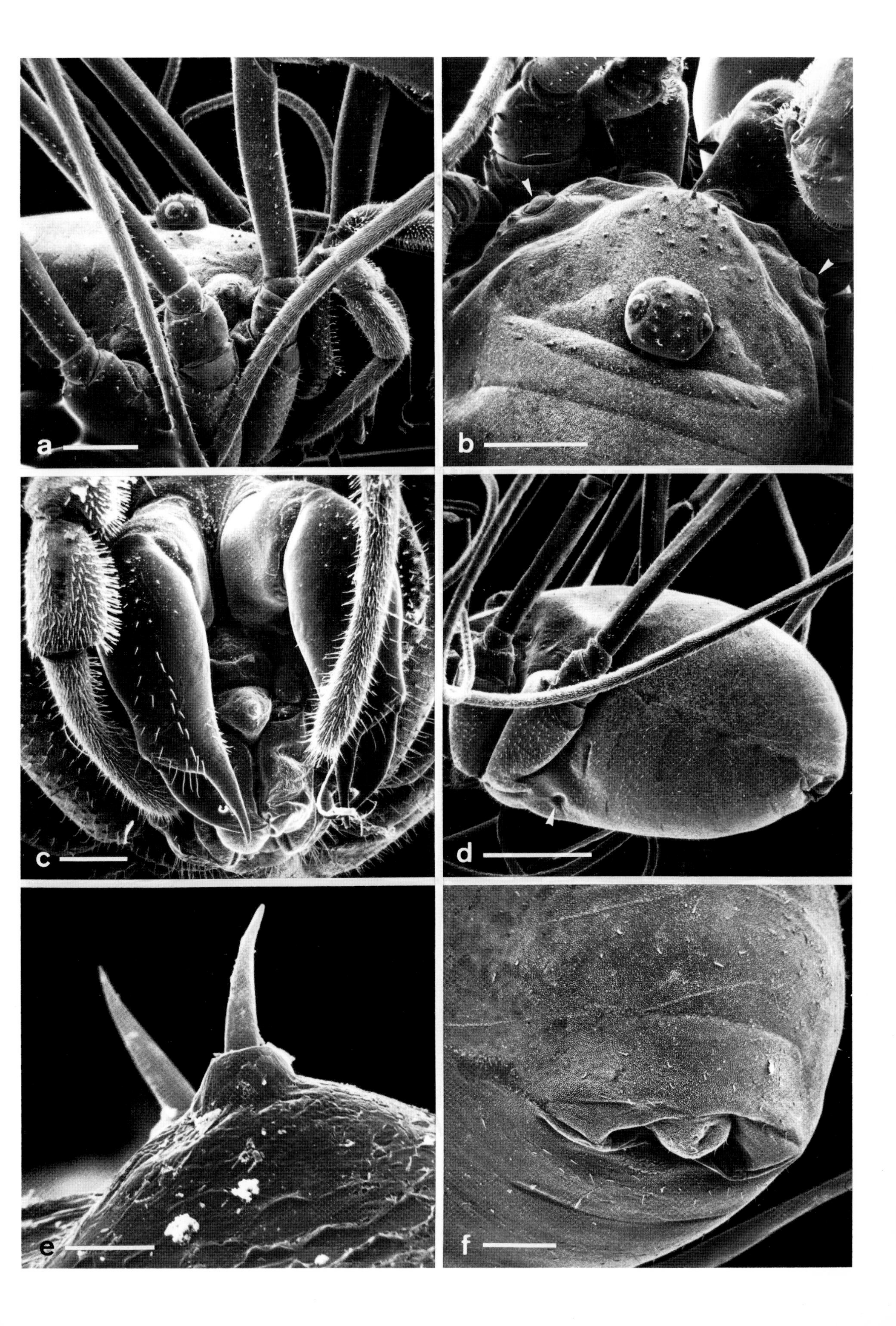
a
b
c
d
e
f

2.3.7 The Development of the Phalangiidae

The life span of many harvestmen is 1 year; but the Sironidae (Juberthie, 1964), Nemastomatidae (Immel, 1954) and Trogulidae families (Pabst, 1953) live several years. The females of *Phalangium opilio* deposit up to 200 eggs in their first batch, followed at intervals of a few days by one or two additional batches with distinctly less eggs (Martens, 1978). Oviposition begins in June, though the peak time is in late summer, with a gradual reduction until late autumn (Pfeifer, 1956). After embryonic development the young animals hatch at intervals according to the time of oviposition. The duration of embryonic development progessively increases as the cold season approaches. It is largely dependent on temperature, taking 21 days under optimal conditions (25 °C), but often 4–6 weeks or longer in the natural biotope till finally both juveniles and the egg batches overwinter (Rüffer, 1966; Juberthie, 1964). In spring juveniles and the newly-hatched animals resume their postembryonic development until sexual maturity, which is attained by summer at the latest.

The oviposition period of *Oligolophus tridens* is limited to the late summer and autumn. However, the eggs overwinter, and the young only emerge in May of the following year (Martens, 1978).

Plate 24. *Oligolophus tridens* (Phalangiidae).
a Tarsal segments. 50 µm.
b Femur-patella joint, lateral, with a single slit sensory organ (*arrow*). 100 µm.
c Tarsus, distal, with claw. 25 µm.
d Articulation at the femur-patella joint. A single slit sensory organ (*arrow*) is situated beside the joint socket to measure the torsion of the joint. 25 µm.
e Outlet of the stink gland with lid, on the prosoma. 50 µm.
f Ventro-lateral glandular pit on the opisthosoma (cf. Plate 23d). 50 µm

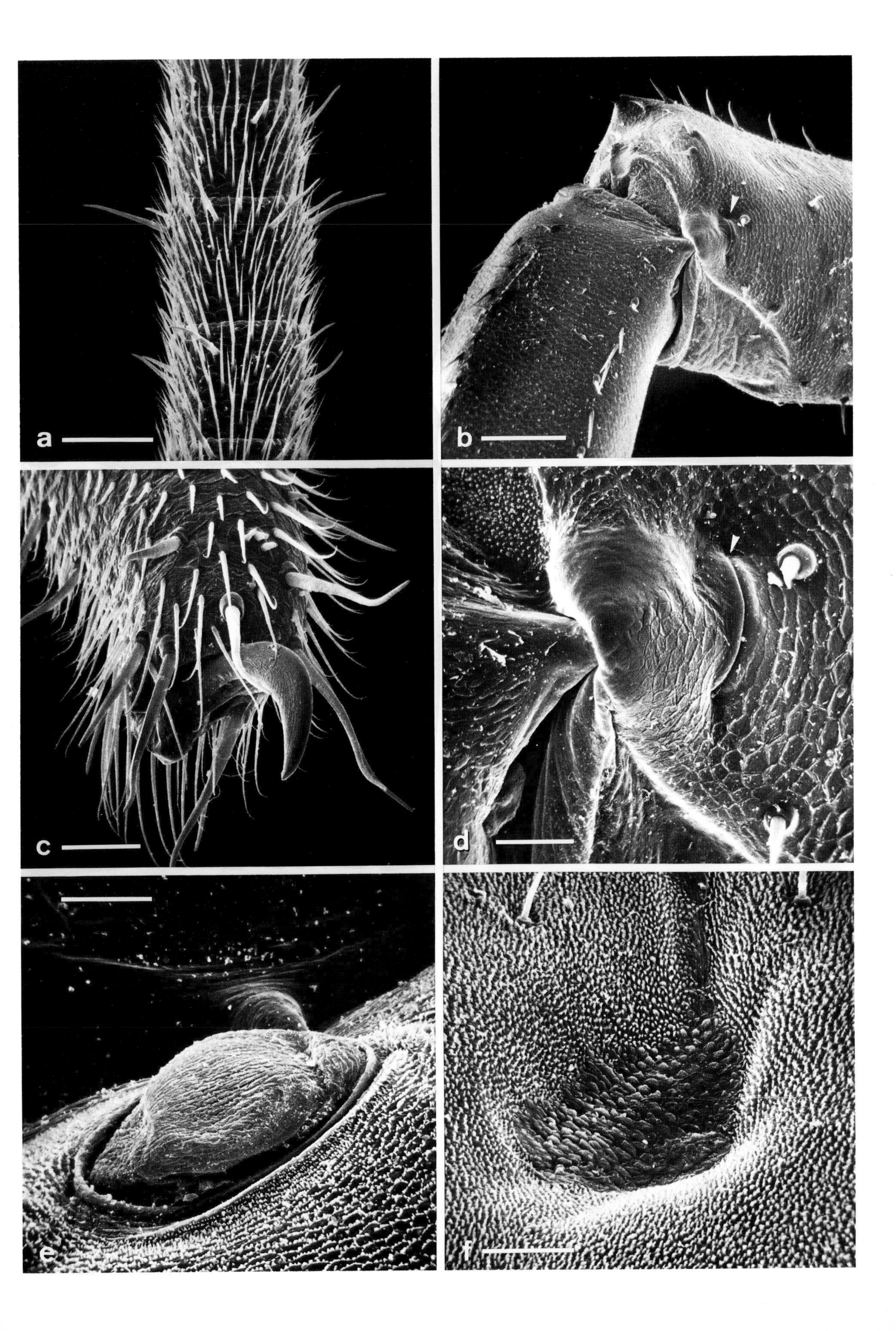
a
b
c
d
e
f

2.3.8 The Ovipositor of the Harvestmen

The ovipositor, with which the female lays her eggs, is homologous to the male's penis. During copulation sperm are transferred directly from the penis to the ovipositor and stored in the receptaculum seminis until oviposition.

The two distinct types of ovipositor (Fig. 51; Plate 25) are considered to be adaptations of the biotope in which oviposition occurs. The non-segmented, short ovipositor is typical for soil-dwelling harvestmen which lay their eggs directly, and often without any protection, on moist soil (e.g. *Ischyropsalis*) or in snail shells (e.g. *Trogulus*). A segmented ovipositor, which can be extended to a considerable length, is found in harvestmen which are not themselves restricted to the damp soil (e.g. *Phalangium*), but which deposit their eggs with their long ovipositors in moist cracks and crevices to protect them from desiccation (Martens et al., 1981; Hoheisel, 1983).

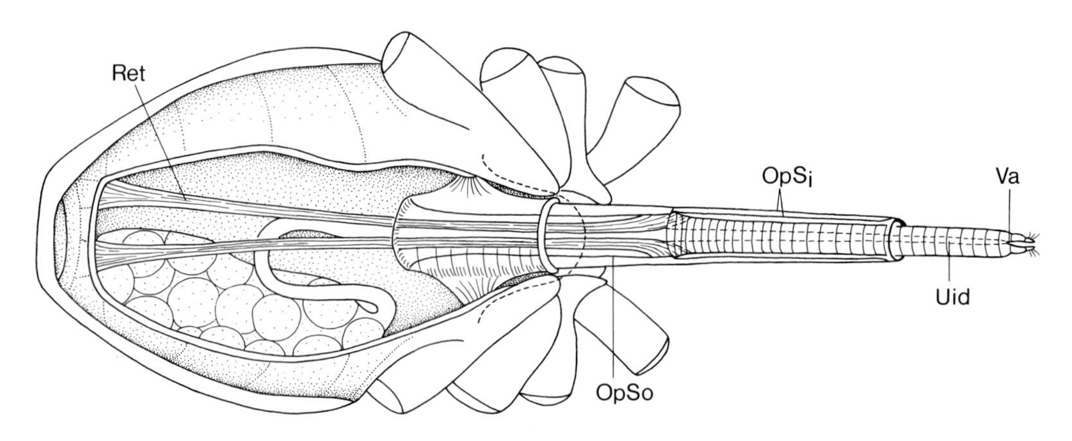

Fig. 51. *Phalangium opilio* (Phalangiidae) with its ovipositor extended for oviposition. The evagination is achieved chiefly by means of hydraulic pressure. (After Martens et al., 1981). *OpS_i* Inner ovipositor sheath; *OpSo* outer ovipositor sheath; *Ret* retractor muscle; *Va* vagina; *Uid* uterus internus, distal part

Plate 25. *Nemastoma dentigerum* (Nemastomatidae) (**a, c, e**), *Phalangium opilio* (Phalangiidae) (**b, d, f**).
a Ovipositor with non-segmented base, lateral. 0.5 mm.
b Ovipositor with segmented base. 200 μm.
c Ovipositor, latero-distal. 100 μm.
d Ovipositor, distal. 50 μm.
e Ovipositor, apical, with partly opened valves. 25 μm.
f Cupula with sensory bristles. 20 μm

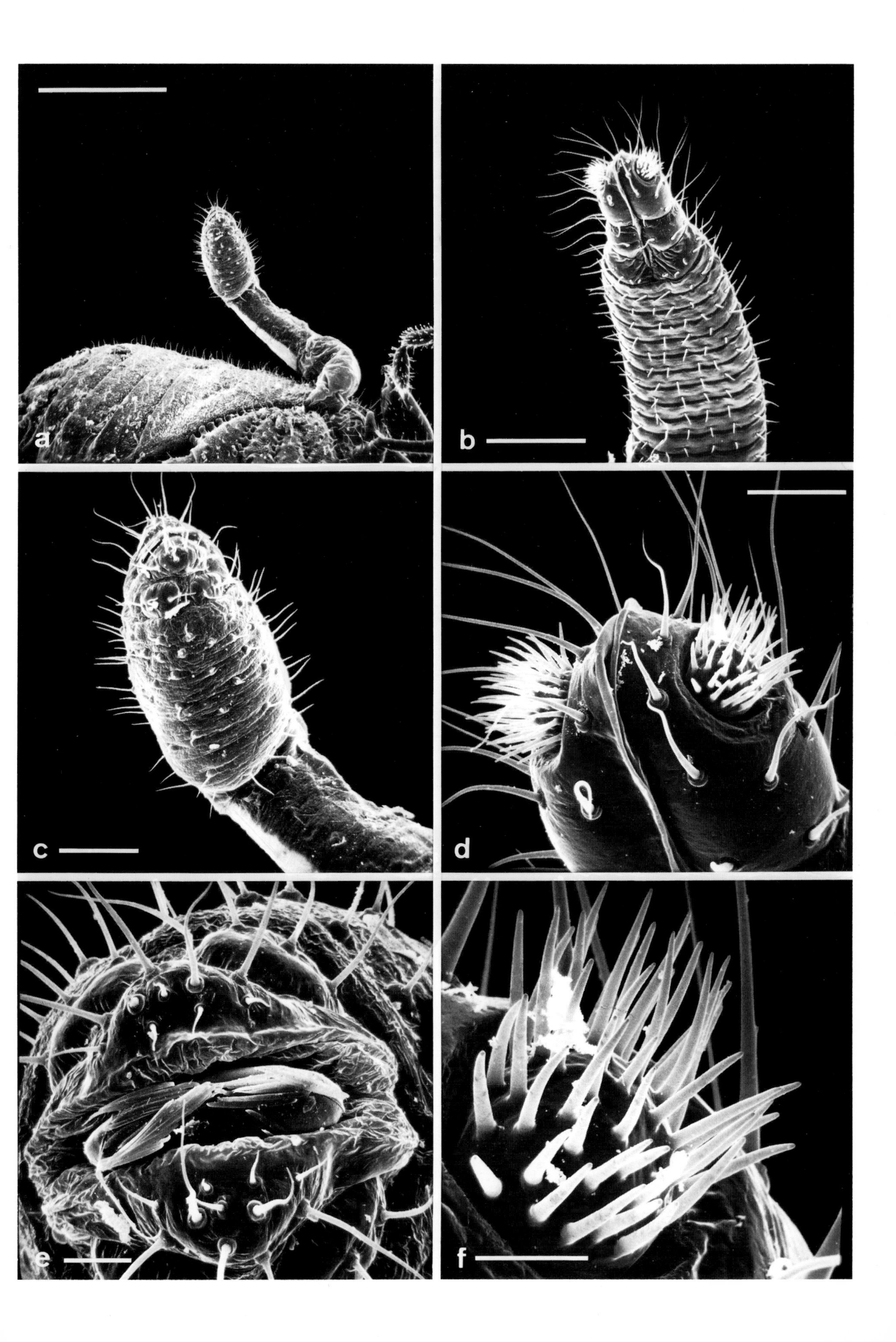

a
b
c
d
e
f

2.3.9 The Habitat of the Hemiedaphic *Siro duricorius*

The Sironidae, *Siro duricorius* (Fig. 52; Plates 26 and 27) and *S. rubens* live stenotopically in the damp ground of thick deciduous woodland, where a uniform soil moisture is maintained by the shade of the tree canopy. During temporary aridity these harvestmen burrow into the soil, sometimes to a depth of 1 m. Similarly, these species construct moulting chambers in deeper layers. Thus, they live in the hemiedaphic manner for short periods at regular intervals. Otherwise they live as part of the epedaphon in the moist, upper soil layers and prey on mites and Collembola (Martens, 1978)

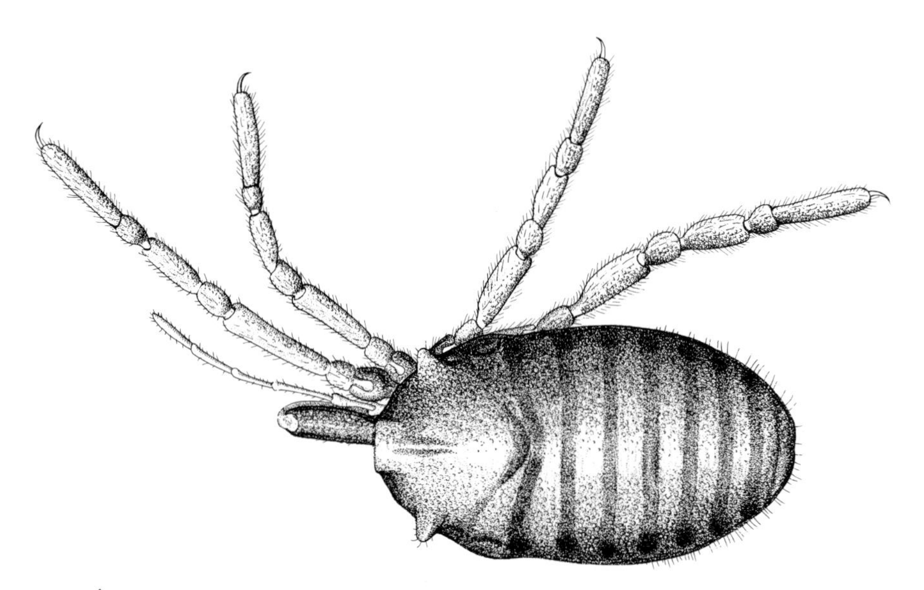

Fig. 52. Habitus of *Siro duricorius*, dorsal. The humps on the prosoma contain the stink glands. (Martens, 1978)

Plate 26. *Siro duricorius* (Sironidae).
a Overview, dorsal. 1 mm.
b Overview, ventral, with female genital aperture. 1 mm.
c Prosoma, lateral, with humps of the stink glands (*arrows*). 200 µm.
d Chelicerae, ventral. 200 µm.
e Special joints between coxa (*below,* integrated into the side of the sternum) and trochanter of the third and fourth walking legs. *Above* lies the hump of a stink gland between the legs. 100 µm.
f Pedipalp, distal, with rudiment of a claw. 20 µm

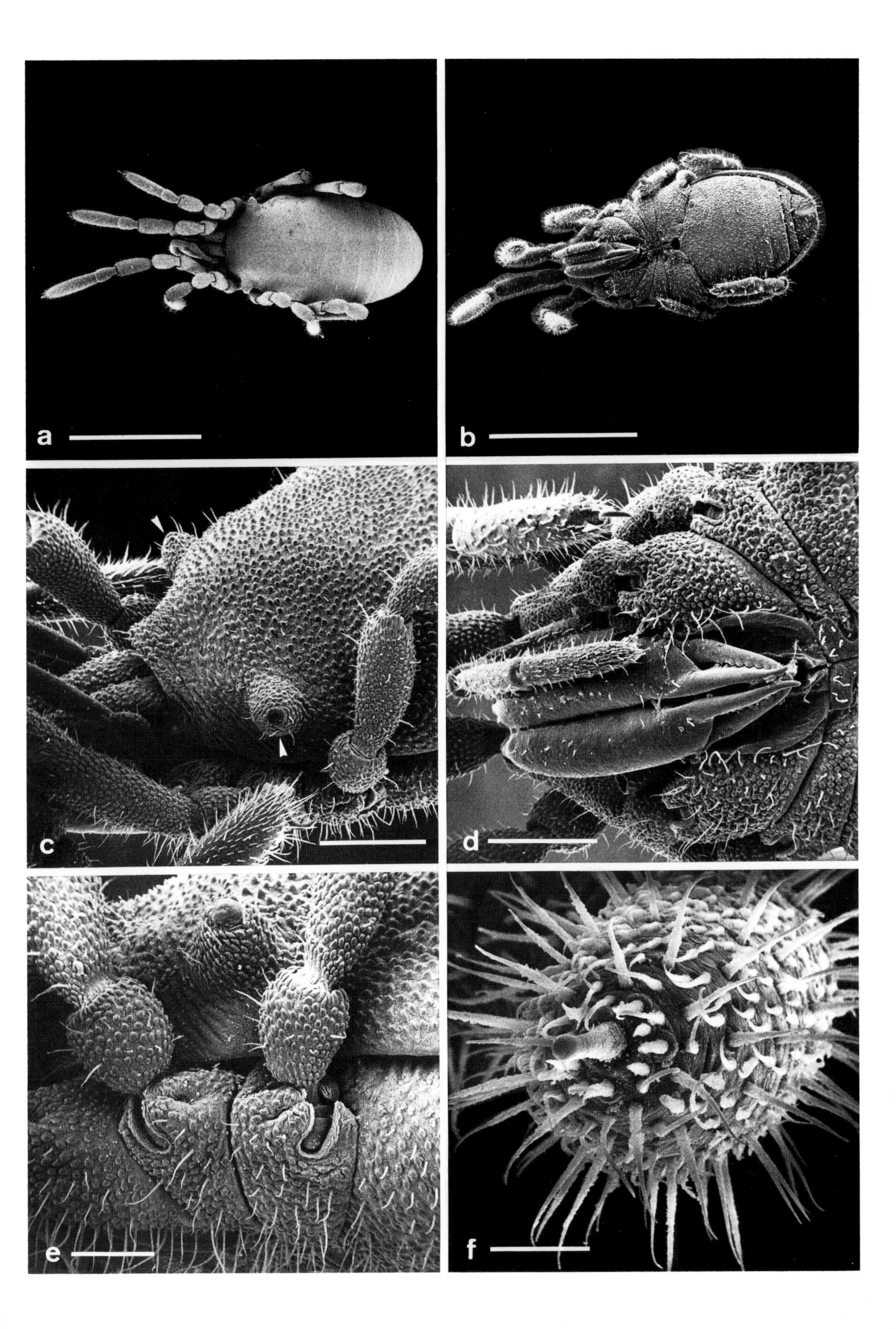
a
b
c
d
e
f

2.3.10 The Defence Mechanisms of Harvestmen

Although harvestmen are predators they are often themselves preyed on by other predatory arthropods. Their principal defence mechanism is a secretion of their stink glands, which quickly repels attackers and can prove deadly to many soil arthropods. *Siro rubens* has stink glands opening directly onto the prosoma humps (cf. Plate 26) from which the secretion is transferred to the tarsi of the walking legs. Then they try to contaminate the attacker's surface with the poison (Juberthie, 1961, 1976). When endangered, the phalangiids – *Leiobunum formorum* and *L. speciosum* – press their extremities against their bodies, which are covered by moist secretion. If, for instance, an ant seizes a leg contaminated by the glandular secretion, it immediately releases its hold and retreats (Blum and Edgar, 1971; Martens, 1978).

Another effective defence mechanism of the phalangiids is the autotomy of the legs. Between the trochanter and the femur there is a potential fracture point which breaks as soon as an enemy seizes the leg. Subsequent healing is rapid but the leg is not regenerated (Wasgestian-Schaller, 1968).

Plate 27. *Siro duricorius* (Sironidae).

a Prosoma with chelicerae and genital aperture, ventral. *Above* on the *right* lies a tracheal spiracle. 200 µm.

b Opisthosoma with anal segments. A complete sclerite ring surrounds the anal operculum. 100 µm.

c Female genital aperture. 100 µm.

d Female genital aperture with partly extended ovipositor. 50 µm.

e Microstructures of the cuticle on the opisthosoma. 5 µm.

f Tracheal spiracle from the latero-ventral region of the opisthosoma. 25 µm

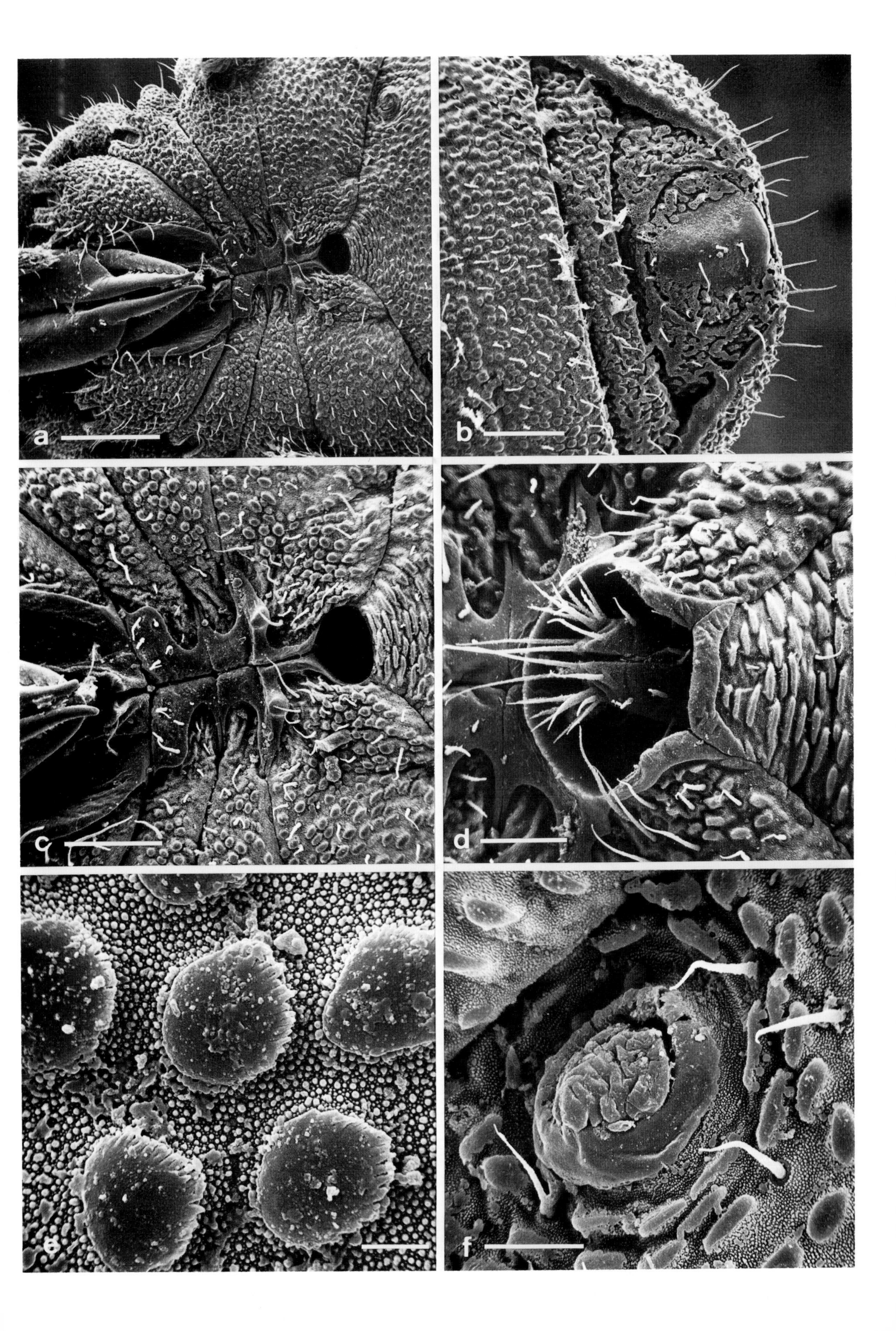
a
b
c
d
e
f

2.4 Order: Acari – Mites (Arachnida)

General Literature: Krantz, 1978; Butcher et al., 1971

2.4.1 Characteristics of the Acari

Approximately half of the 30 000 known species of mites are soil-dwellers. This great diversity of form is combined with high population densities in many cases. Dunger (1983) reports a density of 100 000 to 400 000 mites per m^2 in a moist woodland soil. At least 70% of these are usually "moss" or "horn" mites (Cryptostigmata, Oribatei) and up to 10 000 individuals/m^2 are predatory mites (Parasitiformes). Among the soil arthropods, the mites are the order with the largest number of species, and often also the highest abundance. Their size and shape permits them to penetrate deep into the soil. The Nematalycidae are known for their special adaptation to the pore system of deeper soil layers. Their striking reduction of the four pairs of legs and lengthening of the trunk give them a worm-like shape (Fig. 53) (Coineau et al., 1978). The various suborders of the Acari are united to form two groups, the Parasitiformes and Acariformes.

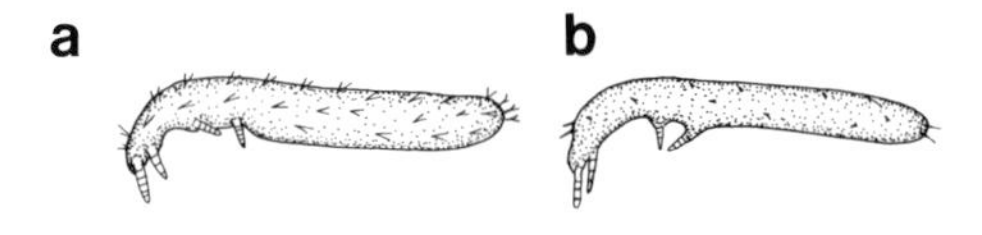

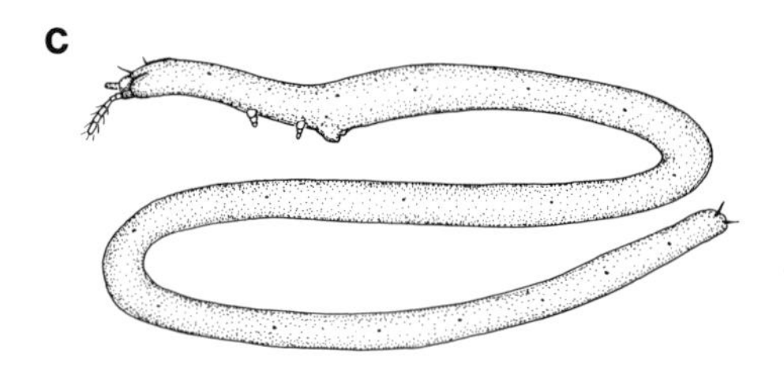

Fig. 53. Evolution of a worm-shaped habitus with reduced extremities in the Nematalycidae family as an extreme adaptation to deep interstitial habitats. (After Coineau et al., 1978).
a *Psammolycus delamarei*. **b** *Nematalycus nematoides*. **c** *Gordialycus tuzetae*. It was found in sandy soils at a depth of 3 m

---

Plate 28. *Uropoda (Cilliba) cassidea* – Uropodidae (Mesostigmata, Uropodina, "tortoise mites").
a Dorsal view of the idiosoma. 200 µm.
b Frontal view. 200 µm.
c Venter of female. The epigynial shield lies between the 3rd to 4th pair of legs. 200 µm.
d Venter of male. Between the three to four pairs of legs lies the genital aperture with operculum. 250 µm.
e Idiosoma, dorsal, edge region. 50 µm
f Male genital aperture with operculum. 25 µm

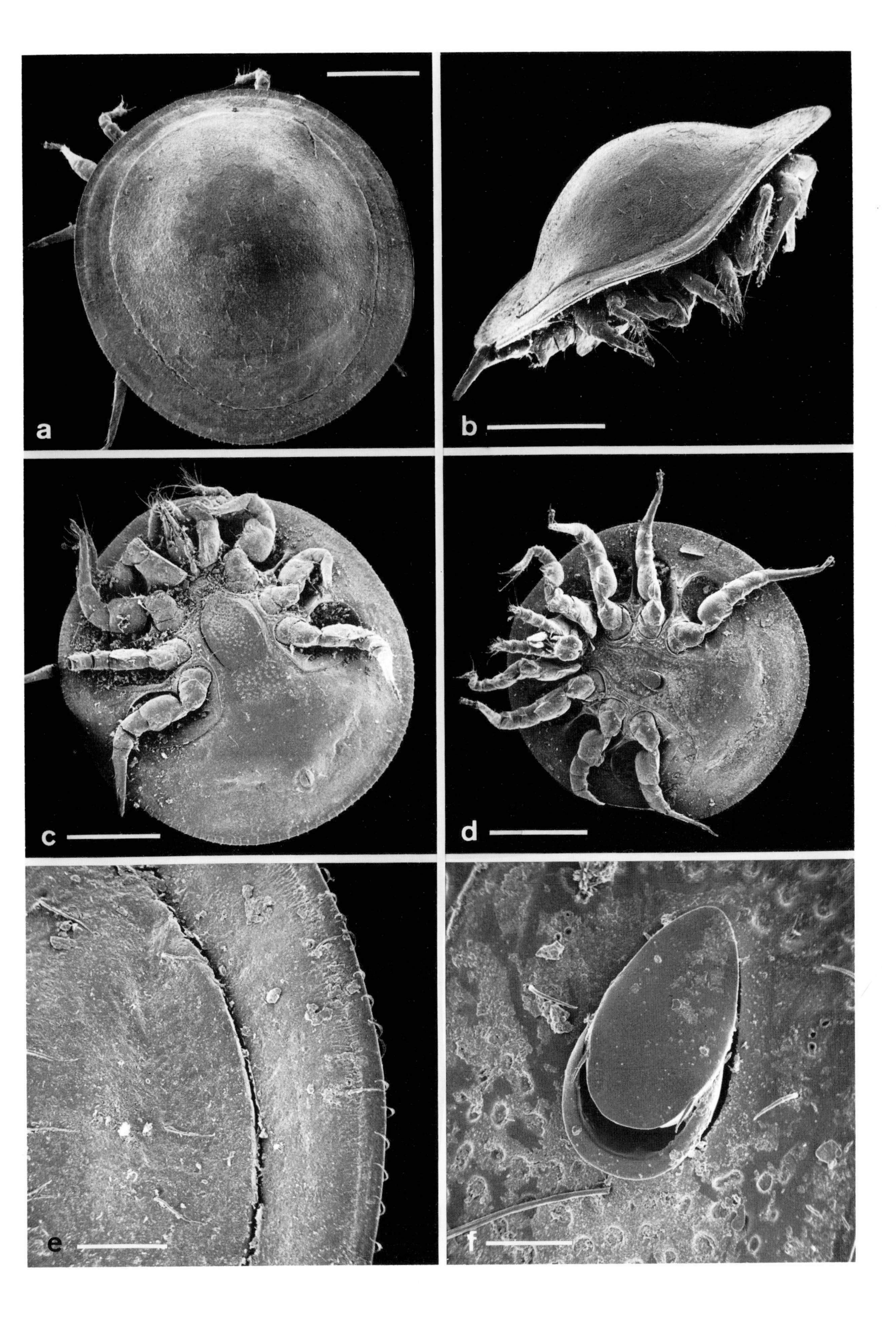
a
b
c
d
e
f

2.4.2 Parasitiformes of the Soil

The free-living Gamasina (Gamasides) and the Uropodina belong to the Parasitiformes of the soil (Karg, 1962, 1971). Scanning electron microscope pictures of members of the genus *Uropoda* and *Urodiaspis* (Uropodidae) (Plates 28 and 29) and the genus *Pergamasus* (Eugamasidae) (Plates 30–33) are presented as typical examples of the Uropodina, "tortoise mites" and the Gamasina, "predatory mites", respectively. Figure 54 shows a schematic diagram of the eidonomy of the Gamasina.

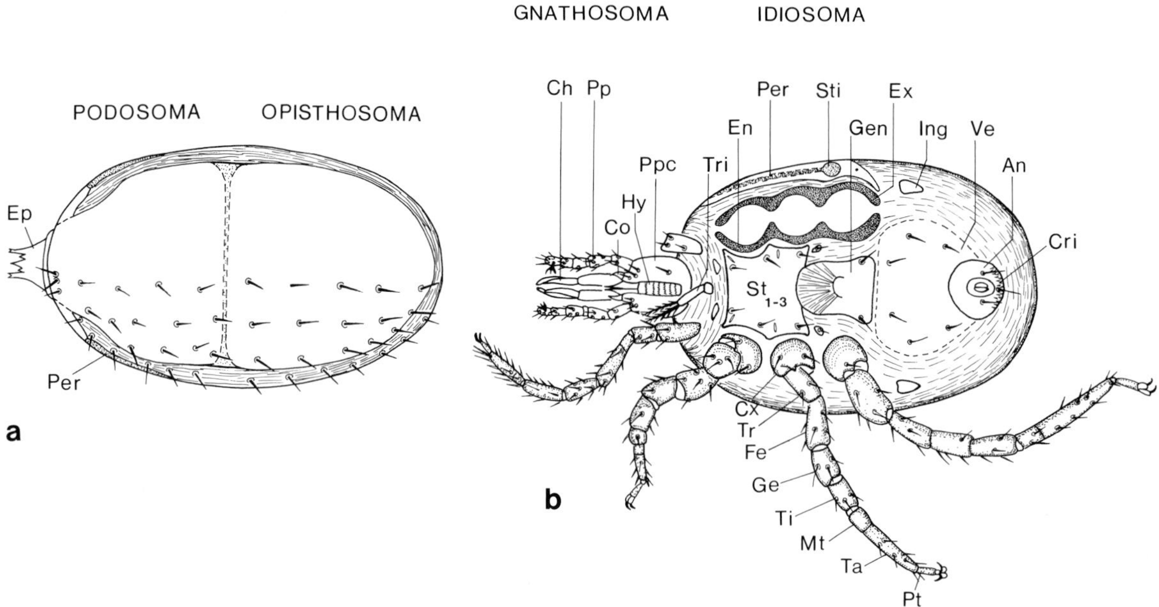

Fig. 54. Arrangement of the body structures of the Gamasina. (Karg, 1971).
a Dorsal view. **b** Ventral view of a female.
An Anal shield with anal valves; *Ch* chelicerae; *Co* corniculi; *Cr* cribrum; *Cx* coxa; *En, Ex* endo-, exopodalia; *Ep* epistome ("Randfigur"); *Fe* femur; *Ge* genu; *Gen* genital shield (operculum of the epigyne); *Hy* hypostome; *Ing* inguinalia; *Mt* metatarsus; *Per* peritrema; *Pp* pedipalp; *Ppc* coxa of pedipalp; *Pt* pretarsus; *Ste* sternum; *Sti* spiracle; *Ta* tarsus; *Ti* tibia; *Tri* tritosternum; *Tro* trochanter; *Ve* ventral shield

Plate 29. *Urodiaspis tecta* – Uropodidae (Mesostigmata, Uropodina, "tortoise mites").
a Idiosoma and gnathosoma, frontal. 100 μm.
b Front of body with gnathosoma, ventral. 50 μm.
c Idiosoma, dorsal. 200 μm.
d Podosoma of a female, ventral. The legs are folded into concavities, the fovae pedales, between them lies the epigynial shield. The gnathosoma touches the left edge of the picture. *Arrow* marks a pit behind coxa 4. 100 μm.
e Gnathosoma, ventral. 50 μm.
f Infolded leg. 50 μm

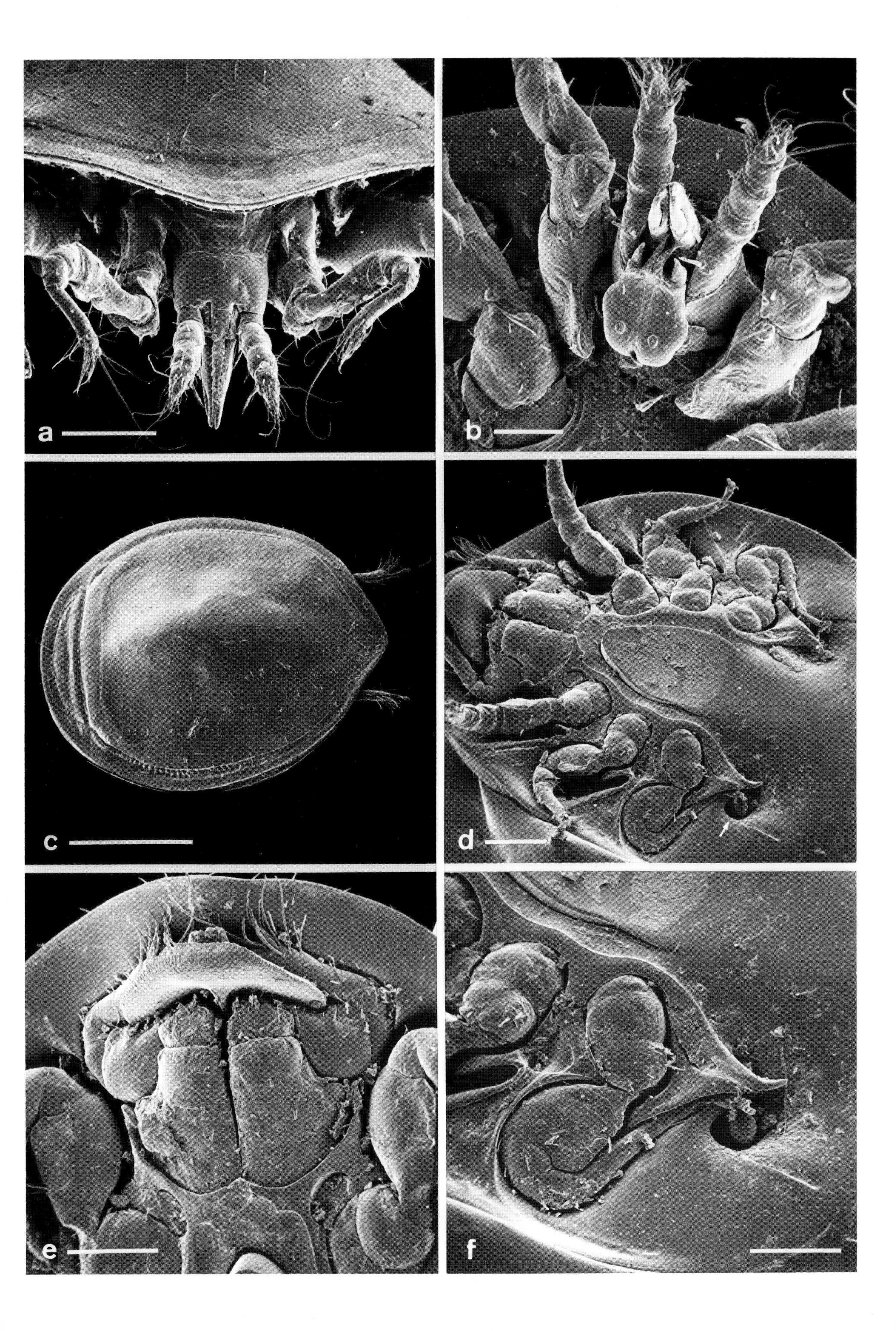

a
b
c
d
e
f

2.4.3 Free-Living Gamasina (Gamasides)

The oval (often twice as long as broad), but only 0.2–2-mm-long Gamasinae are predatory mites, characterized by their mobility, with relatively long legs, and by their efficient seizure of prey in their mouthparts (Fig. 54; Plates 30–33). The mouthparts consist of the chelicerae, of major importance for predatory mites, and the pedipalps (feelers), divided into five segments. The chelicerae are also segmented and consist of a movable and an immovable digit which function like shears (chelae). The inside edges of both digits can be variously toothed. These teeth are distinctly different and characteristic for each species. The base of the chelicerae are inserted into a pocket above the buccal cavity of the gnathosoma (Fig. 55), from which they shoot out by means of hydrostatic pressure to seize prey. They are withdrawn by a retractor muscle.

The narrow mouth opening below the chelicerae only allows the passage of liquid nutrients. These are sucked into the gut with the help of a pharyngal pump. The chelicerae and pharynx are situated in a tube which is open at the front. This tube, known as the gnathosoma, is formed by fusion of the pedipalp coxae (Karg, 1962, 1971; Korn, 1982).

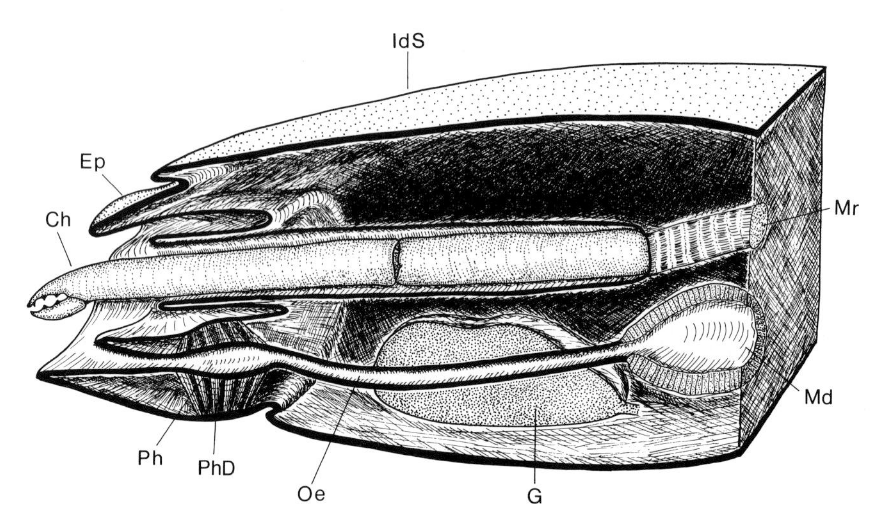

Fig. 55. Section through the gnathosoma of a parasitiform mite with cheliceral pocket. (After Karg, 1971, and Krantz, 1978). *Ch* Chelicerae; *Ep* epistome ("Randfigur"); *G* brain ganglion composed of upper and lower pharyngal ganglia; *IdS* idiosoma; *Md* midgut; *Mr* cheliceral retractor; *Oe* oesophagus; *Ph* pharynx; *PhD* pharyngal dilators.

Plate 30. *Pergamasus* sp. – Eugamasidae (Mesostigmata, Gamasina, "predatory mites") female.

a, b Overview, dorsal and ventral. 0.5 mm, 0.5 mm.

c, d Overview, frontal and caudal. 250 µm, 250 µm.

e Anterior of the body with gnathosoma, frontal. 100 µm.

f Idiosoma, oblique lateral. Spiracle and peritrema (*arrow*) are situated in front of the soft skin of the opisthosoma. 250 µm

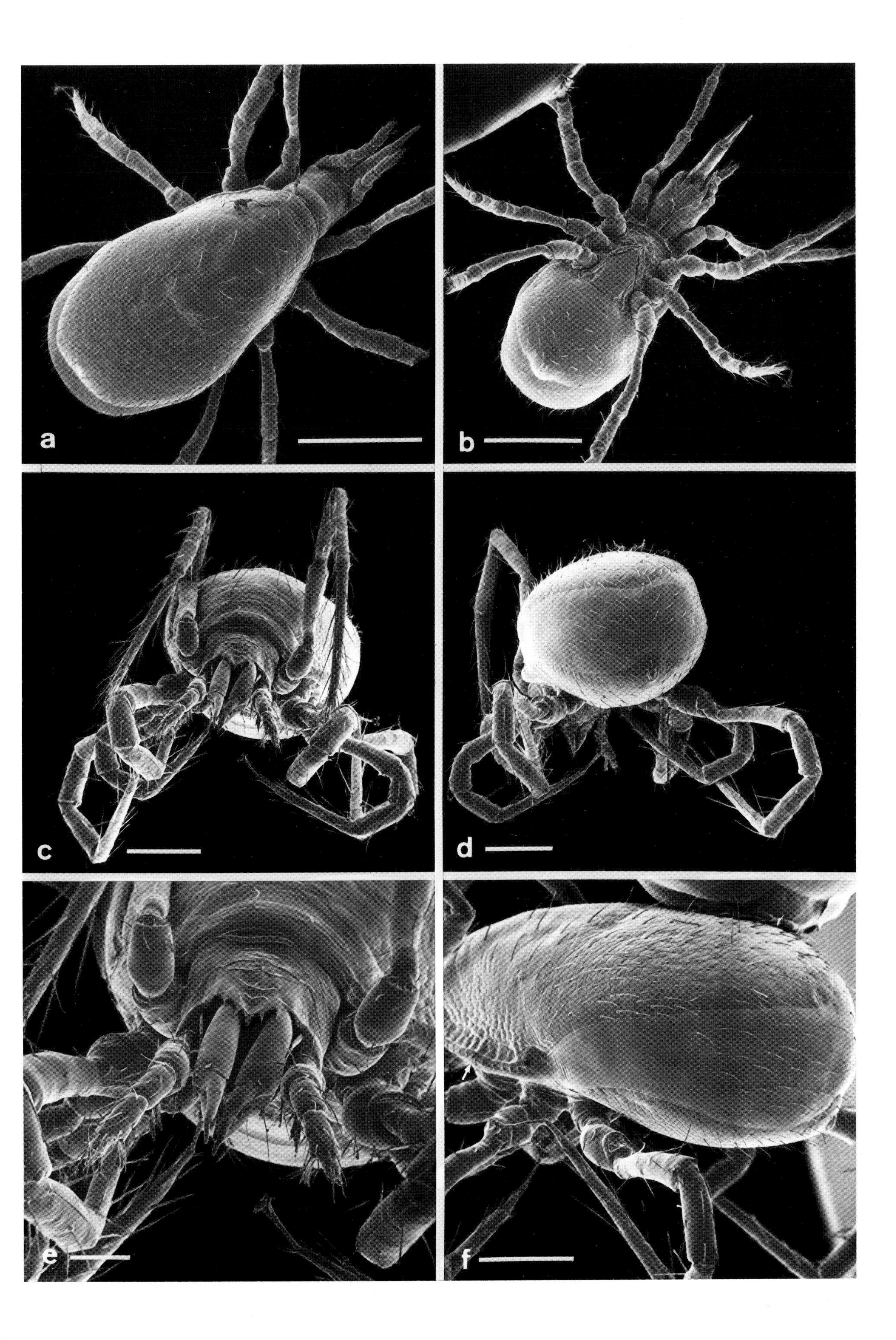
a
b
c
d
e
f

2.4.4 The Prey of Predatory Mites (Gamasina)

Predatory mites prey on animals, occurring in the same biotope, which are mostly smaller and less agile than themselves. Thus, the Gamasina catch nematodes, Collembola and mites, according to their preferences, in their respective characteristic biotopes. Moreover, they feed on the eggs and young larvae of various insects.

The chelicerae are well adapted for catching food. Their shape and the arrangement of the teeth on the digits often indicate food preferences. Thus, short digits with staggered teeth which shut without a gap are tyical of nematode specialists. They are clearly distiguishable from the cheliceral digits found in Collembola- and mite-eating as well as the polyphagous predatory mites. Chelicerae with single protruding teeth on their digits are suitable for piercing insect eggs (Fig. 56).

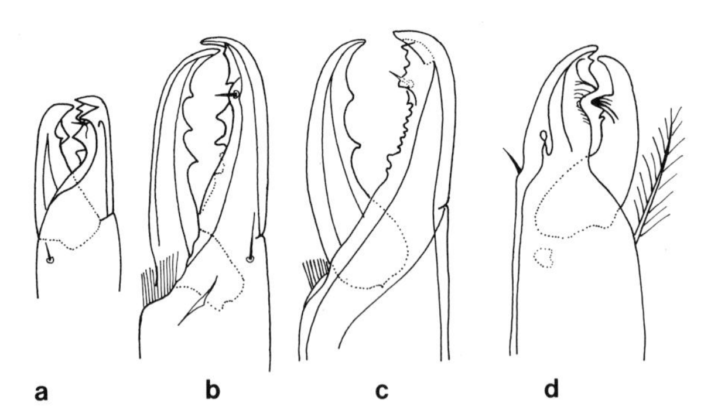

Fig. 56. Chelicerae of various feeding specialists of the Gamasina. (After Karg, 1971).
a Nematode-eating species, *Alliphis siculus.* **b** Predator of Collembola and mites, *Pergamasus misellus.* **c** Polyphagous species, *Hypoaspis aculeifer.* **d** Specialist feeding on insect eggs and larvae as well as nematodes, *Macrocheles insignitus*

Plate 31. *Pergamasus* sp. – Eugamasidae (Mesostigmata, Gamasina, "predatory mites"), female.
a Frontal view. 200 μm.
b Lateral view. 250 μm.
c Pretarsus, dorsal. 20 μm.
d Pretarsus, ventral. 20 μm.
e Spiracle with peritrema (*arrows*). 50 μm.
f Pretarsus, distal, with claws and adhesive lobes (pulvilli). 5 μm

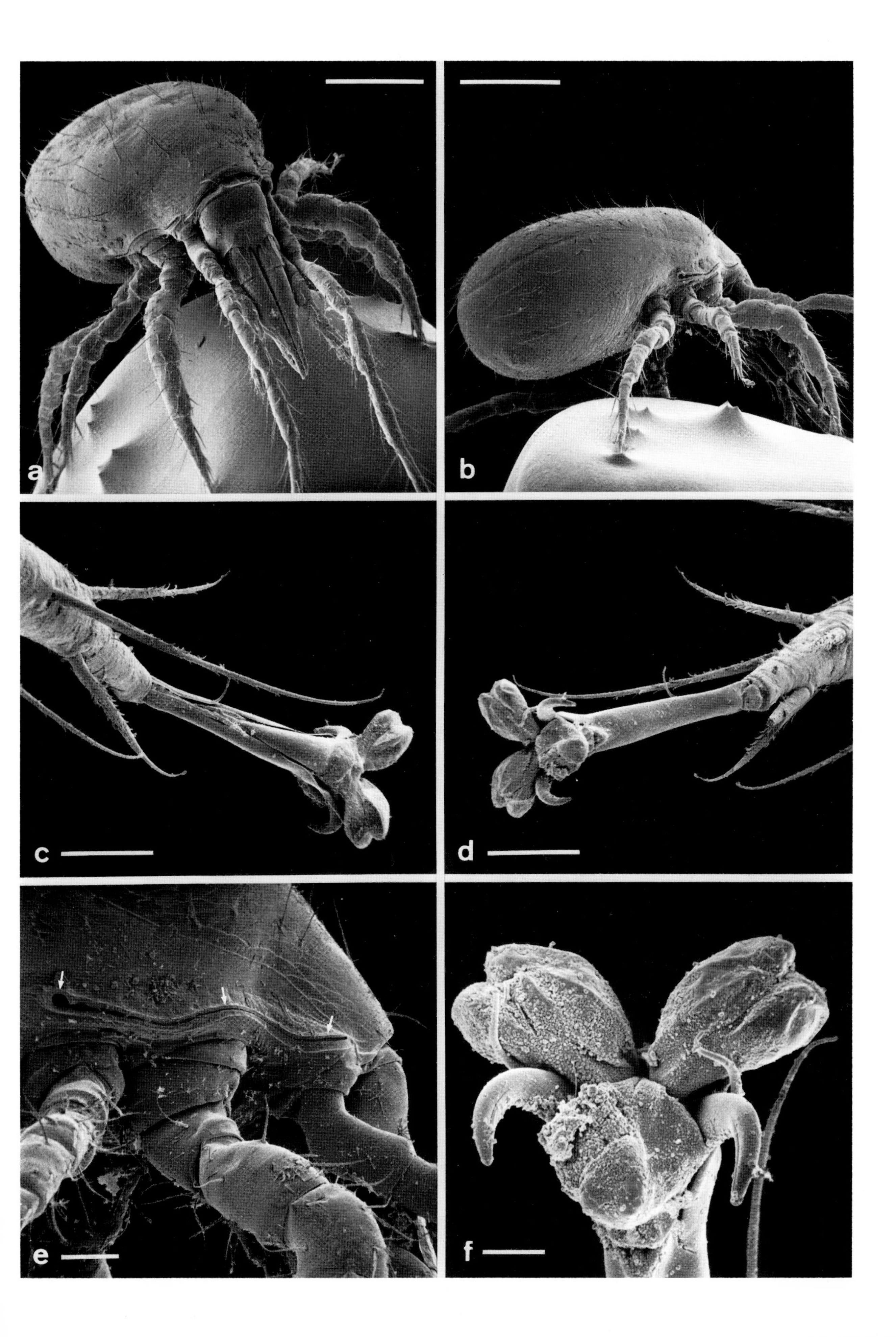
a
b
c
d
e
f

2.4.5 Vertical Distribution of Predatory Mites

According to Karg (1961, 1962, 1971) the Gamasina with their subfamilies are the most common mites in the soil. Their original biotope is regarded as the pore system, from which predatory mites colonized new biotopes above ground by undergoing appropriate adaptations, leading to new life forms. From the soil the Gamasina have colonized the upper layers in several independent developmental steps (Table 3).

The starting point of this development is the euedaphic way of life. The species are smaller than 0.5 mm and with their short extremities they fit into the narrow interstitial spaces. Their mobility is limited, but so is that of the prey animals in this biotope. Thus, *Alliphis siculus* lives well on nematodes.

The epedaphic life forms live on the soil surface, their wedge-shaped, strongly sclerotized bodies permitting them to penetrate into densely packed litter. Their extremities are frequently longer than their body so that fast moving predators such as *Pergamasus crassipes*, can easily subsist on Collembola and other mites.

Table 3. Colonization of a variety of habitats by Gamasina. (Karg, 1971)

Various strata	Plants, wood	Store-rooms	Dead trees	Dung, excrement	Ant nests	Parasites
Atmobiotic	*Amblyseius* *Typhlodromus*	*Blattisocius*	*Dendrolaelaps*	*Macrocheles* *Scarabaspis* *Parasitus* *Halolaelaps*	*Pseudoparasitus*	*Laelaps*
Litter epedaphic	*Lasioseius* *Ameroseius*	*Pergamasus* *Veigaia* *Gamasellus*	*(Pachylaelaps)*	*Asca* *Sejus* *Zercon*		
Soil euedaphic	*Proctolaelaps*	*Rhodacarus* *Rhodacarellus*	*Eviphis* *(Pachylaelaps)* *Alliphis*	*Leioseius* *Arctoseius*	*(Eulaelaps)* *Hypoaspis*	
Superfamily	Phytoseioidea	Eugamasoidea	Eviphidoidea	Ascoidea	Dermanyssoidea	

Plate 32. *Pergamasus* sp. – Eugamasidae (Mesostigmata, Gamasina, "predatory mites"), male.

a Overview, dorsal, with clasping legs (second pair of legs). 0.5 mm.
b Overview, ventral. 0.5 mm.
c Anterior of body, ventral, with gnathosoma and clasping legs. 100 μm.
d Anterior of body, dorsal, with clasping legs. *Right* the thinner, first pair of legs, the pedipalps and the toothed epistome. 200 μm.
e Clasping leg, ventral. 100 μm.
f Overview, oblique frontal. 250 μm

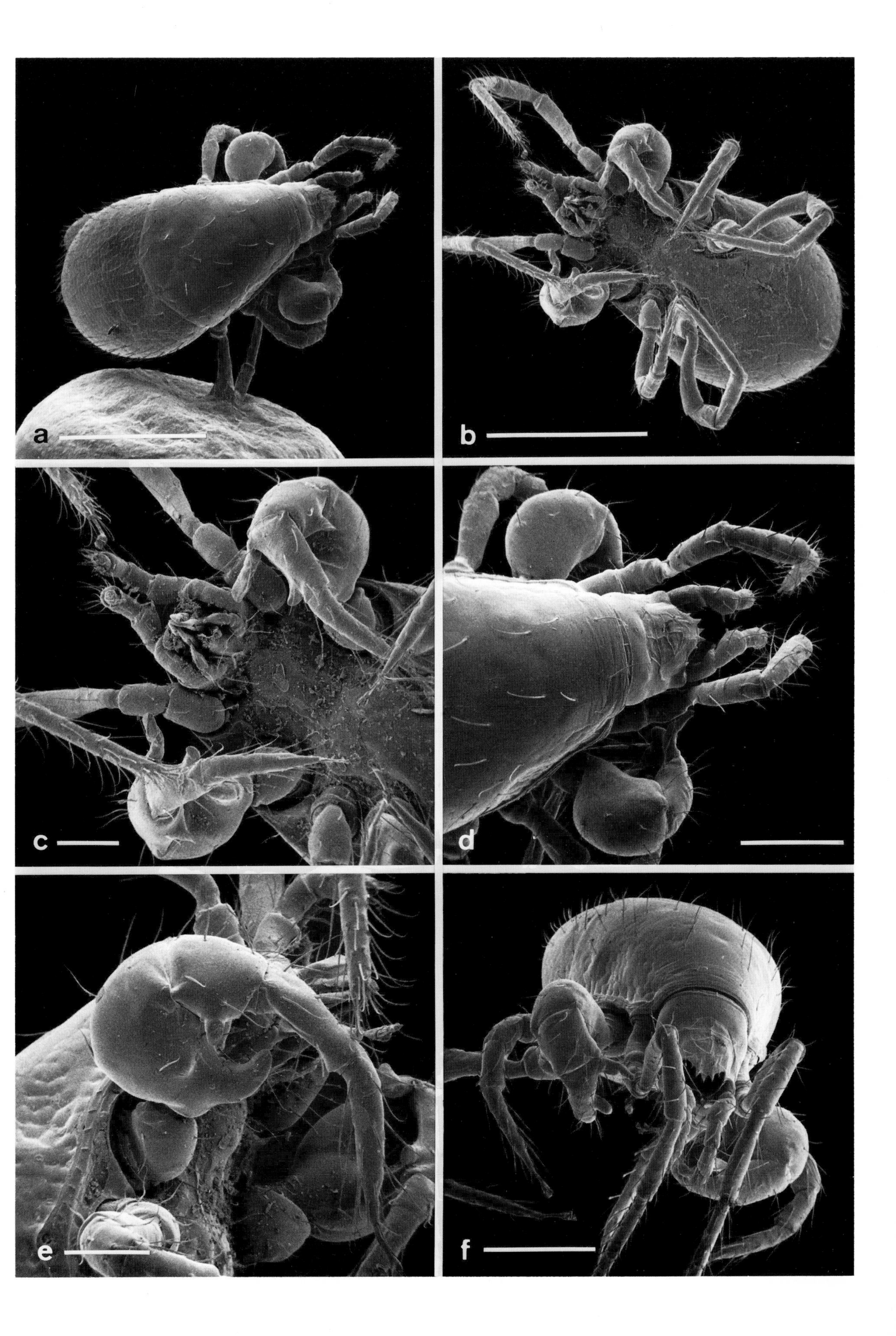

2.4.6 Body Segmentation of the Acari

The body of a typical mite is divided into an anterior gnathosoma and a posterior idiosoma. The gnathosoma bears two pairs of extremities, the chelicerae and pedipalps, the idiosoma has four pairs of legs. Figure 57 shows the hypothetical arrangement of an original, acariform mite, simplified according to a suggestion from Coineau (1974) after Krantz (1978). According to this plan, the body is composed of 16 real segments (somites) with an anterior precheliceral segment (acron). These segments can be grouped into different tagmata (segmental groups). The part of the idiosoma without extremities, the opisthosoma, only attains its complete set of segments in the course of postlarval moulting. Present-day mites show very little evidence of this original arrangement, as the boundaries between segments have either partly disappeared or have undergone a secondary shift. A primitive segmentation is even more difficult to interpret in the parasitiform mites than in the acariform mites. The idiosoma is possibly composed of 12 segments with a terminal anal segment (telson).

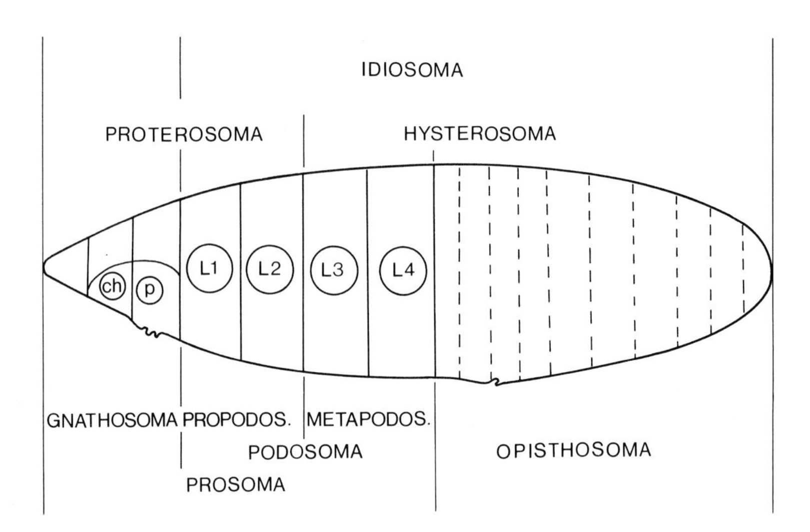

Fig. 57. Hypothetical schematic plan of the original segmentation and tagma arrangement of an acariform mite. Accordingly the adult body is composed of 16 segments and the acron. (After Coineau, 1974 from Krantz, 1978). *ch* Chelicere; *p* pedipalp; *L1–L4* legs

Plate 33. *Pergamasus* sp. – Eugamasidae (Mesostigmata, Gamasina, "predatory mites").
a Pedipalp, distal. 25 μm.
b Venter of a female with genital shield and anal zone. 250 μm.
c Gnathosoma, ventral, with pedipalps, chelicerae and corniculi (*arrows*). 100 μm.
d Anal shield with anal valves and cribrum (*arrow*). 30 μm.
e Cheliceral digits, basal, with proprioreceptive hairs; *right*, the corniculi. 50 μm.
f Cuticle of the opisthosoma, dorsal. 25 μm.
g Cuticle of the soft pleural region. 10 μm

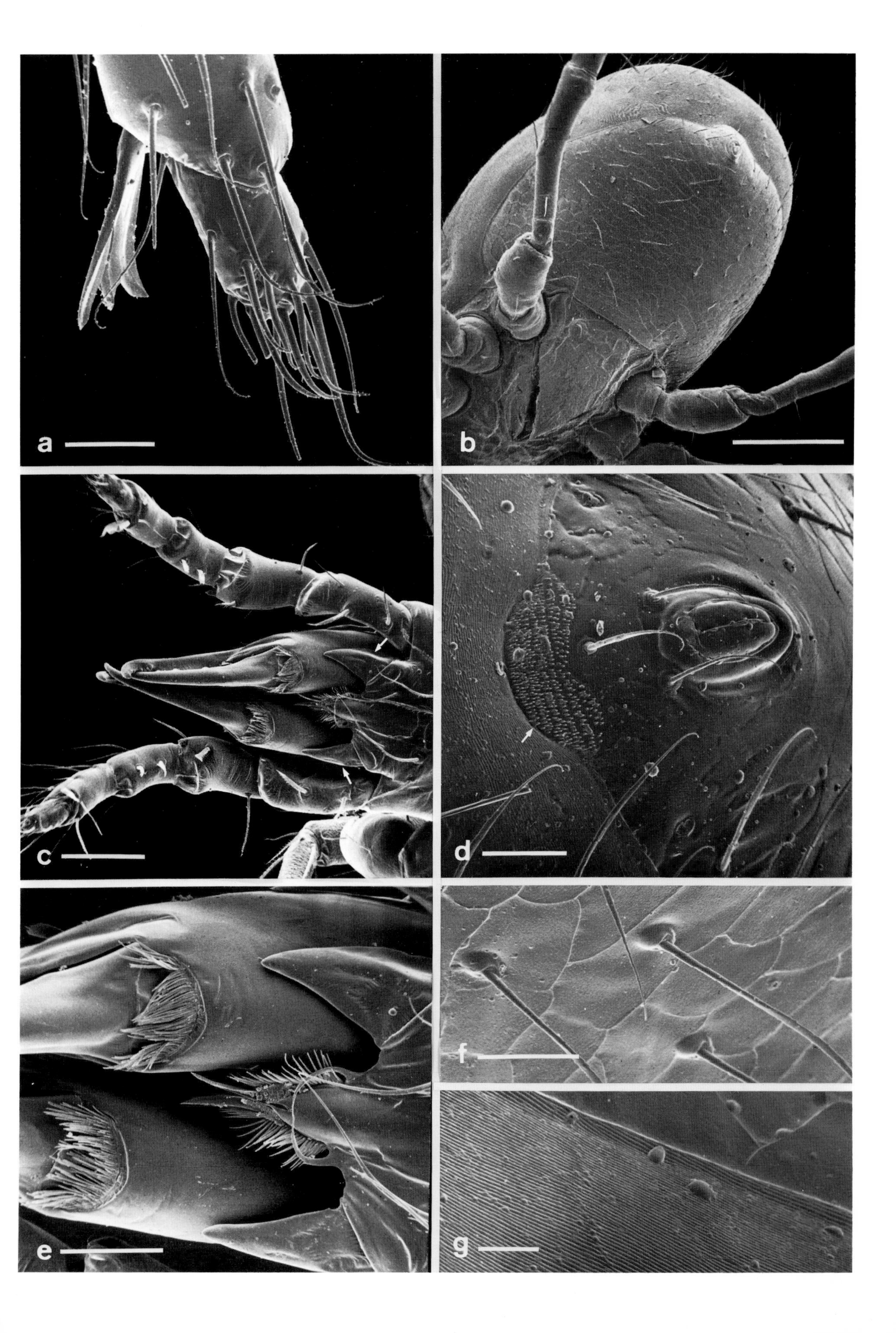
a
b
c
d
e
f
g

2.4.7 Cryptostigmata (Oribatei) "Horn, Moss or Beetle Mites"

The Oribatei (Fig. 58; Plates 34–41) (Willmann, 1931; Sellnick, 1960; Dindal, 1977) have common names derived from their strongly chitinous exoskeleton and their frequent occurrence in moss cushions. They do not live exclusively in moss, but are widely distributed, showing a distinct preference for the soil. Due to their high individual density they are of importance to soil biology and make a considerable contribution to the decomposition of vegetable matter (Forsslund, 1938, 1939; Luxton, 1972; Schuster, 1956; Mittmann, 1980; Wallwork, 1983). Many species of the Oribatei are acidophilic and are extremely abundant in acid soils (Hågvar and Abrahamsen, 1980; Hågvar and Amundsen, 1981; Hågvar and Kjøndal, 1981).

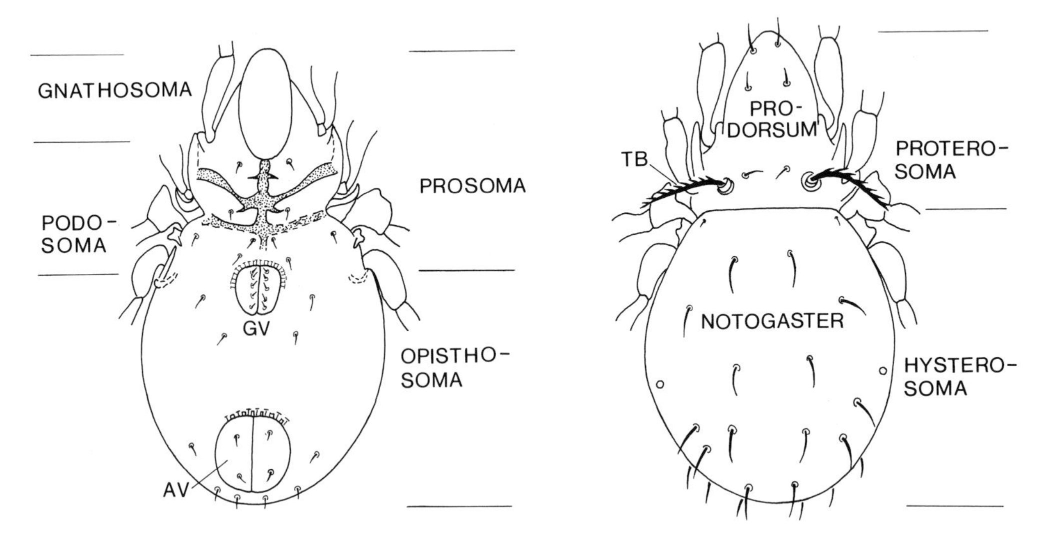

Fig. 58. Arrangement of the body of the Cryptostigmata (Oribatei). (Wallwork, 1969). Venter of a "higher" oribatid (*left*). *AV* Anal valves; *GV* genital valves. Dorsum with paired trichobothria (*TB*) on the prodorsum (*right*)

Plate 34. Damaeidae, Cryptostigmata (Oribatei), tritonymph.

a, b Animal with calyptra encrusted with earth particles, lateral and frontal. 0.5 mm, 300 μm.

c Edge of the calyptra above the leg bases. 100 μm.

d Proterosoma, frontal. Above the leg bases lie the cupulate sockets of the trichobothria (pseudostigmatic organs). 100 μm.

e Opisthosoma, ventral with genital and anal valves. 50 μm.

f Proterosoma, oblique frontal. 200 μm

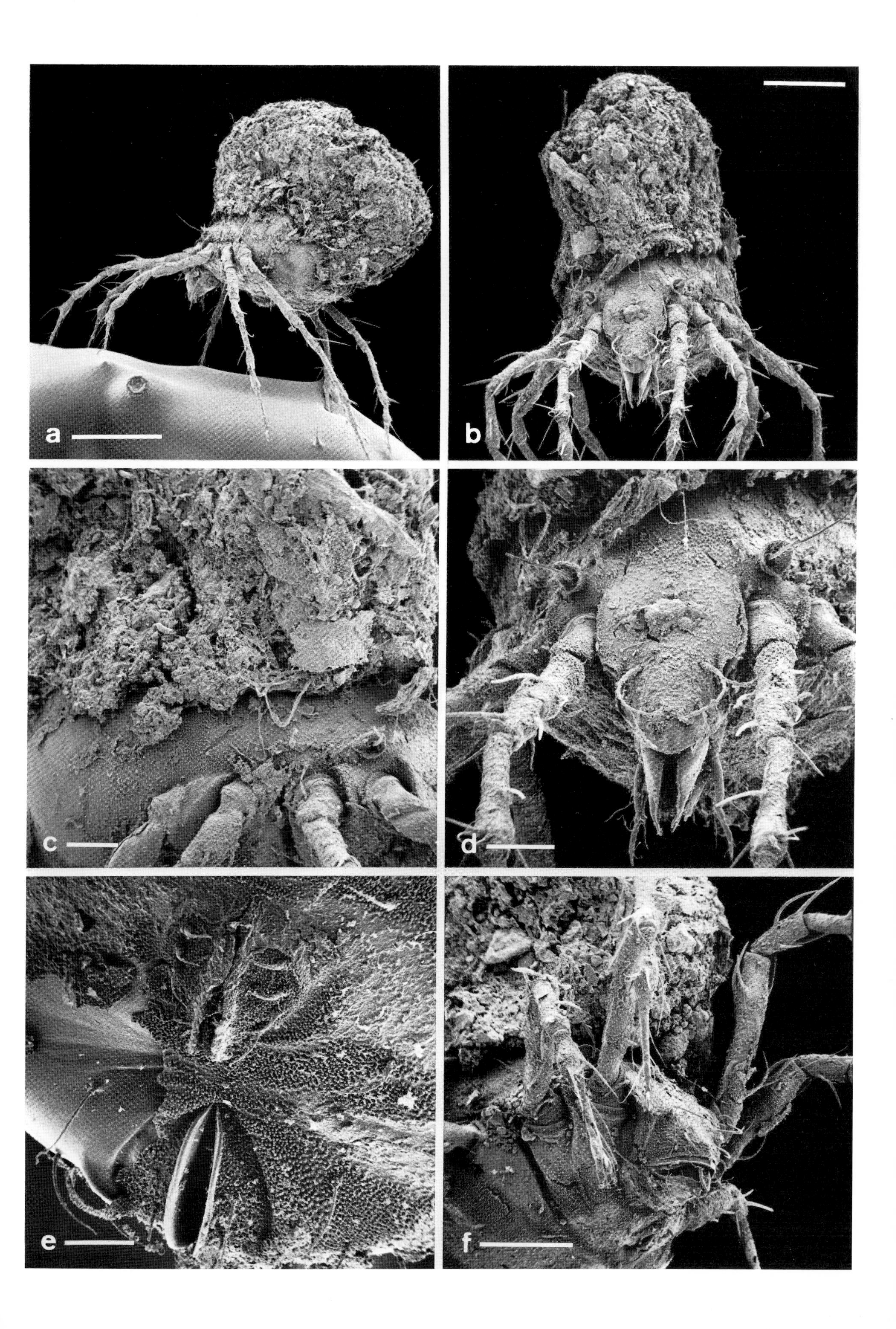
a
b
c
d
e
f

2.4.8 Horizontal Distribution and Aggregation of the Cryptostigmata (Oribatei)

The distribution of the soil-dwelling Oribatei ranges from dry coniferous woods, through dry and damp deciduous woodland, fen woodland and marshes to salty meadows at the boundary between land and sea (Strenzke, 1952; Knülle, 1957; Märkel, 1958; Weigmann, 1967, 1971, 1973; Wink, 1969; Luxton, 1972, 1975, 1979; Usher, 1975; Schatz, 1977; Mitchell, 1977, 1979; Thomas, 1979; Mittmann, 1980, 1983; Wallwork, 1983).

With their high density of individuals in the soil, $6.7-134\times10^3$ ind/m^2 (Berthet, 1964; Berthet and Gérard, 1965; Gérard and Berthet, 1966), the Oribatei show a distinct tendency to congregate into nest-like aggregations in the soil (Usher, 1975 etc.). On the one hand, the density of individuals in an aggregation seems to be in a state of dynamic equilibrium with the number of aggregations in an area. On the other, these aggregations are obviously caused be ecological mechanisms, common to both species and individuals, allowing them to adapt to the various ecological niches and, at the same time, avoiding and withstanding adverse environmental conditions.

Plate 35. *Damaeus* sp. – Damaeidae, Cryptostigmata (Oribatei).

a Overview, dorsal, showing the distinct division into protero- and hysterosoma. 0.5 mm.

b Overview, frontal. 250 µm.

c Anterior body, oblique ventral, with genital valves. The mouthparts have been withdrawn into the hood (rostrum). 200 µm.

d Proterosoma, dorsal. *Arrows* mark the trichobothria (pseudostigmatic organs). 100 µm.

e Rostrum with enfolded gnathosoma. 50 µm.

f Typical cuticular structure on the opisthosoma with trichoid secretion. 10 µm

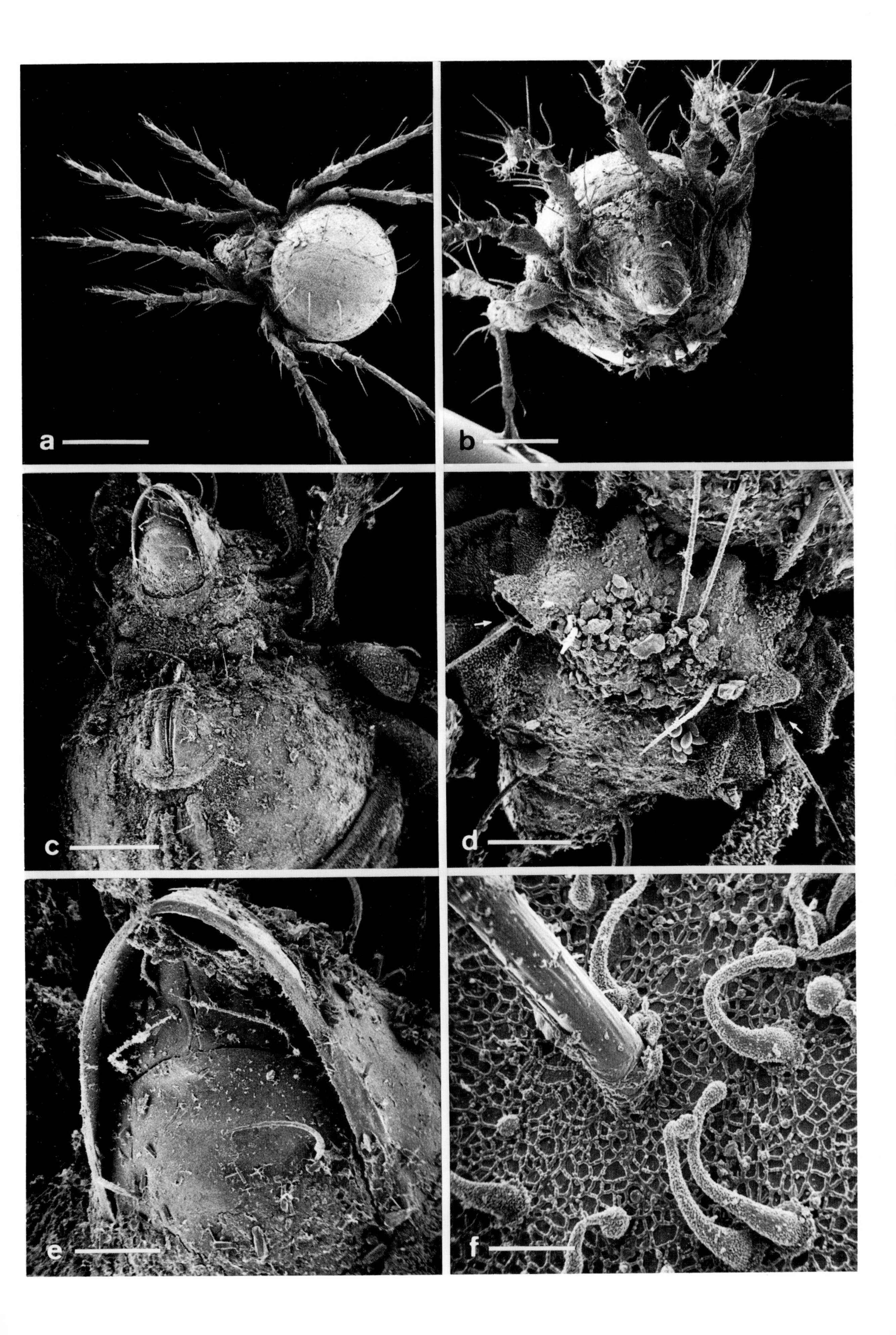

a
b
c
d
e
f

2.4.9 The Feeding Habits of the Oribatei

The feeding habits of the Oribatei are of primary importance for the decomposition of plant debris in the soil. Schuster (1956) (Fig. 59) distinguished between three types of feeding behaviour:

1. Microphytophages, which consume predominantly algae, fungal hyphae and spores (various species of the genera: *Belba, Oppia, Damaeus*, etc.).
2. Macrophytophages, which feed an leaves, conifer needles and wood (various species of the genera: *Carabodes, Oribotritia, Phthiracarus*, etc.).
3. Non-specialists, which are both microphytophagous and macrophytophagous (various species of the genera: *Nothrus, Euzetes, Pelops*, etc.).

No correlation is clearly evident when the nutritional preferences of species are compared with the shape of their chelicerae (Schuster, 1956). However, the diversity of the chelicerae undoubtedly corresponds to the different textures of the nutritional substances (Mittmann, 1980).

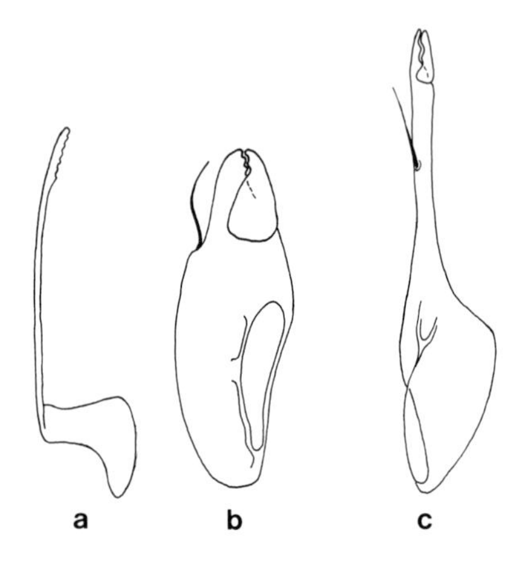

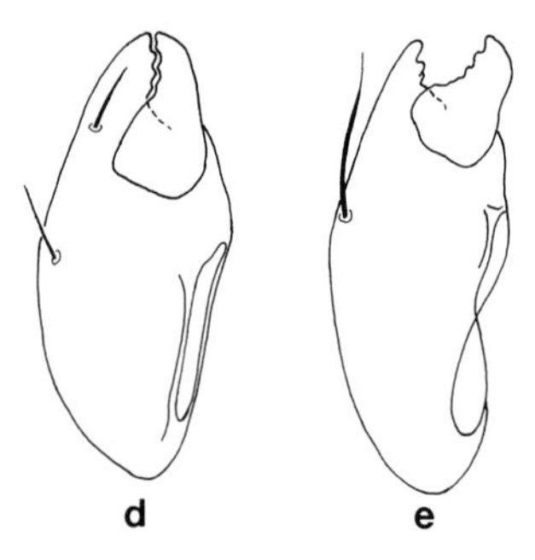

Fig. 59. Cheliceral forms, arranged into feeding types. (Schuster, 1956).
a, b Microphytophages: *Gustavia microcephala, Belba verticillipes.* **c, d** Non-specialists: *Pelops hirtus, Nothrus silvestris.* **e** Macrophytophage: *Hermanniella granulata*

Plate 36. *Porobelba spinosa* – Belbodamaeidae, Crytostigmata (Oribatei).
a Adult mite, oblique lateral. The calyptra is composed of four tiers, one larval and three nymphal exuvia. 200 µm.
b Calyptra with attached eggs and pollen grains. 100 µm.
c Head (proterosoma), oblique frontal. *Left arrow* marks a trichobothrium, the *right arrow* a long bristle hair. 100 µm.
d Trichobothrium with cupulate socket (bothridium). 20 µm.
e Surface of a leg with solidified thread-like secretion. 30 µm.
f Cuticle with secretion formations. 5 µm

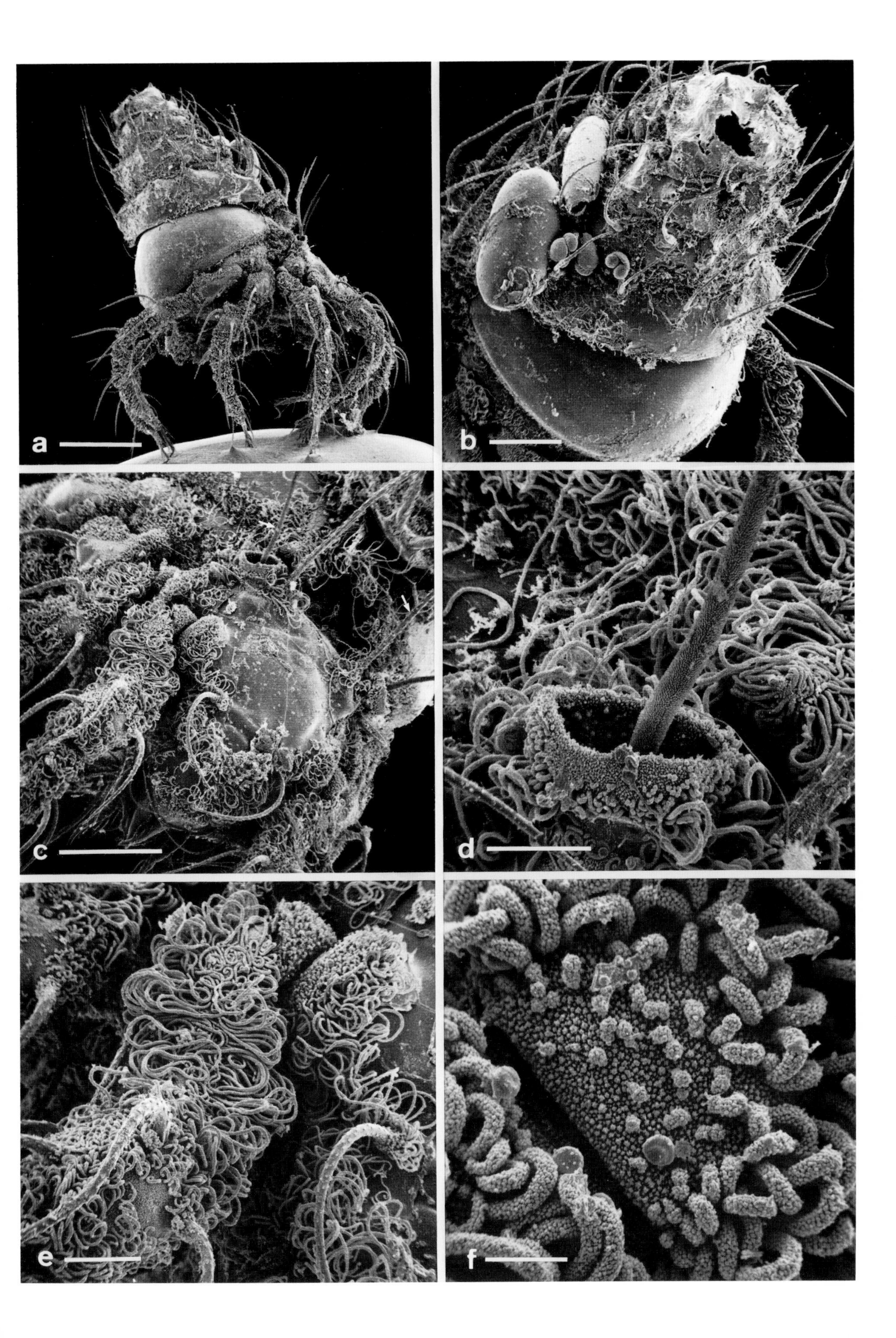
a
b
c
d
e
f

2.4.10 The Significance of the Oribatei to Soil Biology

Micro- and macrophytophagous oribatids influence the decomposition process in woodland soil in different ways (Fig. 60) (Mittmann, 1980, 1983). Those macrophytophages which feed on leaf litter and wood are primary decomposers, playing an essential role in the turnover of soil substances. They are voracious feeders, consuming about 20% of their body weight in beech leaves daily and creating favourable conditions for further degradation by other decomposers. The microphytophages contribute indirectly to the decomposition process. They act as "catalysts", stimulating microbial activity by grazing on fungal hyphae and dispersing spores (Mitchell and Parkinson, 1976; Mittmann, 1980). Their share in the entire turnover of all mycophages in woodland soil is estimated to be 50%.

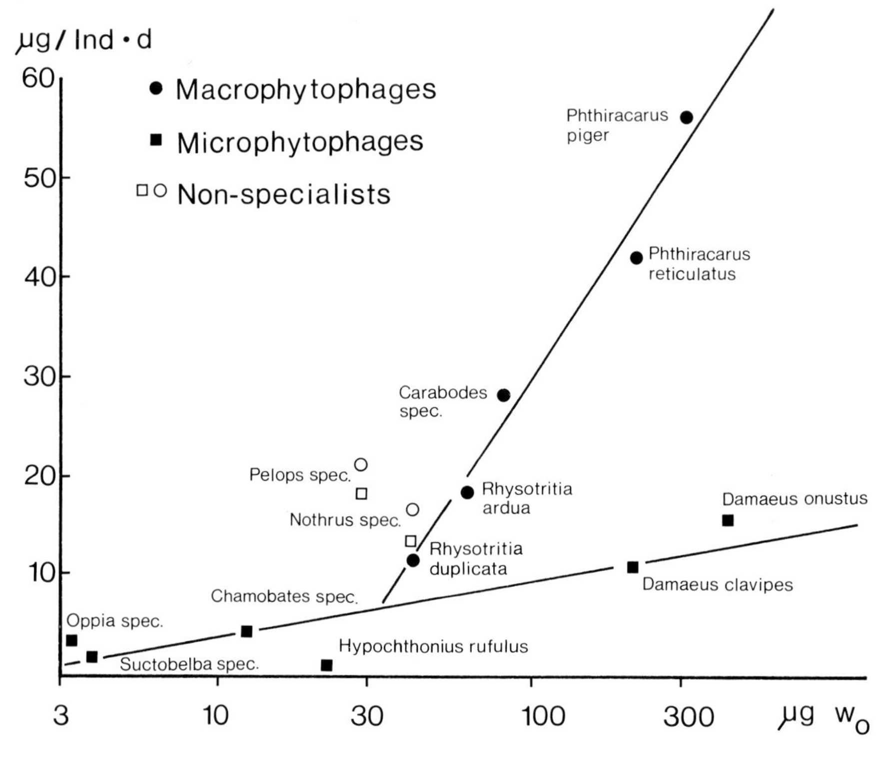

Fig. 60. Relationship between the daily consumption of a variety of oribatid species and their mean body mass w_0. (Mittmann, 1980). *µg/Ind·d* = Microgram per individual per day

Plate 37. *Nothrus* sp. – Nothridae. Cryptostigmata (Oribatei).

a, b Overview, dorsal and lateral. 0.5 mm, 250 µm.
c Ventral view with genital and anal valves. 250 µm.
d Proterosoma, oblique frontal. 100 µm.
e Valve complex. 100 µm.
f Proterosoma, dorsal, with trichobothria pair. 100 µm

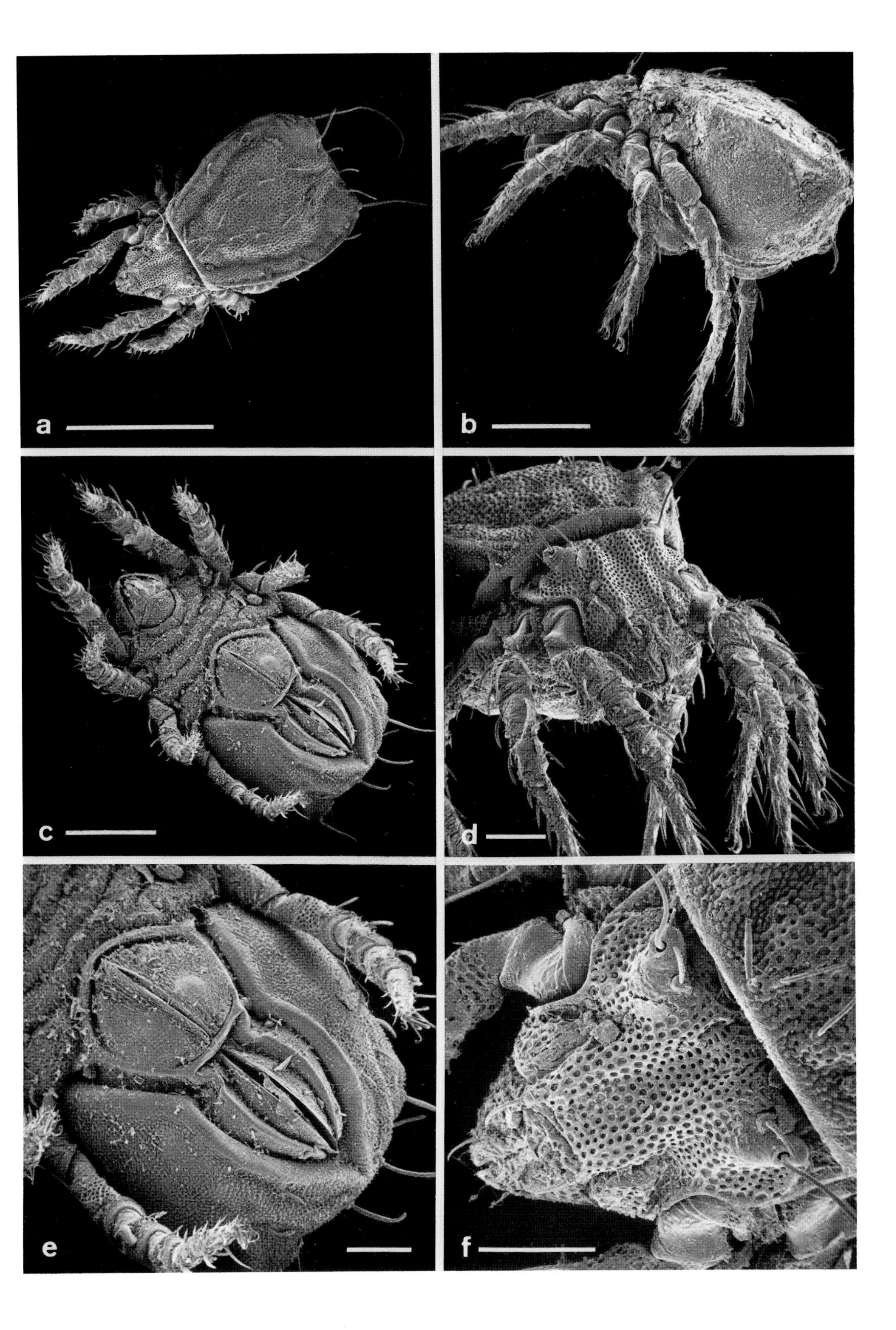
a
b
c
d
e
f

2.4.11 Vertical Distribution of the Oribatei

In accordance with their role as decomposers of plant debris most Oribatei are restricted to the upper 5 cm of the soil where they live in decaying leaf litter. Only a few species occur in deeper layers (Klima, 1956). The few euedaphic species, to which *Nothrus silvestris* and *Pseudotritia minima* belong (Fig. 61) (Lebrun, 1968, 1969, 1970; Märkel, 1958), exhibit hardly any of the typical adaptations shown by other euedaphic soil arthropods. It is primarily their diminutive size – *Pseudotritia minima* is 0.3-mm-long – which allows these Oribatei to invade the fine interstitial spaces of deeper soil layers. Their exoskeleton is hardly sclerotized and pigmented, so that they appear pale. In addition, they have high moisture requirements. Some oribatid mites migrate vertically between the litter layer and the mineral soil depending on the moisture conditions (Metz, 1971). Fluctuations in moisture and temperature in the litter layer evidently cause the seasonal (summer– winter) and diurnal, vertical migrations in the direction of the gradient of the relevant ecological factors (Wallwork, 1959, 1970).

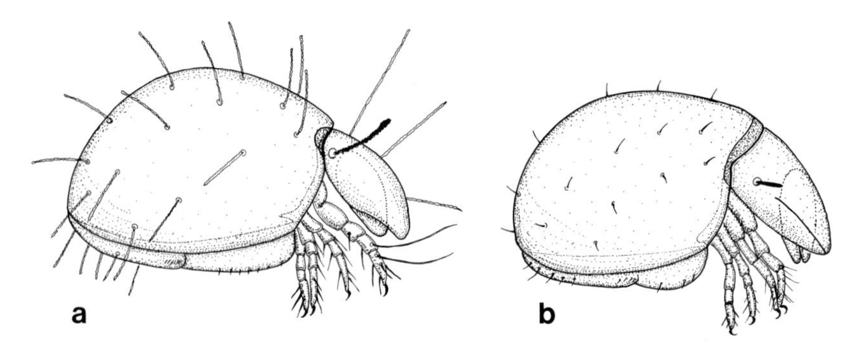

Fig. 61. Oribatid mites of the Phthiracaridae family. (After Märkel, 1958).
a *Pseudotritia duplicata*, a litter-dweller. **b** *Pseudotritia minima*, an euedaphic form with shorter bristles

---

Plate 38. *Steganacarus magnus* – Phthiracaridae, Cryptostigmata (Oribatei).

a, b Overview, lateral and ventral. The prosoma (gnathosoma and podosoma) is freely exposed. 0.5 mm, 300 μm.

c Prosoma, ventral. The "mouthparts" are covered by the hood of the rostrum. 200 μm.

d Proterosoma, with rostrum, dorsal. The *arrow* marks the right trichobothrium. 100 μm.

e Legs, distal. 50 μm.

f Mouthparts with pedipalps (*left arrow*), chelicerae and rutellum (*right arrow*). 50 μm

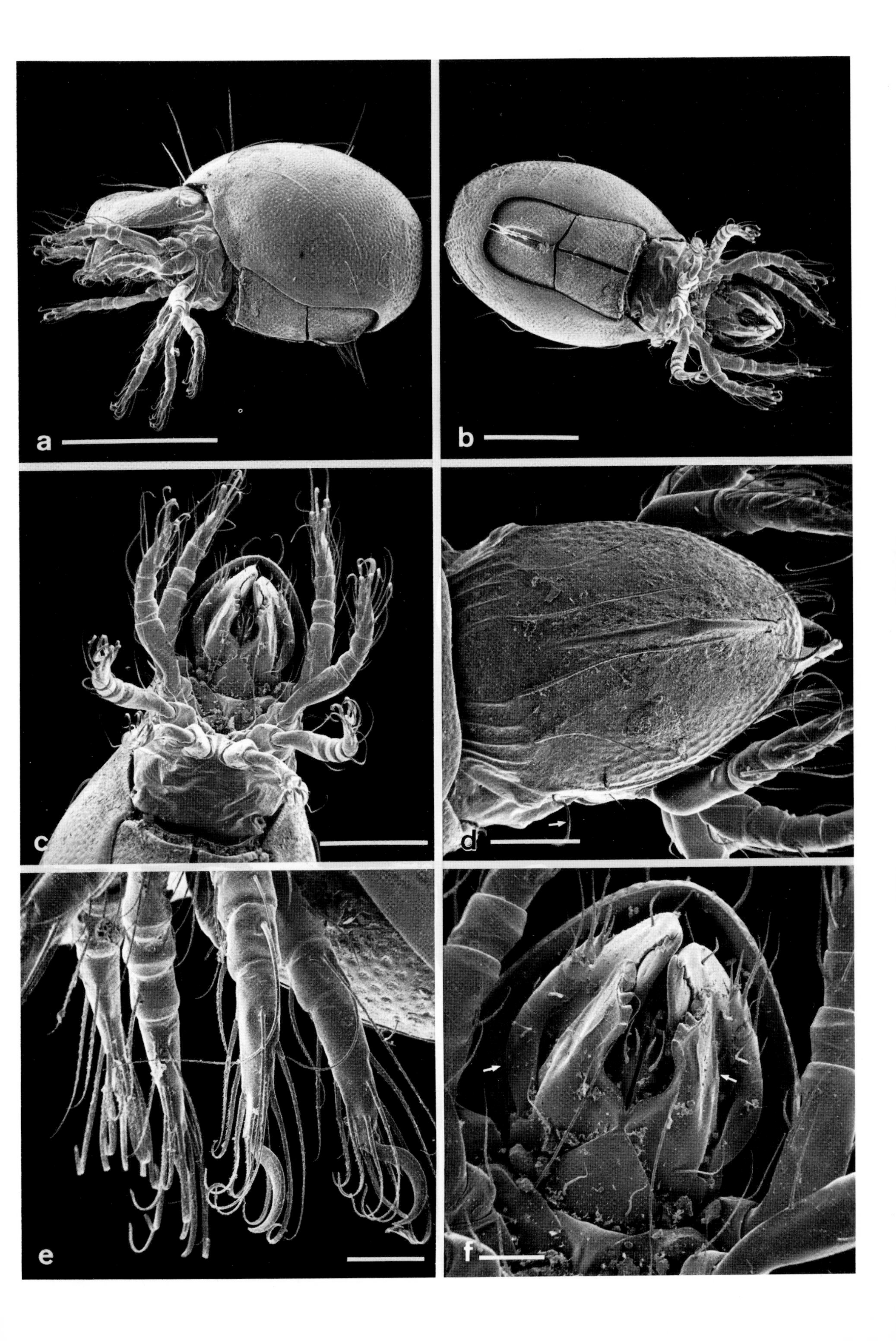
a
b
c
d
e
f

2.4.12 Protective Mechanisms of the Oribatids Against Predation

The soil-dwelling oribatids are frequently preyed upon by other carnivorous arthropods from the same edaphic biotope. As protection from their enemies most "horn" mites have a hard, sclerotized integument and are often spherically-shaped. Their extremities are protected in a variety of different ways, for example, as in the case of *Euzetes globus*, covered laterally by wing-like extensions of the hysterosoma (pteromorphs; Plate 41), or pressed close to the body into hollows in the rounded surface. A further protective device shown by some oribatids, e.g. the *Phthiracarus* species, which they have in common with diplopods (millipedes) and isopods, is their ability to roll themselves into a "ball" (Fig. 62; Plates 38 and 39). When threatened they can enclose their extremities within the articulated hysterosoma and proterosoma by drawing their legs into the former and covering them with the latter. On the other hand, rolling themselves into a ball can prevent desiccation.

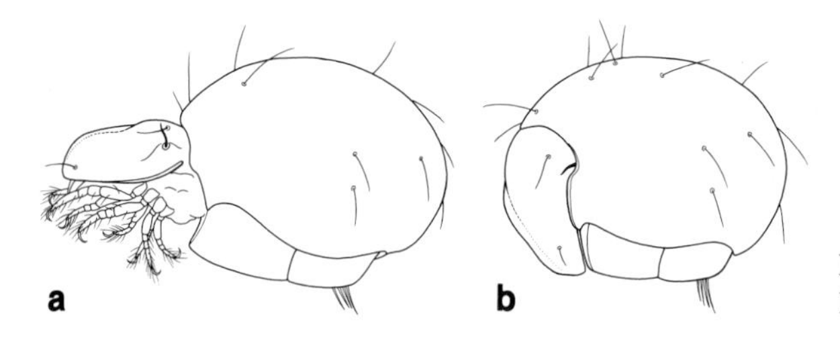

Fig. 62. Rolling up for protection in the Phthiracaridae. Proterosoma in extended (**a**) and "jack-knifed" position (**b**)

Plate 39. *Steganacarus magnus* (**a,c**) and *Phthiracarus piger* – Phthiracaridae, Cryptostigmata (Oribatei).

a Opisthosoma, ventral with genital and anal valves. 100 µm.

b "Rolled up" animal with prosoma shut, lateral. 100 µm.

c Cuticle of the opisthosoma, dorsal, with glandular pits. 10 µm.

d "Rolled up" animal, oblique frontal. 100 µm.

e Cuticle of the opisthosoma. 3µm.

f Edge of the opisthosoma where it overlaps the proterosoma in the rolled-up animal. The bristle is the distal part of the trichobothrium. 10 µm

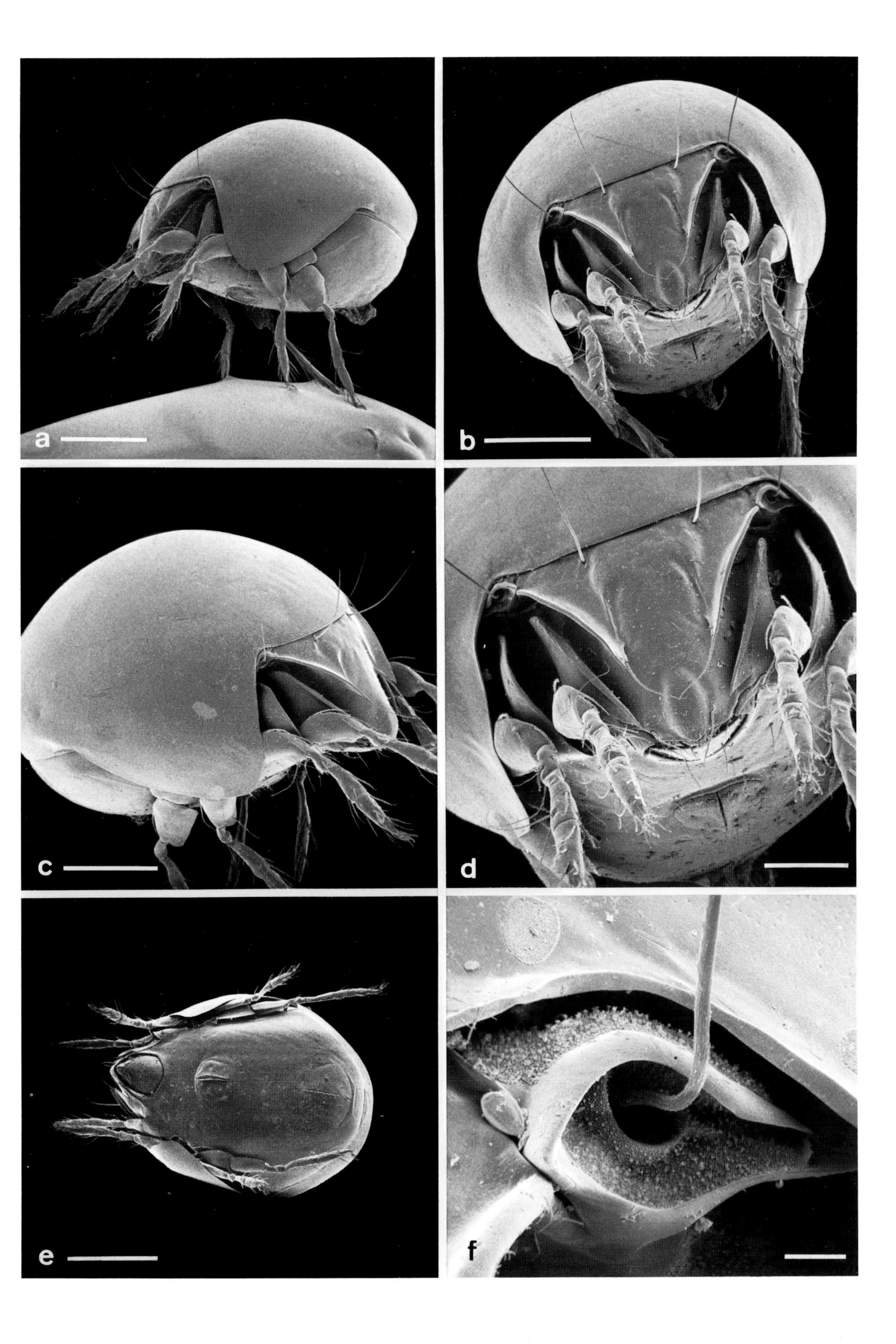

a
b
c
d
e
f

2.4.15 Phoresis in Mites

Certain animal and plant species allow themselves to be transported over some distance for a period of time by so-called carrier species. This phenomenon, known as phoresis, involves a temporary partnership (symphorium) and serves as a dispersal mechanism for the species. The most common passengers are small soil arthropods like the mites, which can thus reach their diverse and often widely separated feeding and breeding sites, e.g. carrion and dung. The host carriers are, in most cases, members of the macrofauna, e.g. isopods, millipedes, beetles, etc. Mites of the Gamasina, Uropodina and Acaridina are most often transported. In many cases a certain developmental stage is specially adapted to phoresis (one-stage rule) in order to survive adverse living conditions, such as desiccation and food shortage. In this case it is the deutonymph, for this reason also known as a "wander" nymph. Well-known examples are the wander nymphs of the Anoetidae, which cling to their host by means of a suction apparatus (Fig. 65; Plate 42). The nymphs of the tortoise mites (Uropodina) (Plates 28 and 29) have also developed an interesting phoretic mechanism. They secrete a solid stalk to attach themselves to their host, allowing space for many mites side by side. Thus, 600 deutonymphs of *Uroobovella marginata* were once encountered attached to a single woodlouse.

The sensory organs for detecting the host carrier are situated on the tarsi on the first pair of legs. Many mites make use of various hosts (polyxenic, heteroxenic), others specialize in one carrier species (monoxenic) e.g. *Coprolaelaps meridionalis* which travels exclusively with the dung beetle *Geotrupes silvaticus* (Hirschmann, 1966). In carabid beetles (especially hibernating carabids) Desender and Vanneechoutte (1984) found members of 17 mite genera mostly at the deutonymph stage. Sometimes they observed damage to beetle hind wings due to phoretic mites, which could reduce the dispersal capacity of species.

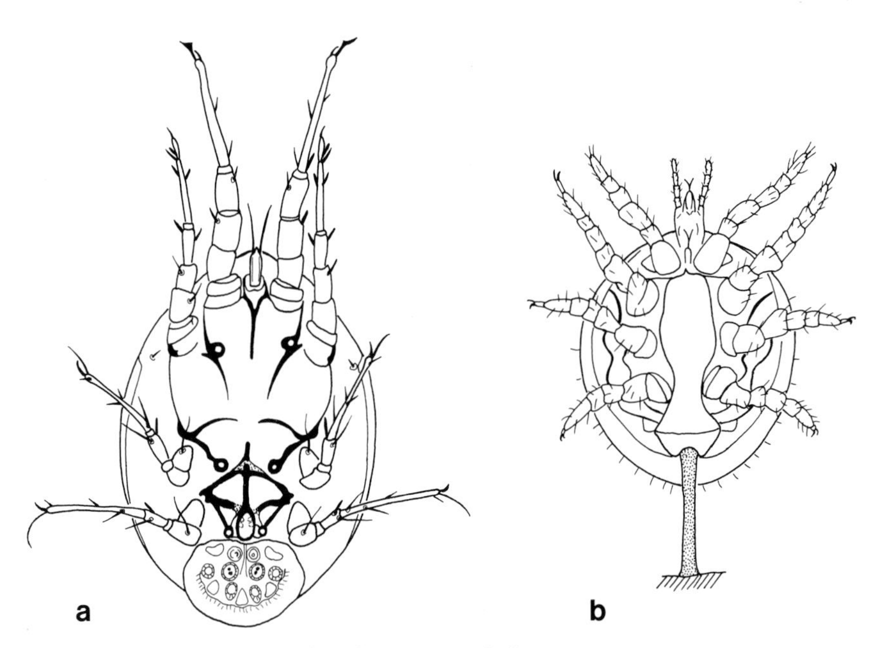

Fig. 65. Phoretic (deuto-) nymphs of the Acari. (After Krantz, 1978).
a Anoetidae nymph, ventral, with caudal attachment disc. **b** Uropodidae nymph on a solidified secretion stalk in the "travelling position". Sometimes the nymphs are totally motionless ("thanatosis"). Then they fold their legs into the fovae pedals (cf. Plate 29)

Plate 42. Phoretic ("wander") nymphs of the Anoetidae (Astigmata).

- **a** A group of "wander" nymphs on the pleurite of a common pill millipede, *Glomeris marginata.* 50 µm.
- **b** Sternal region of *Lithobius* with a "wander" nymph between the sternites. 150 µm.
- **c** "Wander" nymph on the base of a maxillipede of *Lithobius.* 50 µm

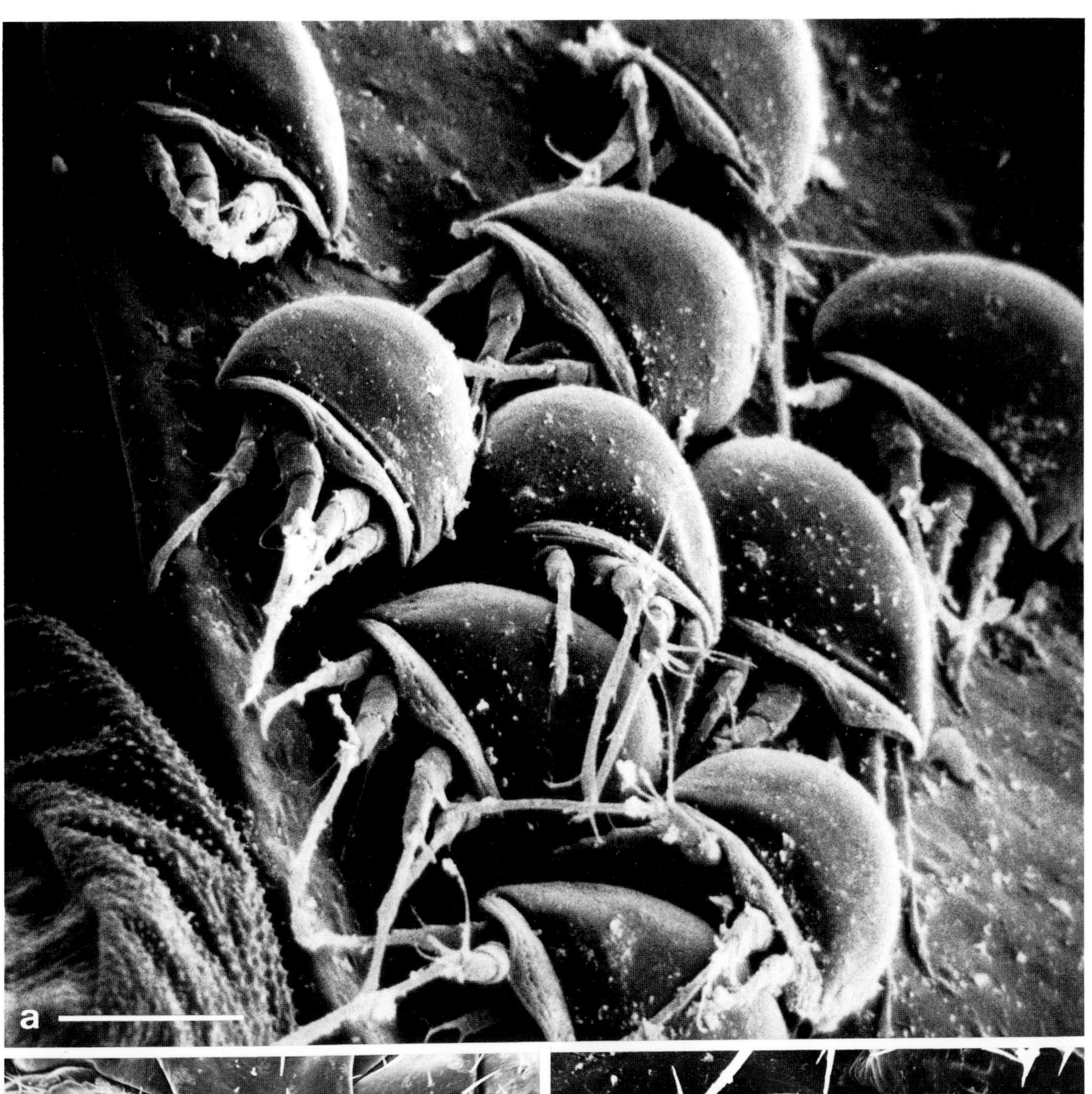
a

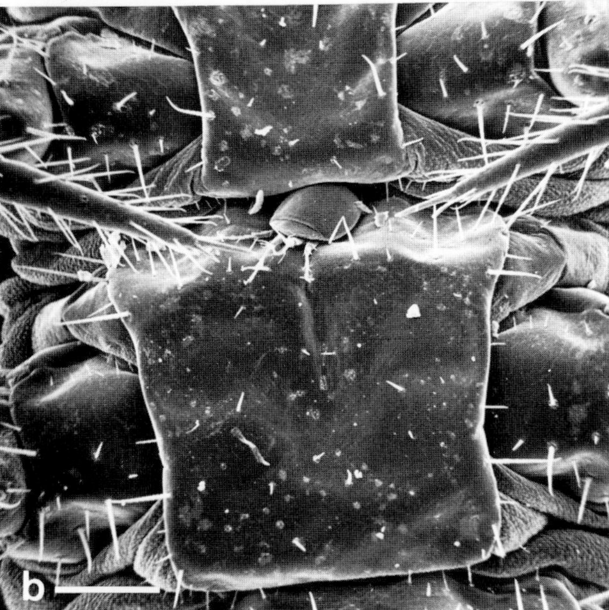
b

c

2.5 Order: Isopoda – Isopods (Crustacea)

General Literature: Wächtler, 1937; Schmölzer, 1965; Gruner, 1965–1966; Kaestner, 1967; Sutton and Holdich, 1984

2.5.1 Oniscoidea – Terrestrial Isopods

Originally crustaceans were exclusively marine animals. Along with the terrestrial copepods, amphipods and land crabs (Gecarcinidae), the isopods of the suborder Oniscoidea are among the crustaceans which have gradually adapted from life in the sea to that on land. The adaptive stages are illustrated by the Oniscoidea families:

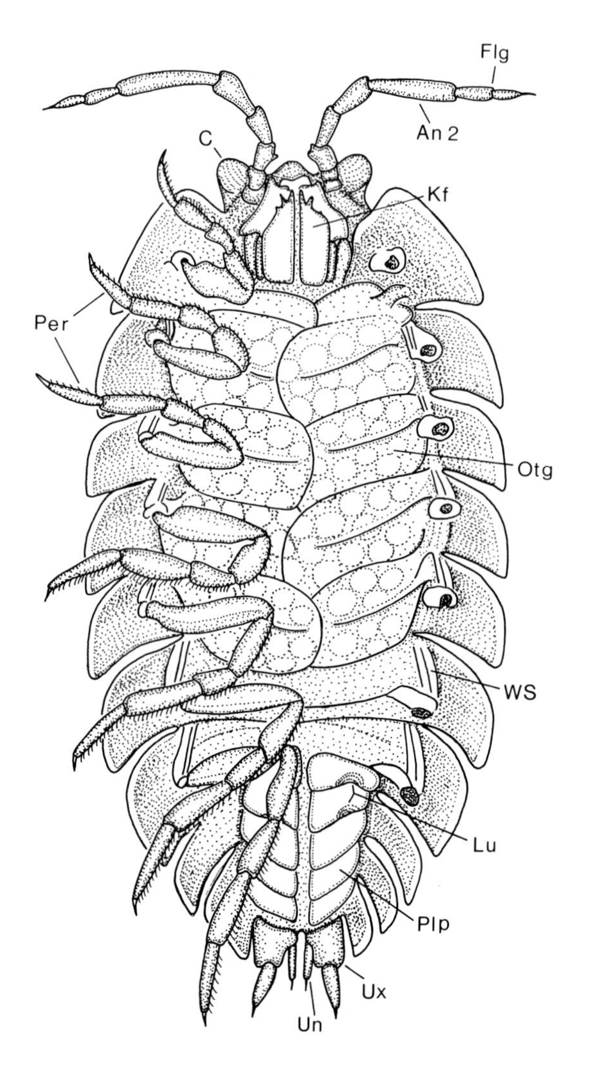

1. Amphibious isopods, e.g. living in wet riparian zones of tide water, running water and woodland pools. — Ligiidae
2. Isopods of moist biotopes, e.g. in the F-horizon of damp deciduous woodland. — Trichoniscidae
3. Isopods of moderately moist biotopes, e.g. under stones, bark, in the litter layer of damp woodland. — Oniscidae
4. Isopods of dry to moderately damp biotopes, e.g. leaf litter, stony rubble, crevices on buildings. — Porcellionidae
5. Isopods of dry and sunny biotopes, semi-deserts — Armadillidiidae

These stages show the development of physiological mechanisms necessary for ecological adaptation, particularly with regard to water balance and respiration.

Fig. 66. Structural characters of a terrestrial isopod exemplified by the woodlouse *Porcellio scaber* (Porcellionidae). (After Kaestner, 1967). Ventral aspect with brood pouch (marsupium).
An 2 Antenna 2; *C* "head" (cephalothorax); *Flg* flagellum with terminal sensory organ (brush organ, often conical in shape); *Kf* maxillipedes; *Lu* lung entrance areas, (white bodies in living animal); *Otg* oostegite (brood pouch plates, with eggs beneath); *Per* pereiopods; *Plp* pleopods; *Un, Ux* uropod endopodite, uropod exopodite; *WS* water-conducting system

Plate 43. *Ligidium hypnorum* (Ligiidae).
a Overview, lateral. The legs have relatively long basal segments. 1 mm.
b "Head" (cephalothorax), frontal. Between the antennal bases of the second antennae lie the small first antennae. 250 µm.
c Compound eye. 100 µm.
d "Lower lip" and upper lip with sensory hairs and cleaning structures. 50 µm.
e Ommatidia of the compound eye. 20 µm.
f Sensory hairs of the "lower lip", distal. 1 µm

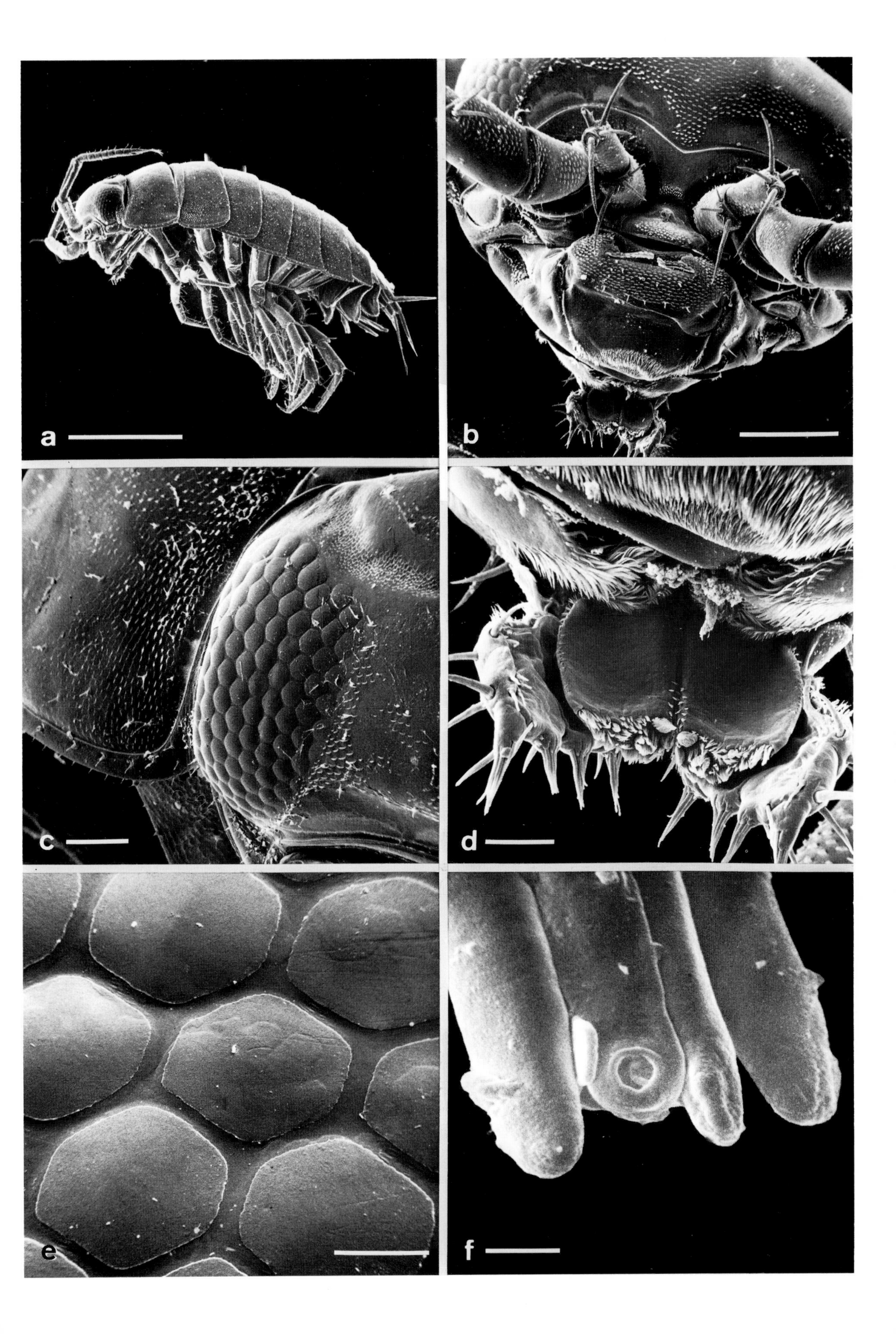
a
b
c
d
e
f

2.5.2 Sensory Organs on the Antennae of the Oniscoidea

Adaptation of the Oniscoidea to life on land is also recognizable in the ultrastructure of their antennal sensory organs, which has been described in *Ligidium hypnorum*. *Porcellio scaber* and *Armadillidium nasutum* (Risler, 1976, 1977, 1978) as well as the Tunisian desert isopod, *Hemilepistus reaumuri* (Seelinger, 1977). The morphology and function of the chemoreceptive organs of the terrestrial isopods show convergence with those of the insects (Risler, 1977).

Sensory physiological investigations suggest that the cone at the tip of the second antenna bears contact chemoreceptors involved in the detection of food (Henke, 1960). Olfactory receptors, which are important for the social behaviour of desert isopods (Linsenmair and Linsenmair, 1971; Schneider, 1971) and the tendency of indigenous isopods to form aggregations (Schneider and Jakobs, 1977), are thought to be located beside them on the second antenna.

Fig. 67. The fine structure of the first antenna of *Porcellio scaber* (Porcellionidae). (After Risler, 1977). At the distal end the three-segmented antenna terminates in a bundle of tubular hairs, known as the aesthetascs (*Ae*). They are presumed to be chemoreceptors. Beside them, in the interior of the antenna, there are four chordotonal organs which function as stretch receptors (*CH 1–4*). Dendrites with ciliary structures on the outer segment emanate from the sensory cells (*Sz*). *Di* Inner dendritic segment with axial filaments for support; *M 1–3* muscles; *N* conducting nerve

Plate 44. *Ligidium hypnorum* (Ligiidae).

a Antenna 1 with terminal cone (*arrow*). 200 µm.

b Leg joint with bristle-like sensory hairs and cuticular scales. 50 µm.

c Antenna 1, terminal cone with aesthetascs (tubular hairs). 10 µm.

d Antenna 2, distal, with its terminal organ (brush organ). 100 µm.

e Venter of the trunk, lateral, with water-conducting system (*arrow*) and cone-shaped processes (stolons). 250 µm.

f Surface of endopodite IV with cellular arrangement. 20 µm

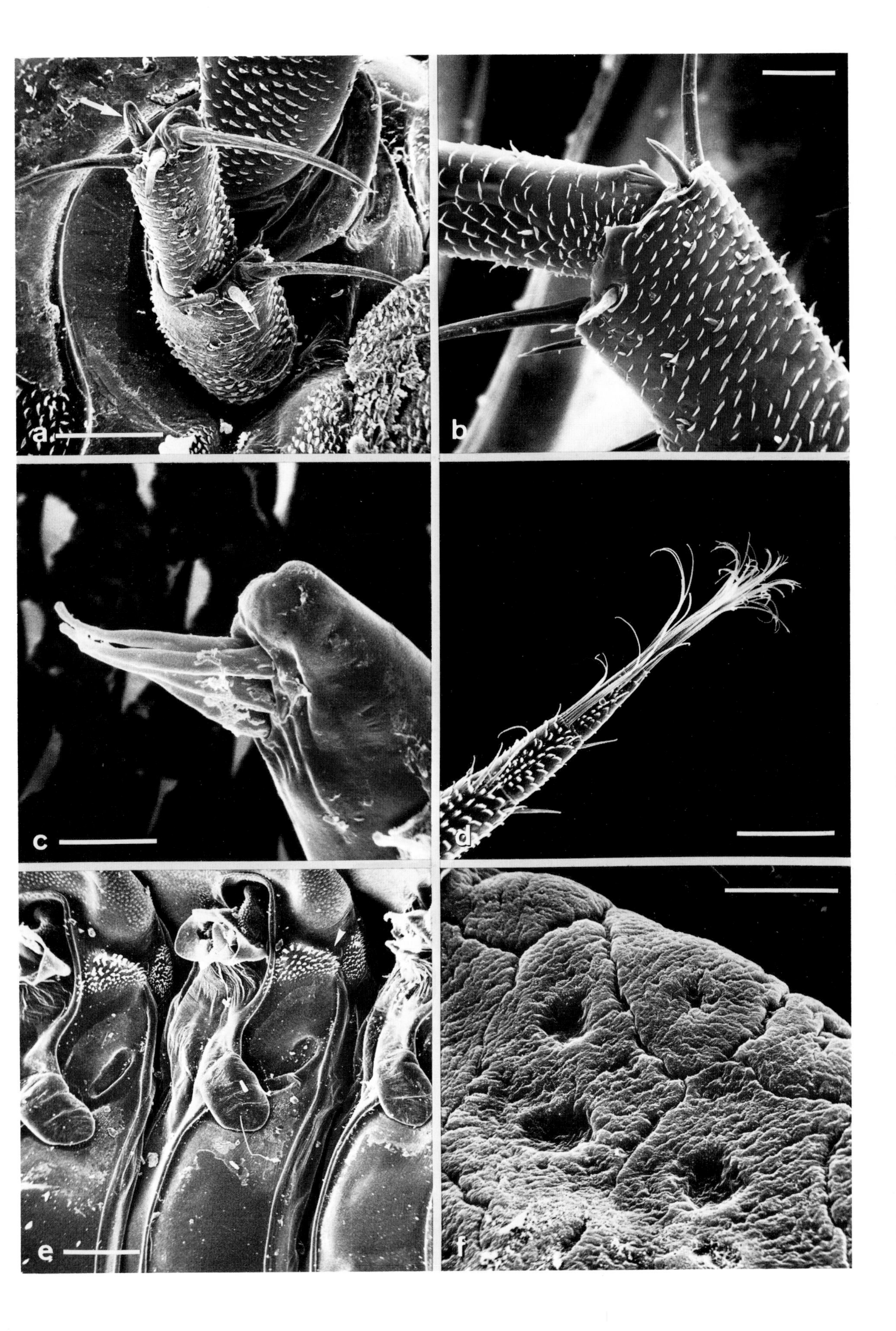
a
b
c
d
e
f

2.5.3 Isopods as Primary Decomposers – Food Preferences

The terrestrial isopods – Oniscoidea – have powerful biting mouthparts, enabling these primary decomposers to contribute significantly to the decomposition and fragmentation of plant debris in the soil. This important participation in soil biology is coupled with a high density of individuals in moist rather than dry, open woodland or open country. The seasonal activity and population dynamics of the isopods is highly dependent on moisture and temperature. In warm, dry summers they are predominantly nocturnal and remain in the hollow spaces in the upper soil layers; but on damp, cool autumn days they become active in the newly fallen leaves on the surface (Herold, 1925; Brereton, 1957; Den Boer, 1961; Beyer, 1964).

When offered a choice of different types of leaves their food preferences are similar to those of other primary decomposers: snails, enchytraeids, earthworms, centipedes and certain insects. The acceptance or rejection of various foliage types has complex causes, influenced by a range of chemical and physical factors. Preference differences between freshly fallen and overwintered leaves are especially noticeable. *Oniscus,* for instance, showed the following sequence of preference for freshly fallen leaves: *Ilex aquifolium* (holly) – *Fagus silvatica* (beech) – *Carpinus betulus* (hornbeam) – *Acer pseudoplatanus* (maple) – *Quercus rubra* (red oak). For old litter leaves, allowed to overwinter, the sequence was: *I. aquifolium* – *A. pseudoplatanus* – *F. silvatica* – *Q. rubra.* Thus, *Ilex* is distinctly preferred, while oak leaves are largely rejected even in a decayed state. In selection experiments to determine whether fresh or overwintered leaves were more palatable, *O. asellus* showed a preference for fresh material (Dunger, 1958, 1962; Beck and Brestowsky, 1980).

Plate 45. *Trichoniscus pusillus* (Trichoniscidae).
a Overview, frontal. 200 μm.
b Pleon, dorsal. 250 μm.
c "Head" (cephalothorax), ventral, with mouthparts. 100 μm.
d Exopodite of a uropod, distal, with tubular hairs. 50 μm.
e Antenna 1, distal, with aesthetascs (*arrow*). 10 μm.
f Antenna 2, distal, with long tubular hairs which form the terminal organ (brush organ). 25 μm

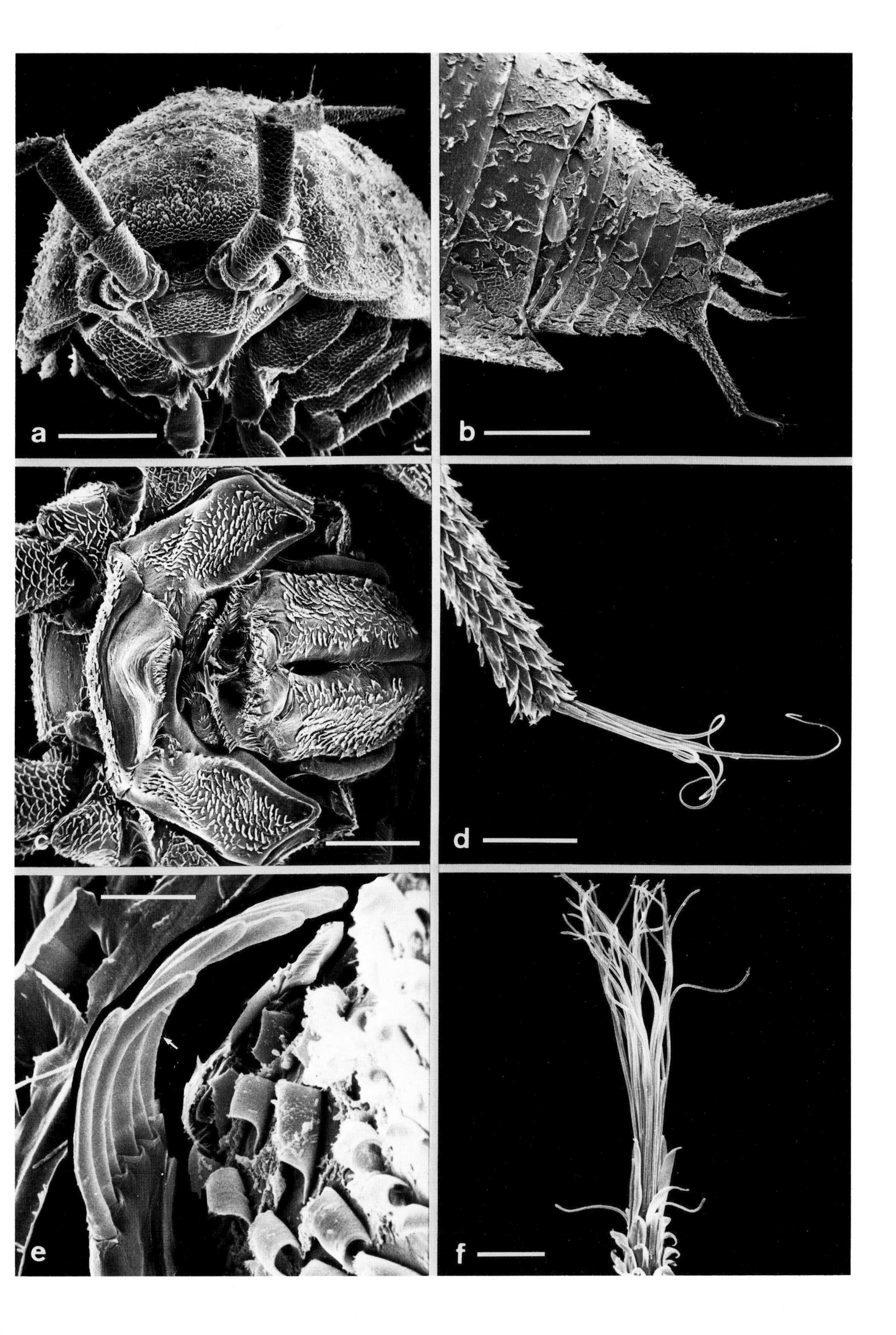

a
b
c
d
e
f

2.5.4 Isopods as Primary Decomposers – Turnover of Organic Matter

The terrestrial isopods, as primary decomposers, are macrophytophagous, consuming freshly fallen leaves. As their feeding activity progresses the leaves are perforated and skeletonized till only the hard vascular ribs remain. Moreover, as saprophytophages they feed on overwintered leaves which have already been degraded by microorganisms. The daily food consumption of an isopod depends on the animal's size and developmental stage. It is also influenced by the leaf species, the degree of decomposition and the moisture content of the nutrition as well as by microclimatic conditions in the habitat (Dunger, 1958). The growth rate of *Oniscus asellus* is considerably higher when the animals are offered fresh instead of overwintered leaves (Beck and Brestowsky, 1980).

Table 4 shows the daily production of excrement by a variety of isopod species as a measure of the turnover of organic matter. Several different leaf species of fresh and overwintered foliage are taken into account.

Table 4. Mass of excrement produced daily (mg) per gram body mass of the isopods *Armadillidium vulgare* (1), *Tracheoniscus rathkei* (2), *Porcellio scaber* (3), *Onsicus asellus* (4) and *Ligidium hypnorum* (5) for fresh leaves (a) and overwintered leaves (b) (Dunger, 1958)

Leaf species		1	2	3	4	5
Tilia cordata	a	–	–	–	–	
	b	57,3	27,3	39,0	34,7	–
Fraxinus excelsior	a	82,9	–	61,1	57,1	11,2
	b	50,7	22,2	27,4	29,3	–
Carpinus betulus	a	18,2	–	27,4	18,9	17,2
	b	56,8	68,7	39,6	44,6	–
Alnus glutinosa	a	52,7	–	38,4	43,7	78,9
	b	43,3	39,6	26,8	36,3	–
Ulmus carpinifolia	a	59,6	–	37,7	31,1	61,7
	b	63,8	–	74,0	54,0	–
Acer pseudo-platanus	a	–	–	–	–	–
	b	17,7	59,7	19,3	39,1	–
Acer platanoides	a	17,8	–	28,0	25,8	13,8
	b	19,1	28,3	21,8	16,9	–
Fagus silvatica	a	–	–	–	–	–
	b	20,1	32,2	20,7	24,6	–
Quercus robur	a	8,1	–	9,1	9,8	4,7
	b	13,6	32,2	16,6	22,6	–

Plate 46. *Trichoniscus pusillus* (Trichoniscidae).

a Ventral side with part of the brood pouch (marsupium). It is sealed by flap-like processes, the oostegites. 100 μm.

b Pleon, ventral, with pleopods and uropods. 200 μm.

c Posterior margin of a segment, dorsal, with cuticular scales removed. 10 μm.

d Cuticular scales on the ventral side of an exopodite. 20 μm.

e Cuticular fold from which cuticular scales originate. 2 μm.

f Surface of an endopodite. 10 μm

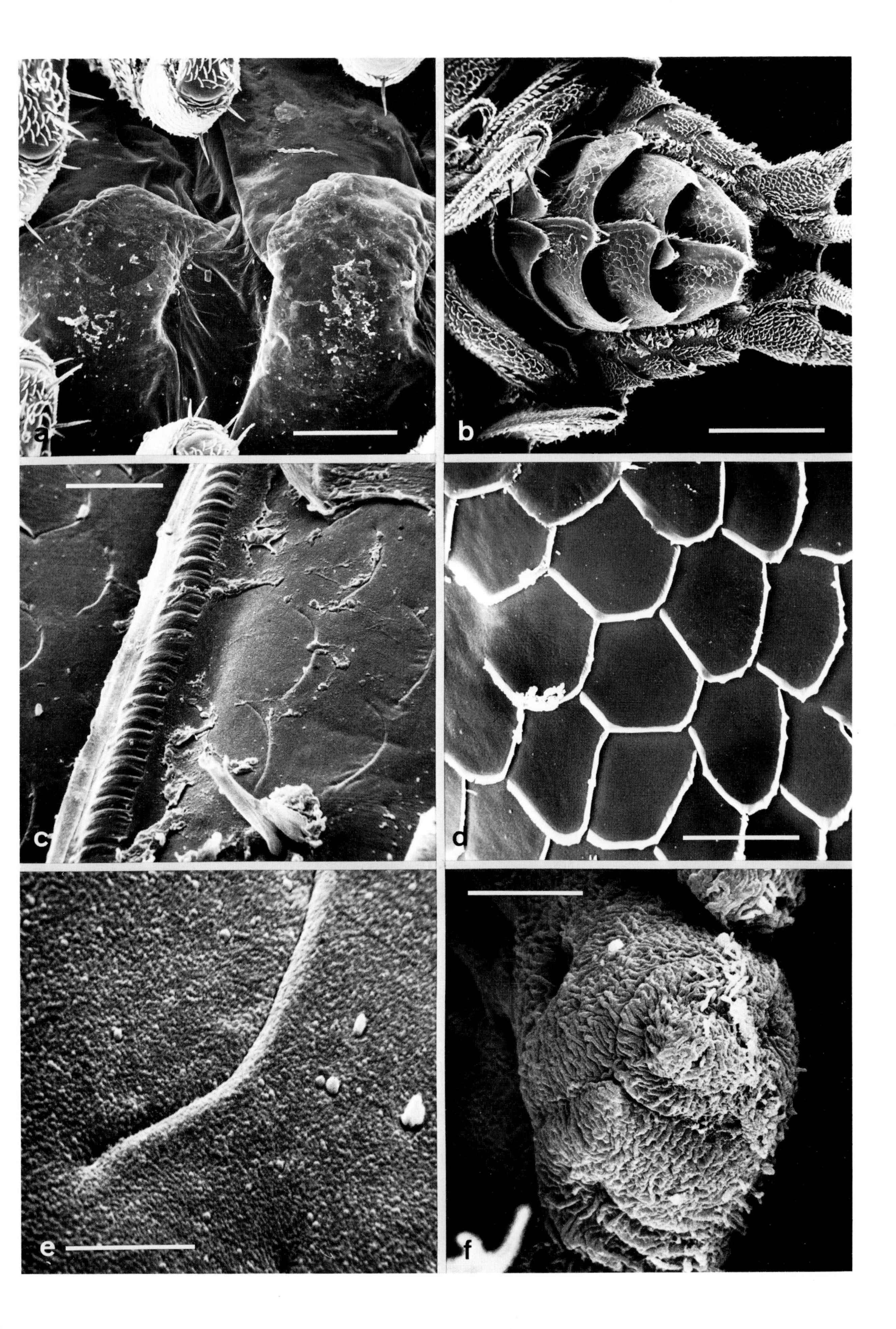
a
b
c
d
e
f

2.5.5 Surface Structures for the Formation of a Plastron

The tergite surface of terrestrial isopods often exhibits diverse hump-like, scale-like and symmetrical or asymmetrical polygonal microstructures which are absent in aquatic isopods. While water-dwellers have a smooth tergite surface, an adaptation to minimize hydrodynamic friction, the bizarre surface structures on the terrestrial isopods can be interpreted as an adaptation to life on land.

Such surface structures frequently occur on euedaphic soil arthropods. These animals have little possibility of escape when the soil becomes waterlogged after rain or during long-lasting flooding in marshy and riparian woodland. Hydrophobic structures on their body surface such as a wax coating, densely crowded hairs or special cuticular structures enable the formation of a plastron between the body and the surrounding water which 1. reduces osmosis and the necessity for osmoregulation, and 2. facilitates unimpeded respiratory exchange.

This adaptive mechanism is irrevelant for most of the terrestrial isopods as they belong to the epedaphon, living in loose, or preferably, in moist litter layers where they can easily escape flooding. However, there are several small, cylindrical, Trichoniscidae about 5-mm-long, partly lacking pigment and eyes, with shortened antennae and pereiopods, which live euedaphically in deeper, moist layers (F- and H-layers), exposed to the danger of flooding, like other euedaphic soil arthropods.

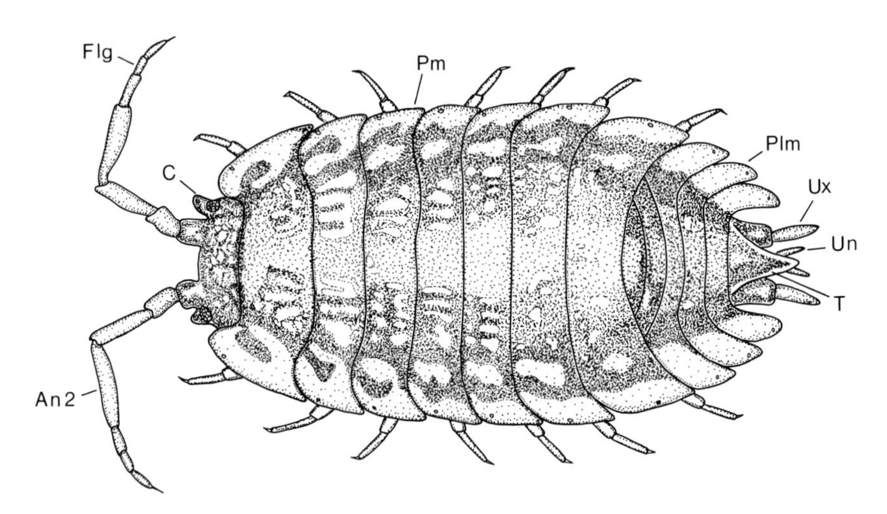

Fig. 68. *Oniscus asellus* (Oniscidae), habitus dorsal. (After Kaestner, 1967). The defensive glands open onto the lateral, flatly extended peripheral plates of the segments, the epimers. *An2* Antenna 2; *C* "Head" (cephalothorax), with lobes in front of the eyes; *Flg* flagellum of antenna 2 with terminal organ (brush organ); *Pm* pereiomer (thoracic segment); *Plm* pleomer (abdominal segment); *T* telson (anal shield); *Un, Ux* uropod endopodite, exopodite

Plate 47. *Oniscus asellus* (Oniscidae).
a Habitus, lateral. 1 mm.
b Anterior of body, lateral. Antenna 2 terminates in the three-segmented flagellum. 250 µm.
c "Head" (cephalothorax), frontal. At the base of the large, second antennae lie the tiny, first antennae. 250 µm.

a

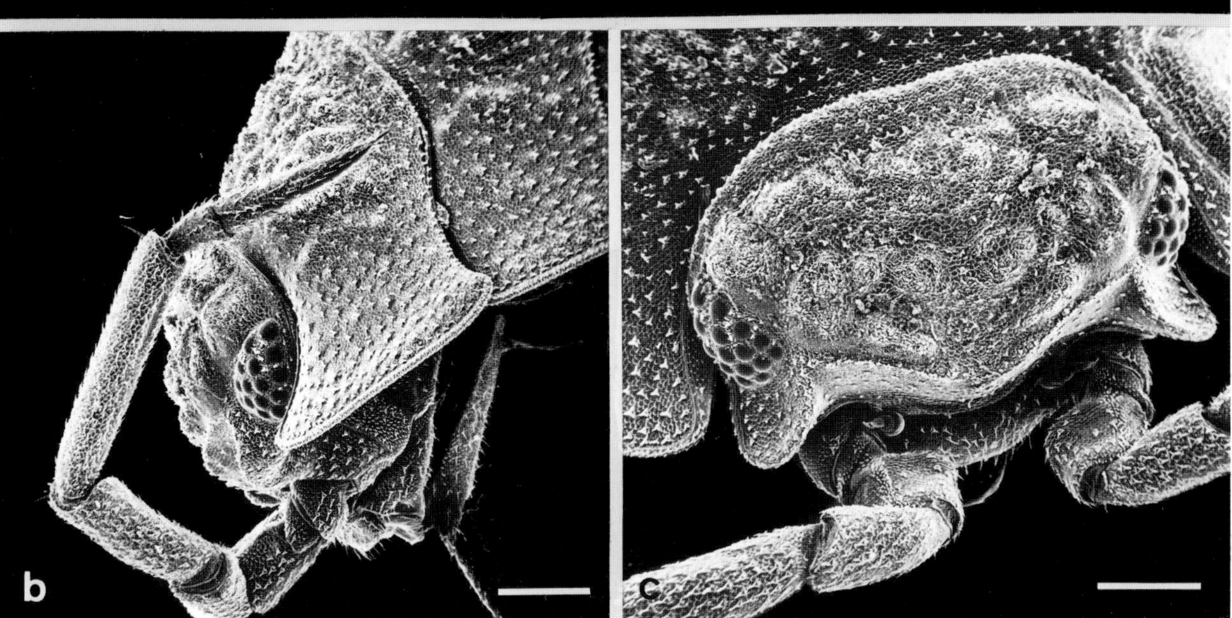
b
c

2.5.6 Cuticular Structures to Minimize Contact with Surrounding Surfaces

The surface structures of soil arthropods can have another function which is particularly important for many terrestrial isopods in very moist habitats (Schmalfuss, 1977, 1978). Their surface is ridged, and roughened by cuticular scales, thus minimizing the area of contact with moist surfaces and particles, which would otherwise adhere to the tergite surface due to the surface tension of water. These surface structures are known as "anti-adhesive structures" according to the function attributed to them (Schmalfuss, 1977). Moist particles in the litter could stick to a smooth surface and greatly impair the animal's mobility and the exchange of substances with its surroundings.

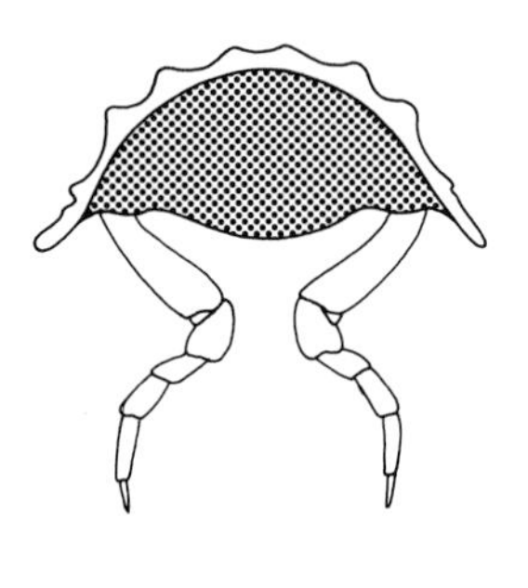

Fig. 69. *Haplophthalmus m.*

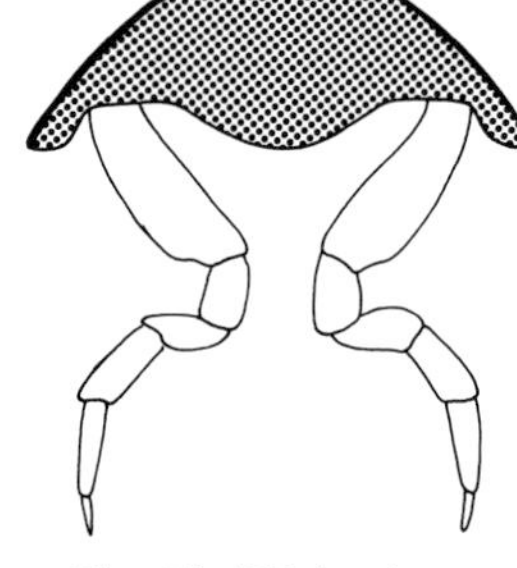

Fig. 70. *Trichoniscus p.*

Figs. 69, 70. Comparison of body cross-sections and walking legs of the smooth, long-legged, mobile *Trichoniscus pusillus* (inhabitant of the F-layers) and the ridged, short-legged, slow *Haplophthalmus montivagus* (euedaphic form). (Schmalfuss, 1977)

Plate 48. *Oniscus asellus* (Oniscidae).
a Trunk, caudal, with telson and uropods. 0.5 mm.
b Dorsal tubercle with cuticular scales. 25 µm.
c Trunk, caudal, with telson and uropods. Secretion is emerging from the uropod gland on the *right,* outer branch of the right uropod. 250 µm.
d Typical cuticular scales in the dorsal region. 10 µm.
e Trunk, caudal, lateral, with telson, uropods and pleopod exopodites. 250 µm.
f Terminal, inner branches of the uropods. 50 µm

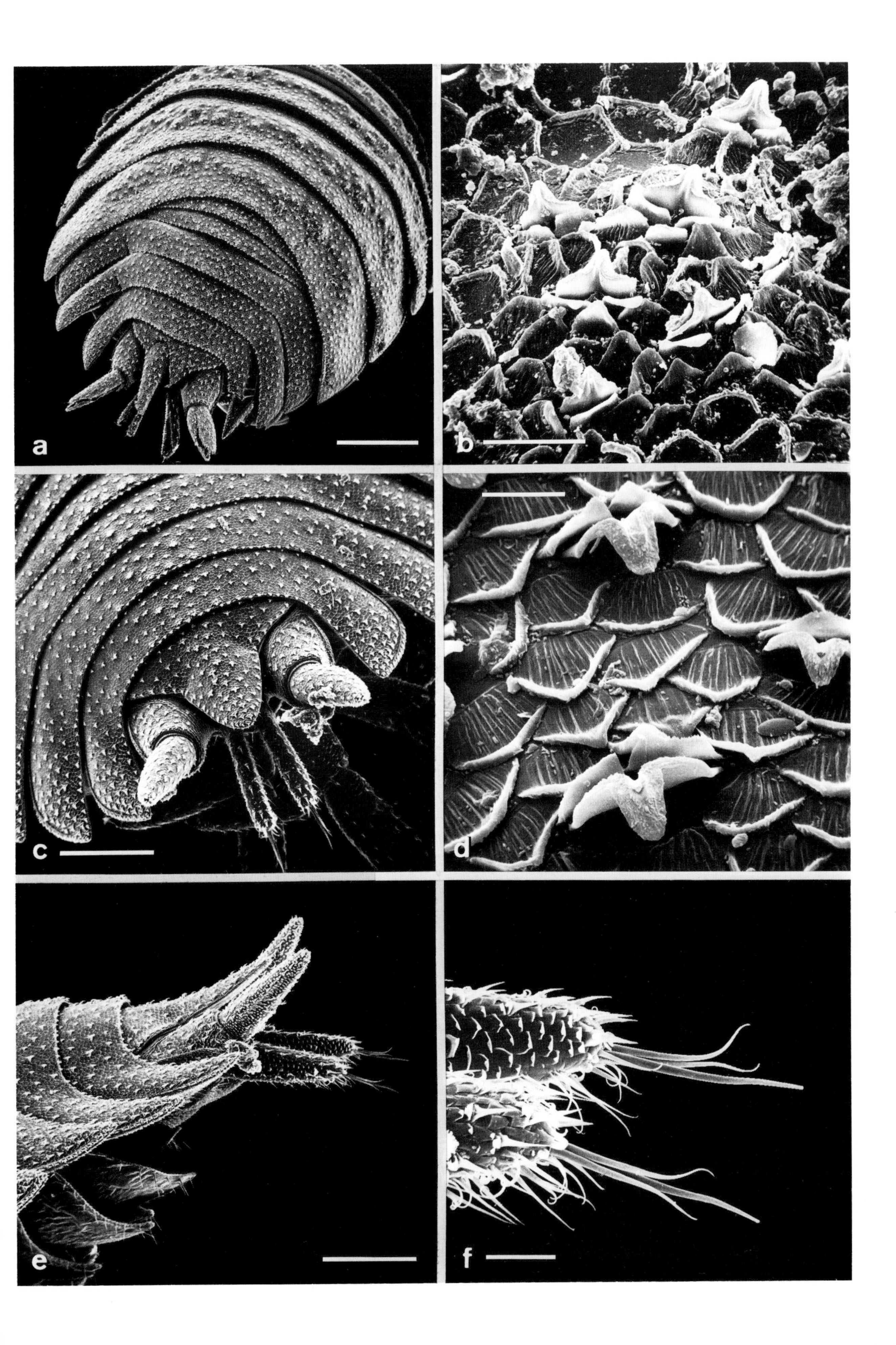
a
b
c
d
e
f

2.5.7 Pleopod-Endopodites as Gills

In the Oniscoidea males the pleopods of the second abdominal segment and the endopodites of the first pleopod are differentiated to form gonopods (Plate 49 a, c) which play an important role in copulation. The endopodites of the remaining male pleopods (Plate 49 a, b) and those of the female have been modified to gills. The flattened endopodite forms the respiratory surface. Oxygen has only a short diffusion path from the exterior, through a thin cuticle and a flat epithelium into the lumen of the gills (haemolymph space). The exopodites serve as protective opercula, covering the delicate skin of the gills like tiles on a roof. In the amphibious terrestrial isopods, e.g. *Ligidium hypnorum*, the exopodites only cover the endopodites loosely so that a constant stream of water can flow between them and gas exchange with the ambient air is enhanced. The exopodites of isopods, which are more closely associated with life on land, bulge outwards, fitting laterally into the epimers and enclosing the endopodites in a pleoventral cavity (Plate 52). Liquid rich in oxygen is supplied via the ventral, water-conducting system. Thus, the endopodites retain their respiratory function (Hoese, 1981).

Plate 49. *Oniscus asellus* (Oniscidae).
a Pleon, ventral, with view of the pleopods of the male. Half of the pleoventral cavity is open. The *arrows* show the water-conducting system. 0.5 mm.
b Endopodites III–V. They serve as gills. Their permanently moist surface is colonized by ciliates. 250 μm.
c Styloid endopodites of a male serve as auxiliary structures for copulation (gonopods). 0.5 mm.
d Surface of an endopodite with ciliates (*arrows*). Matthes (1950) identified the members of an endemic gill fauna: ten ciliate species, one bdelloid rotifer. 25 μm.
e Exopodite III, dorsal, with respiratory surface (respiratory area) (*arrow*). 0.5 mm.
f Respiratory surface (respiratory area), dorsal. 110 μm

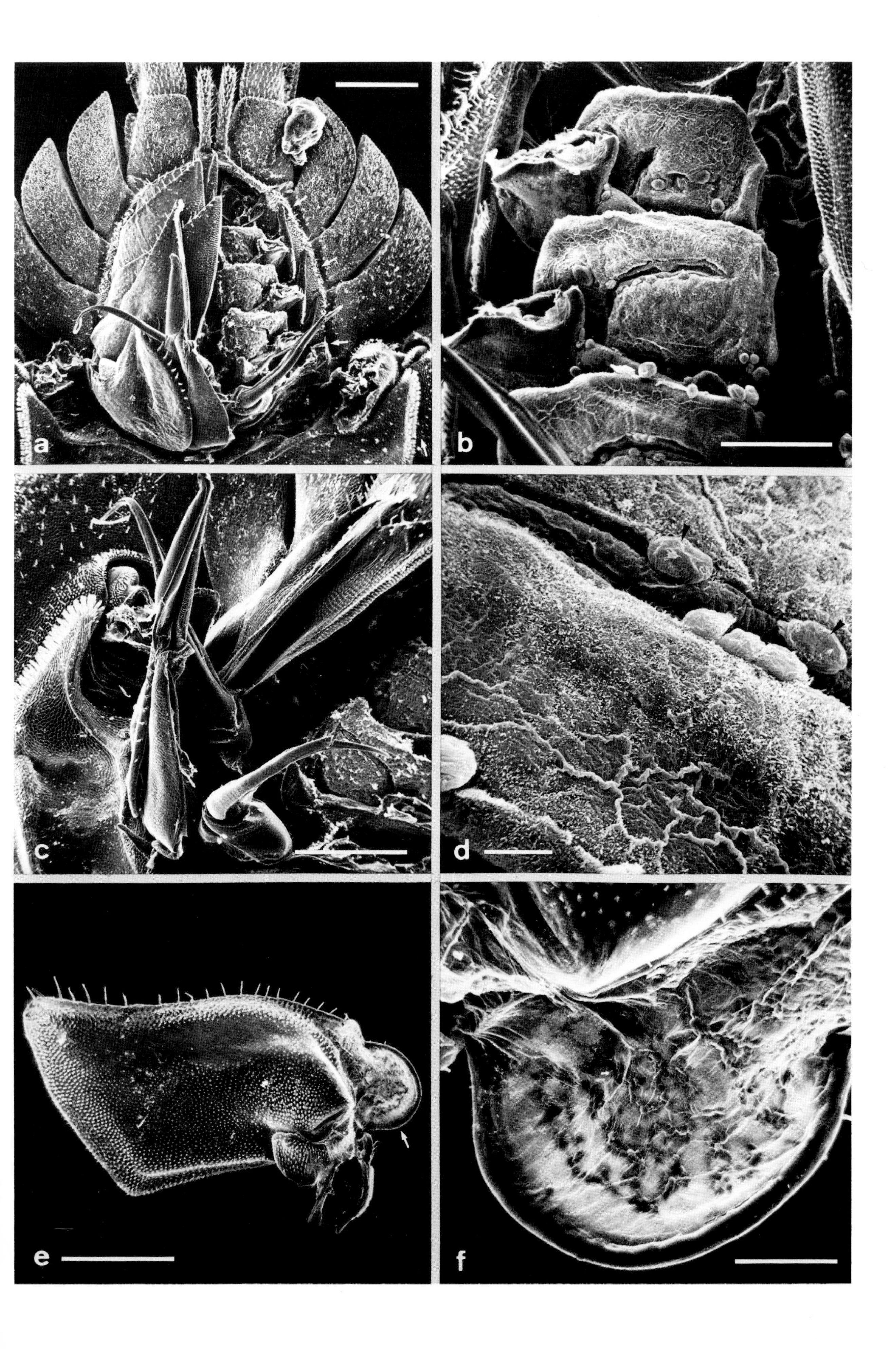
a
b
c
d
e
f

2.5.8 The Water-Conducting System of the Terrestrial Isopods – *Porcellio* Type

The water-conducting system of the terrestrial isopods represents one of the methods of adaptation to drier biotopes developed by animals of marine origin for successful invasion of the land (Hoese, 1981). It originates at the laterally situated urinary opening on the cephalothorax and extends through the ventral conducting structures, interconnected by transverse tergite grooves, and terminates around the pleoventral cavity. Urine from the maxillary nephridia is channelled into the conducting system. Due to its increased exposure as it flows along the conducting system NH_3 evaporates continuously from the urine (ammoniotelic excretion), while oxygen diffuses into it. In the pleoventral cavity the oxygen diffuses in the direction of the concentration gradient, through the respiratory surface of the endopodite gills into the body. After flowing through the conducting system and passing through the anus, the NH_3-free water is reabsorbed in the rectum.

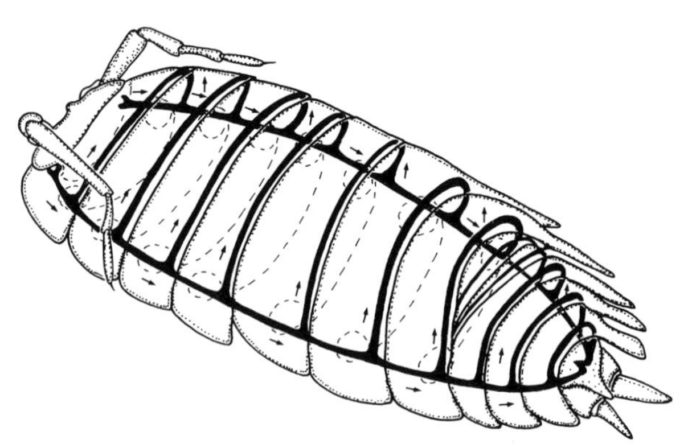

Fig. 71. The closed, circulatory water-conducting system of *Porcellio scaber* (Porcellionidae) *Porcellio* type. (After Hoese, 1981). The urine excreted from the maxillary nephridia is partly distributed over the body surface. In the open *Ligia* type external water, e.g. from the soil surface, is taken up by the pereiopods

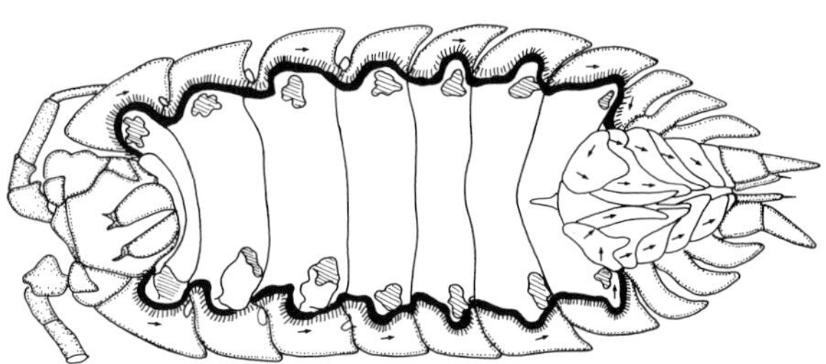

Fig. 72. The course of the ventral channels of the water-conducting system of *Porcellio scaber* (Porcellionidae). (After Hoese, 1981). The *arrows* mark the direction of flow from the cephalothorax to the pleopods.

Plate 50. *Porcellio scaber* (Porcellionidae).

a Habitus, lateral. 1 mm.

b Anterior of the body, lateral. The flagellum of antenna 2 is, in contrast to that of *Oniscus asellus,* divided into two segments. 250 µm.

c Anterior of the body, dorsal. 0.5 mm.

d Antenna 1 with aesthetascs on the third segment. 50 µm.

e Tubercle with cuticular scales on the dorsal region of a segment. 25 µm.

f Compound eye. 100 µm

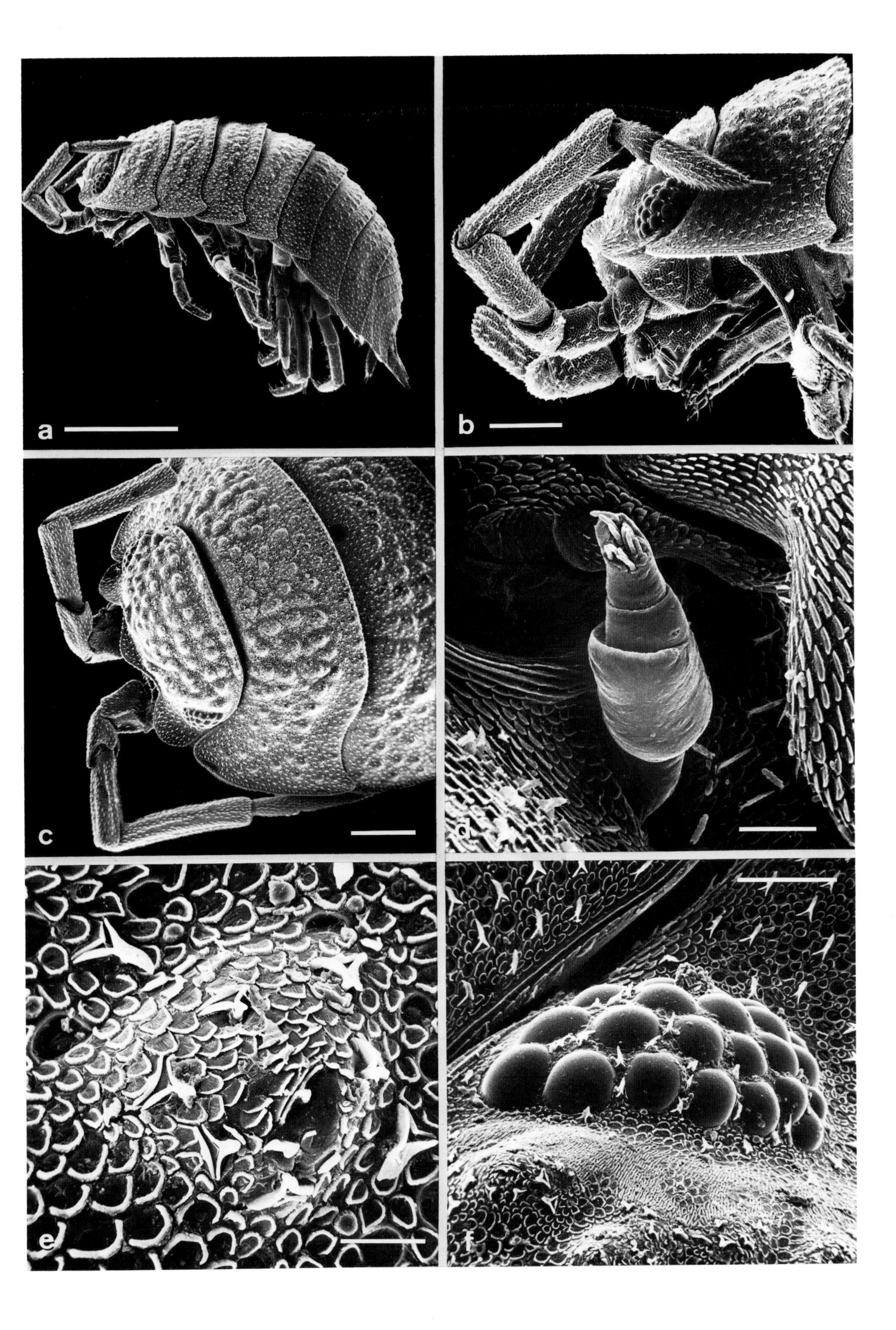

a
b
c
d
e
f

2.5.9 The Water-Conducting System of the Terrestrial Isopods – *Ligia* Type

The *Ligia* type water-conducting system is similarly structured to that of the *Porcellio* type, both having ventral and dorsal channels (Hoese, 1982b). In contrast to the *Porcellio* type (closed recycling system), the *Ligia* type functions as an open system. These hygrophilic isopods can take in water from the moist substrate of their environment. The isopod places its sixth and seventh pereiopods together in a drop of water which is drawn into the conducting system by means capillary suction.

Due to the long passage through the channels of the conducting system and the increased exposure of its surface, some of the liquid evaporates, thereby lowering the body temperature (evaporative cooling). Thus the water-conducting system of the terrestrial isopods not only serves to maintain appropriate moisture conditions, but is also involved in excretion, respiration, thermoregulation and possibly in osmotic and ionic regulation.

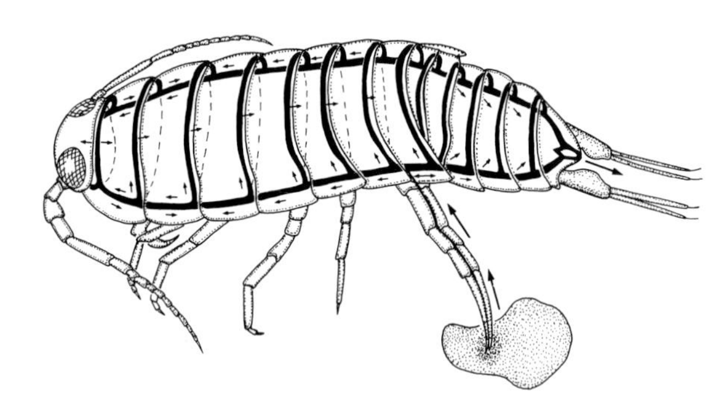

Fig. 73. The *Ligia* type of water-conducting system of the terrestrial isopods. Urine already present in the conducting system is diluted when water from the environment is drawn into the system, using the sixth and seventh pereiopods placed together. The liquid is then distributed over the body surface as in the *Porcellio* type. (After Hoese, 1982b)

Plate 51. *Porcellio scaber* (Porcellionidae).

a Anterior of body, ventral. 0.5 mm.

b Second antenna, distal, with terminal organ (brush organ). The sensory hairs, which are separate in *Ligidium* and *Trichoniscus* are fused together to form a compact sensory stylus. 50 µm.

c Cuticle of the ventral side along the sagittal plane. 50 µm.

d Proximal surface of the terminal organ of antenna 2. 5 µm.

e Leg bases with water-conducting system (*arrow*). 200 µm.

f Water-conducting system, formed by cuticular scales. 25 µm

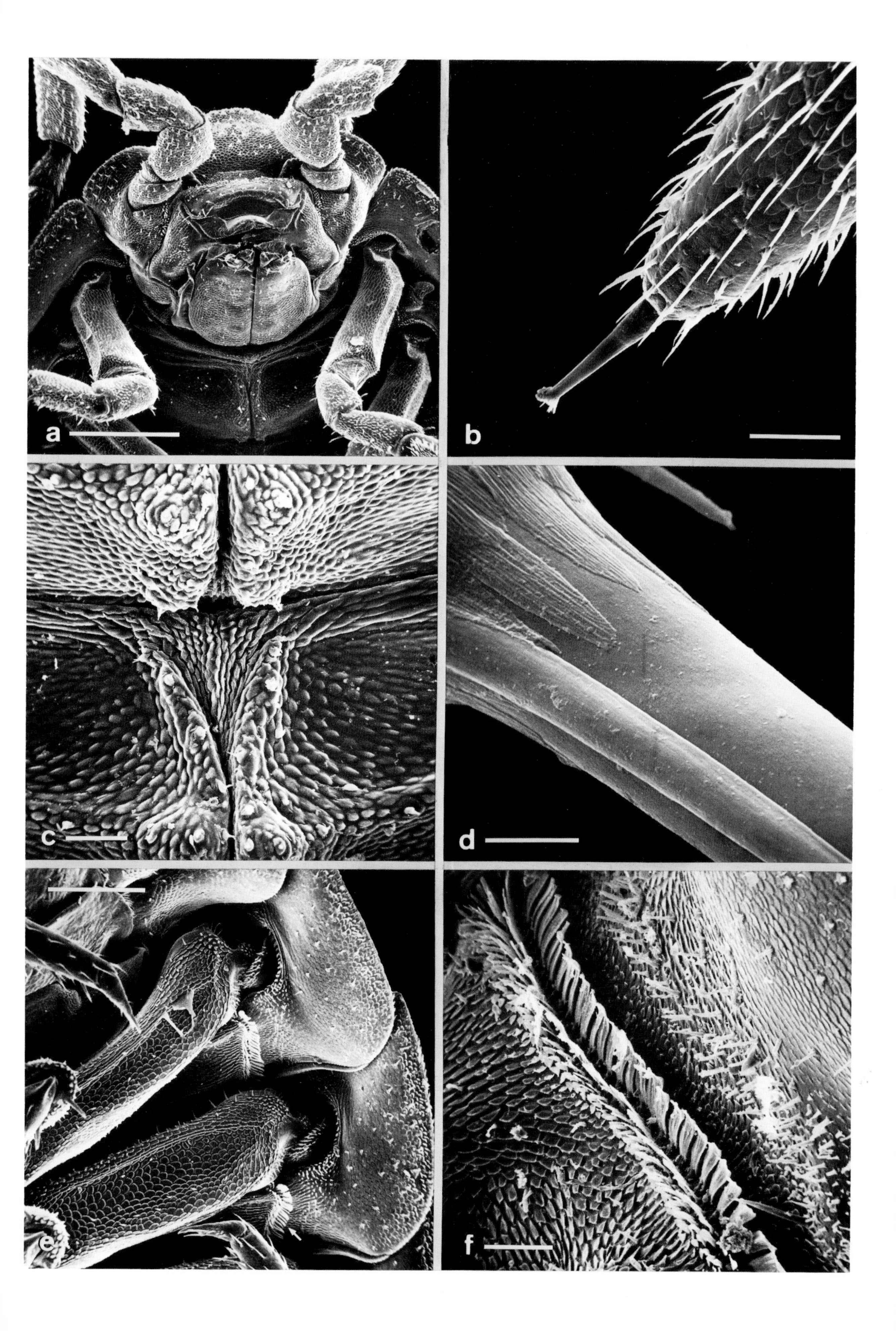
a
b
c
d
e
f

2.5.10 Lungs in the Pleopod-Exopodites of the Terrestrial Isopods

Terrestrial isopods with more advanced adaptation to atmospheric respiration possess appropriately differentiated lungs in the exopodites of the pleopods, in addition to the gills on the pleopod-endopodites which are supplied with oxygen-enriched water by the water-conducting system. The gills of isopods, which are entirely restricted to moist biotopes (*Ligia, Ligidium*) ensure an adequate supply of oxygen, but larger terrestrial isopods in semiarid and arid biotopes (*Hemilepistus*) require highly developed lungs. The morphological prerequisites for functionally effective lungs are an enlarged respiratory surface due to membrane invagination, a shortened diffusion path through thin lung epithelia and the formation of a respiratory cavity to reduce transpiration rates. Gradual evolution of these structural requirements is found in the diffentiation stages of the lungs of progressively adapted species. Whereas *Oniscus asellus* only has exposed respiratory fields on the exopodites (Plate 49), *Porcellio scaber* possesses a well-developed pair of lungs in exopodites 1 and 2 (Plate 52) (Hoese, 1982a, 1983).

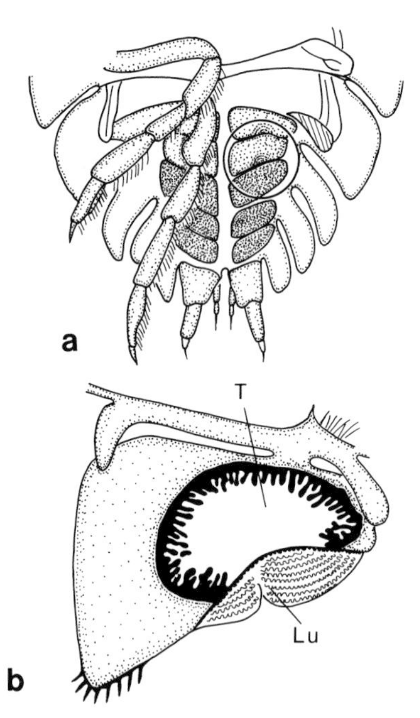

Fig. 74. Position of the tracheal lungs of *Porcellio scaber* (Porcellionidae). (After Kaestner, 1967 from Topp, 1981). **a** Overview of the pleopods. The *circle* shows the position of exopodites 1–3 on the left half of the body. As a result of the air within the lungs, pleopods 1–2 appear white (white bodies) in the living animal. **b** Exopodite showing the position of the lung (*T*) and its entrance area (*Lu*)

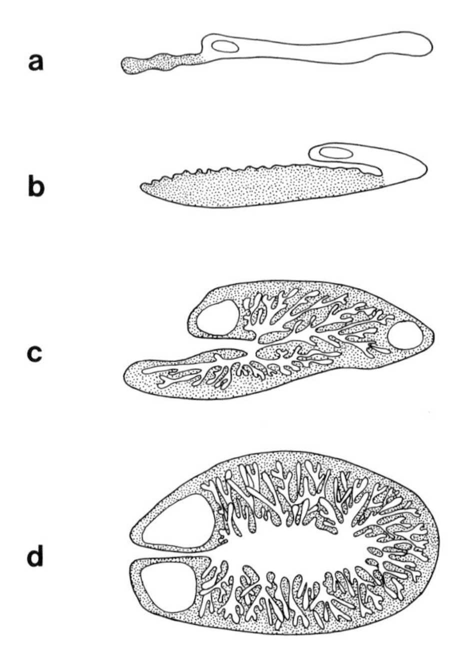

Fig. 75. Evolution of the respiratory-active surfaces on the exopodites of several terrestrial isopods. (After Hoese, 1981 from Topp, 1981). Cross-section through the exopodites. The respiratory areas or lung portions are *dotted*. **a** *Oniscus asellus,* **b** *Trachelipus ratzeburgi,* **c** *Porcellio scaber,* **d** *Hemilepistus reaumuri*

Plate 52. *Porcellio scaber* (Porcellionidae).

- **a** Trunk, caudal, with telson, uropods and the plate-shaped exopodites of the pleopods. 200 μm.
- **b** Trunk, caudal, ventral, with pleopods, uropods and telson. The *arrow* marks a lung entrance on exopodite II. 0.5 mm.
- **c** Pleoventral cavity with exposed endopodites. The *arrows* mark the water-conducting system. 0.5 mm.
- **d** Endopodite I and II with the lung entrance areas. 250 μm.
- **e** Lung entrance area. 50 μm.
- **f** Cuticle of a lung entrance area. 5 μm.
- **g, h** Comb scales on the dorsal side of exopodite IV. 10 μm, 10 μm

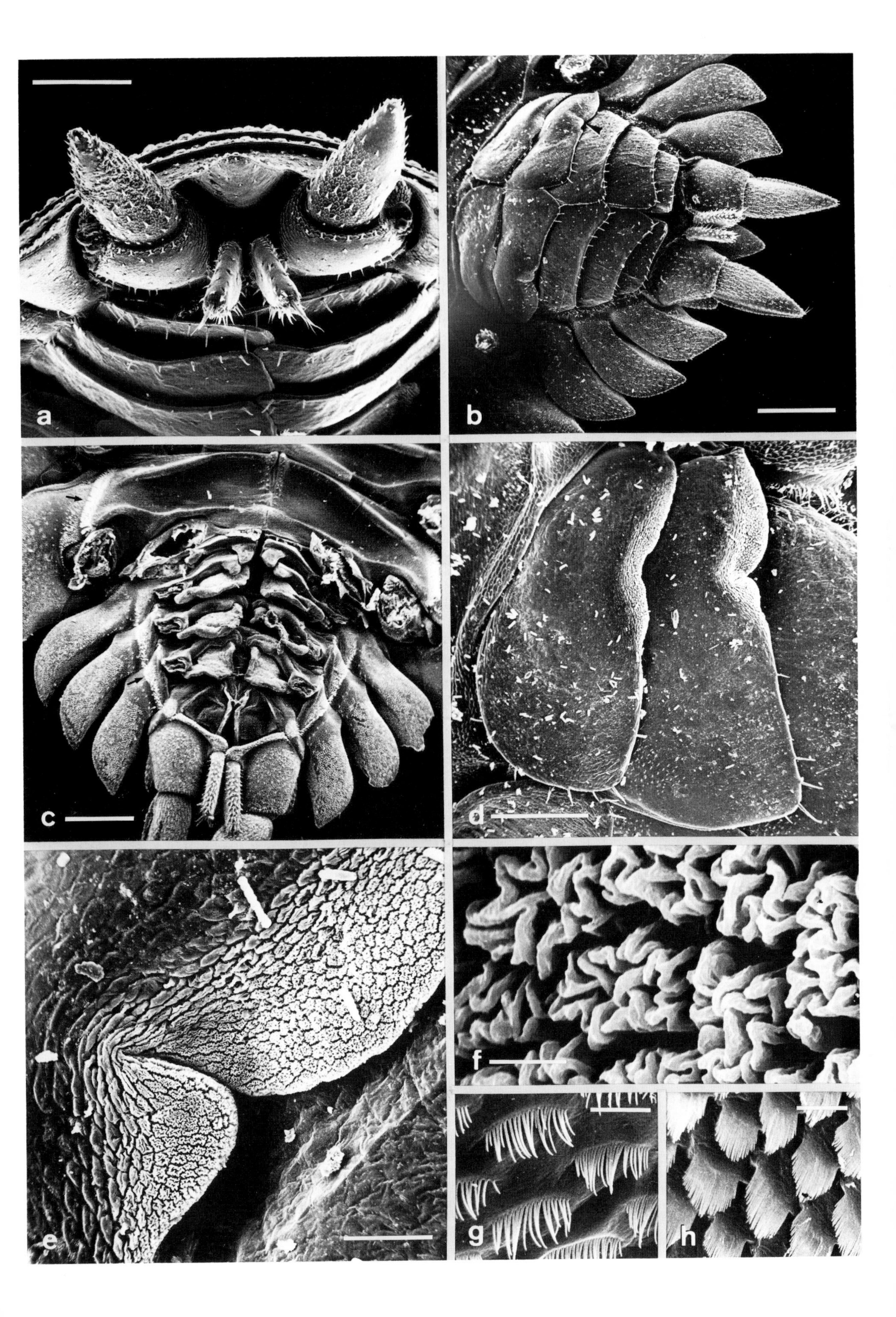
a
b
c
d
e
f
g
h

2.5.11 Reabsorption by the Endopodites of the Terrestrial Isopods

In addition to their respiratory role, the thin-walled endopodites of the pleopods possibly function as water reabsorptive organs. The urine-water mixture (*Ligia* type) or the urine (*Porcellio* type) flows through the water-conducting system, during which NH_3 is continuously released and O_2 taken up into the liquid, until it finally reaches the endopodites in the pleoventral cavity. The ultrastructure of the epithelium of the endopodites exhibits apical and basal surface enlargement in the form of invaginations as well as numerous mitochondria, rich in cristae, which are typical features of a transport epithelum (Babula and Bielawski, 1976; Kümmel, 1981, 1984). Thus, it can be postulated that water from the conducting system is not only reabsorbed through the gut in the rectum, but also through the transport epithelium of the endopodites by means of active transport of electrolytes into the lumen of the endopodites followed by an osmotic movement of water. Similar structures and mechanisms for maintaining body water balance have been described for many soil animals with high transpiration rates (Edney, 1954, 1960, 1977; Den Boer, 1961; Lindqvist, 1972; Cloudsley-Thompson, 1977; Coenen-Stass, 1981).

Plate 53. *Armadillidium vulgare* (Armadillidiidae).

a Anterior of the body, oblique dorsal, with neck shield. 1 mm.

b Trunk, caudal, with telson and plate-shaped uropods. 0.5 mm

c Anterior of the body, ventral. 1 mm.

d Anterior of the body, lateral. 1 mm

e Leg bases under the epimers. The *arrow* marks structures for linking the segments in the rolled-up position. 250 μm.

f Cuticle on the dorsal surface of the trunk segments with small scales. 25 μm

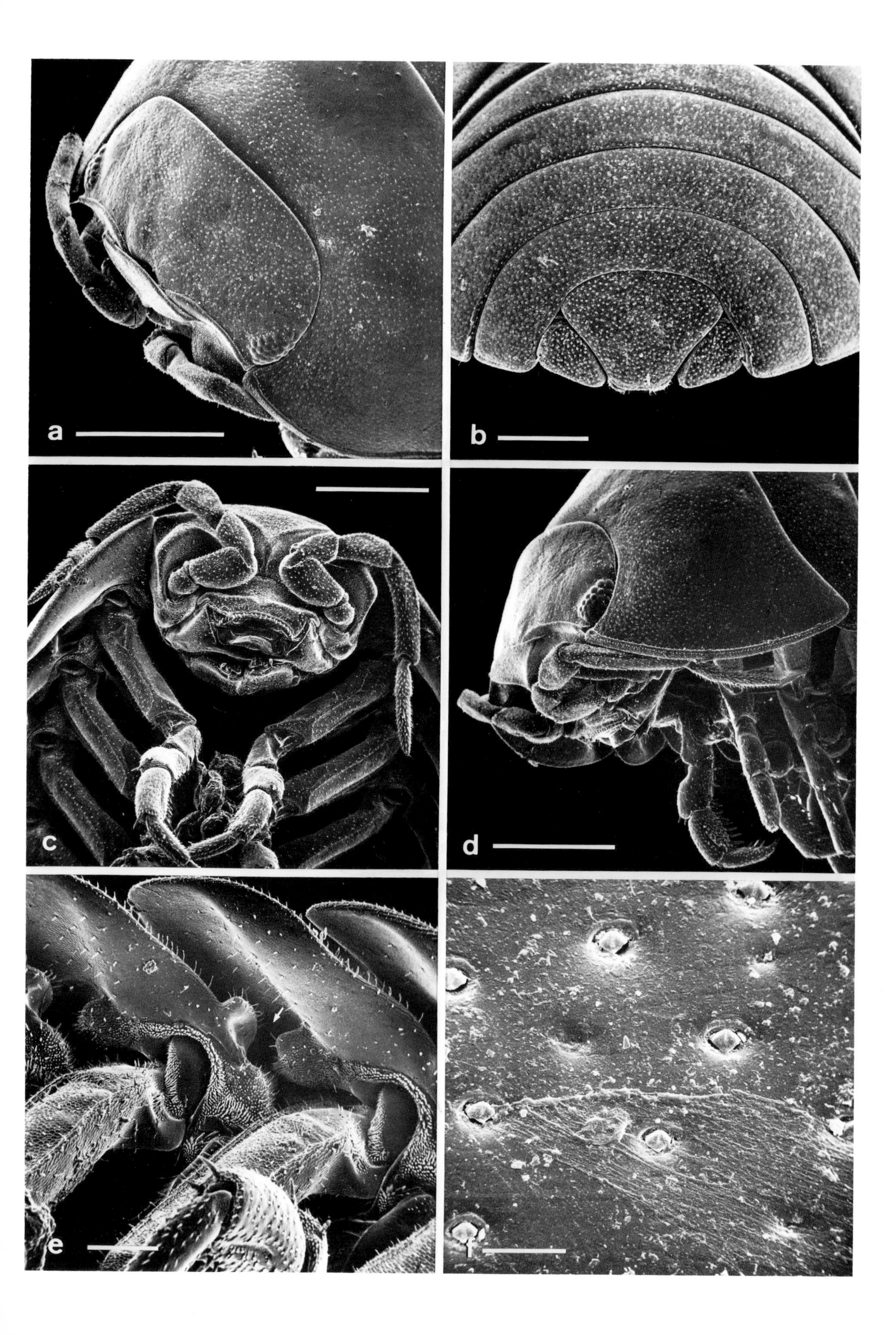

a
b
c
d
e
f

2.6 Subclass: Chilopoda – Centipedes (Myriapoda)

General Literature: Verhoeff 1925, Dobroruka, 1961; Camatini, 1979; Lewis, 1981

2.6.1 Geophilomorpha and Lithobiidae

Together with the soil-dwelling Diplopoda, Pauropoda and Symphyla, the centipedes or Chilopoda belong to the class of the Myriapoda. Their various life forms are well adapted to life in the soil. Centipedes exhibit two distinctly differing life forms. The epedaphic way of life has been adopted by the surface-dwelling Lithobiidae, while the Geophilomorpha live a purely euedaphic life. Both groups, are carnivores, the primary disparities between them being in their anatomy, the diversity of their sensory organs and their adaptations to different moisture conditions.

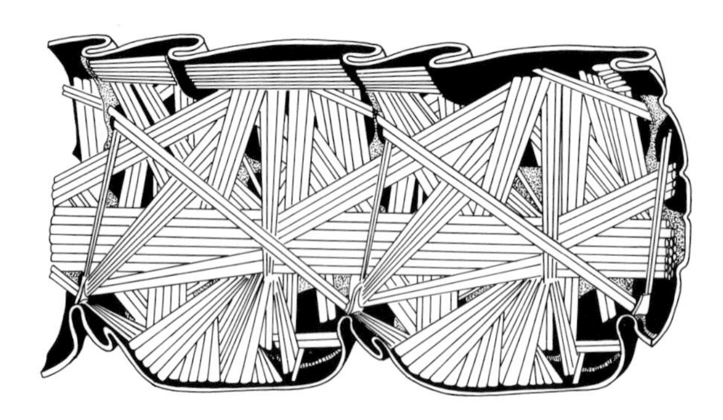

Fig. 76. Musculature of the Geophilidae. (After Füller, 1963). View of part of the muscle system showing the longitudinal, dorso-ventral and diagonal musculature at different levels. The abundance of muscle types and the flexible intersegmental membranes lend great mobility to their segments. This is a prerequisite for their remarkable adaptation to life in the interstitial spaces of the soil

Plate 54. *Scolioplanes acuminatus* (Scolioplanidae).
a Anterior of the body, oblique dorsal. 0.5 mm.
b Anterior of the body, oblique ventral. 0.5 mm.
c Trunk segments in the front third of the body. They taper towards the head. 250 µm.
d Trunk segments, dorsal, with pretergites and metatergites. 100 µm.
e Region on the flank of a trunk segment with base of a leg and spiracle. 100 µm.
f Ventral region of trunk segments. In the caudal portion of the sternites there are areas with pores of defensive glands (*arrow*). 100 µm

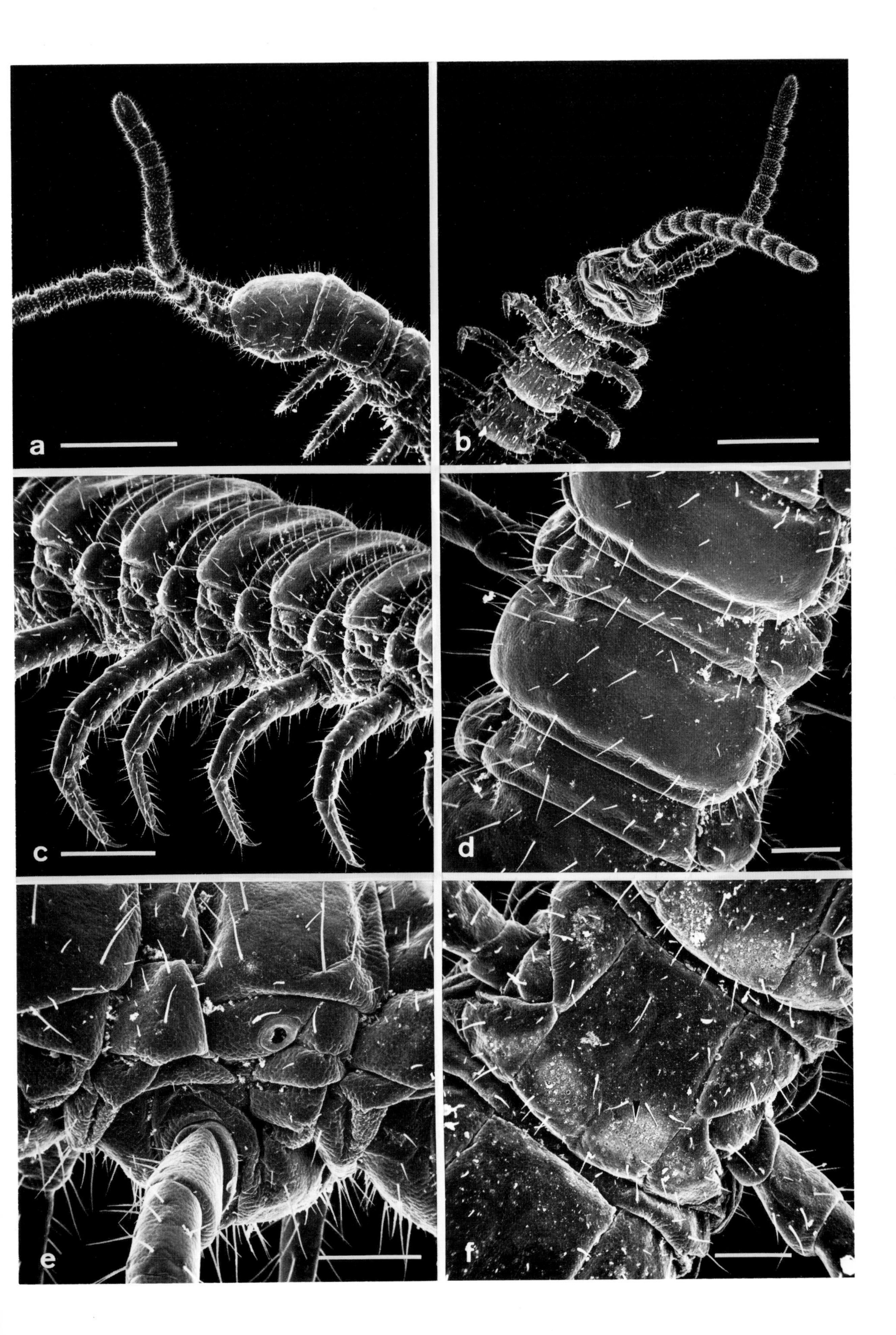
a
b
c
d
e
f

2.6.2 Euedaphic Life in the Geophilomorpha

The geophilomorph centipedes are worm-shaped and consist of 35 to 175 trunk segments. Their length ranges from 9 to 200 mm. With short legs lying close to their body they can crawl through the soil spaces to a depth of more than 30 cm, preying almost exclusively on earthworms and other soft-skinned animals from deeper soil layers.

An additional indication of a subterranean way of life is the absence of eyes. Instead the Geophilomorpha bear sensory hairs spaced at almost regular intervals over the surface of their usually pale bodies. These hairs function as tactile mechanoreceptors for exploring the narrow soil channels and their immediate surroundings. Defensive glands, opening onto glandular areas on the sternites of the trunk segments (Plate 54), protect them from predators. *Geophilus* repels approaching attackers by pointing its ventral side towards the enemy and spraying it with glandular secretion (Dobroruka, 1961).

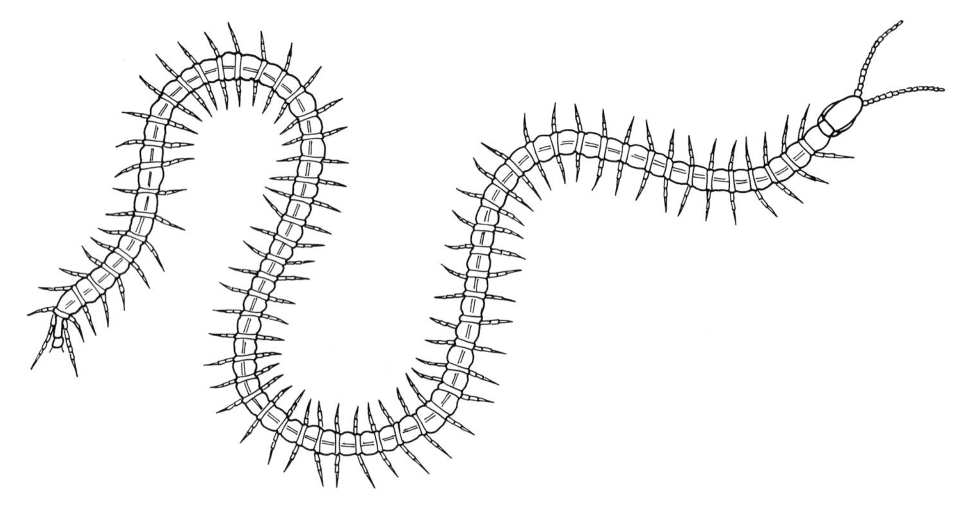

Fig. 77. *Geophilus electricus* (Geophilidae) in its walking posture. (After Kaestner, 1963)

Plate 55. *Scolioplanes acuminatus* (Scolioplanidae).
- **a** Head with maxillipede segment and antennal bases, dorsal. 200 μm.
- **b** Head with maxillipede segment, lateral. 200 μm.
- **c** Head, oblique ventral, with maxillipedes. 200 μm.
- **d** Base of antenna, lateral. Eyes are lacking. 50 μm.
- **e** Maxillipedes and mouthparts. The latter are largely hidden. 100 μm.
- **f** Tip of forceps of a maxillipede with the poison gland outlet. Conical sensilla located in shallow depressions are distributed over its surface. 10 μm

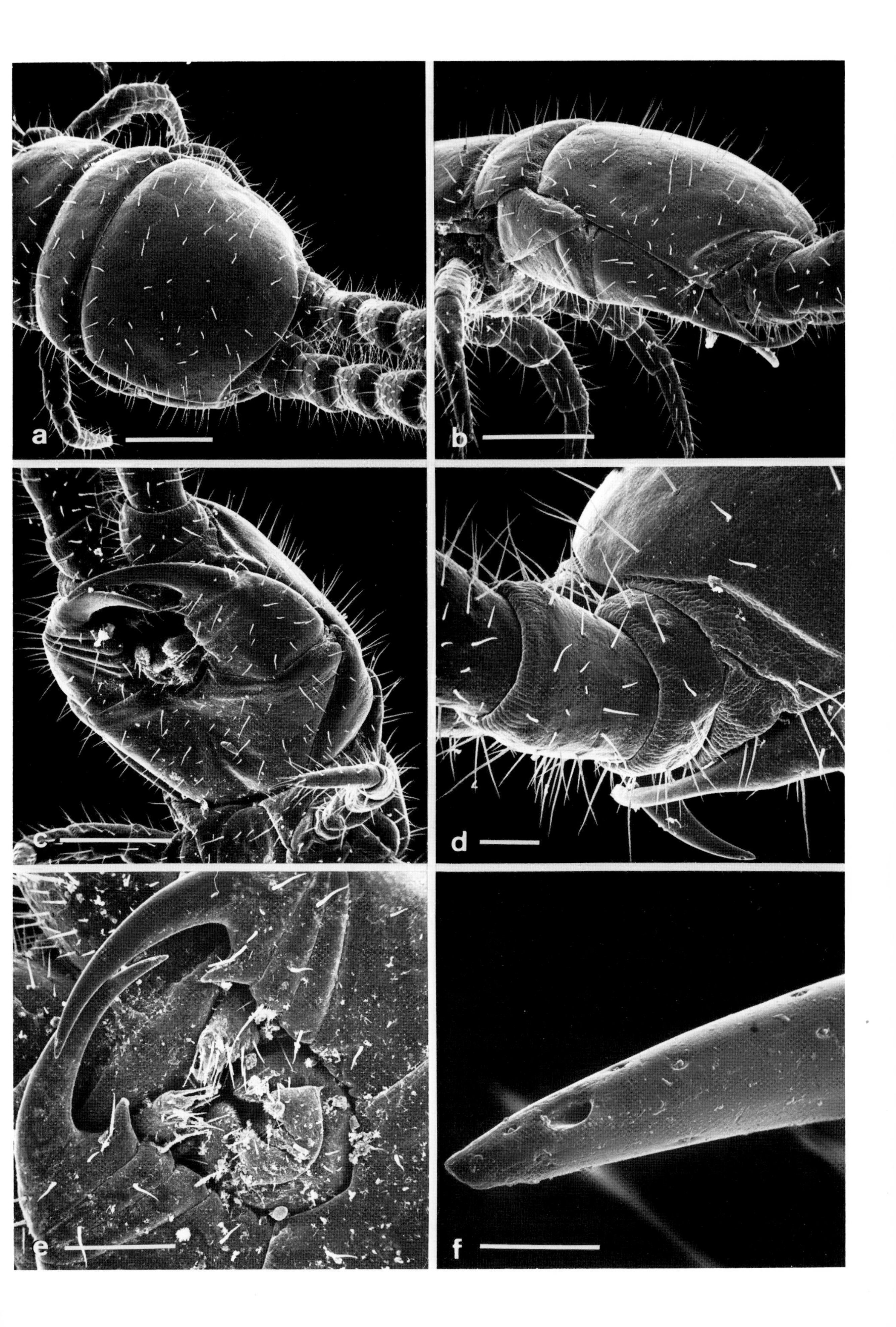

a
b
c
d
e
f

2.6.3 Antennal Sensory Organs in the Geophilomorpha

Like the insects, the antennae of the Chilopoda bear chemosensitive and mechanosensitive sensilla which are important for detecting prey, recognizing the opposite sex and for orientation in the soil. There are various types of sensilla which not only show structural differences, their number and distribution on the antennae also varies according to their function. Ernst (1976, 1979, 1981) described three types of sensilla on the distal segments of the 14-segmented antennae of *Geophilus longicornis*:

1. The sensilla trichodea (620–660) are the most numerous type of sensilla. They are pointed, slightly curved hairs with a shaft length of 50–75 µm. Their ultrastructure suggests that they function as contact chemoreceptors.

2. The cone-shaped sensilla basiconica are situated in two lateral grooves in the distal antennal segments. Their wall is perforated by numerous pores. Chemo-, hygro- and thermoreception are all discussed as functions of these sensilla.

3. The pointed, conical sensilla brachyconica are found at the apex of the terminal antennal segment. Seven sensilla form a group, with one central sensillum surrounded by an almost circular ring of the six others. This arrangement of sensory cones is, in turn, encircled by a border of numerous sensilla trichodea. In this case the possibility of a double function as thermoreceptors and hygroreceptors is also discussed.

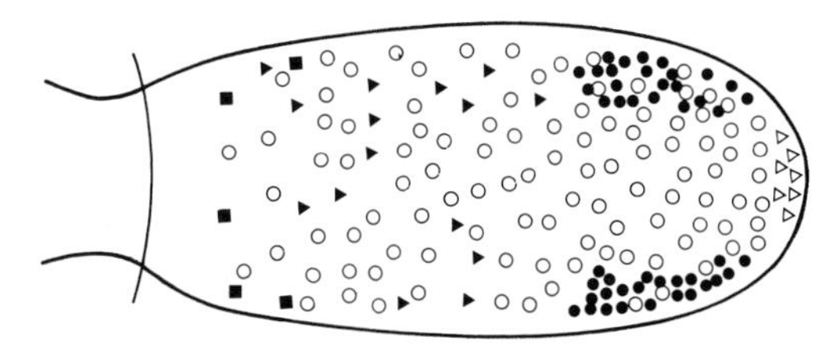

Fig. 78. *Geophilus longicornis* (Geophilidae). Distribution of hair sensilla on the dorsal side of the terminal antennal segment. (Ernst, 1979, 1981). △ Small, thick bristles (sensilla brachyconica) 17–20 µm, ○ Pointed hairs of medium length (sensilla trichodea) 50–75 µm, ● Short, leaf-shaped sensilla (sensilla basiconica 10–14 µm, ▲ Small sensory hairs 3.5–7 µm, ■ Long, pointed hairs 170–270 µm

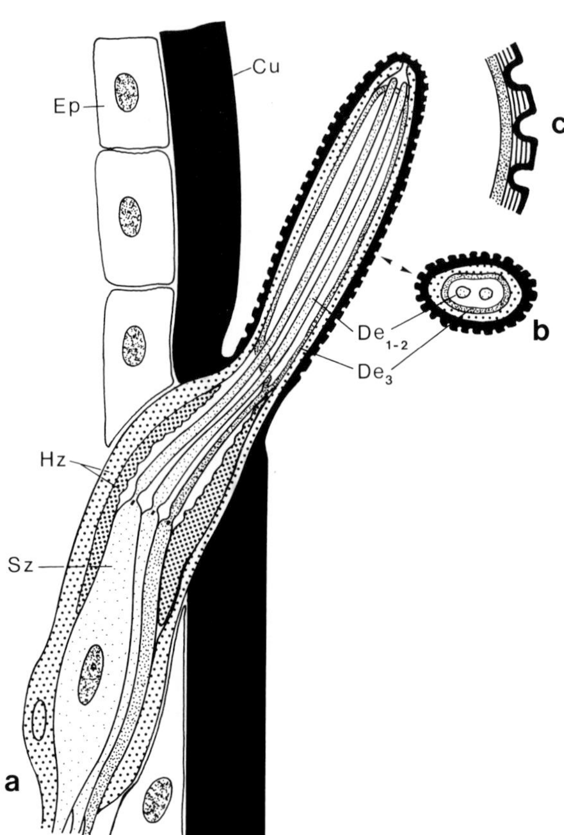

Fig. 79. *Geophilus longicornis* (Geophilidae). Sensory structures on the antenna. (After Ernst, 1979).
a Sensillum basiconicum, schematic, with three sensory cells (*Sz*) and two envelope cells (*Hz*). The hair-shaft cuticle is thin and perforated, at least in the outer layer. Three dendrites extend through the hair: two slim (De_{1-2}) and one broader sheath-like sensory process (De_3). **b** Cross-section through the hair shaft. **c** Fine structure of the hair-shaft cuticle.
Ep Epidermis; *Cu* cuticle

Plate 56. *Scolioplanes acuminatus* (Scolioplanidae).
a Antenna, distal with sensilla trichodea, s. brachyconica and s. basiconica. The latter extend in a broad row (*arrow*) to the apex of the antennal segment. 50 µm.
b Middle portion of the antenna. 100 µm.
c Three s. basiconica and one s. trichodeum, distal. 2 µm.
d Segment in the middle portion of the antenna. 25 µm.
e Spiracle. 10 µm.
f Cuticle with cellular pattern and outlets of the epidermal glands (*arrows*). 3 µm

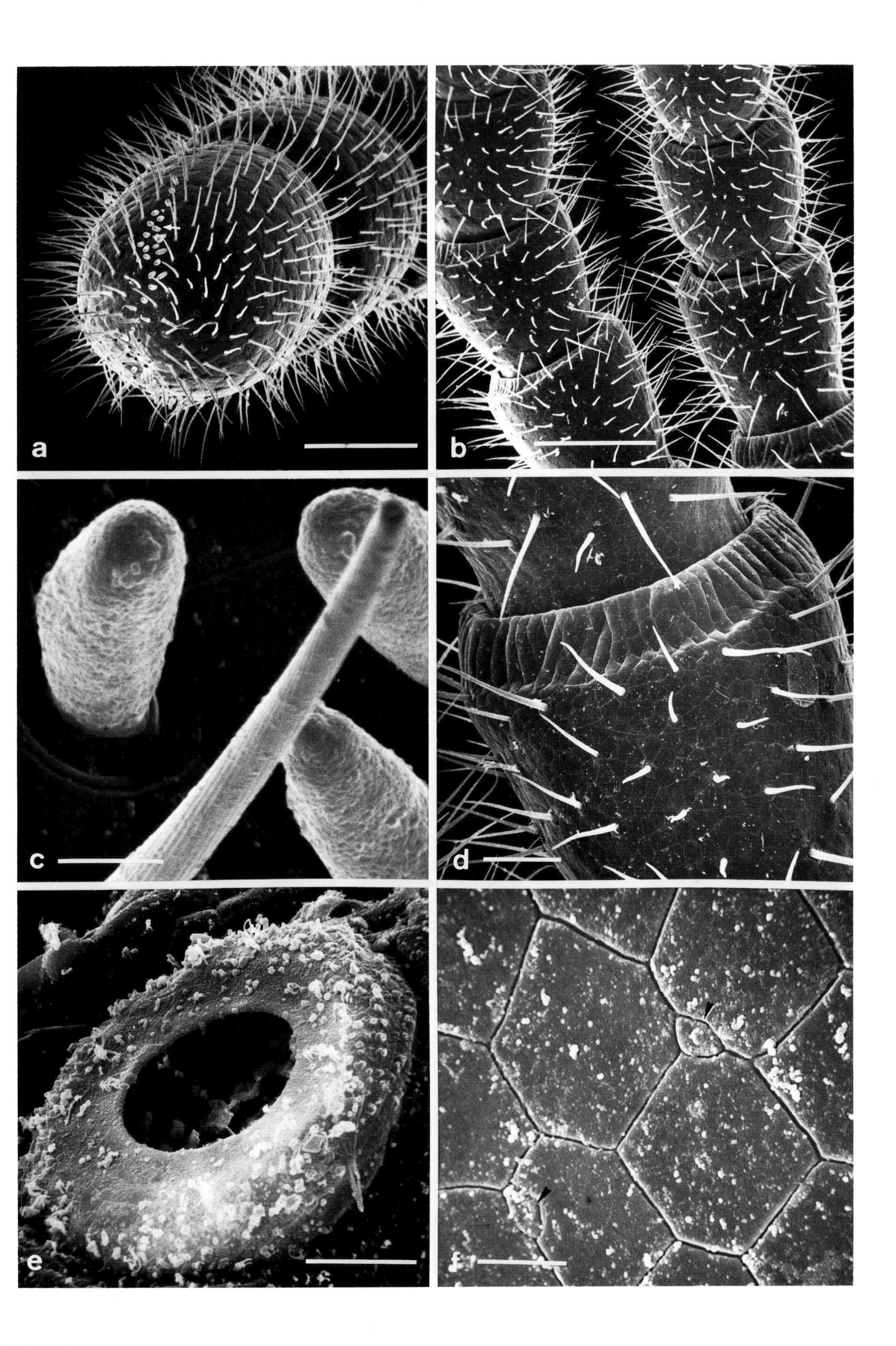
a
b
c
d
e
f

2.6.4 Coxal Organs of the Geophilomorpha

The broad, flattened coxae of the legs on the terminal segment of the Geophilomorpha bear pore fields which have been described as coxal organs (Rosenberg and Seifert, 1977; Rosenberg, 1982). These pores have a diameter of 15–20 µm and open into cylindrical cavities, sunk deeply into the coxae. Beneath the thin cuticular lining of these cavities lie large transport cells with their nuclei situated in the basal region. The basal and apical cell membranes are remarkable for their surface enlargement due to infoldings. Above all, the deep, narrow infoldings of the basal membrane are associated with a large number of cristae-rich mitochondria running parallel to them (cf. Fig. 83).

The organs are most probably involved in water balance, as many hygrophilic soil animals possess similar structures. It can be postulated that water drops, drawn into the pore cylinders by capillary suction, are absorbed into the haemolymph by means of existing transport mechanisms.

Rosenberg and Bajorat (1983) proved that the coxal pores have a higher permeability than the normal cuticle in *Lithobius forficatus.* When the pores are blocked the influx of tritiated water from the atmosphere is decreased by 40%. This cannot, however, be regarded as proof of active uptake of water vapour, as shown by other arthropods (Rudolph and Knülle, 1982). *Lithobius* does not gain weight in an atmosphere saturated with water vapour. Even when it has a deficit of body water it continues to suffer heavy water losses.

Plate 57. *Scolioplanes acuminatus* (Scolioplanidae).

a Trunk, caudal ventral, with anal region. The coxae of the defensive legs have perforated areas. 200 µm.
b Trunk, caudal, lateral. 250 µm.
c Coxae with "coxal gland" areas. The anus is bordered by three valves. 200 µm.
d Pore of a "coxal gland". 10 µm.
e Distal segmentation of a leg. 50 µm.
f Leg, distal, with claw. 20 µm

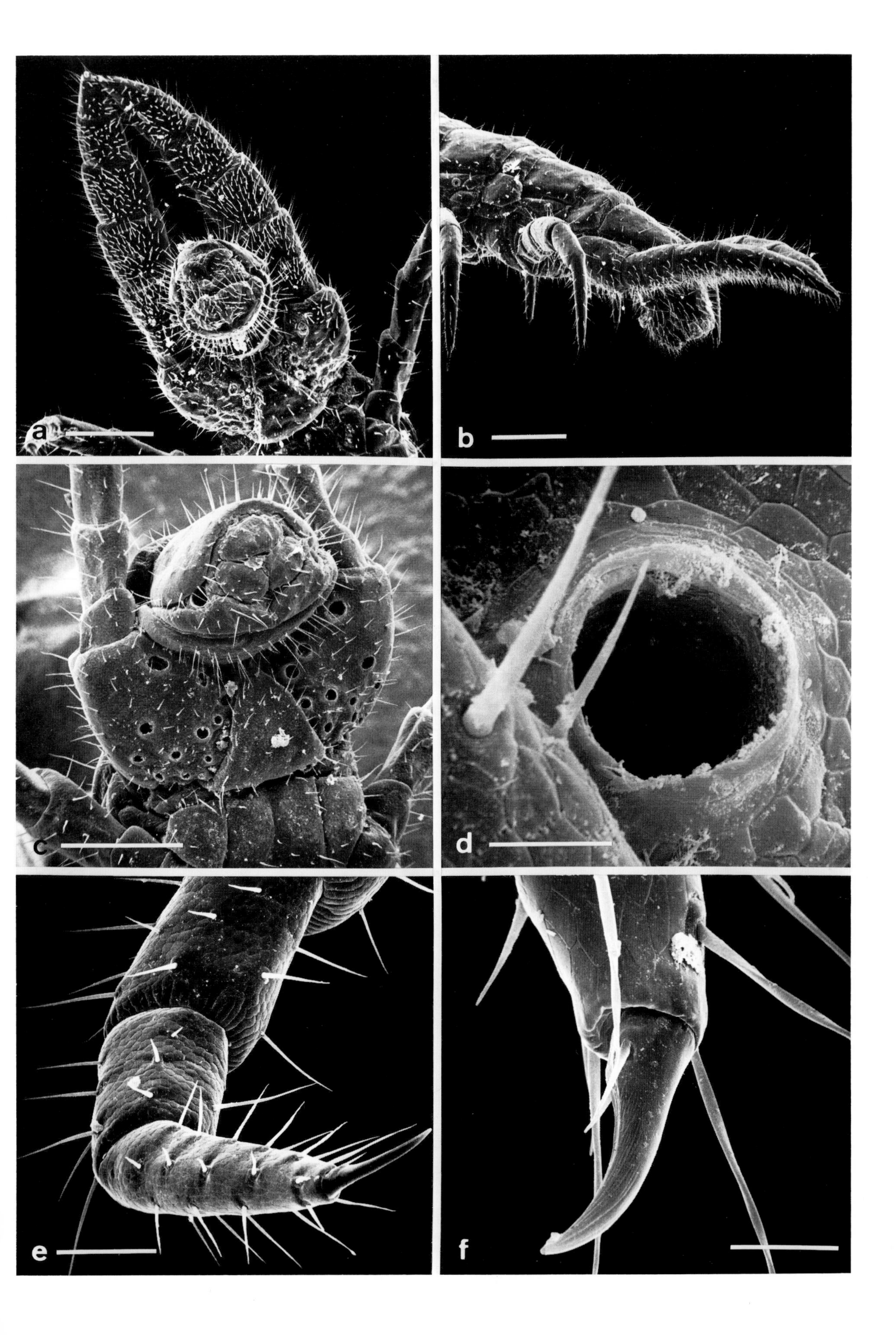

a
b
c
d
e
f

2.6.5 Epedaphic Life in the Lithobiidae

Lithobius forficatus is a fast and agile soil carnivore, living predominantly in deciduous and mixed woodland. The 15 pairs of legs sticking straight out from its sides emphasize the dorso-ventral flattening of its body (Fig. 81). Although this shape does not permit *Lithobius* to colonize the deeper layers of the soil, it favours life between the loose layers of fallen leaves and enables rapid discovery of a safe refuge in flat hiding places under stones and loose bark when danger threatens. In a 130-year-old beech wood (Luzulo-Fagetum) in the Solling area, West Germany, the average annual abundance and biomass was determined as 41 ind/m^2 and 113.1 mg dry body mass/m^2 for *Lithobius mutabilis* and 32 ind/m^2 and 38.5 mg dry body mass/m^2 for *Lithobius curtipes* (Albert, 1977).

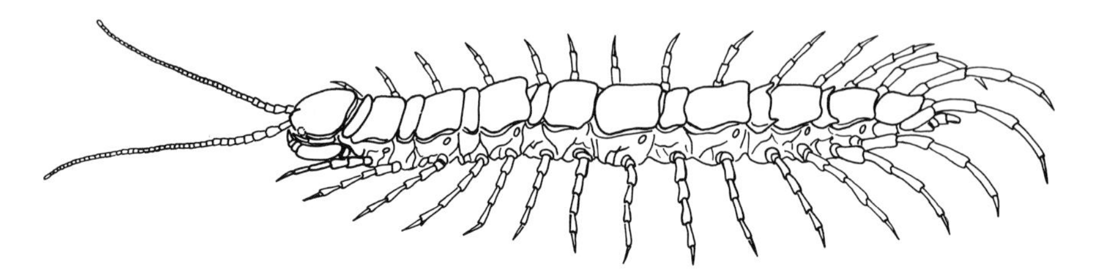

Fig. 80. *Lithobius forficatus* (Lithobiidae) lateral view, oblique dorsal. (After Kaestner, 1963)

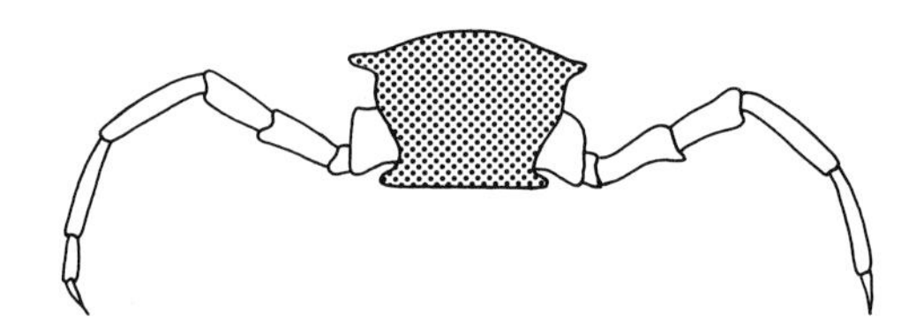

Fig. 81. Diagrammatic cross-section of the trunk of a lithobiid centipede. (After Dunger, 1963)

Plate 58. *Lithobius* sp. (Lithobiidae).
a Anterior of the body, dorsal. 1 mm.
b Head, oblique ventral, with antennal base, simple compound eye, Tömösváry organ (*arrow*) and maxillipedes. 250 μm.
c Anterior of the body, lateral. 1 mm.
d Anterior of the body, ventral. 1 mm.
e Trunk, caudal, lateral, with defensive legs. 1 mm.
f Trunk, caudal, dorsal. 0.5 mm

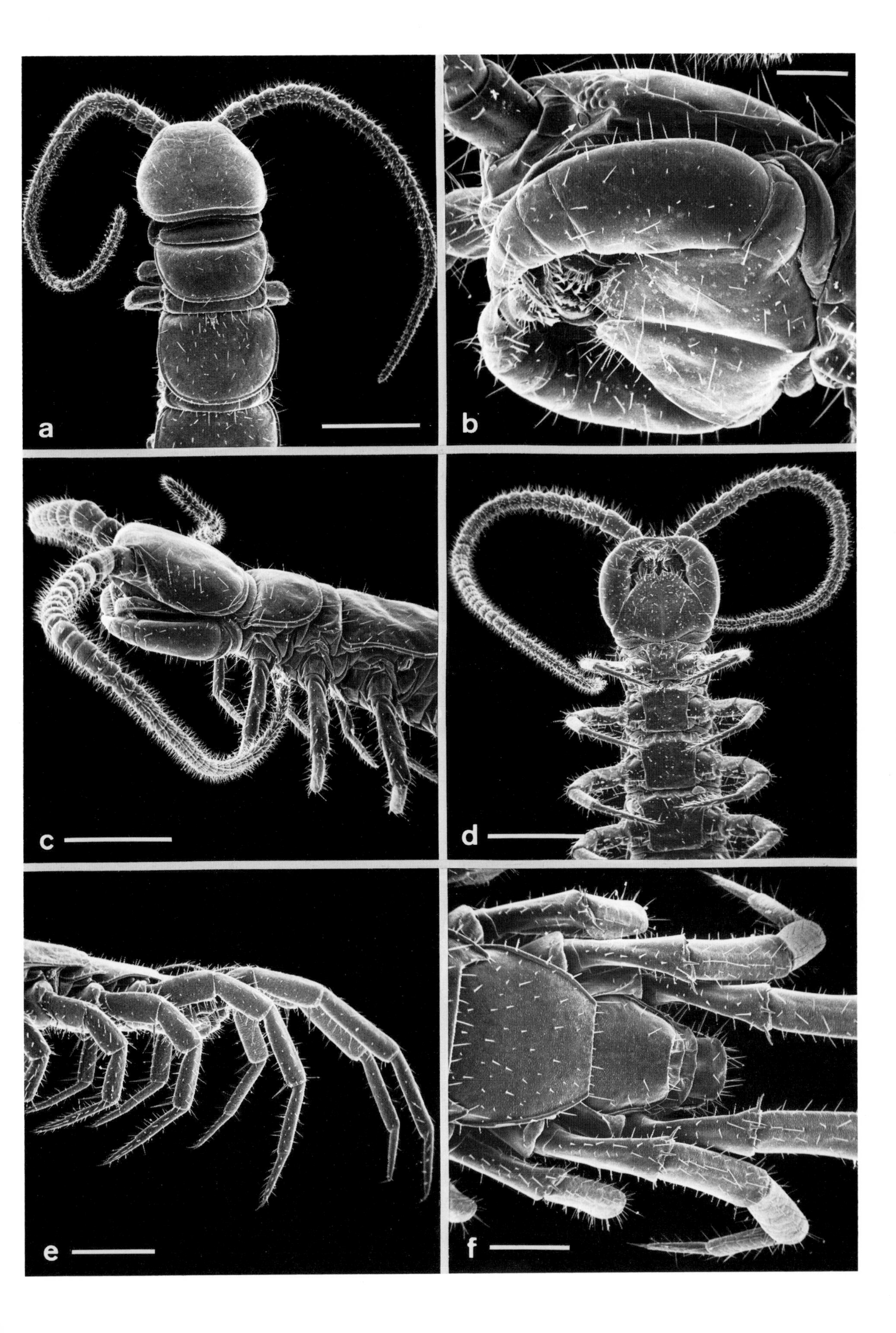
a
b
c
d
e
f

2.6.6 Predatory Feeding Habits of the Lithobiidae

Lithobius forficatus mainly preys on small insects, catching them as soon as its antennae come in contact with them. In addition to the role they play in perceiving the immediate proximity, the antennal sensory organs are obviously also important for detecting prey. Each antenna of a mature lithobiid centipede bears approximately 2000 sensilla trichodea which fulfil combined functions as contact chemoreceptors and mechanoreceptors (Keil, 1976). These sensilla evidently play a role in perceiving prey by means of tactile and chemical stimuli.

Lithobius forficatus seizes its prey with its maxillipedes. These maxillary feet have evolved by modification of the first pair of walking legs. The syringe-like claws contain poison glands with a narrow secretory duct leading to the opening at their tip (Fig. 82; Plate 59). The prey is bitten by the maxillipedes and killed or immobilized by an injection of poison. Then it is torn open by the mandibles, covered in digestive fluid and sucked out until only the hollow chitin skeleton of the victim remains (Rilling, 1960).

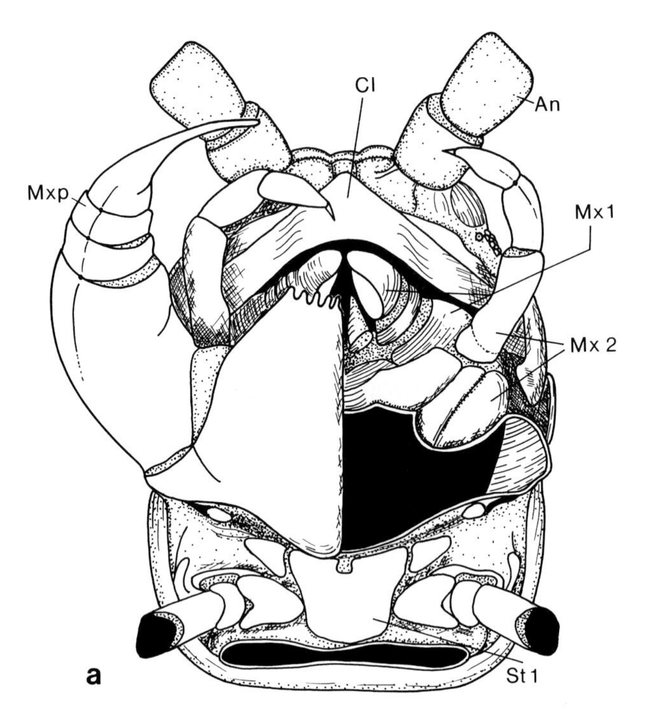

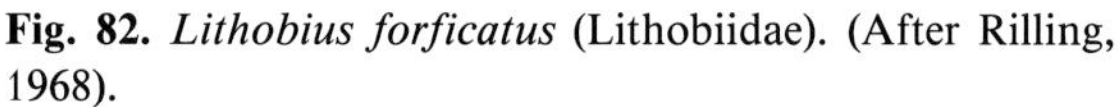
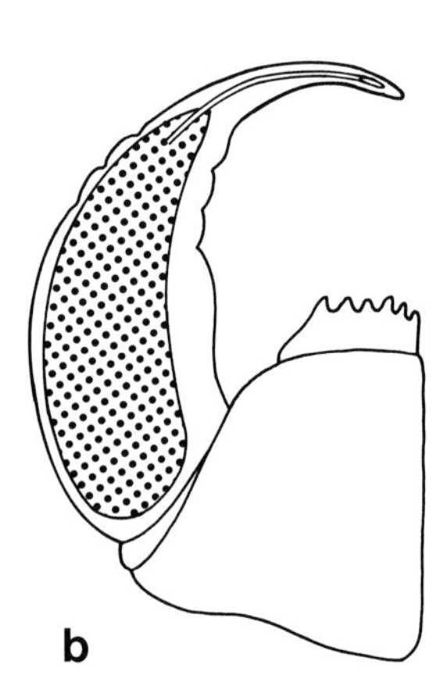

Fig. 82. *Lithobius forficatus* (Lithobiidae). (After Rilling, 1968).
a Head, ventral, the maxillipede on the left side of the body has been removed. The upper lip, mandibles and the hypopharynx are hidden by the first maxilla. *An* Antenna; *Cl* clypeus; *Mx 1,2* maxilla 1 and 2; *Mxp* maxillipede; *St1* sternum 1. **b** Schematic diagram of a maxillipede with poison gland

Plate 59. *Lithobius* sp. (Lithobiidae).
a Mouthparts with maxillipedes and maxillae. 250 μm.
b Cleaning hairs on the second maxilla. 20 μm.
c Dorsal aspect of the maxillipede with poison gland outlet. Small sensilla in shallow depressions are distributed over the surface of the claw. 50 μm.
d Surface of a maxillipede claw with sensillum. 4 μm.
e Middle portion of the antenna. 100 μm.
f Antennal surface with hair base of a sensillum trichodeum. 5 μm

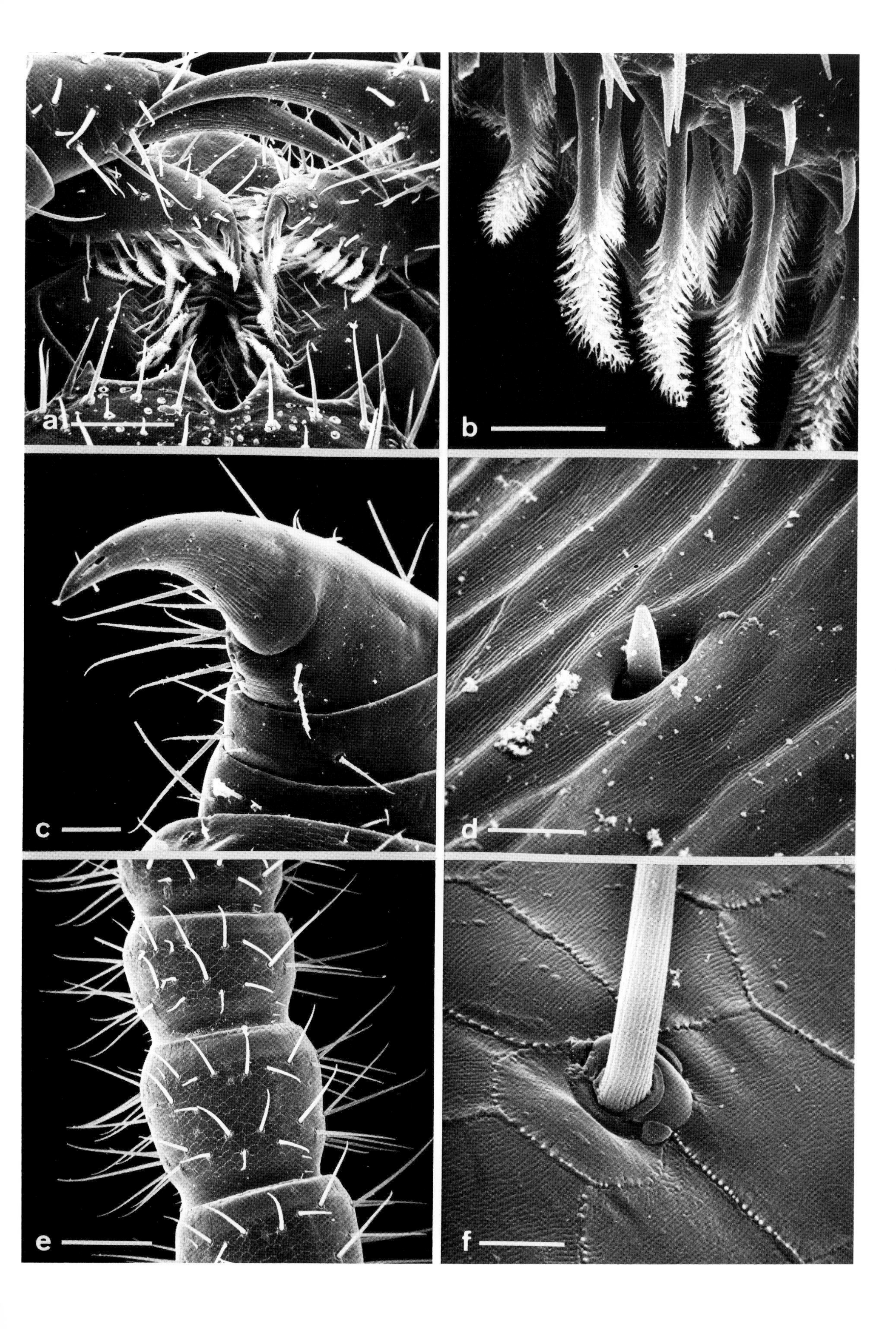
a
b
c
d
e
f

2.6.7 Moisture Requirements of the Chilopoda

The epedaphic and euedaphic ways of life give rise to a variety of ways of adapting to the various moisture conditions of different soil layers. As soon as air humidity falls to levels below almost complete saturation, the epedaphic lithobiids suffer severe water loss. Besides the continual efflux of water vapour from the open spiracles, the permeable exoskeleton is largely responsible for the high losses, as the thin, water-repellent lipid film is insufficient to reduce cuticular transpiration (Curry, 1974). As fast and mobile runners the Lithobiidae are able to compensate for this deficit under normal conditions in woodland soil by moving to moister microhabitats, e.g. damp piles of leaves. The Tömösváry organs situated in the ocular region were recognized as hygroreceptors which play a directing role in the detection of moisture (Tichy, 1971, 1972; Haupt, 1979). Their absence in the Geophilomorpha is noteworthy. However, in contrast to the freely mobile lithobiids, the geophilomorphs are more confined to soil tunnels in horizons of uniform moisture with only slight fluctuations in temperature and evaporation. In both cases, the animals have coxal organs on the terminal trunk segments (Fig. 83; Plates 57 and 60), which obviously help to compensate for water loss by the absorption of water, e.g. from dew.

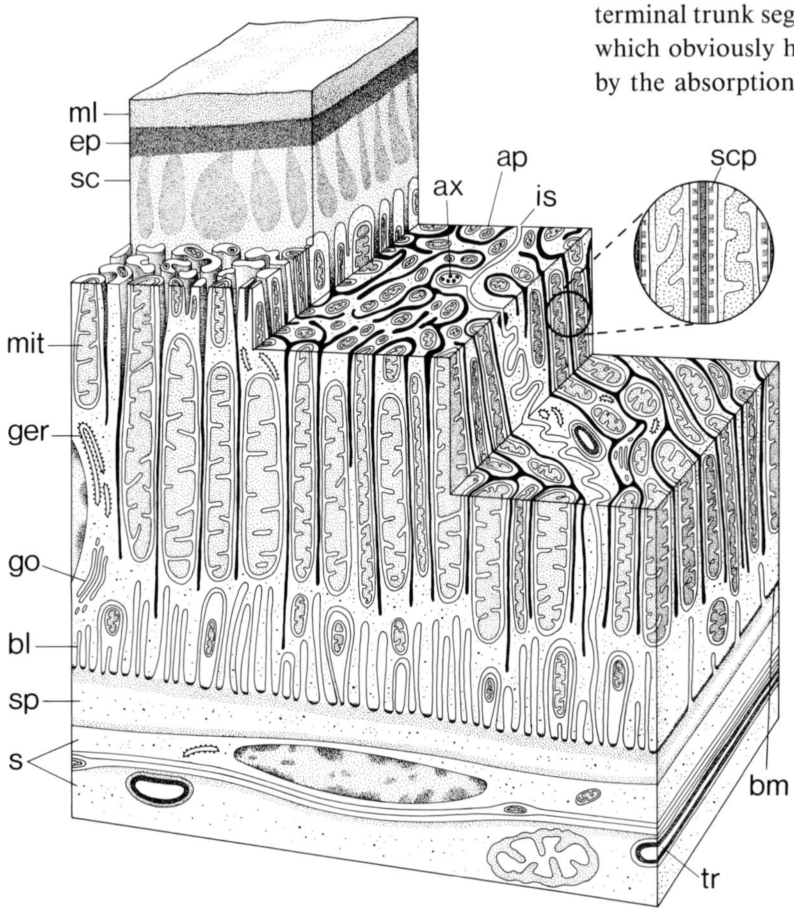

Fig. 83. *Lithobius forficatus* (Lithobiidae). Sections through the transport epithelum of a coxal gland. (After Rosenberg, 1983). The main function of the coxal pore structures is probably water and ion transport from the substrate into the haemolymph space of the animals. *ap* Apical outfolding; *ax* axon-containing neurosecretory granules; *bl* basal labyrinth; *bm* basal membrane; *ep* epicuticle; *go* Golgi apparatus; *ger* granular endoplasmic reticulum; *is* intercellular slit; *ml* mucous layer; *mit* mitochondria; *sc* subcuticle; *scp* surface coat particles; *sp* subepithelial slit leading to haemolymph space; *s* cellular basal sheath; *tr* tracheae

Plate 60. *Lithobius* sp. (Lithobiidae).
a Trunk segments, lateral. 250 μm.
b Trunk, caudal, ventral, with coxal gland outlets. 250 μm.
c Spiracle, 50 μm.
d Coxa with glandular pores. 100 μm.
e Cuticle and intima at the edge of the spiracle. 5 μm.
f View into the interior of a coxal pore. 20 μm

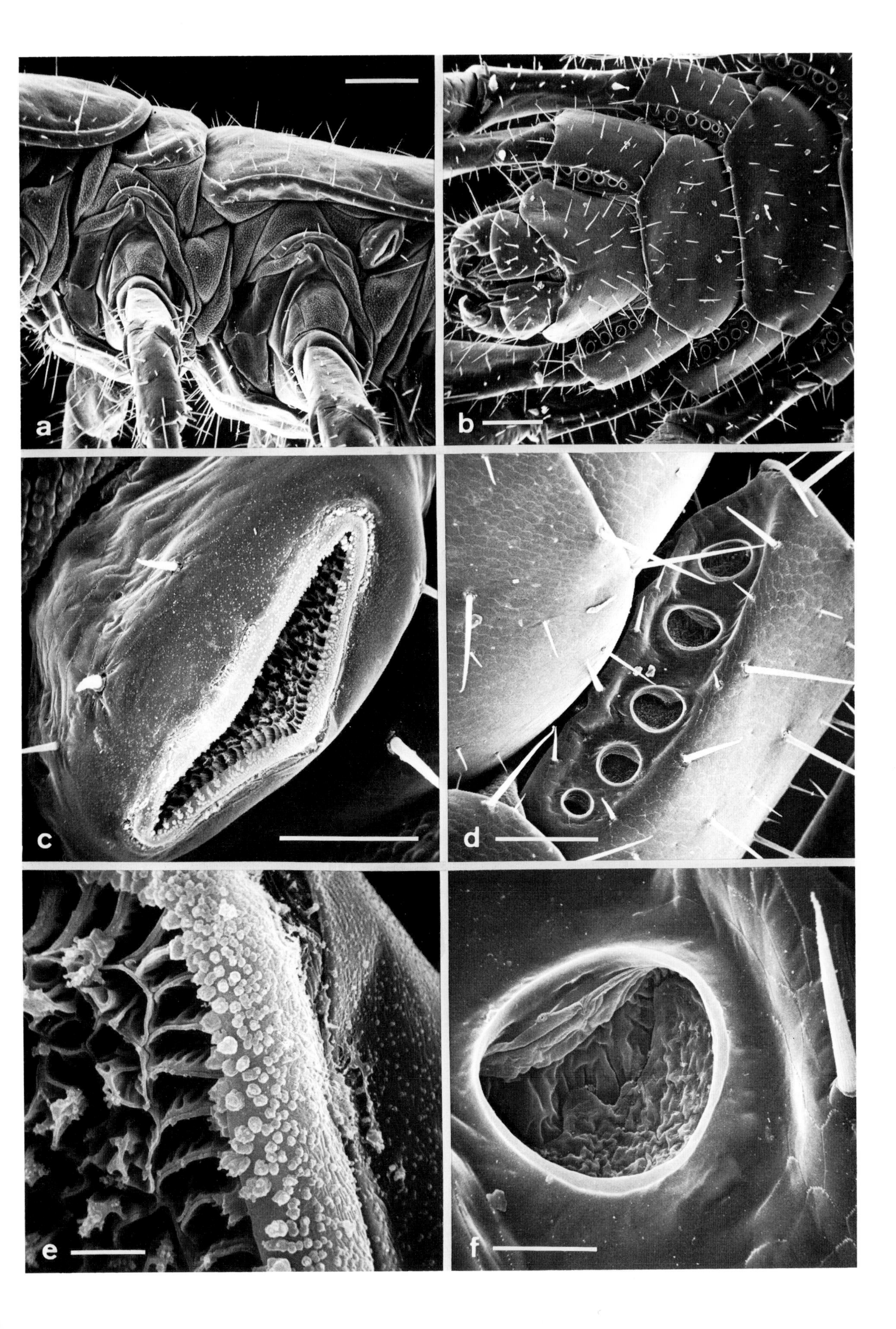
a
b
c
d
e
f

2.7 Subclass: Diplopoda – Millipedes (Myriapoda)

General Literature: Verhoeff, 1932; Schubart, 1934; Seifert, 1961

2.7.1 The Epedaphic and Hemiedaphic Diplopoda

The millipedes or Diplopoda make a major contribution to the decomposition of plant debris in the soil. They are exclusively herbivorous, gnawing plant material and fragments of decaying vegetation, especially fallen leaves and detritus in the litter layer of woodland. Thus, they belong to the macrophytophagous and saprophytophagous primary decomposers in the soil (Striganova, 1967; Marcuzzi, 1970).

The shape of their body and their function are well adapted to both the litter and to deeper layers. Dunger (1983) and Manton (1977) presented the following five types of epedaphic and hemiedaphic diplopod life forms in the soil:

1. Life form: rammer (bulldozer) type (Iulidae)
2. Life form: globular type (Glomeridae)
3. Life form: borer type (Polyzoniidae)
4. Life form: wedge type (Polydesmidae)
5. Life form: bark-dweller (Polyxenidae)

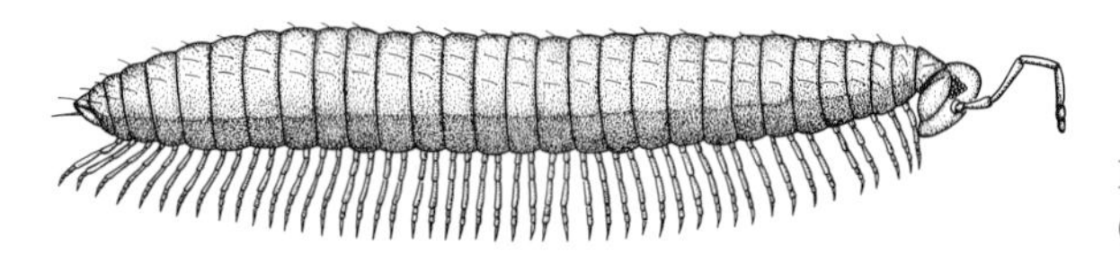

Fig. 84. Habitus of a diplopod, *Chordeuma silvestre* (Chordeumidae). (After Verhoeff, 1932)

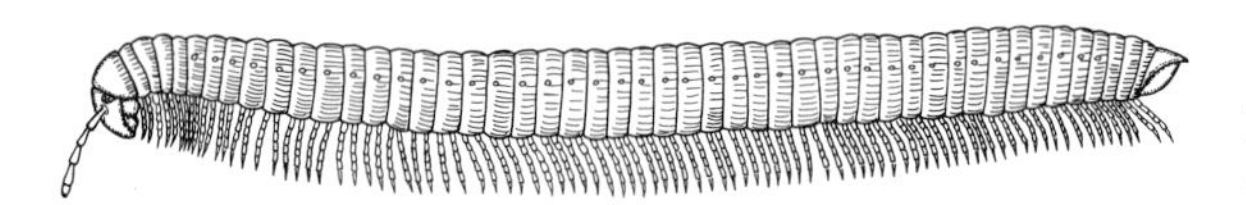

Fig. 85. Habitus of a diplopod, *Tachypodoiulus albipes* (Iulidae). (After Verhoeff, 1932)

Plate 61. *Orthochordeuma germanicum* – Chordeumidae (**a, b**); Iulidae (**c–f**).
a Anterior of the body, lateral. 200 μm.
b Compound eye and opening of the Tömösváry organ. 50 μm.
c Anterior of the body, lateral. 1 mm.
d Head and neck shield (collum), lateral. 0.5 mm.
e Trunk segments, ventral. 250 μm.
f Leg bases. 100 μm

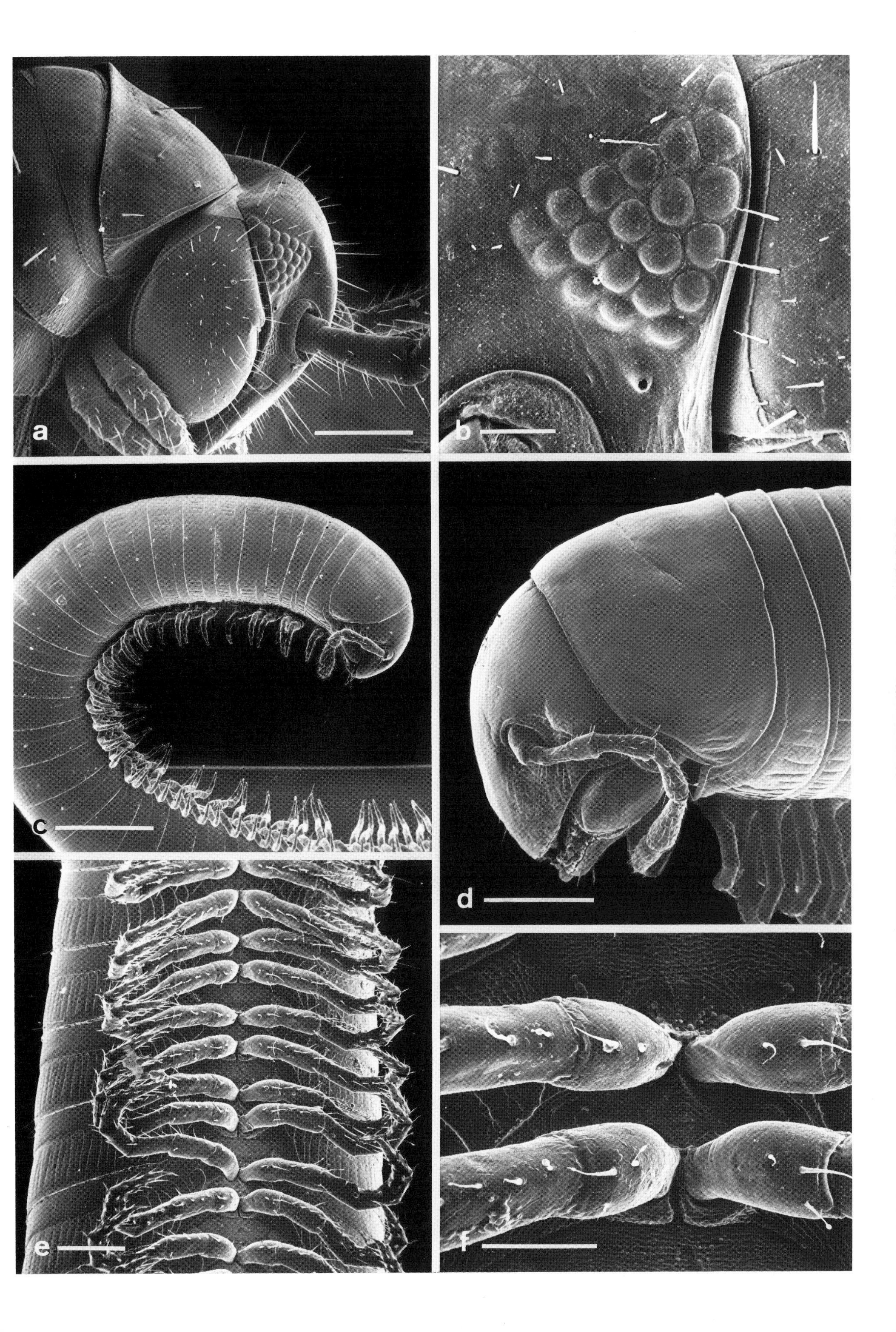
a
b
c
d
e
f

2.7.2 Life Form – Rammer (Bulldozer) Type (Iulidae)

In central Europe the iulid millipedes represent the hemiedaphic way of life (Fig. 85; Plates 61–63). While some species live under tree bark, the majority prefer to inhabit the upper soil layer, but like earthworms they are also prepared to excavate deep subterranean tunnels. Accordingly the animals are worm-shaped with round, elongated bodies consisting of at least 35 double segments, each bearing two pairs of legs. This abundance of legs arranged in rows gives them the necessary power to push aside the earth in their path like a bulldozer, using their broad head and the legless neck segment (collum) as a ram. Even when the iulids excavate a curved tunnel, full power transmission is ensured by the intersegmental connections which are similar to ball joints. The dorsal shields (tergites) are fused with the ventral shields (sternites) to form rigid segmental rings (Fig. 86), making the elongate animals less susceptible to lateral pressure. When the earth is too hard, the iulid millipedes eat their way through the substrate like earthworms.

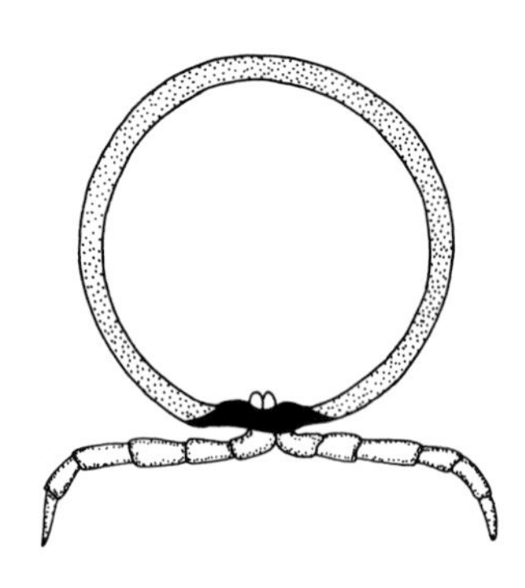

Fig. 86. Schematic cross-section through the body of a rammer-type diplopod (Iulidae) with its rigid body ring. (After Manton, 1977 and Topp, 1981)

Plate 62. Iulidae.

a Trunk, caudal, dorsal. 0.5 mm.

b Trunk segments with furrowed metazonites. Laterally the prozonites and the defensive glands (*arrow*) are recognizable. 250 μm.

c Trunk, caudal, lateral, with preanal segment and anal valves. 0.5 mm.

d Trunk segments, lateral, with furrowed metazonites and smooth prozonites. The *arrow* marks a defensive gland outlet. 0.5 mm.

e Trunk, caudal, with preanal segment and anal valves. 0.5 mm.

f Metazonite and prozonite, latero-dorsal, with defensive gland outlet (*arrow*). 100 μm

a
b
c
d
e
f

2.7.3 Protection Against Attackers – Defensive Glands

As well-adapted herbivores and primary decomposers in the soil, the millipedes do not exhibit the active defensive mechanisms shown by predatory soil animals, but rely more on protective devices. Besides the thick cuticle of the exoskeleton, hardened by deposits of calcium, they have defensive glands opening through a series of small lateral pores starting from the fifth or sixth diplosomite (Fig. 87; Plates 62 and 63). When threatened, small drops of secretion are expelled from the defensive glands by means of haemolymphatic pressure and muscle power. The secretion either has a repugnant smell or contains toxic substances. The effective substances in the defensive secretion of the spirostreptids, spirobolids and iulids have been identified as substituted p-benzoquinones (= quinones) alone or as a mixture (2-methylquinone, 2-methyl-3-methoxyquinone). Hydroquinone was shown to be their precursor. The secretion of *Polyzonium germanicum* (Diplopoda, Colobognatha) contains a volatile, defensive substance smelling like camphor and a milky, sticky component (protein) (Röper, 1978).

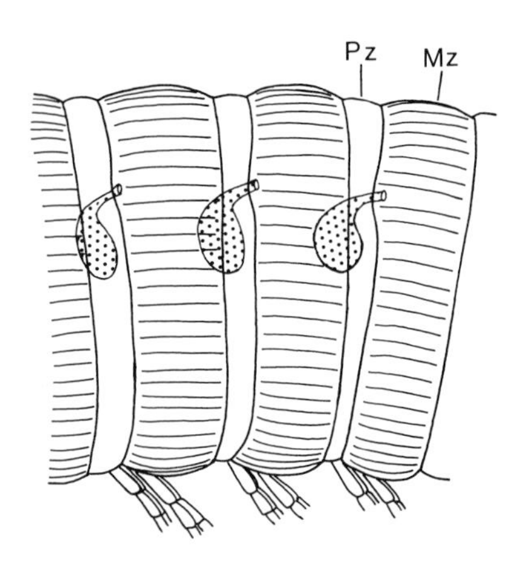

Fig. 87. Division of the trunk segments of a iulid millipede into prozonites and metazonites (*Pz, Mz*). The defensive gland outlets are in the dorso-lateral region of the segmental rings. (After Blower, 1955 from Dunger, 1983)

Plate 63. Iulidae.

a	Prozonite and metazonite with a defensive gland outlet. 50 μm.
b	Base of a bristle at the boundary between a prozonite and a metazonite. 10 μm.
d	Furrowed metazonite. 50 μm.
g	Outlet of a defensive gland. 10 μm.
c, e, f, h, i, j	Various cuticular patterns on the prozonites and metazonites. 10 μm, 5 μm, 10 μm, 3 μm, 2 μm, 3 μm

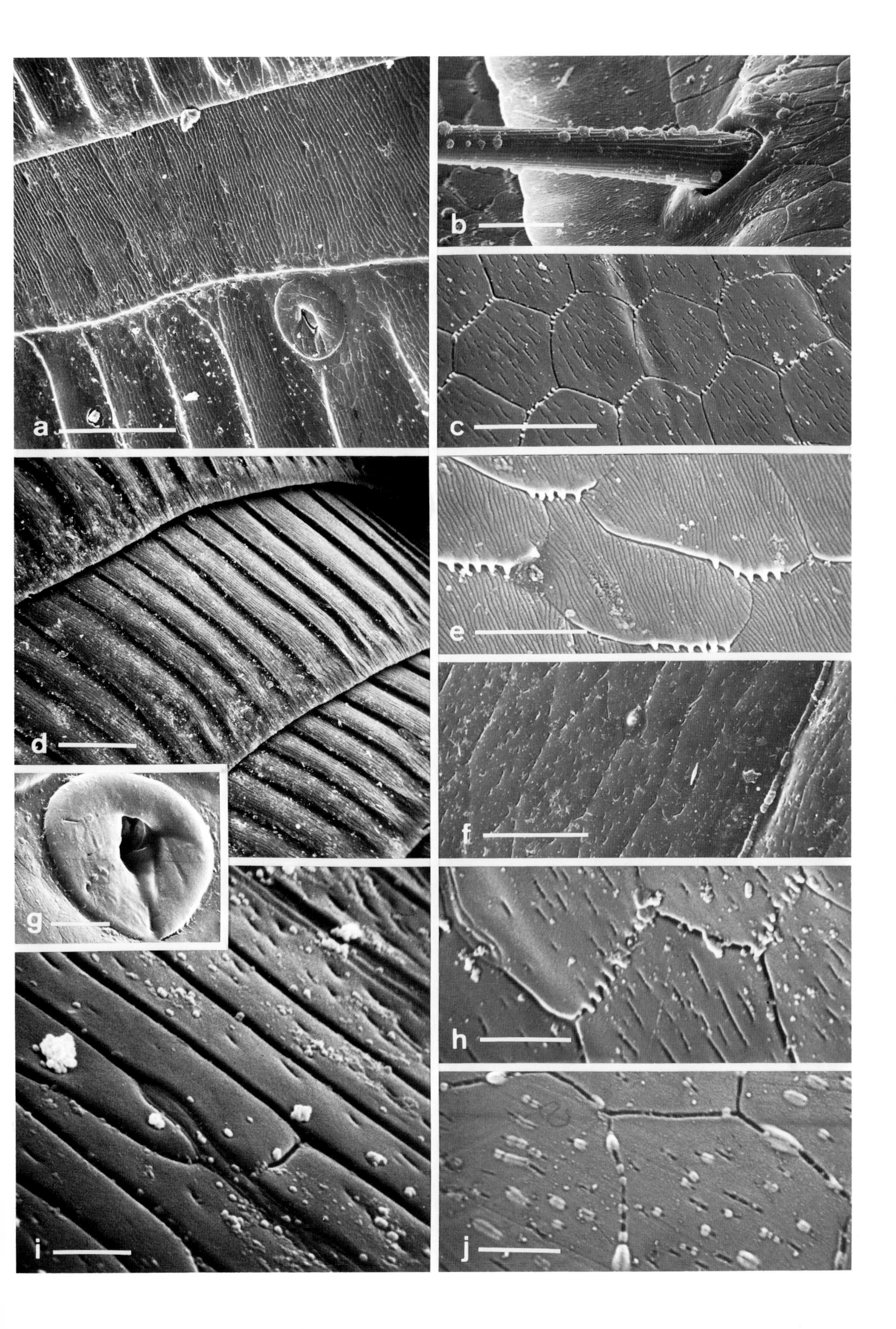
a
b
c
d
e
f
g
h
i
j

2.7.4 Life Form – Globular Type (Glomeridae)

Like certain isopods, these Diplopoda, to which the pill millipede, *Glomeris*, belongs, can roll themselves into a spherical ball enclosing their head, collum, pleurites and sternites together with the extremities. The tergites, which form the capsule, are not rigidly fixed to the sternites, but are so articulated that the ventral plates (sternites) and the lateral plates (pleurites) can be folded into the larger, vaulted tergites. The sphere is sealed by a large dorsal plate which has evolved by fusion of the second and third tergites.

The "globular diplopods" also like to burrow in the upper layers of the soil. In contrast to the "bulldozer type" they use their dorsal shield as a ram, since their head and collum are considerably smaller. They do not generally occur in the lower soil layers. With only 13 body segments and correspondingly fewer pairs of legs they have a comparatively stocky appearance and are not equipped for excavating tunnels deeper into the earth.

They are therefore more endangered by the carnivorous soil animals of the litter layer. The resulting adaptation of rolling up to escape danger is augmented by the emission of clearly visible droplets from the defensive glands as soon as they are threatened (Fig. 88b).

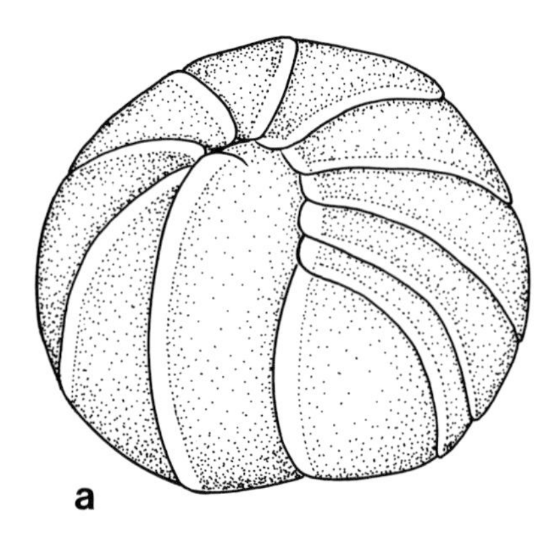

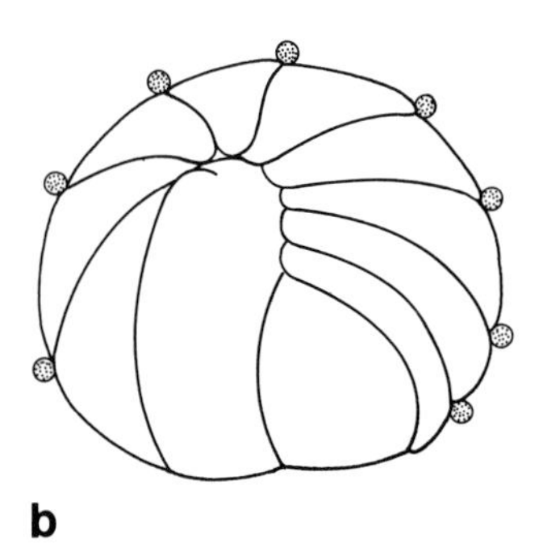

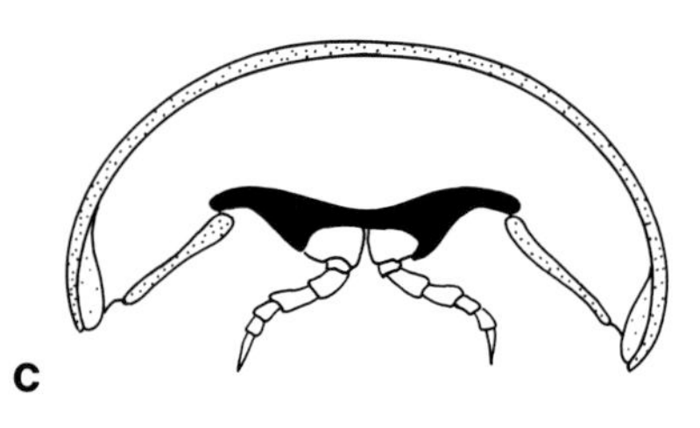

Fig. 88. *Glomeris marginata* (Glomeridae), pill millipedes. **a** Animal rolled up. **b** Rolled-up animal after provocation. Drops of a defensive secretion are exuded by the intersegmental glands on the back. **c** Schematic cross-section through the body. The sclerites of the ventral side form a flexible connection between the tergites and the rigid region of the leg bases. (After Dunger, 1983)

Plate 64. *Glomeris* sp. (Glomeridae), pill millipedes.
a Rolled-up animal, dorsal aspect. 0.5 mm.
b Animal partly unrolled, with a view of the dorsal shield (*above*), the neck shield (collum), the head with the antennae and the anal shield (*below*, surrounded by segments). 0.5 mm.
c Head with collum. 0.5 mm.
d Antenna. 0.5 mm.
e Mouthparts. 200 μm.
f Antenna, distal, with four conical sensilla. 100 μm

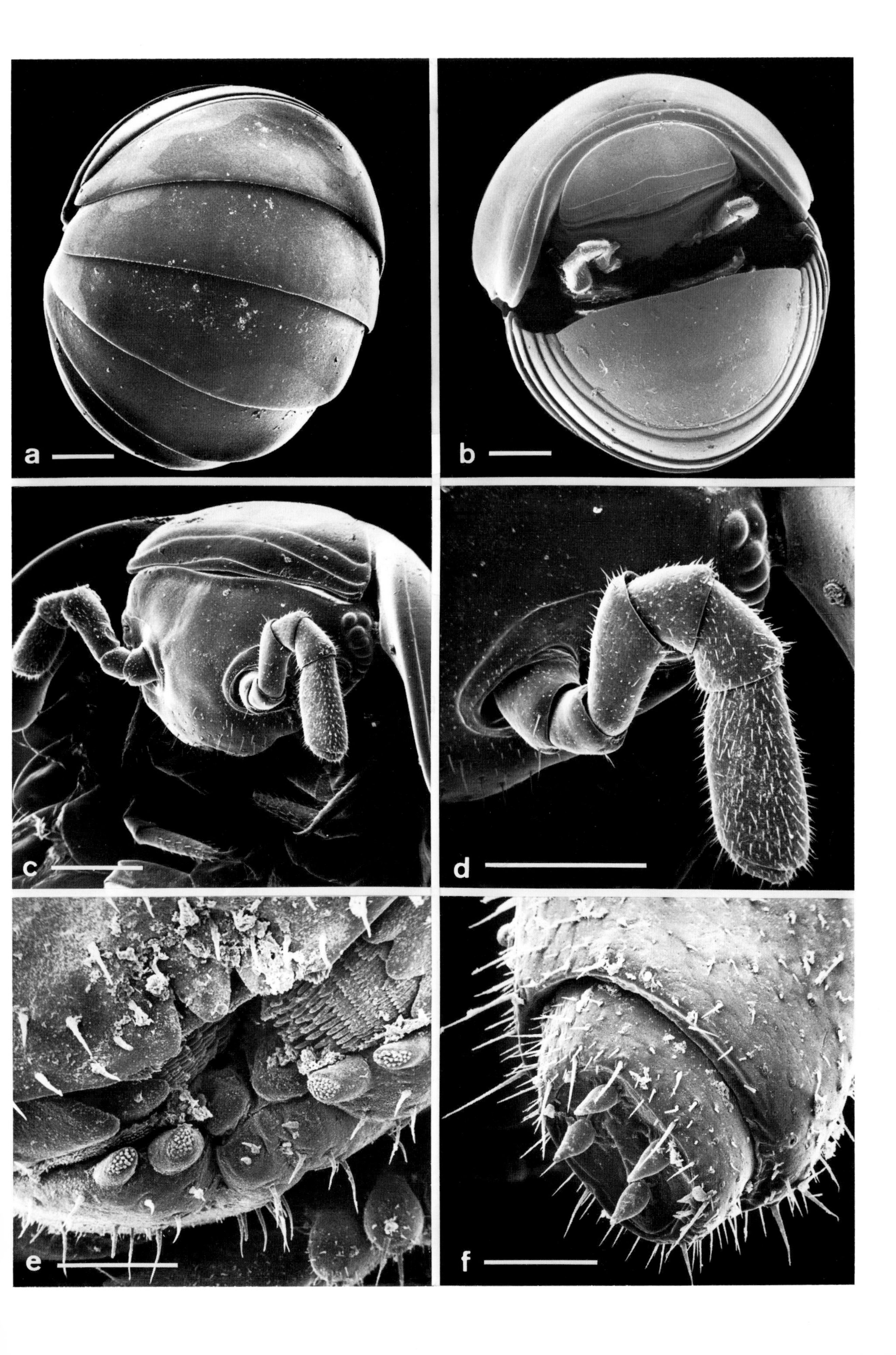
a
b
c
d
e
f

2.7.5 Moisture Requirements of the Diplopoda

Like most soil arthropods adapted to life under conditions of relatively high air humidity, the Diplopoda have high moisture requirements due to the absence of a fully developed lipid layer on the epicuticle for reducing evaporation. These requirements are closely related to the avoidance of bright light (scototaxis). In general, the diplopods seek shelter from desiccation and direct sunlight under stones, in litter or in rotting tree stumps. Suitable habitats are detected by a hygroreceptor which is obviously identical to the temporal organ described by Bedini and Mirolli (1967), which Haupt (1979) classified according to morphological and functional considerations as belonging to the Tömösváry group of organs.

Possible water losses are compensated by oral water and food intake (Edney, 1951, 1977). Like primitive insects, the Callipodidae can obviously transport water by way of transport cells on the coxal sacs. New findings show the high efficiency of anal water uptake in the diplopods. After a period of desiccation iulids and species belonging to other families resorb water by bladder-like eversion of their anal sac tissue, which they press against moist substrate enabling a rapid uptake of water (Meyer and Eisenbeis, 1985).

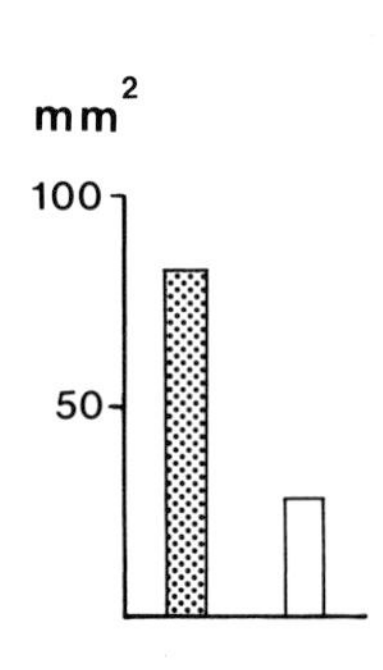

Fig. 89. *Glomeris marginata* (Glomeridae), pill millipedes. Reduction of the body surface area to about a third of its normal value due to rolling up (*white column*) for an individual weighing 20 mg

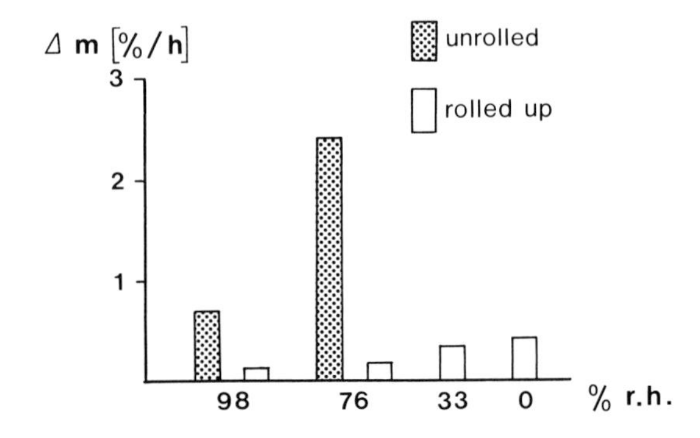

Fig. 90. *Glomeris marginata* (Glomeridae), pill millipedes. Comparison of transpiration rates expressed as a percentage of the change in water content m_0 of normal hydrated animals in variable ambient humidities at 22 °C. The animals remain unrolled only in high ambient humidities (approx. 75–100% r.h.)

Plate 65. *Glomeris* sp. (Glomeridae), pill millipedes.

a Mouthparts and genital aperture at the base of the second pair of legs. 0.5 mm.
b Eye with postantennal organ. 100 µm.
c Trunk, ventral. 1 mm.
d Trunk, ventral, with the movable ventral sclerites. 0.5 mm.
e Cuticle of the dorsal plates. 20 µm.
f Leg, distal, with claw. 100 µm

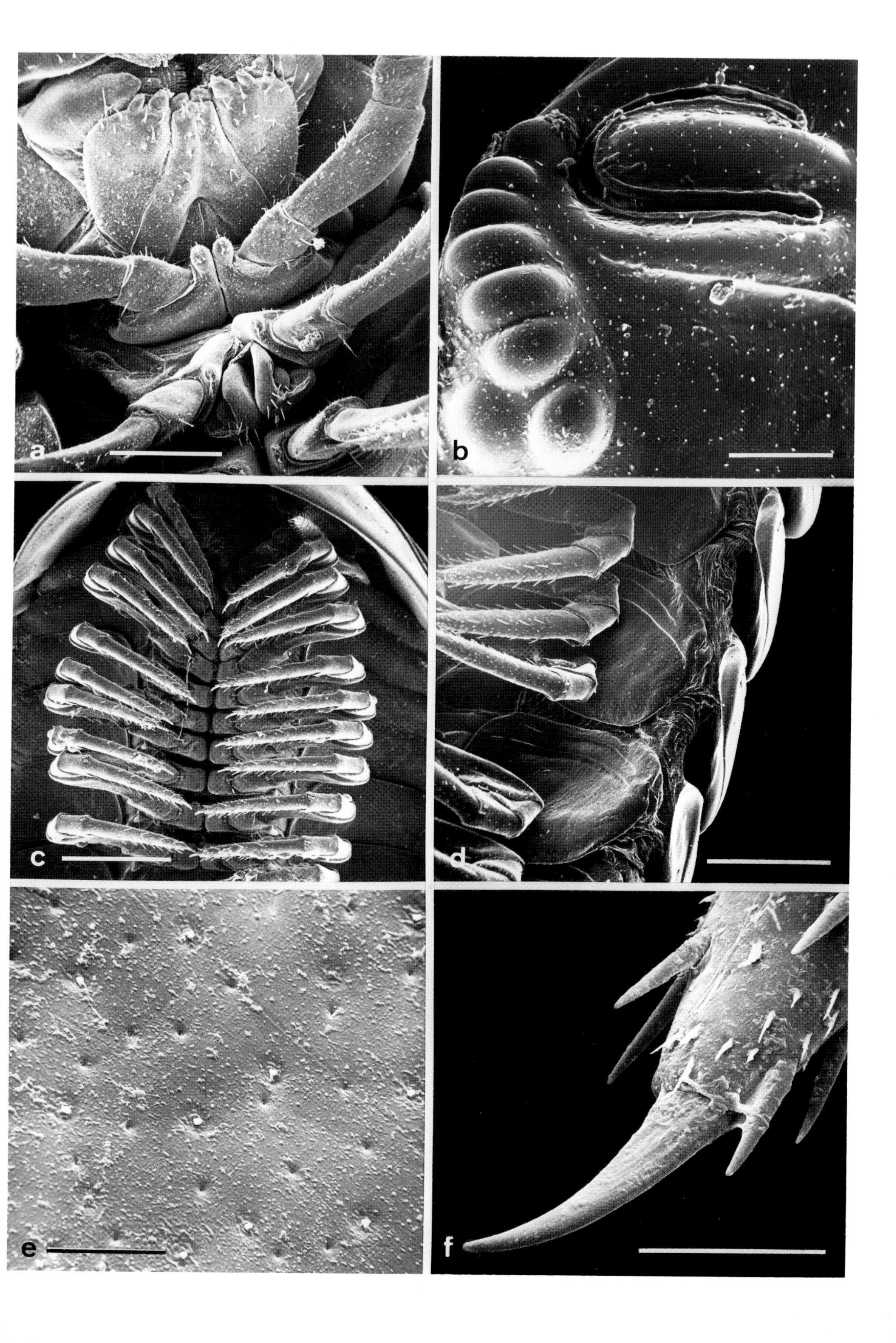
a
b
c
d
e
f

2.7.6 Life Form – Wedge Type (Polydesmidae)

The polydesmids, which embody the epedaphic life form, live in litter and in the upper soil layer. Their conspicuously wedge-shaped bodies are well adapted to life in this biotope (Figs. 91 and 92; Plates 66 and 67). The constructive elements are: (1) the tapering of their body towards the front, the head and collum being extremely small and (2) the construction of the following body segments. Like the bulldozer diplopods, their tergites and sternites are fused to form a rigid segmental ring. However, the tergites are leveled on top with wing-like lateral extensions, the paratergites, which increase their surface area (Fig. 92; Plate 66).

The tapering anterior of the body and the flattened back enable the animals to thrust their way between leaves and force themselves under stones like a wedge. The resulting external pressure is exerted on the dorsal surface and absorbed by the powerful legs. However, the enlargement of the dorsal surface through the paratergites prevents the animals from penetrating into deeper layers.

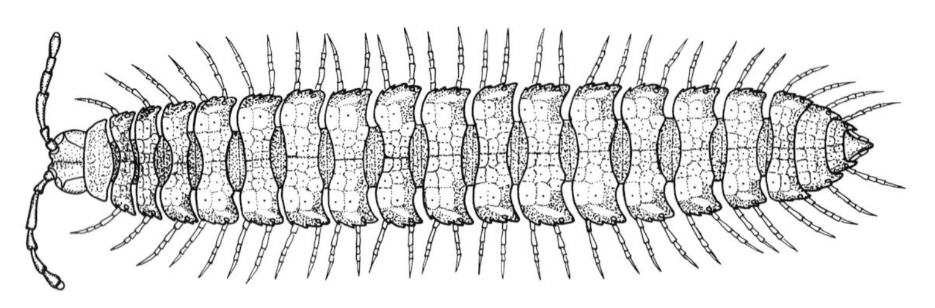

Fig. 91. *Polydesmus angustus* (Polydesmidae). Habitus, dorsal. The tergites of the segments are winged with longitudinal ridges. In their biotope the animals are reddish-brown in colour

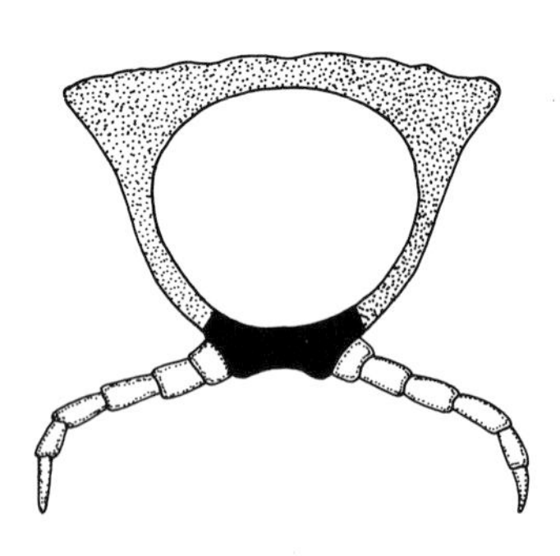

Fig. 92. Cross-section through the body of a polydesmid (wedge type) with rigid body ring. (After Manton, 1977 and Topp, 1981)

Plate 66. *Polydesmus angustus* (Polydesmidae).
a Anterior of the body, lateral. 1 mm.
b Head, ventral. 250 µm.
c Trunk segment with winged tergites. 0.5 mm.
d Trunk segments, ventral with leg bases. 250 µm.
e Trunk, caudal, lateral. 1 mm.
f Trunk, caudal, dorsal. The *arrow* marks the opening of a defensive gland. 1 mm

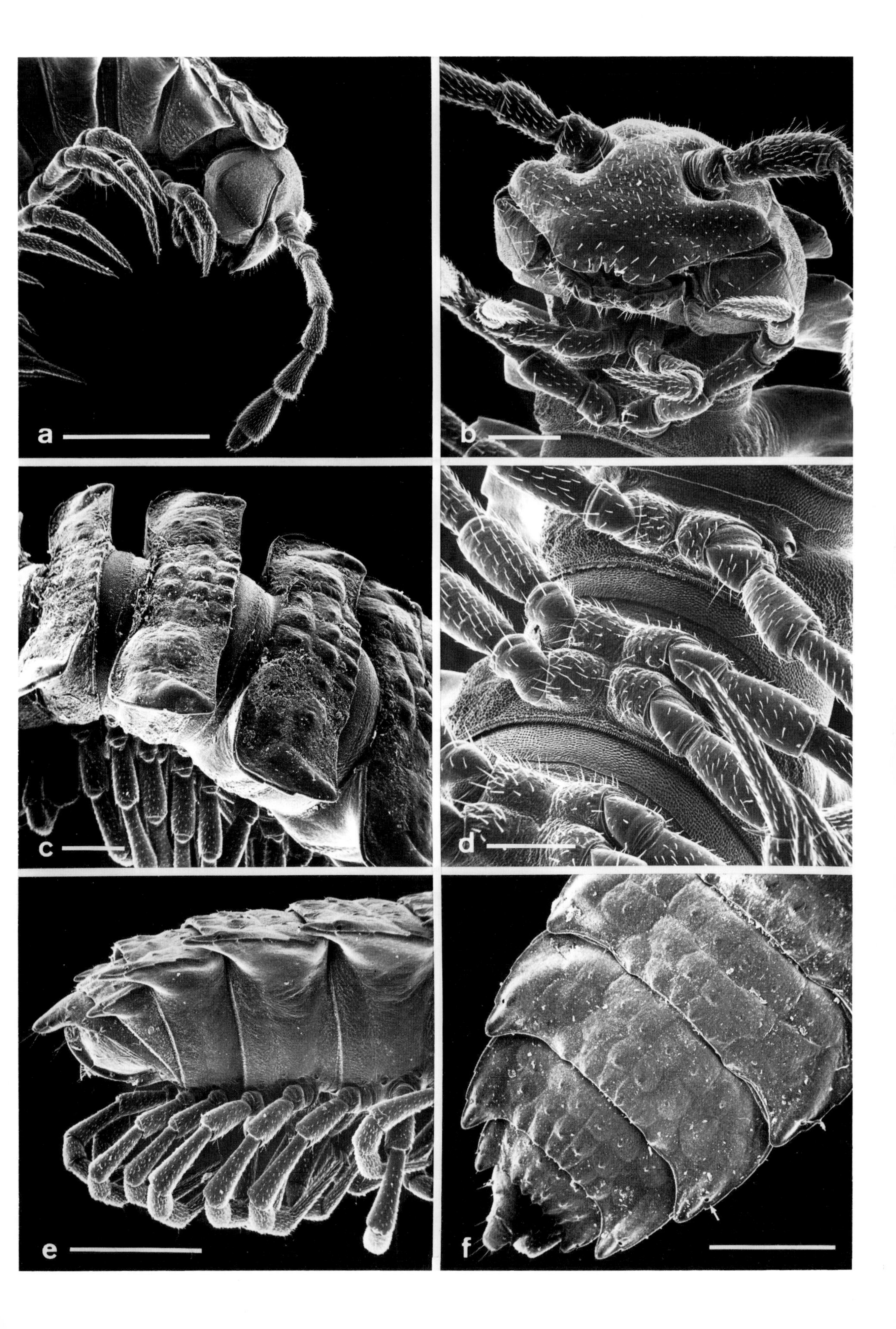
a
b
c
d
e
f

2.7.7 Respiratory Organs of the Diplopoda

Together with the insects, the myriapods belong to the Tracheata, which possess tracheae as respiratory organs. Within the Myriapoda the diplopods are characterized by the fusion of the trunk segments to form double segments or diplosomites. While the second to the fourth segment each have only one pair of legs and spiracles and the first segment (collum) has neither organ, the posterior diplosomites each possess two pairs of legs and spiracles.

The spiracles are situated on the sternites immediately beside the coxae of the legs. As a rule their opening is protected from the intrusion of extraneous particles by a cuticular lattice or network (Plates 66 and 67). Moreover, they can be closed by laterally placed, movable processes (*Glomeris*). Inside the spiracles the space opens out to an atrium from which bundles of tracheae carry oxygen to the body cells.

During rain and waterlogging of the soil the diplopods have to come up to the surface to breathe. Unlike the euedaphic soil arthropods, the diplopods are adapted to life in the upper layers of the soil. Thus, they lack hydrophobic structures on their body surface to retain a film of air round their body and spiracles as in the case of gas gills.

Plate 67. *Polydesmus angustus* (Polydesmidae).
a Trunk segments, lateral, with spiracles and leg bases. 0.5 mm.
b Leg bases with coxa, trochanter and spiracle opening. 100 µm.
c Leg segmentation. 250 µm.
d Spiracle with sieve plate. 20 µm.
e Anal segment, dorsal. 250 µm.
f Trunk, caudal, with anal valves. 250 µm

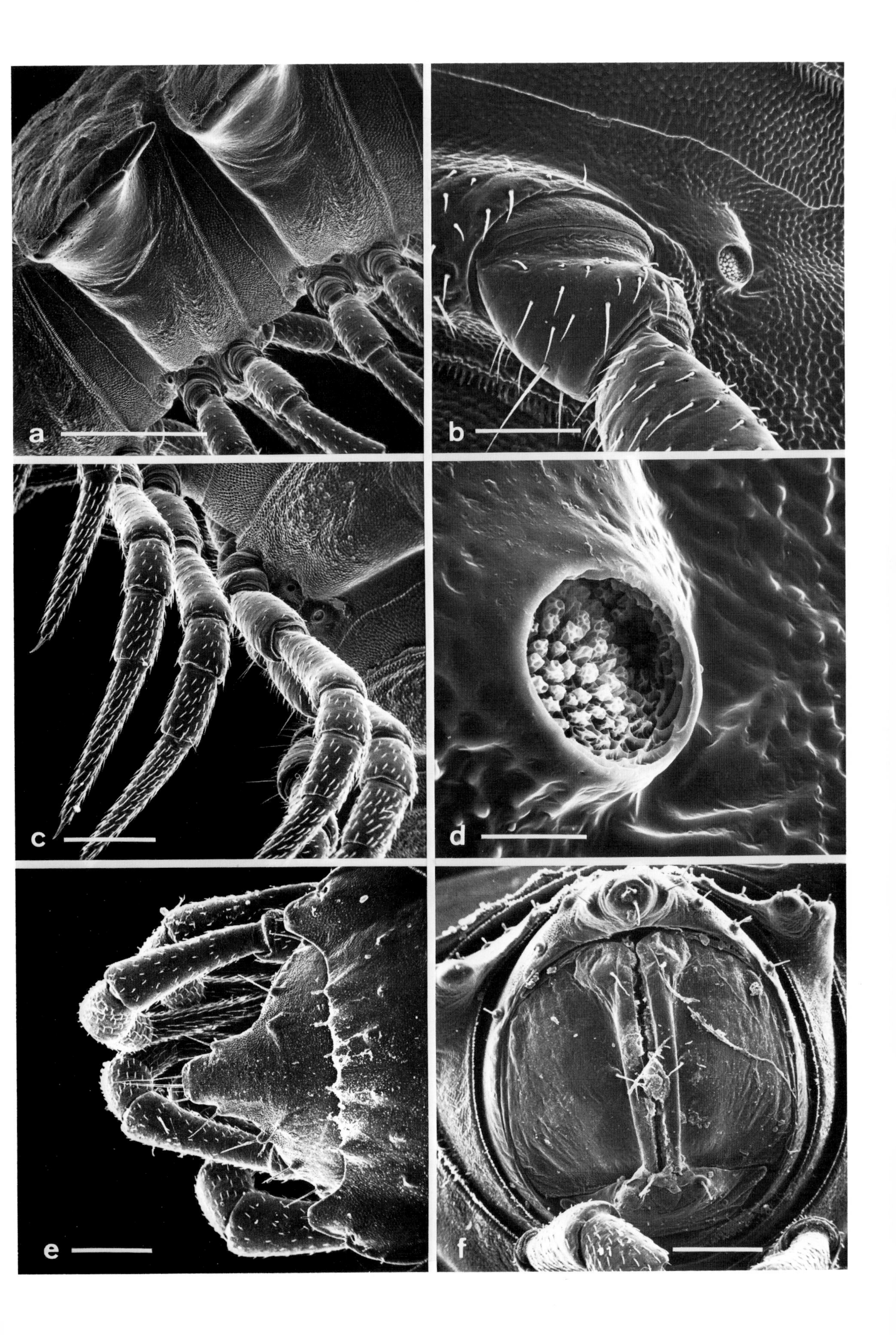
a
b
c
d
e
f

2.7.8 Life Form – Bark-Dweller (Polyxenidae)

The final diplopod life form presented by Dunger (1974, 1983) is the bark-dweller, which shows no typical adaptation mechanisms to life in the soil. The 2–3-mm-long, bizarre polyxenid millipedes belong to this group. Their appearance is impressive, their body being covered by decoratively distributed trichomes (Fig. 93; Plate 68). *Polyxenus lagurus* lives in loose humus and under leaves, but especially under the bark of trees, ranging from soil height to the 10–15-m-high treetops, and feeds on unicellular algae on the bark. Sometimes the animals are also encountered under stones in aggregations of individuals of all ages.

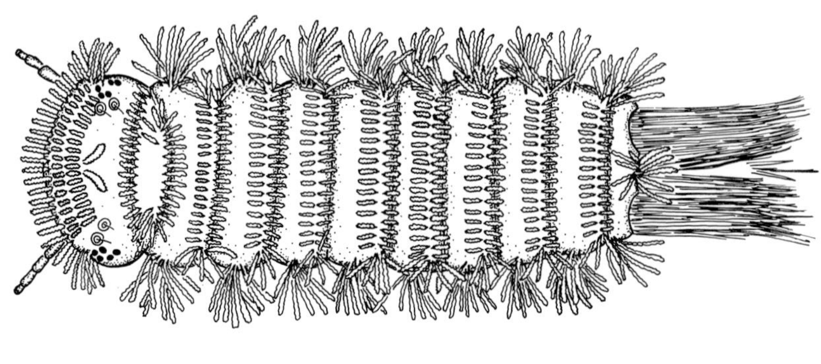

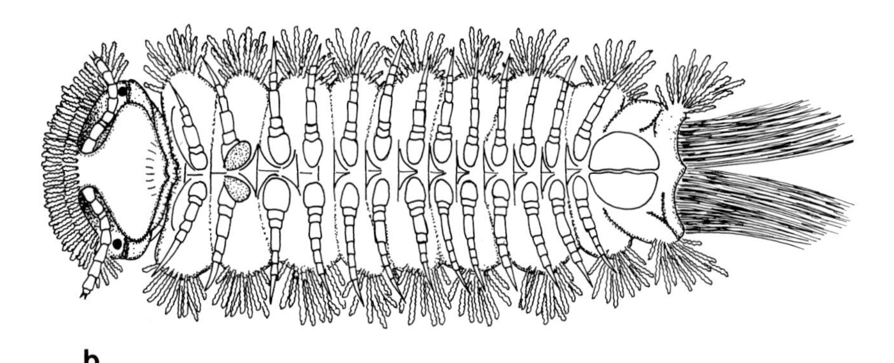

Fig. 93. *Polyxenus lagurus* (Polyxenidae). (After Verhoeff, 1932).
a Habitus, dorsal, with groups of ocelli and trichobothria on the head. The body is covered by trichomes, while the brush is composed of bundles of longer hairs which are grouped together on plates. **b** Habitus, ventral. The head is capped by an apron-like shield (clypeus). The genital papillae are situated medially on the second trunk segment. In front of the brush lie the anal opercula, consisting of two movable valves

Plate 68. *Polyxenus lagurus* (Polyxenidae).
a Habitus, oblique ventral. 0.5 mm.
b Habitus, oblique dorsal. 0.5 mm.
c Head region, oblique dorsal, with ocelli-bearing tubercle and trichobothria. 100 µm.
d Dorsal side of the head, lateral with ocelli-bearing tubercle, trichobothria and antenna. 50 µm.
e Head, ventral, with head shield (clypeus) and antennae. 200 µm.
f Head, lateral, with clypeus, antenna and ventral ocellus. 100 µm

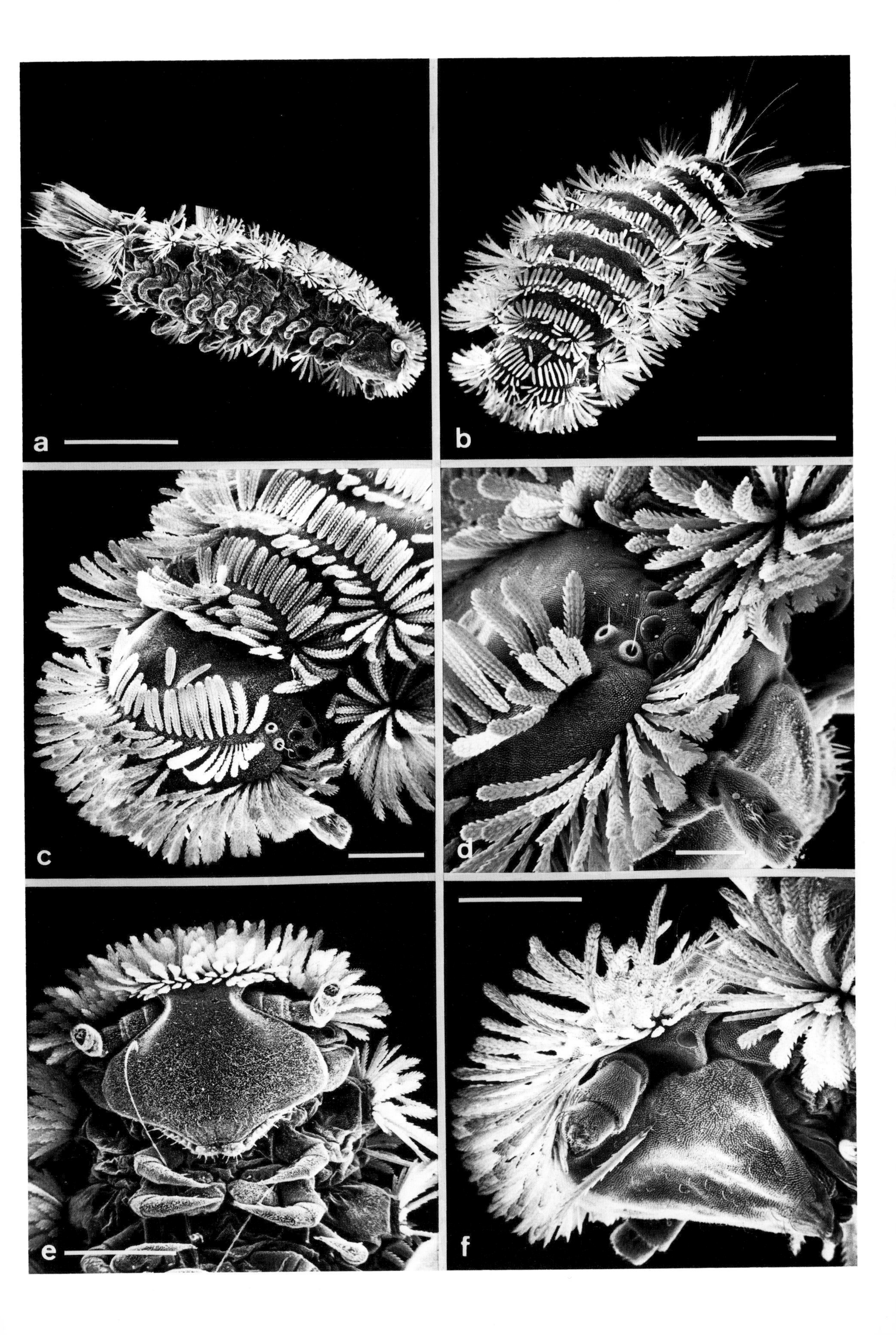
a
b
c
d
e
f

2.7.9 Sensory Hairs of *Polyxenus lagurus*

Sensory hairs permit the spatial orientation of these animals, with mechanoreceptors, as well as chemo-, hygro- and thermoreceptors playing a prominent role. Tufts of feathery trichomes on the flanks and arranged in double rows on the tergites not only give *Polyxenus* an extraordinary appearance, but also ensure that these bark-dwellers are in contact with their immediate surroundings.

Further sensory organs with differing perceptive functions are found on the head and antennae (Plates 68 and 69). On the head beside the eyes there are trichobothria which perceive the slightest air disturbances (Tichy, 1975). Seven different cuticular sensory organs have been identified on the antennae alone (Schönrock, 1981): tactile bristles, contact chemoreceptors and chemoreceptive, cylindrical bristles, two types of antennal cones, one tubercular organ and the distal olfactory cone, with its sensilla at each corner of a square.

The ultrastructure of the four conical sensilla at the tip of the antennae and of several basiconical sensilla below the apex has been described by Nguyen Duy-Jacquemin (1981, 1982). In the first case, two types of innervation of the conical sensilla are recognizable according to the tergal or sternal position of the sensilla. All sensilla have a differentiated innervation, so that more than one function is probable. The conical sensilla are regarded as mechanoreceptors and contact chemoreceptors and the basiconical sensilla as olfactory sensory organs.

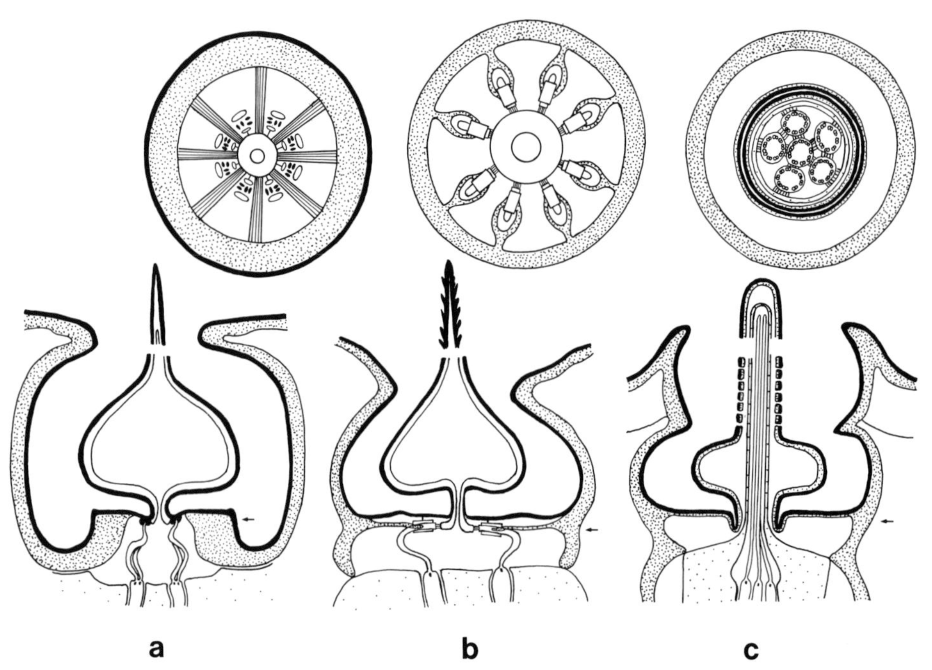

Fig. 94. Modifications of trichobothria (cupulate hairs) of progoneatal Myriapoda. (After Haupt, 1979).
a *Scutigerella immaculata* (Symphyla), **b** *Allopauropus* sp. (Pauropoda), **c** *Polyxenus lagurus* (Diplopoda).
The ciliary processes of the sensory cells are either connected to the base of the hair shaft by a complicated suspension system (**a, b**) or they penetrate into the hair shaft (**c**). Epicuticle is *black*, endocuticle is *dotted*

Plate 69. *Polyxenus lagurus* (Polyxenidae).
a Ocular tubercle with trichobothria. 25 µm. *Inset*: cupulate socket of a trichobothrium. 3 µm.
b Antenna, distal, with four conical sensilla. 10 µm.
c Tufts of trichomes from the flank region of the trunk. 50 µm.
d Antenna, distal, conical sensilla. 5 µm.
e Microstructures on the trichome surface. 3 µm.
f Brush organ, insertion of branched hairs in socket cups arranged to form a plate. 10 µm

a
b
c
d
e
f

2.7.10 Reproduction of *Polyxenus lagurus*

Both parthenogenetic and sexual reproduction, geographically separated, are possible in *Polyxenus lagurus*. Schömann (1956) described the sexual behaviour. Before the male of *P. lagurus* deposits spermatophores, he spins a mesh of threads in a zigzag pattern in a small crevice and then affixes two droplets of sperm onto one of the stretched threads (Fig. 95). He subsequently stretches a conspicuous, 1.5-cm-long, double thread vertically downwards. A passing female, perceiving the thick double threads, follows them and is led directly to the two sperm droplets, which she takes into the valves on her second pair of legs. This indirect sperm transfer requires no contact between male and female.

The eggs of *Polyxenus lagurus* are stuck together like beads by a secretion and deposited in a spiral string to form a disc. The hair of the tail brush is pressed around the still sticky eggs to encircle them in a ventilated protective sheath which keeps them away from the substrate (Seifert, 1960).

Iulids, glomerids and polydesmids build chambers by eating earth, then mixing it with a consolidating secretion as it emerges from the anus and depositing it to construct the nest. In this manner protective bell-shaped nests are made, as by *Ophyiulus falax*. Its eggs are protected by a bell-shaped cover of earth with a central chimney to regulate internal air humidity.

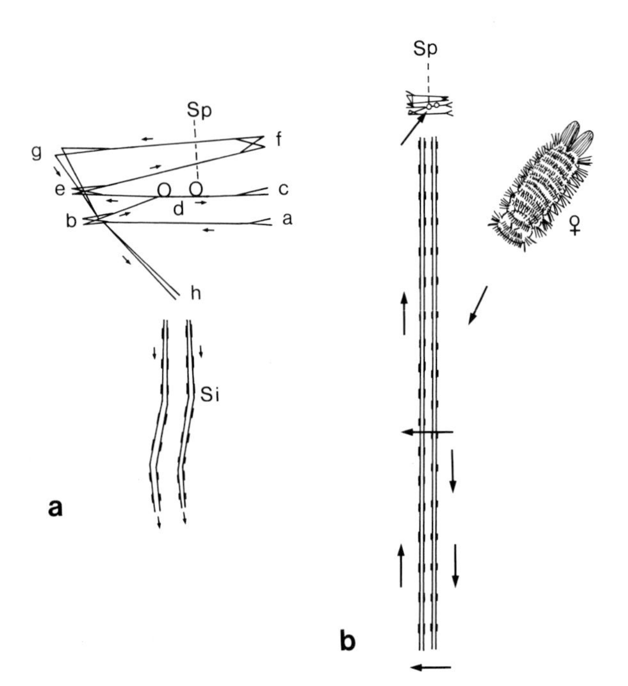

Fig. 95. *Polyxenus lagurus* (Polyxenidae), method of indirect sperm transfer. (After Schömann, 1956).
a Two droplets of sperm (*Sp*) are deposited on a web of threads (secreted by penis glands) spun with zigzag movements by the male. In addition, an approximately 1.5-cm-long trail of signal threads (*Si*) is secreted by the glands on the eighth and ninth pair of legs. **b** A mature female recognizes the signal threads with her antennae and either crosses or walks round to the other side of the signal trail, which leads her to the sperm web. When she finds the mesh of threads, she takes the sperm droplets into her genital valves

---

Plate 70. *Polyxenus lagurus* (Polyxenidae).
a Head region, ventral. 50 μm.
b Palpal structures in the mouth region. 20 μm.
c Flank region of the trunk. 50 μm.
d Microstructures on the cuticle near the leg bases. 10 μm.
e Glandular areas on the leg surface. 10 μm.
f Cuticle on the ventral side in front of the brush organ. 10 μm

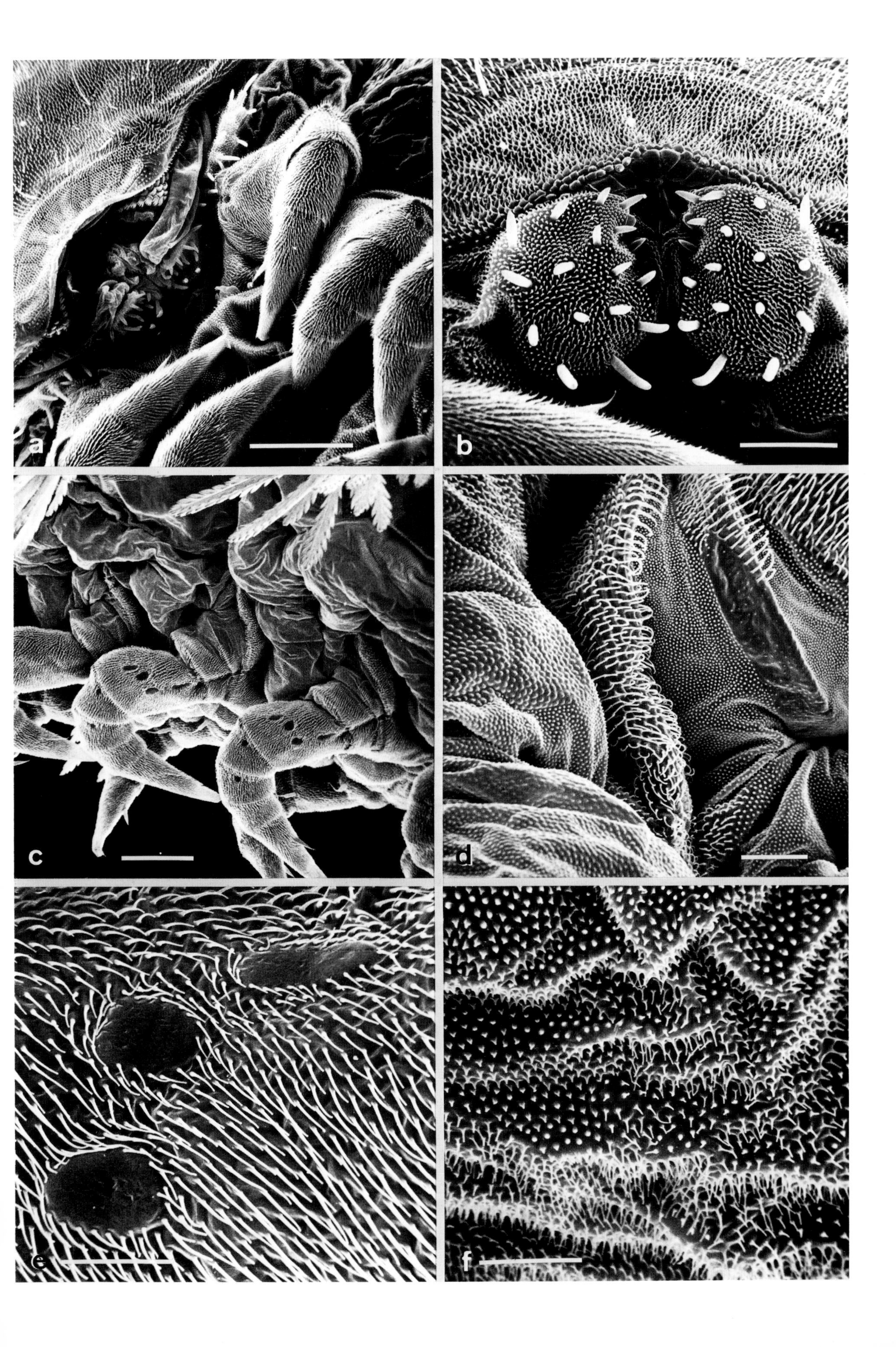
a
b
c
d
e
f

2.7.11 Water Balance in *Polyxenus lagurus*

As expected in a bark-dwelling life form, appropriate adaptations with regard to water balance have evolved in *Polyxenus lagurus*, on the one hand, to reduce water loss through transpiration and, on the other, to utilize water vapour from the atmosphere. Thus, water loss is extremely low even when the animals are kept in completely dry air for several days. In 0% r.h./22 °C loss of water mass per hour remains distinctly below 1% (Table 1; Fig. 96), while it is only 0.08 or 0.28% at 98 and 76% r.h./22 °C respectively.

Under experimental conditions the animals show a constant rate of transpiration over a longer period of time even at 98% r.h./22 °C. However, a sudden, rapid weight increase, a rise of more than 3%/h with respect to the total water mass, was observed in several individuals. This absorption phase was completely linear and sometimes lasted more than an hour. In order to exclude the possibility that the increase in weight was purely a surface adsorption effect, the animals were subsequently tested at 0% r.h. The weight increase was still observable, restoring the former weight of the animal after a transpiration phase lasting several hours. The timing of this absorption behaviour is remarkable; all individuals absorbed in the early morning between 5 and 6 a.m. This is ecologically meaningful, as ambient moisture can be expected to be high because of dew formation at this time. The site of water vapour uptake is unknown.

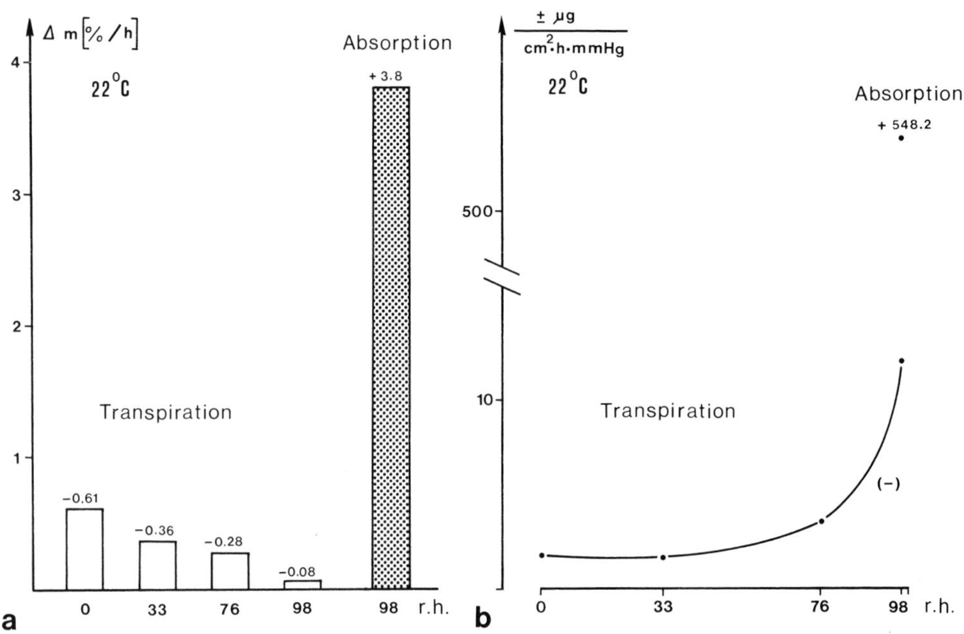

Fig. 96. *Polyxenus lagurus* (Polyxenidae), data on water balance.
a The percentage change in the animal's existing water content m_0 at variable relative humidities. The negative rates are a measure of transpiration, the positive rate indicates how quickly the animals could regain their original water mass by absorption of water vapour at 98% r.h. **b** Comparison of the relative transpiration rates at variable relative humidities. As aridity increases there is a relative decrease in the transpiration rate. The positive value represents the absorption rate relative to the body surface (theoretical value)

Plate 71. *Polyxenus lagurus* (Polyxenidae).
a Brush organ, lateral. 200 µm.
b Brush organ, oblique caudal, with plate into which the brush hairs are inserted (*arrow*). 100 µm.
c Brush organ, distal. 50 µm.
d Tergum near the brush plates. 50 µm.
e Anal plate with two anal valves. 50 µm.
f Hindgut intima between anal valves. 10 µm

a
b
c
d
e
f

2.8 Subclass: Pauropoda – (Myriapoda)

General Literature: Verhoeff, 1937; Hüther, 1974

2.8.1 Characteristics of the Pauropoda

The body of the small, only a few millimetres long Pauropoda consists of the head, 11 segments and a short telson. In contrast to the other members of the myriapods, pauropods use only nine pairs of legs for walking. Their antennae are also very different from those of the other Antennata. The four basal segments of the antennae bear two branches, the upper branch having one flagellate sensillum and the lower two. An additional club-shaped antennal globulus (clavate sensillum) is situated between both flagella (Fig. 97; Plates 72 and 73).

More than 500 pauropod species have been described, of which at least 50 occur in central Europe.

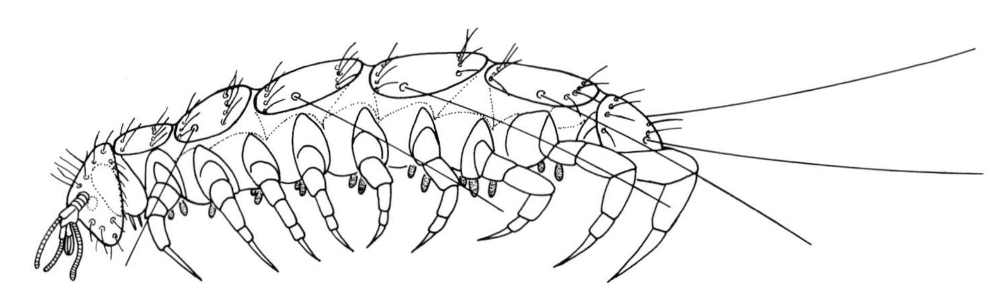

Fig. 97. *Allopauropus* sp. (Pauropodidae), lateral view. The large *dotted area* on the head denotes the area occupied by the pseudoculus (cf. Fig. 99)

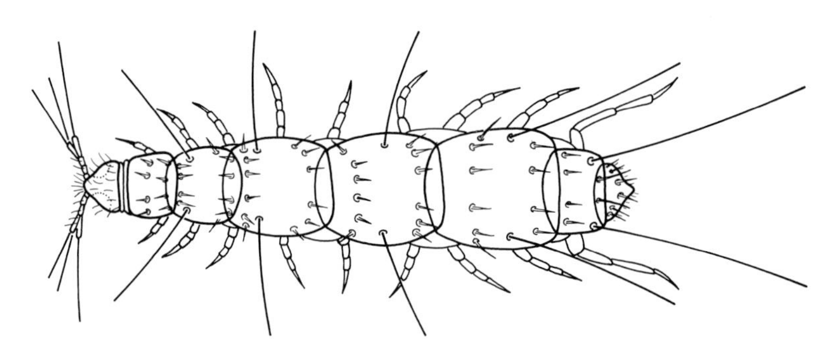

Fig. 98. *Pauropus huxleyi* (Pauropodidae), dorsal view. (After Verhoeff, 1937)

Plate 72. *Allopauropus* sp. (Pauropodidae).
a Habitus, lateral. 200 µm.
b Anterior trunk, lateral. 50 µm.
c Head with first trunk segment, oblique dorsal. 50 µm.
d Head with first trunk segment, ventral. 10 µm.
e Head with conical arrangement of the mouthparts, ventral. 20 µm

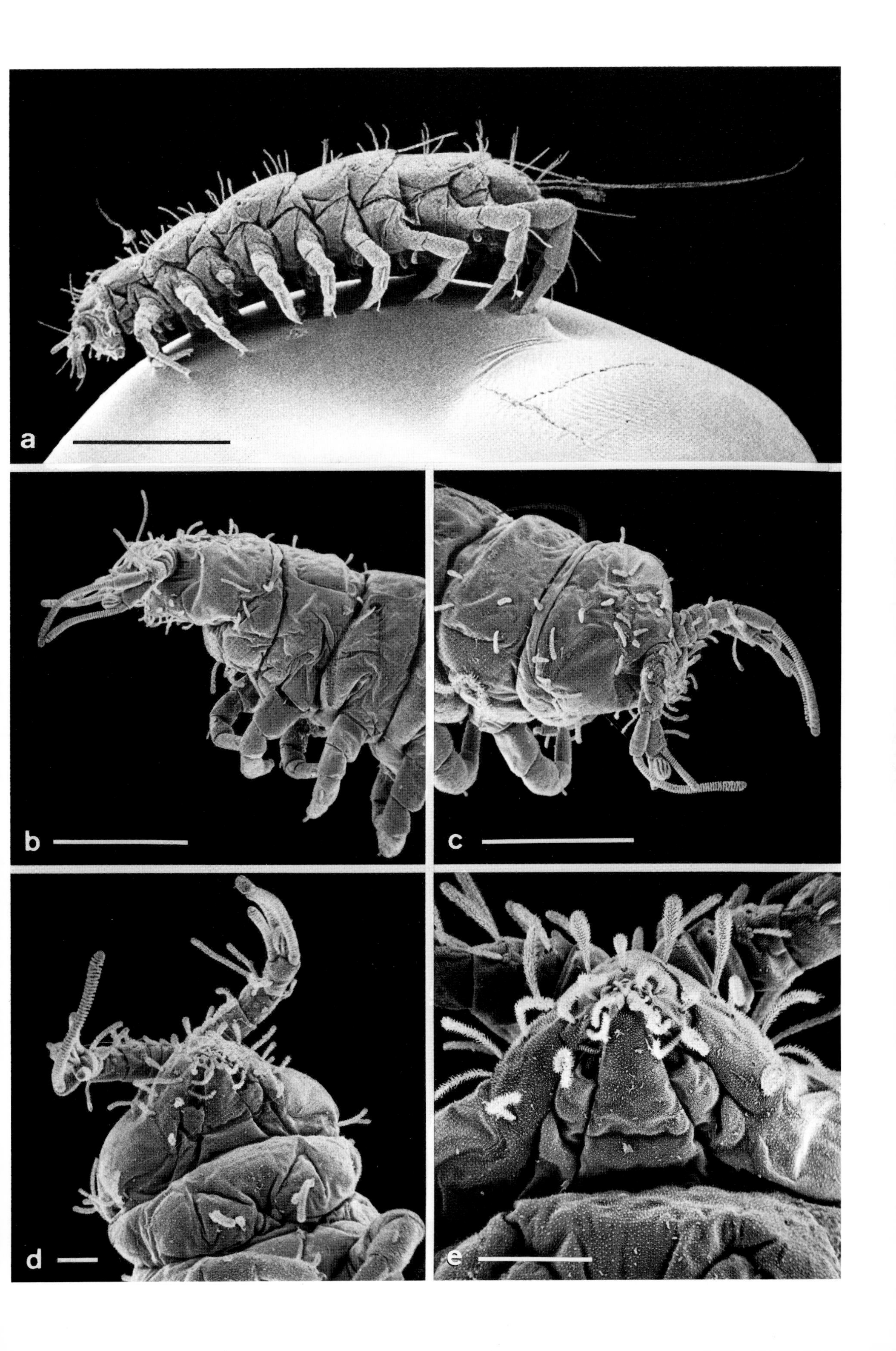
a
b
c
d
e

2.8.2 Adaptation of Sensory Perception – Pseudoculus

As in most euedaphic soil arthropods, the sensory organs of the pauropods have undergone differentiation, adapting to life in the soil. Photoreceptors can hardly play a practical role in the dark soil and are therefore greatly reduced or completely absent. They are replaced by chemoreceptive and mechanoreceptive organs (Haupt, 1976).

The genal organs near the bases of the antennae or in the cheek region of the head, described as Tömösváry organs (in Chilopoda and Symphyla), as temporal organs (in Diplopoda) or as pseudoculus (in Pauropoda) belong to the complex chemoreceptors (Haupt, 1973) (Fig. 105). The abundance of sensory cells in the pseudoculus indicates its complex function. Hygro-, CO_2- and thermoreception are all discussed. Furthermore, olfactory perception is suspected. Fine pores in the cuticle allow contact between the dendritic processes and the exterior surroundings (Fig. 99). Conclusions about both the effectiveness and the significance of the pseudoculi can be drawn from the large area occupied by the pore fields.

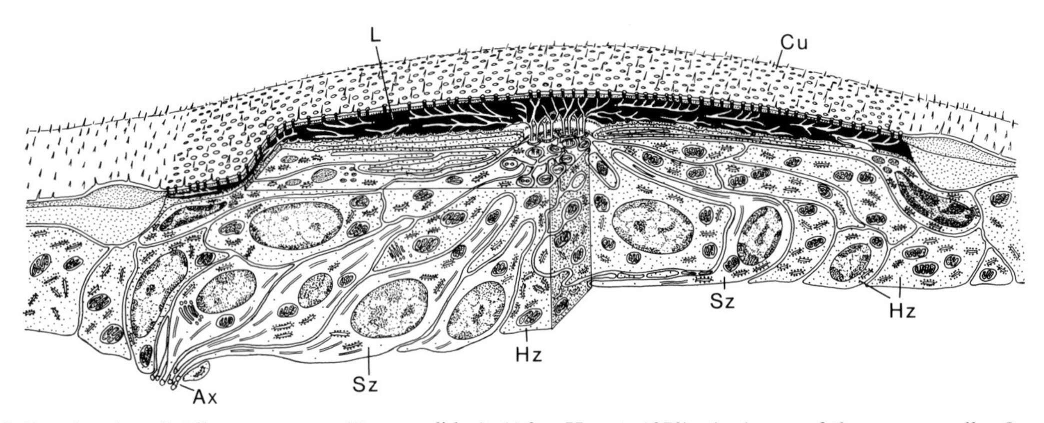

Fig. 99. Pseudoculus of *Allopauropus* sp. (Pauropodidae). (After Haupt, 1973). *Ax* Axons of the sensory cells; *Cu* cuticle, perforated; *Hz* envelope cells; *L* outer receptor lymph cavity with dendrites; *Sz* sensory cells

Plate 73. *Allopauropus* sp. (Pauropodidae).
a Branched antennae, dorsal. 20 μm.
b Branched antenna, dorsal, with antennal globulus. 4 μm.
c Antennal base, lateral. 10 μm.
d Antennal surface. 1 μm.
e Tergum of the trunk, lateral. Cupulate socket of a long (trichobothrium, *left*) and a short bristle hair. 10 μm.
f Anal region, ventral. 25 μm

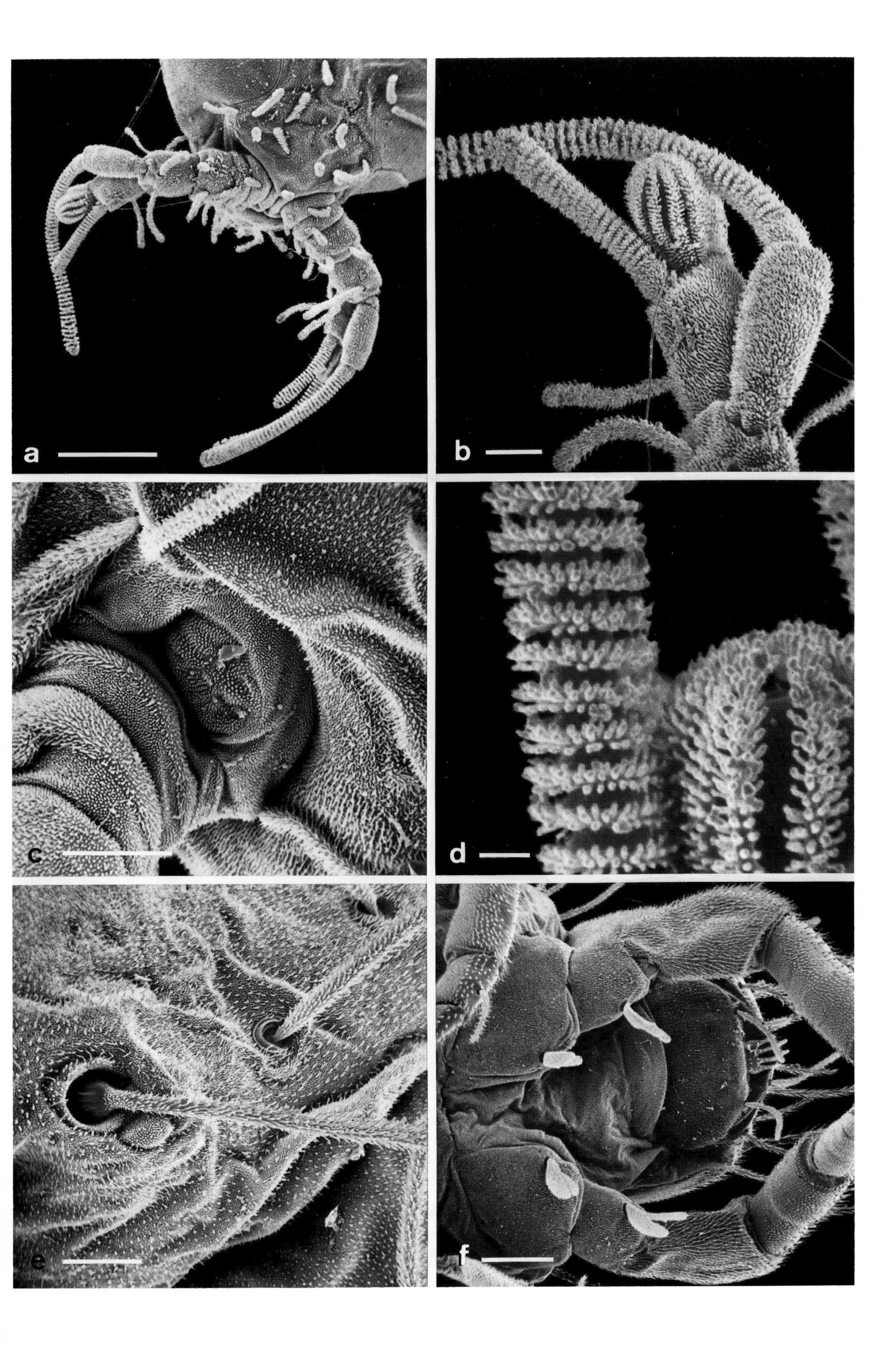

a
b
c
d
e
f

2.8.3 Adaptation of Sensory Perception – Trichobothrium

Paired trichobothria, sensitive to mechanical stimuli, are inserted laterally in the second to the sixth tergites of *Allopauropus* (Fig. 100; Plates 72–74). The remarkably long sensory hairs of the trichobothria perceive the slightest directional changes in air currents, thus facilitating the orientation of these soil-dwellers within the interstitial system. Eight sensory cells are associated with the trichobothria of *Allopauropus*. Their dendrites are linked to the disc-shaped basal plate of the sensory hair (Fig. 100). The sensory hair originates from a bulb, which is attached to the discoid base by a short stalk (Haupt, 1976, 1978).

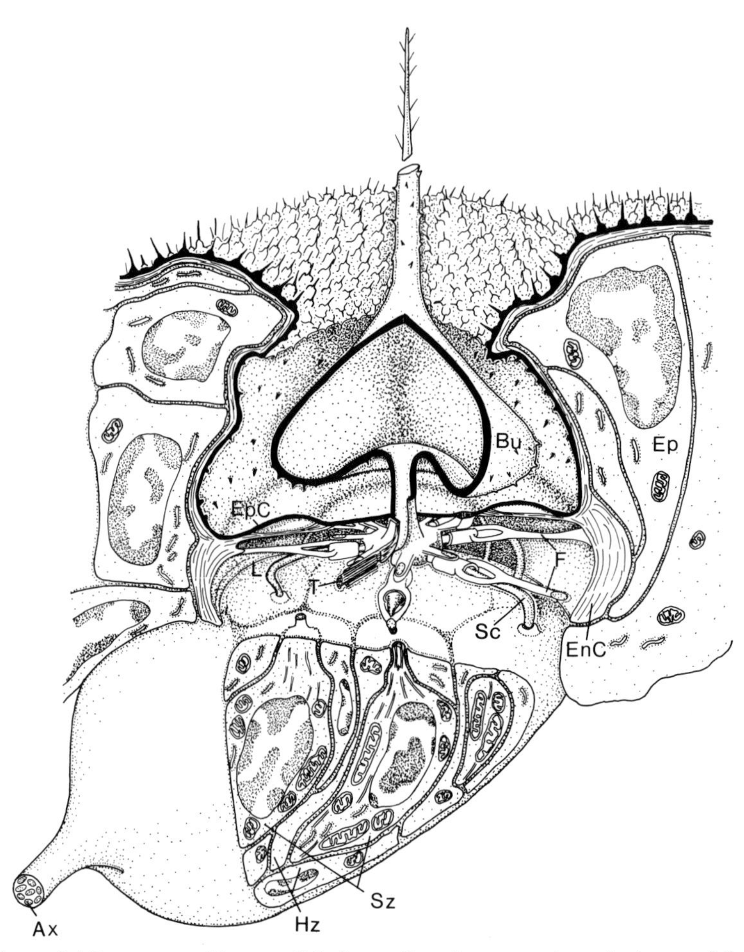

Fig. 100. Trichobothrium of *Allopauropus* (Pauropodidae). (After Haupt, 1976). The hair with its bulbous shaft is sunk into a cupulate socket and is innervated by eight dendritic cilia which are attached via tubular bodies to the hair base by a complicated suspension system, an arrangement typical of mechanoreceptors. *Ax* Axons of the sensory cells; *Bu* hair bulb; *EnC* endocuticle; *EpC* epicuticle; *Ep* epidermis; *F* suspension fibrils; *Hz* envelope cells; *L* receptor lymph cavity; *Sc* sensory cilia (dendrites); *Sz* sensory cells; *T* tubular body

Plate 74. *Allopauropus* sp. (Pauropodidae).
a Trunk segments, dorsal. 50 μm.
b Trunk segments, ventral. 100 μm.
c Arrangement of the flank region of the trunk. 25 μm.
d Trunk, ventral, sternal region. 25 μm.
e Coxa with finely structured club-shaped hair. 10 μm.
f Anal region, caudal, with anal plate (*above*) and telson (*arrow*). 25 μm

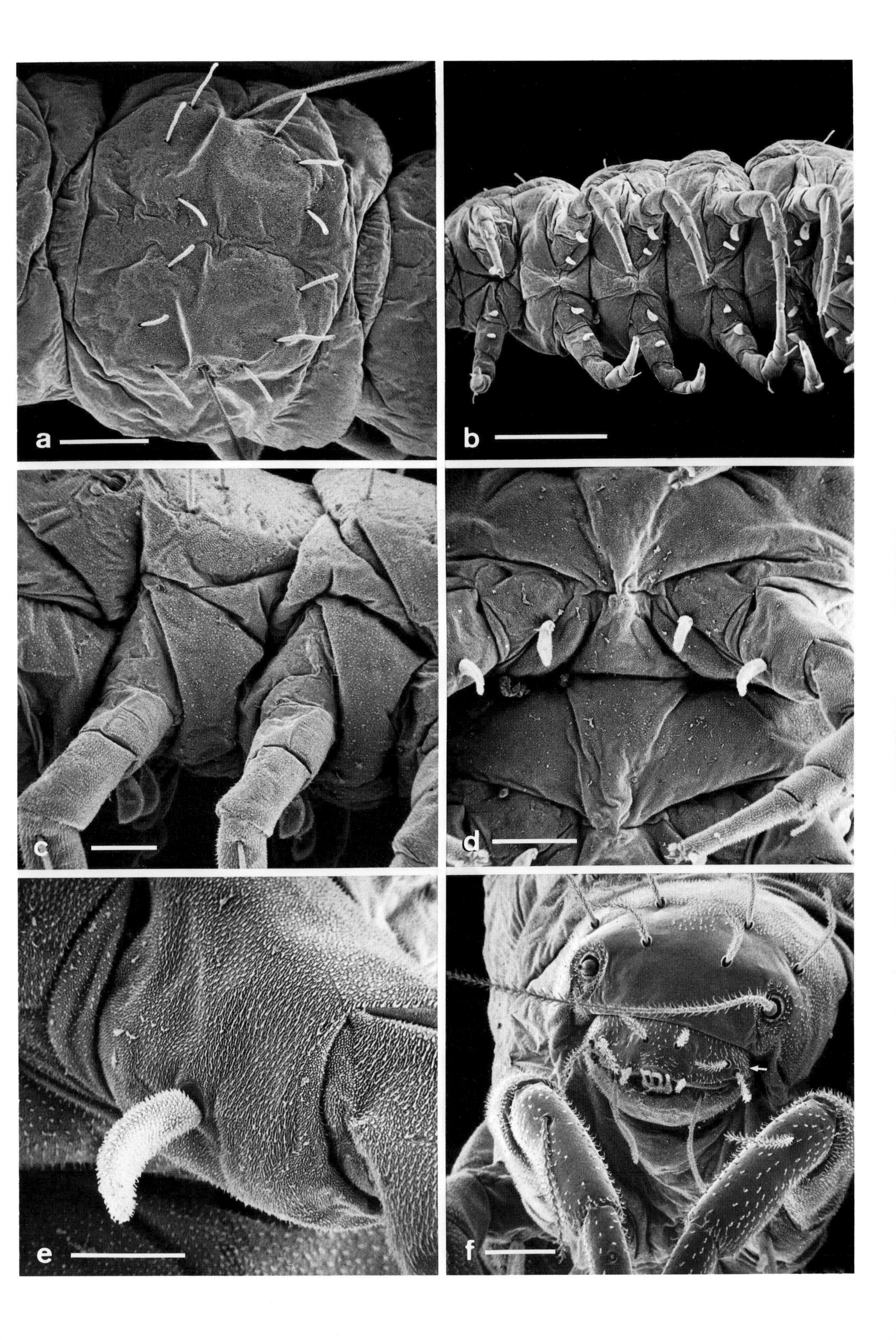
a
b
c
d
e
f

2.9 Subclass: Symphyla – (Myriapoda)

General Literature: Verhoeff, 1934

2.9.1 Characteristics of the Symphyla

The symphylans, like the pauropods, are members of the euedaphon. Their size and shape, with an average length of 4–5 mm and breadth of 0.5 mm, are well suited to subterranean life in soil pore spaces (Fig. 101; Plate 75). They are often encountered along the course of roots. The dark conditions of their biotope make eyes unnecessary, instead there are numerous mechanoreceptors distributed over their usually colourless bodies. These are supplemented by chemoreceptors and hygroreceptors.

The diet of the symphylans partly consists of dead vegetation fragments in the humus layer. In addition, they like to feed on the fine roots of young plants and are thus regarded as agricultural and horticultural pests.

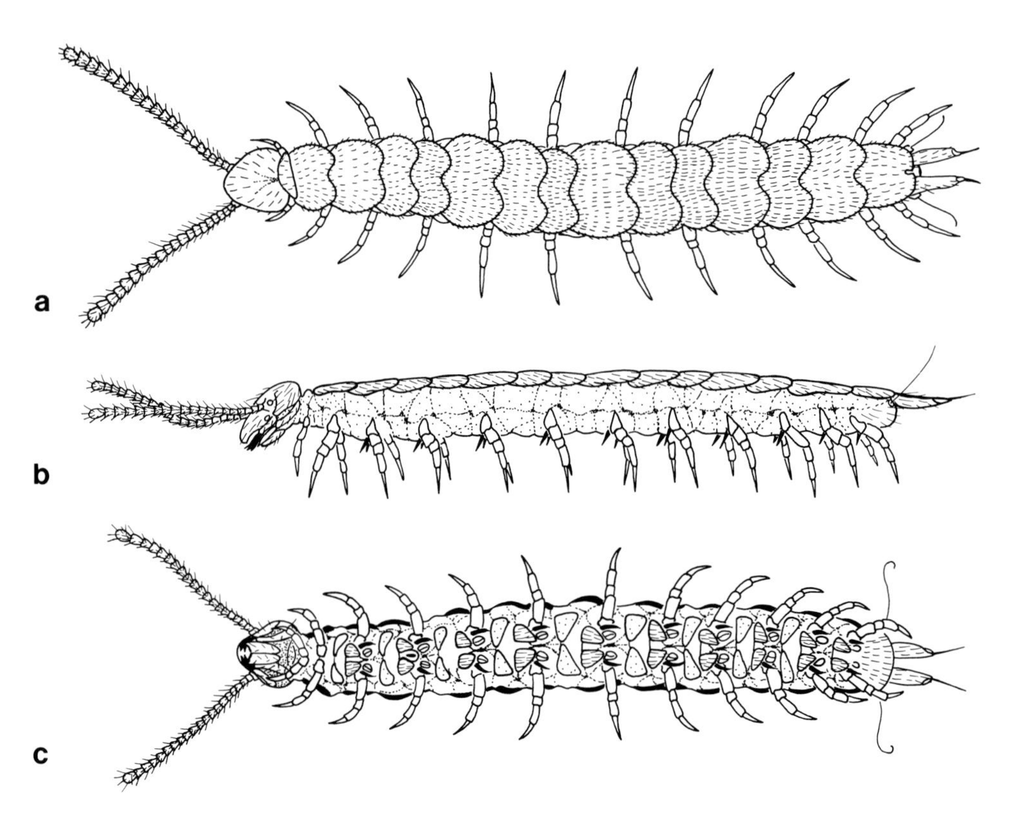

Fig. 101. *Scutigerella immaculata* (Scutigerellidae). (After Kaestner, 1963).
a Habitus, dorsal. **b** Habitus, lateral. **c** Habitus, ventral. The ventral side shows the characteristic features: 12 pairs of legs, the sternites, styli, coxal vesicles, spinning styli and trichobothria

Plate 75. *Scutigerella immaculata* (Scutigerellidae).
a Overview, lateral. 1 mm.
b Head, frontal. Eyes are absent. 100 μm.
c Trunk, lateral, with leg base and stylus. 100 μm.
d Trunk, ventral, with sternites (hairy), styli and retracted coxal vesicles (*arrow*). 100 μm.
e Trunk, dorsal, with tergites. 100 μm.
f Leg segments with scales and hair texture. 10 μm

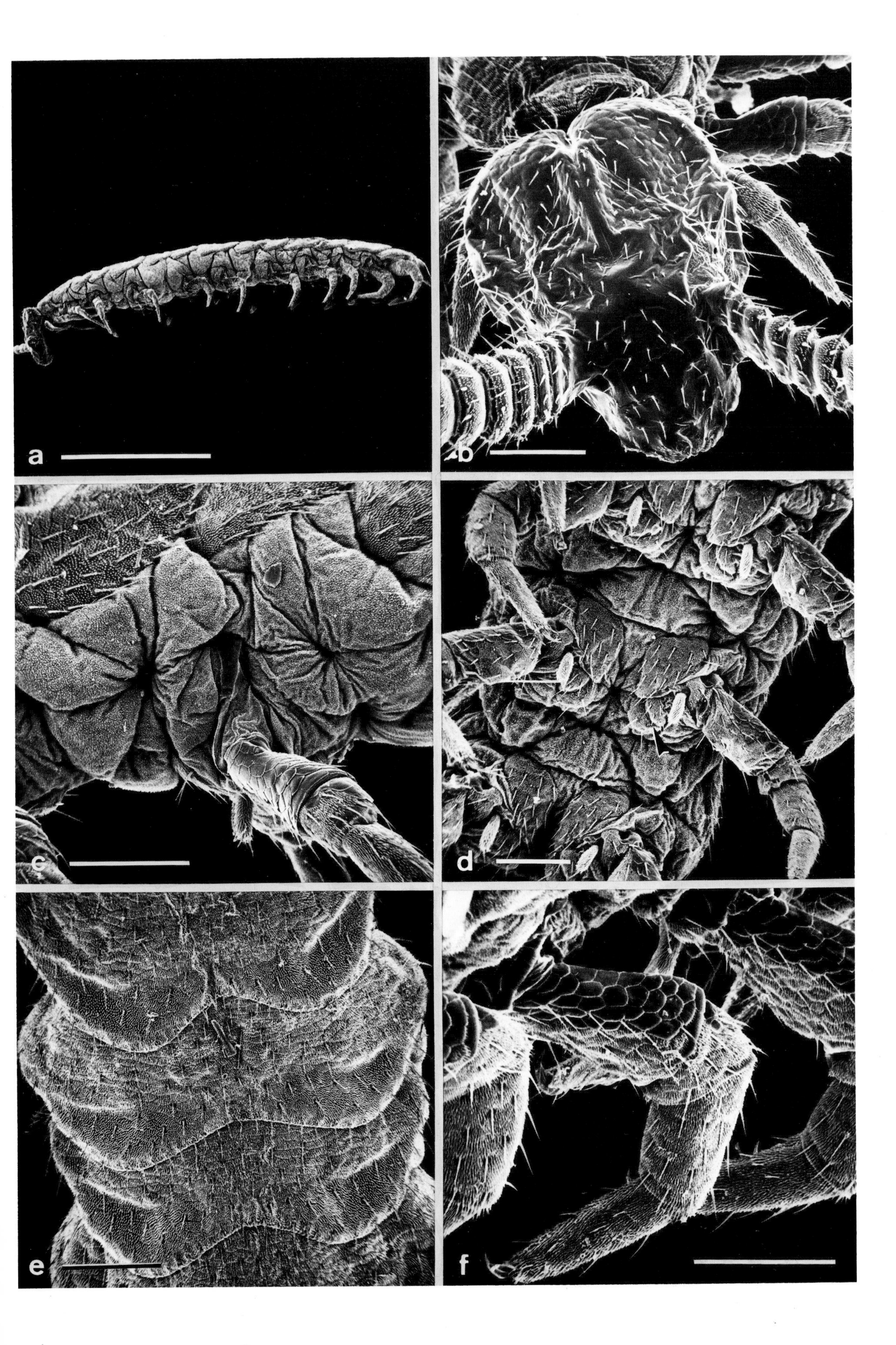
a
b
c
d
e
f

2.9.2 Moisture as an Ecological Factor – Tömösváry Organ

The Symphyla reach a high population density in humus-rich soil with sufficient moisture content. Seasonal differences in soil moisture as well as temperature fluctuations result in vertical migrations of the animals to reach favourable conditions. Warm humidity in spring and autumn favours a high population density in the upper soil layers. In the dry summer months, however, the optimal vertical distribution lies deeper the soil (Friedel, 1928; Michelbacher, 1938).

As this vertical distribution can be attributed to the active perception of the prevailing humidity conditions, a receptor sensitive to moisture was sought. Friedel (1928) believed he had found a hygroreceptor as a result of experimental investigations of the antennae and coxal vesicles. However, he overlooked the Tömösváry organ in his experiments. The ultrastructure of this organ, situated behind the antennae, makes it very probable from a functional-morphological viewpoint (Haupt, 1971) that it is the long sought moisture receptor. This conclusion is supported by comparisons with the so-called genal (temple), temporal or postantennal organs of the Myriapoda (Fig. 105), the Tömösváry organs of epedaphic chilopods, the temporal organ of the diplopods and the pseudoculi of the pauropods (Haupt, 1979). This sensory organ was first described by Tömösváry (1883) and is known as the Tömösváry organ (Fig. 102; Plate 76) after Hennings (1904, 1906).

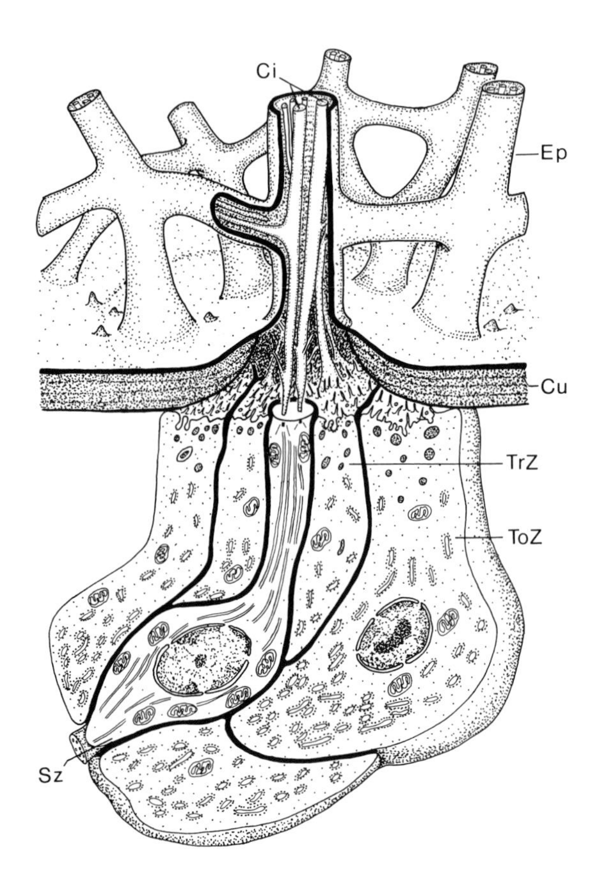

Fig. 102. Tömösváry sensory organ of *Scutigerella immaculata* (Scutigerellidae), schematic diagram of a sensory unit. (After Haupt, 1971). Two dendritic cilia of a sensory cell penetrate into the finely porous tubular system of the epicuticle (*Ep*) and divide into branches. Two envelope cells, the trichogen and tormogen envelope cell (*TrZ, ToZ*), enclose the sensory structures. Moreover, the former possesses apical processes. *Sz* Sensory cell

---

Plate 76. *Scutigerella immaculata* (Scutigerellidae).
a Head, frontal. 100 μm.
b Head, lateral, with sclerites, mouthparts and antennal base. 100 μm.
c Mouth cavity with mandibles. 30 μm.
d Antennal base with Tömösváry organ (*above*) and cranial spiracle. 50 μm.
e Tömosváry organ. 10 μm.
f Cranial spiracle. 5 μm

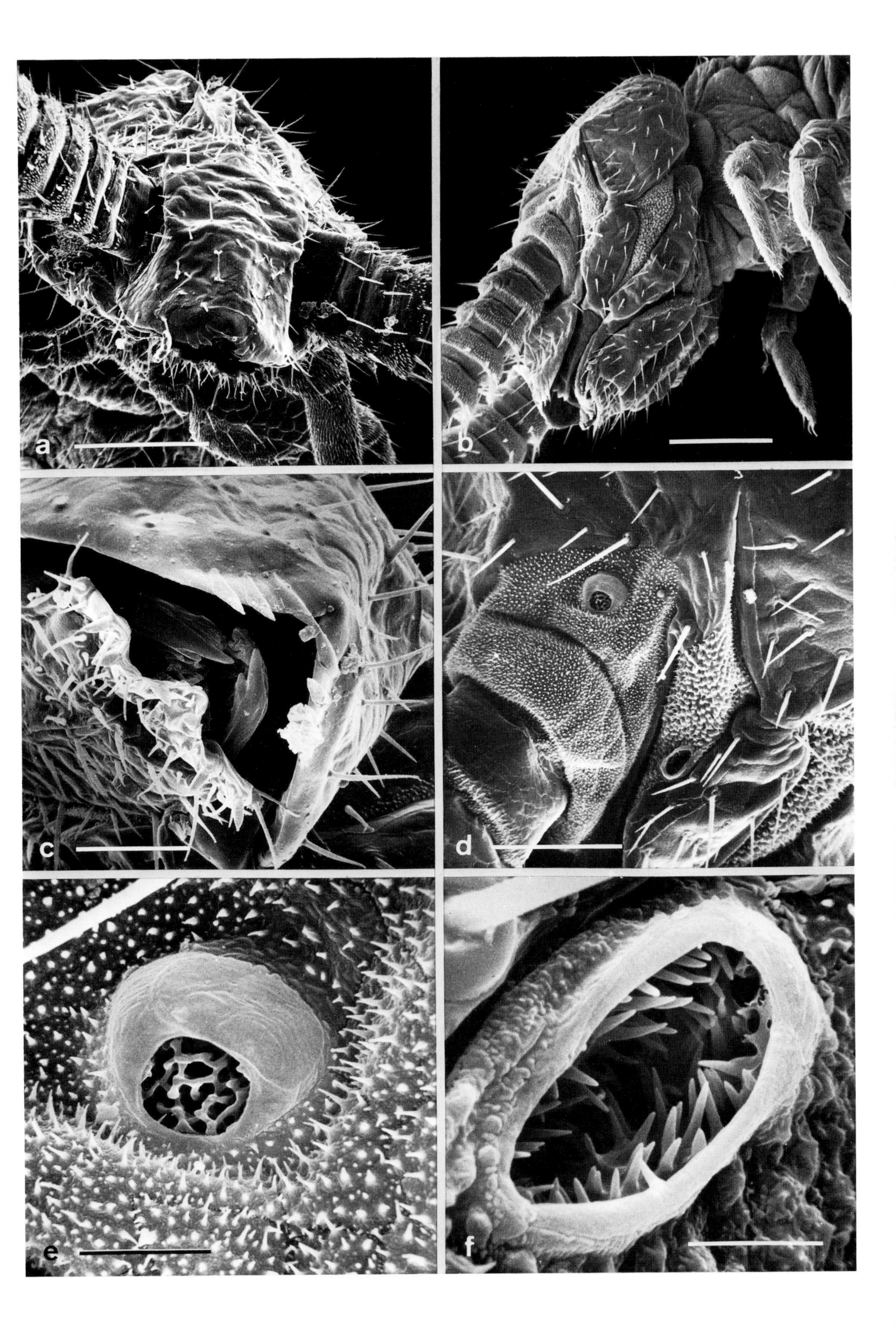
a
b
c
d
e
f

2.9.3 Moisture as an Ecological Factor – Coxal Vesicles

Each of the third to the eleventh trunk segments in *Scutigerella immaculata* bears a pair of coxal vesicles (coxal sacs) mediad from the coxae of the legs and the styli (Fig. 103; Plate 77). In their retracted state they are covered by two semi-circular flaps on which there are three, or more often four, long sensory hairs. These hairs are believed to be mechanoreceptors (Gill, 1981). Opening of the two flaps exposes the underlying vesicle epithelium covered by a thin, barely structured cuticle. A hydraulic mechanism, by which muscle power raises the haemolymphatic pressure in the body, causes eversion of the vesicles.

The fine structure of the epithelium cells, investigated by Gill (1981), indicates an active transport function. The epithelium is composed of three highly differentiated cells, each having a lobed nucleus and a pronounced basal labyrinth with rows of mitochondria parallel to it. The cell apex, however, hardly shows any infoldings. These transport cells are probably involved in ion and consequently water uptake to balance the water losses constantly suffered by the animals even in high ambient humidity and function within the framework of adaptation to seasonal variation in soil moisture.

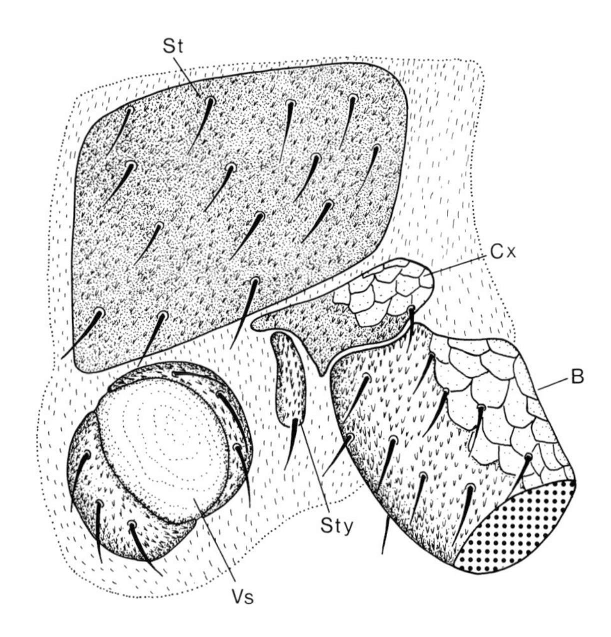

Fig. 103. *Scutigerella immaculata* (Scutigerellidae). The region around the leg base (*B*) with sternite (*St*), coxal organ with everted vesicle (*Vs*), stylus (*Sty*) and coxa (*Cx*). Beneath the thin vesicle surface lie transport cells, typical for water and ion transport. At the edge of the flaps there are glandular openings

Plate 77. *Scutigerella immaculata* (Scutigerellidae).
a Genital aperture in anterior portion of trunk. 30 µm.
b Ventral cuticle with hairy sternite and spiny membrane surface. 20 µm.
c Base of leg with stylus and coxal vesicle (*arrow*). 25 µm.
d Coxal vesicle, open, with shrivelled cuticle. 20 µm.
e Stylus. 10 µm.
f Marginal region of a coxal vesicle. 5 µm

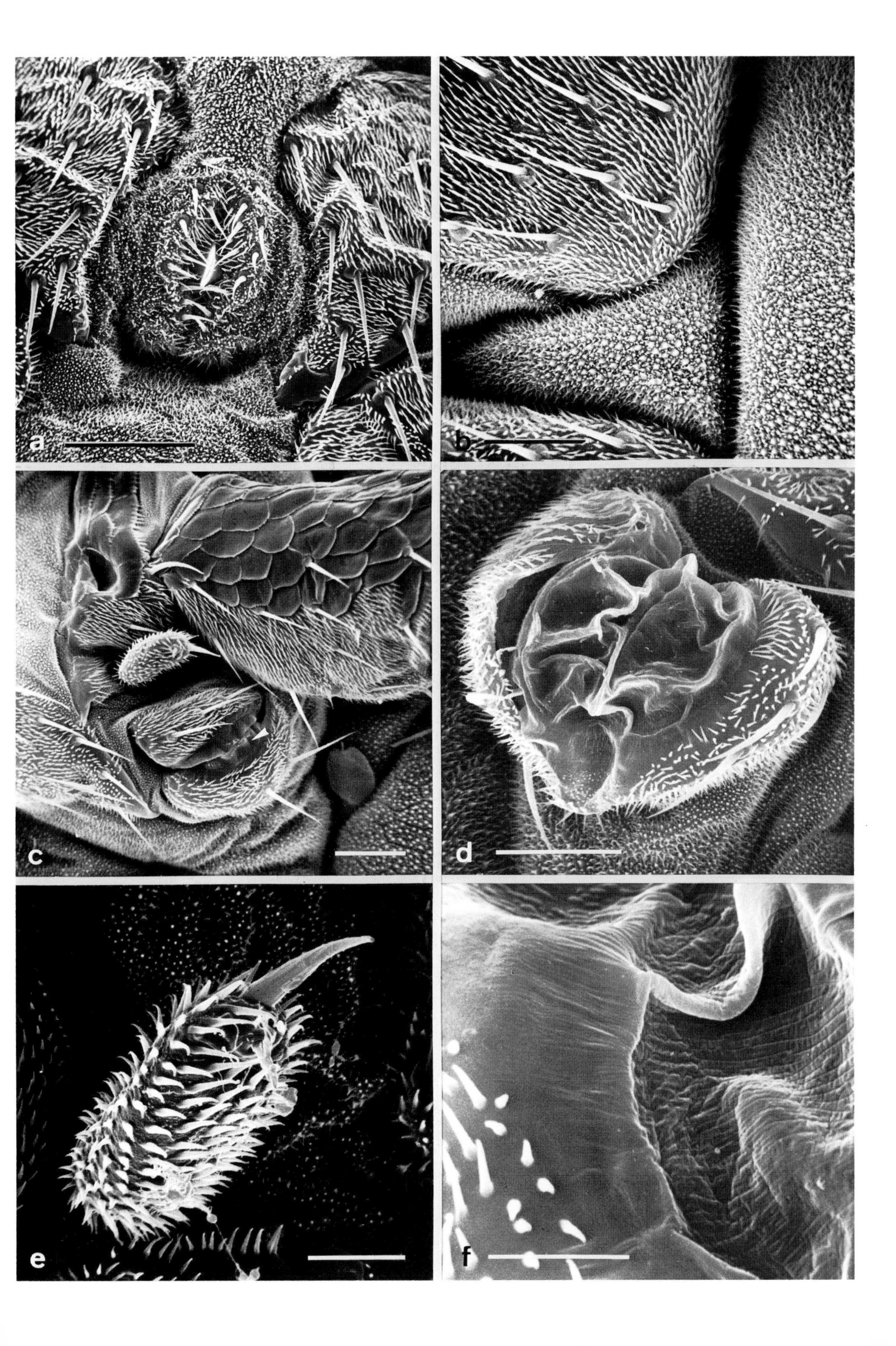
a
b
c
d
e
f

2.9.4 Indirect Transfer of Spermatophores

As in other euedaphic soil arthropods *Scutigerella* has developed a method of reproduction which ensures fertilization of the eggs without copulation. Juberthie-Jupeau (1956, 1959a, b) succeeded in analyzing the complicated sequence of behaviour shown by both sexes.

In the interval between two moults the *Scutigerella* male deposits about 150–450 spermatophores in the absence of the female. From the genital opening on the fourth trunk segment (Plate 77a) a drop is extruded onto the substrate. By lifting the front of his body he stretches it to a stalk, on the tip of which the actual spermatophore remains (Fig. 104a). Females feed on a large number of the sperm capsules (approx. 18 daily). Some are swallowed and digested and others are stored in buccal pouches. The insemination takes place during oviposition (Fig. 104b). An egg, expelled out of the genital aperture, is seized in the mouthparts and attached to the substrate. The female appears to chew at the surface of the egg, thereby inseminating the egg by transfer of sperm from the pouches.

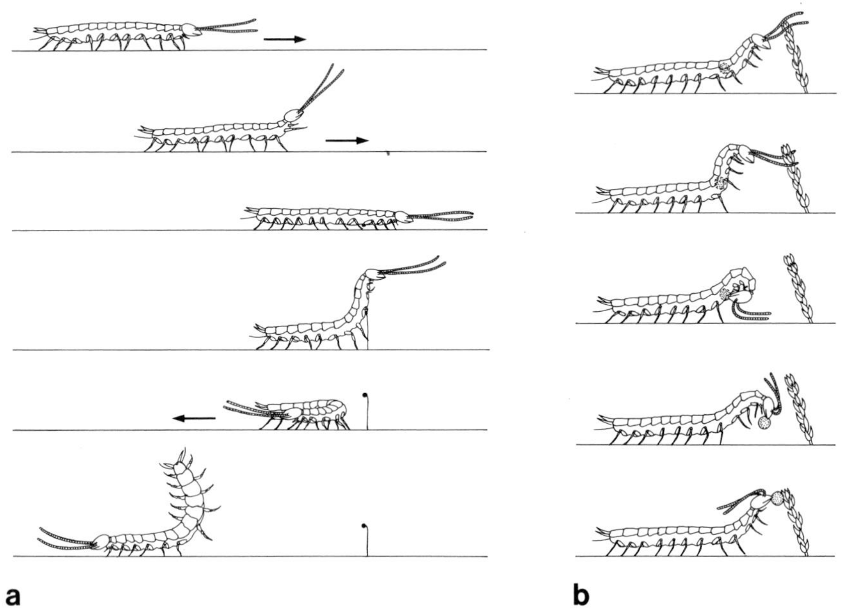

Fig. 104. Reproduction biology of *Scutigerella.* (Juperthie-Jupeau, 1959).
a Male, deposition of a stalked spermatophore from the genital aperture on the fourth trunk segment. Later the droplet of sperm is taken up by the mouthparts of the female and stored in buccal pouches. **b** Female depositing eggs. The egg from the genital aperture is received by the mouthparts and attached to the substrate. At the same time insemination is achieved by transferring sperm from the buccal pouches onto the egg surface

Plate 78. *Scutigerella immaculata* (Scutigerellidae).
a Antenna. 200 µm.
b Antenna, distal. 100 µm.
c Terminal antennal segment, sensilla. 20 µm.
d Middle portion of the antenna. 5 µm.
e Terminal antennal segment, sensilla. 5 µm.
f Middle portion of the antenna with sensilla chaetica and small trichomes. 5 µm

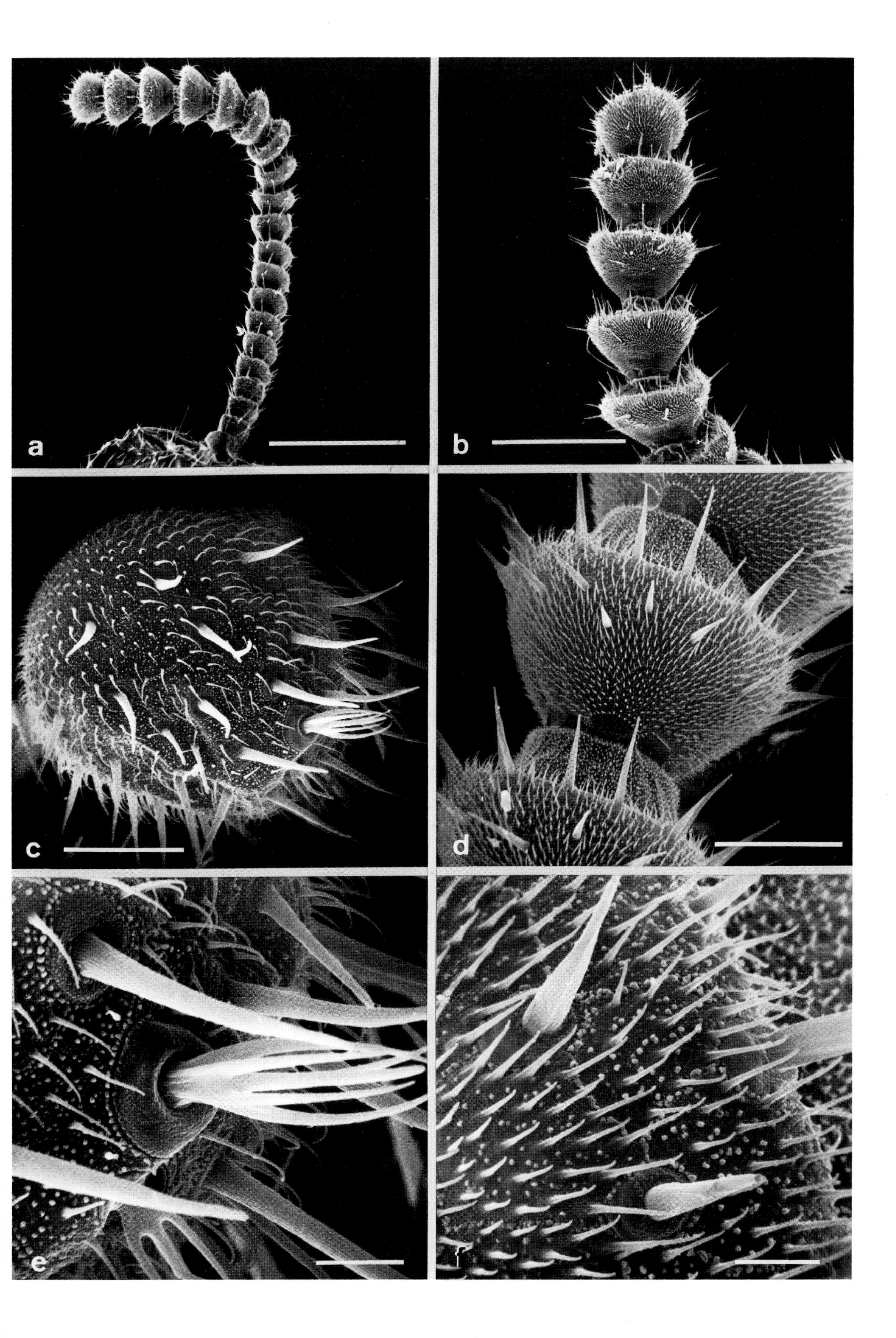
a
b
c
d
e
f

2.9.5 Protection Against Predators – Spinning Styli

Euedaphic animals have only limited freedom of movement in the pore spaces of the soil, making them easy prey for predators. They often possess defensive glands with poison or a repugnant secretion with which to repel attackers. Symphylans have silk glands which open at the tip of the abdominal spinning styli (Plate 79). The paired spinning styli, which are derived from the thirteenth trunk segment, each terminate in a spinning hair on the base of which there is a small glandular area. When threatened and hunted by predators they spin threads which evidently hinder the pursuer, at the same time offering the possibility of letting themselves drop rapidly into another soil crevice.

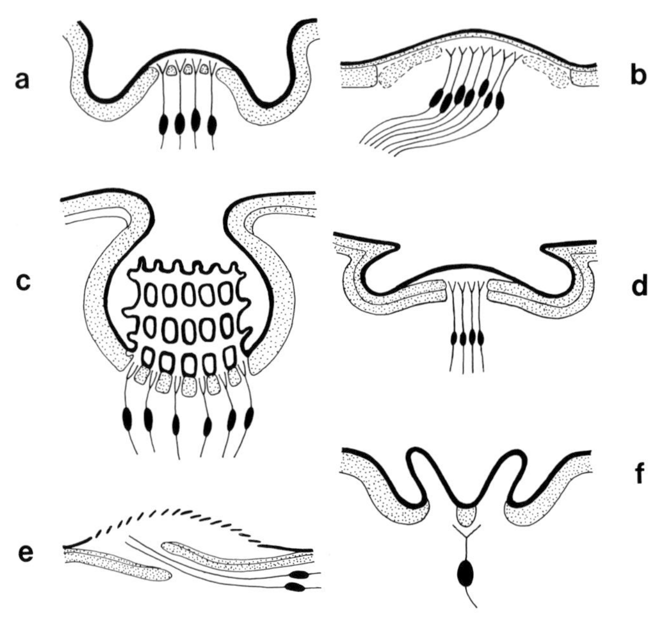

Fig. 105. Structure of the temporal sensory organs of a variety of soil arthropods. (After Haupt, 1979). The processes >—— of the sensory cells extend to just below the thin epicuticle (*thick black line*). In addition to variations in the outer structure to increase the surface area, the number of sense cells and cilia also varies. The micromorphology of these sensory organs suggests that they function as chemoreceptors in the broadest sense.
a *Glomeris* (Diplopoda); **b** *Allopauropus* (Pauropoda); **c** *Scutigerella* (Symphyla); **d** *Lithobius* (Chilopoda); **e** *Eosentomon* (Protura); **f** *Onychiurus* (Collembola)

Plate 79. *Scutigerella immaculata* (Scutigerellidae).
a Trunk, caudal, lateral, with spinning styli. Lateral from the base of each stylus lies a trichobothrium, recognizable as a long fine thread. 100 µm.
b Spinning styli with glandular areas. 30 µm.
c Spinning stylus, terminal, with glandular area. 10 µm.
d Spinning stylus, terminal, glandular area. 3 µm.
e Leg, distal, with claw. 10 µm.
f Leg surface with scales and hair texture. 20 µm

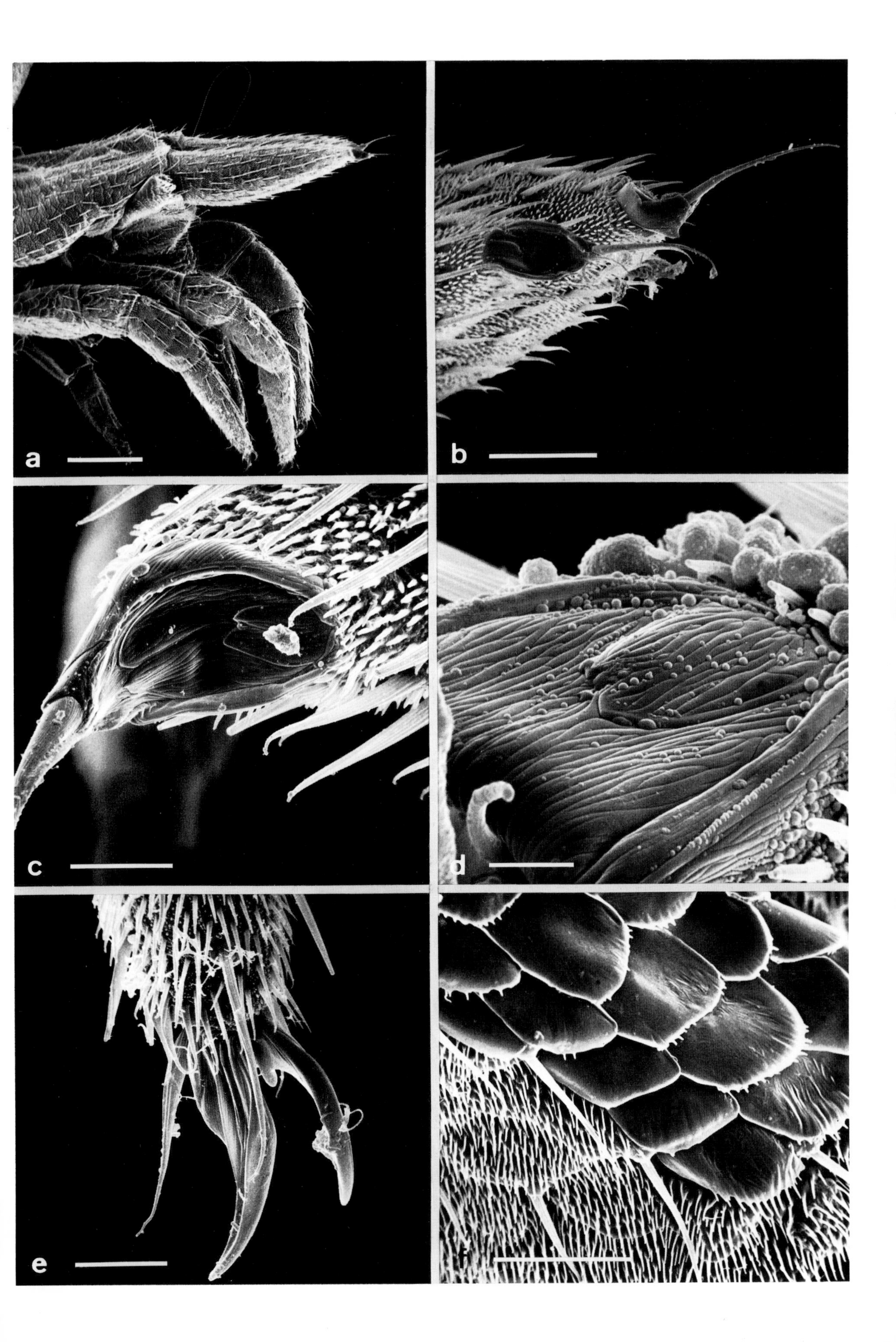

a
b
c
d
e
f

2.10 Order: Diplura – Doubletails (Insecta)

GENERAL LITERATURE: Paclt, 1956; Palissa, 1964

2.10.1 Characteristics of the Diplura

The Diplura or doubletails were formerly assigned together with the Protura, Collembola and Thysanura to the Apterygota, regarded as primitive insects at the beginning of the class Insecta. They do not, however, belong to a natural systematic unit. According to the phylogenetic system of classification the primary wingless insects of the orders Diplura, Protura and Collembola are grouped together to form the subclass of the Entognatha.

The soil is the preferred biotope of the Entognatha. The (mostly 3–10-mm) elongate Diplura lack eyes, but are well equipped with sensory hairs. They live in moss, under bark or stones, in deeper litter, but also in the moist pore spaces of the lower soil layers. They are often encountered together with symphylans. The dipluran family of the Campodeidae is widely distributed. Just over ten campodeid species reach the far north of the temperate regions of northern and central Europe. The Japygidae family prefers the warmer zones of southern Europe as well as subtropical and tropical climates (Paclt, 1956; Palissa, 1964; Bareth, 1986).

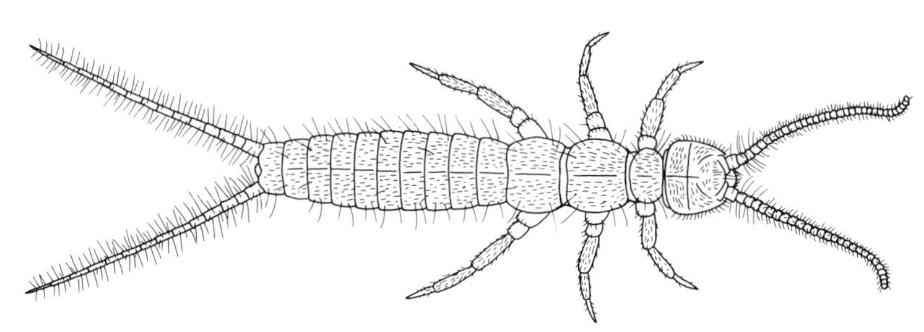

Fig. 106. Habitus of a *Campodea*, dorsal aspect, with its abundantly segmented antennae and cerci. Both pairs of appendages are held in constant oscillating movement. Eyes are lacking. (After Handschin, 1929)

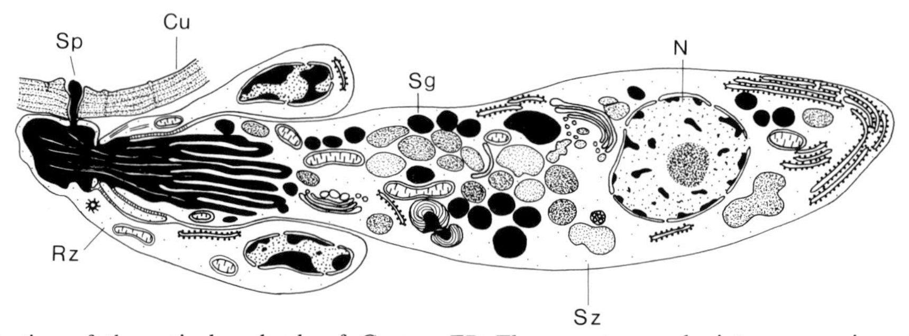

Fig. 107. Organization of the cuticular glands of *Campodea*. (After Juberthie-Jupeau and Bareth, 1980a). The glandular cells (*Sz*) are rich in dictyosomes and granular ER. They secrete granules into a reservoir, out of which the secretion emerges from the cuticle (*Cu*) through a pore (*Sp*). *Rz* reservoir cell (cf. Plate 84f)

Plate 80. *Campodea* sp. (Campodeidae).
a Overview, lateral. 1 mm.
b Head with antennae, dorsal. 0.5 mm.
c Head, pro- and mesothorax, dorsal. 200 μm.
d Cerci, proximal, dorsal. 200 μm.
e Head capsule and antennal bases, dorsal. Eyes are lacking. 100 μm.
f Head, pro- and mesothorax, ventral. 200 μm

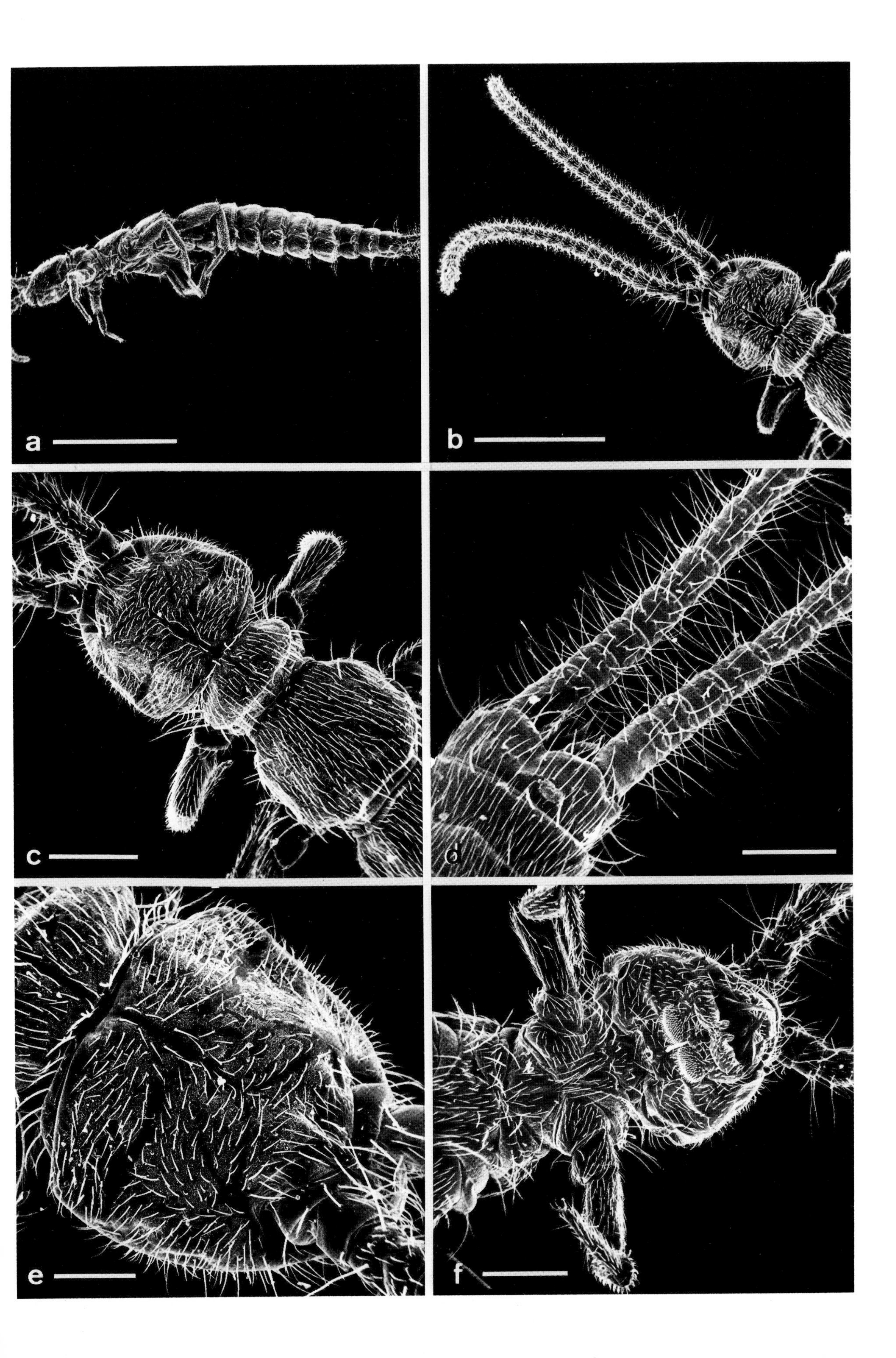
a
b
c
d
e
f

2.10.2 Sensory Organs on the Antennae of *Campodea*

As appropriate for their thigmotactic habits, the campodeids have antennae bearing abundant sensilla. On one antenna, comprising an average of 25 segments, a total of around 2.800 sensilla were determined (Endres, 1980).

The basal segments of the antennae, scapus and pedicellus, are relatively poorly endowed with sensilla. The medial segments exhibit a uniform pattern, while the terminal segment bears the especially striking pit organ with its cupuliform sensilla (Plate 81). Besides the trichobothria and macrochaeta in the region of the antennal base and the cupuliform sensilla, 12 further sensilla types are recognizable when size, surface structure, shape and articulation of the hair shaft are considered (Fig. 108). About 90% of them are classified as sensilla trichodea, presumably with a mechanoreceptive or multi-modal function. The others can be considered as chemoreceptors. Fine structural results indicate that the cupuliform pit organ (Plate 81f, g) is an olfactory receptor (Juberthie-Jupeau and Bareth, 1980b; Bareth, 1983). The fine feathery trichobothria at the base of the antennae (Plate 81b–d) are regarded as receptors for the perception of slight air movements. Disturbance of the air causes *Campodea* to retreat immediately.

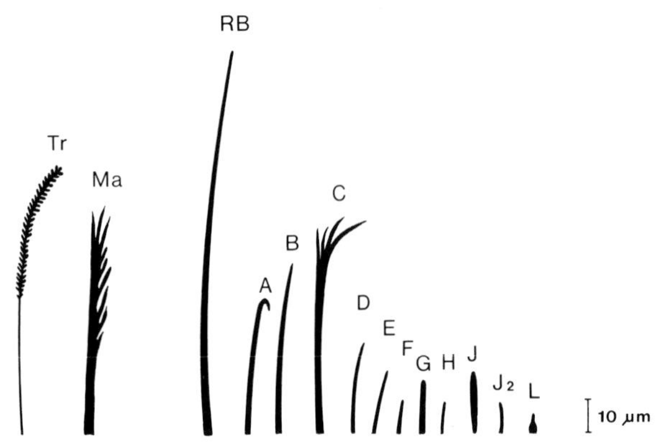

Fig. 108. Shape of hairs and sensilla on the antenna of *Campodea* (Campodeidae). (Endres, 1980). *Tr* Trichobothria; *Ma* macrochaeta (on the basal segments of the antenna, presumably vibrational and tactile (mechano-)receptors; *RB* (giant bristles), types *A, B, C, D* (sensilla trichodea distributed over the antenna, presumably mechanoreceptors); *E, F* (small sensilla trichodea, especially at joint membranes, presumably postural hairs); *G, J,* J_2*, L* (sensilla basiconica, few in number, situated in exposed places, presumably chemoreceptors); *H* (sensilla trichodea on the scapus and pedicellus, presumably identical to type *D*, see above). The apical pit organ was not included

Plate 81. *Campodea* sp. (Campodeidae).
a Antennal bases, dorsal. 50 µm.
b Antennal segments 4–6. The *arrows* point to the trichobothria. 50 µm.
c Trichobothrium, distal. 2 µm.
d Trichobothrium proximal. 2 µm.
e Sensillum trichodeum (thread hair), proximal, from the base of the cercus. The bristle cup is narrowed to allow a directional hair movement. Beside the hair base lies the outlet of a typical dermal gland (*arrow*). 5 µm.
f Antenna, distal, with pit organ. 20 µm.
g Pit organ with four cupuliform sensilla. 3 µm

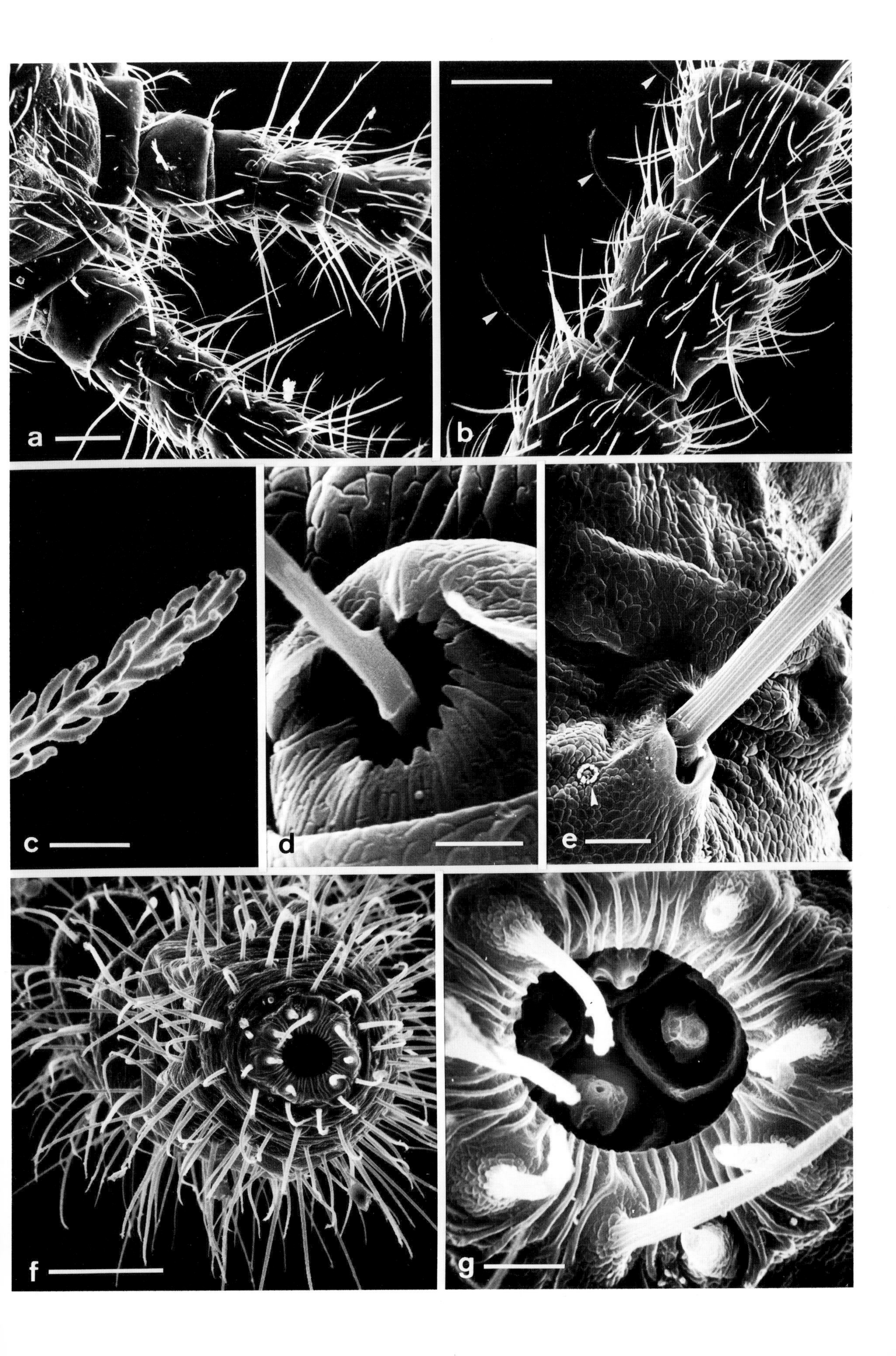
a
b
c
d
e
f
g

2.10.3 Morphology of the Head of *Campodea*

As in the Collembola and Protura the mandibles and maxillae are inserted in buccal pouches on the eyeless head capsule so that only the apical part protrudes (Plate 82). Between them lies a tongue-like structure, the lingua, which is regarded as part of the hypopharynx (Francois, 1970). Its apical surface is covered by cuticular scales, reminiscent of the radula of a snail.

Of the mouthparts only the lower lip (labium) is freely exposed. It connects the oral folds at the sides of the head which form the buccal pouches (Fig. 109). The labium is greatly modified and divided into several parts. Most striking are two flat areas covered with bristle-like hairs, which are interpreted as labial palps (Plate 82). The fine structure of these palp hairs was described by Bareth and Juberthie-Jupeau (1977). Each hair is innervated by seven to ten sensory cells. The typical structural elements of mechanoreceptors and contact chemoreceptors are combined in each hair.

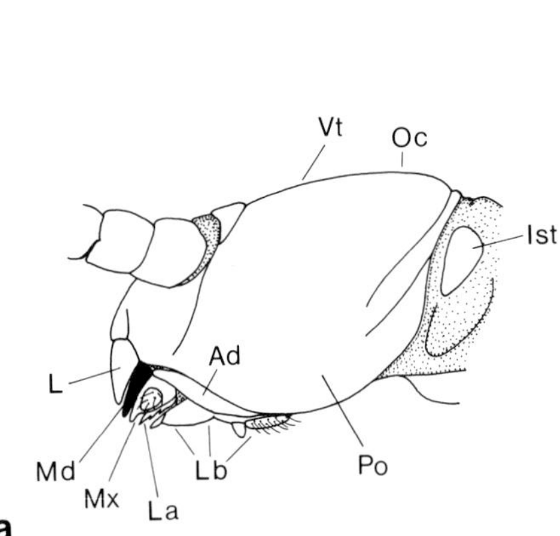

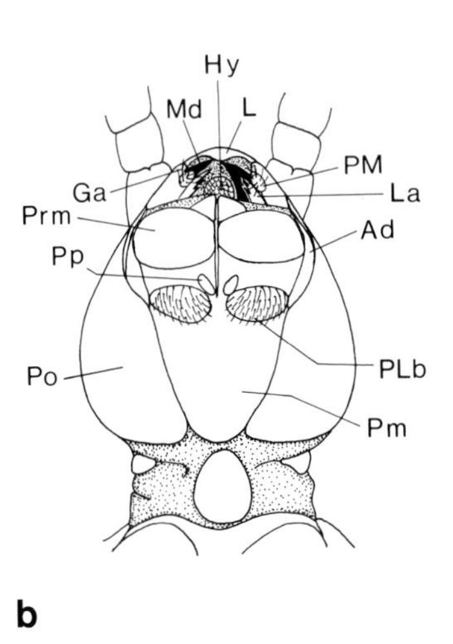

Fig. 109. Morphology of the head capsule of *Campodea chardardi* (Campodeidae). (After Francois, 1970).
a Side view; **b** ventral view. *Ad* Admentum; *Ga* galea; *Hy* lingua of the hypopharynx; *Ist* intersternite; *L* labrum; *La* lacinia; *Lb* labium; *Md* mandible; *Mx* maxilla; *Oc* occiput (posterior of head); *Po* plica oralis (oral fold); *PLb* palpus labialis; *Pm* postmentum; *PM* palpus maxillaris; *Pp* processus palpiforme; *Prm* prementum; *Vt* vertex

Plate 82. *Campodea* sp. (Campodeidae).

a Mouth cone, frontal. The *arrow* marks the lingua of the hypopharynx. Beside it lie the maxillae, above the upper lip (labrum). 50 μm.

b Mouth cone, oblique ventral, the toothed laciniae (*arrow*) are easily recognizable. 50 μm.

c Mouth cone with stiletto-shaped laciniae. 50 μm.

d Lingua of the hypopharynx, distal, with texture of scales. 5 μm.

e Brush-like organ of the processus palpiforme on the lower lip (labium) between the labial palps. 25 μm.

f Surface of labial palp with sensory hairs and dermal gland outlet (*arrow*) (cf. Fig. 107). 5 μm.

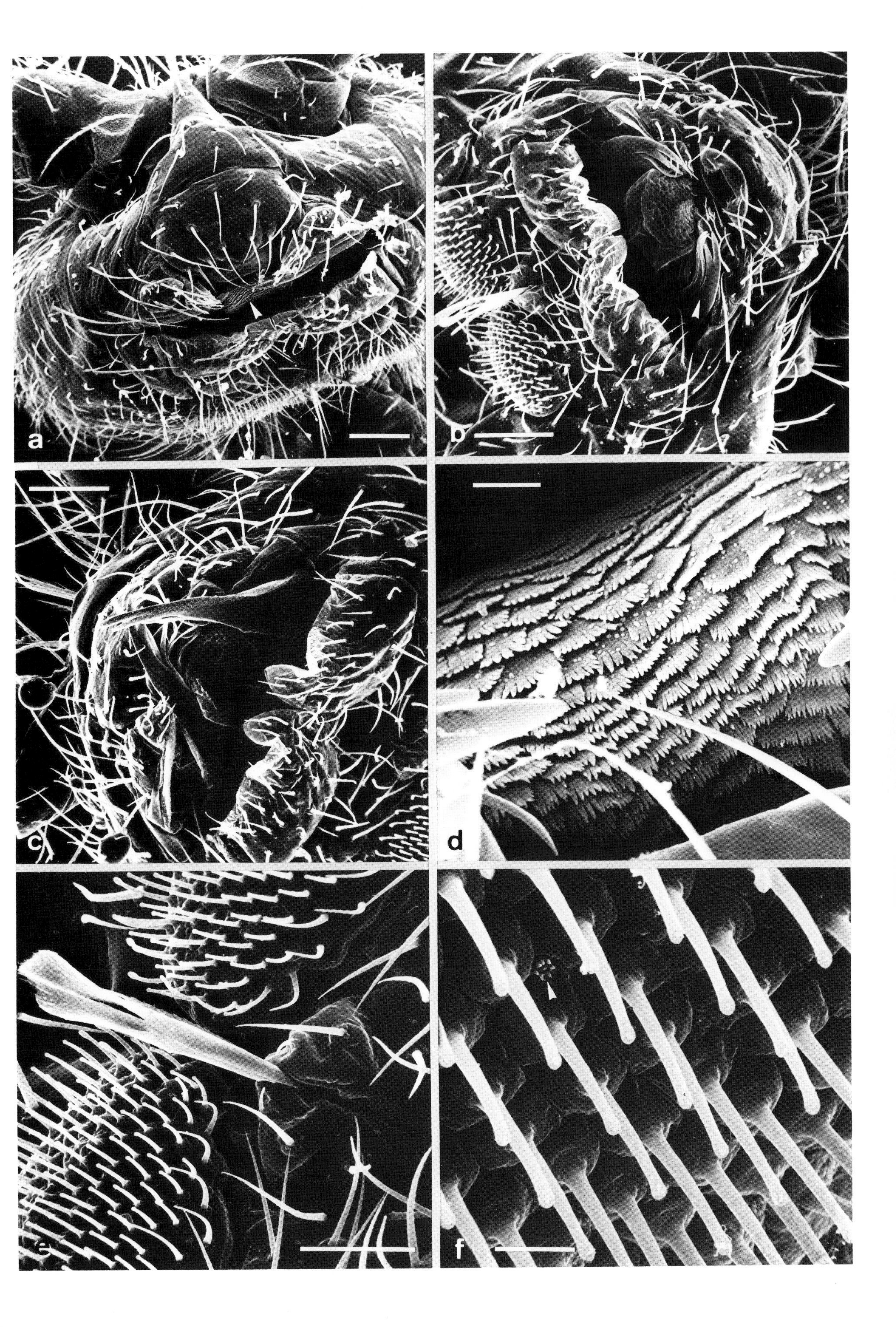

a
b
c
d
e
f

2.10.4 Respiration of the Diplura

Air is the respiration medium of the soil-dwelling animals. When the ground is waterlogged the soil arthropods are either forced to leave their biotope or they possess hydrophobic structures which enable the formation of a plastron to keep the spiracles of the open tracheal system water-free and ensure temporary gas exchange at the air-water boundary. The body surface of the euedaphic soil arthropods is frequently completely or partly covered with hydrophobic structures with emerging microstructures for the formation of a plastron.

The Diplura have an open tracheal system, *Campodea* possesses three pairs of thoracic spiracles, while *Japyx* has four pairs on the thorax and seven on the abdomen (Fig. 110). Tracheal branching shows primitive characters with initial and simple anastomosis in the head of *Campodea* and between the the eighth and ninth abdominal segments of *Japyx* (Paclt, 1956).

As *Campodea* has so few spiracles, cuticular exchange presumably makes a considerable contribution to respiration. They are considered extremely hygric on account of their high transpiration rate (average loss of water mass 77.4%/h at 0% r.h./22 °C; turnover rate for the exchange of body water in high ambient humidity 6–7 h) (Eisenbeis, 1983a, b). This is linked to a high cuticular permeability. *Japyx*, on the other hand, is considered to be more thermophilic and to be mesic in its humidity requirements.

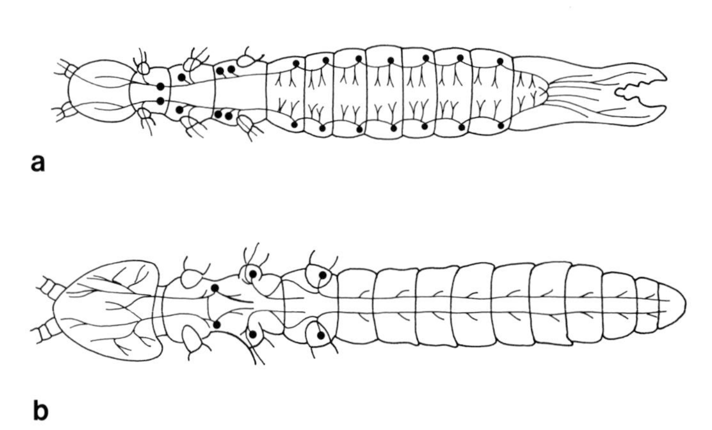

Fig. 110. The tracheal system of *Japyx* (**a**) and *Campodea* (**b**). (After Paclt, 1956)

Plate 83. *Campodea* sp. (Campodeidae).

a Pleural region between the pro- and mesothorax with spiracle (*arrow*). The ventral sclerites are covered with macrochaeta. 100 µm.

b Ventro-lateral thoracic sclerites with macrochaeta and filiform hairs, between them lie dermal glands. 25 µm.

c Spiracle on the mesothorax. 25 µm.

d Thorax, ventral. 100 µm.

e Open spiracle with view into the tracheal intima. Near the rim of the spiracle lies the ring structure of a dermal gland. 5 µm.

f Claw. 10 µm

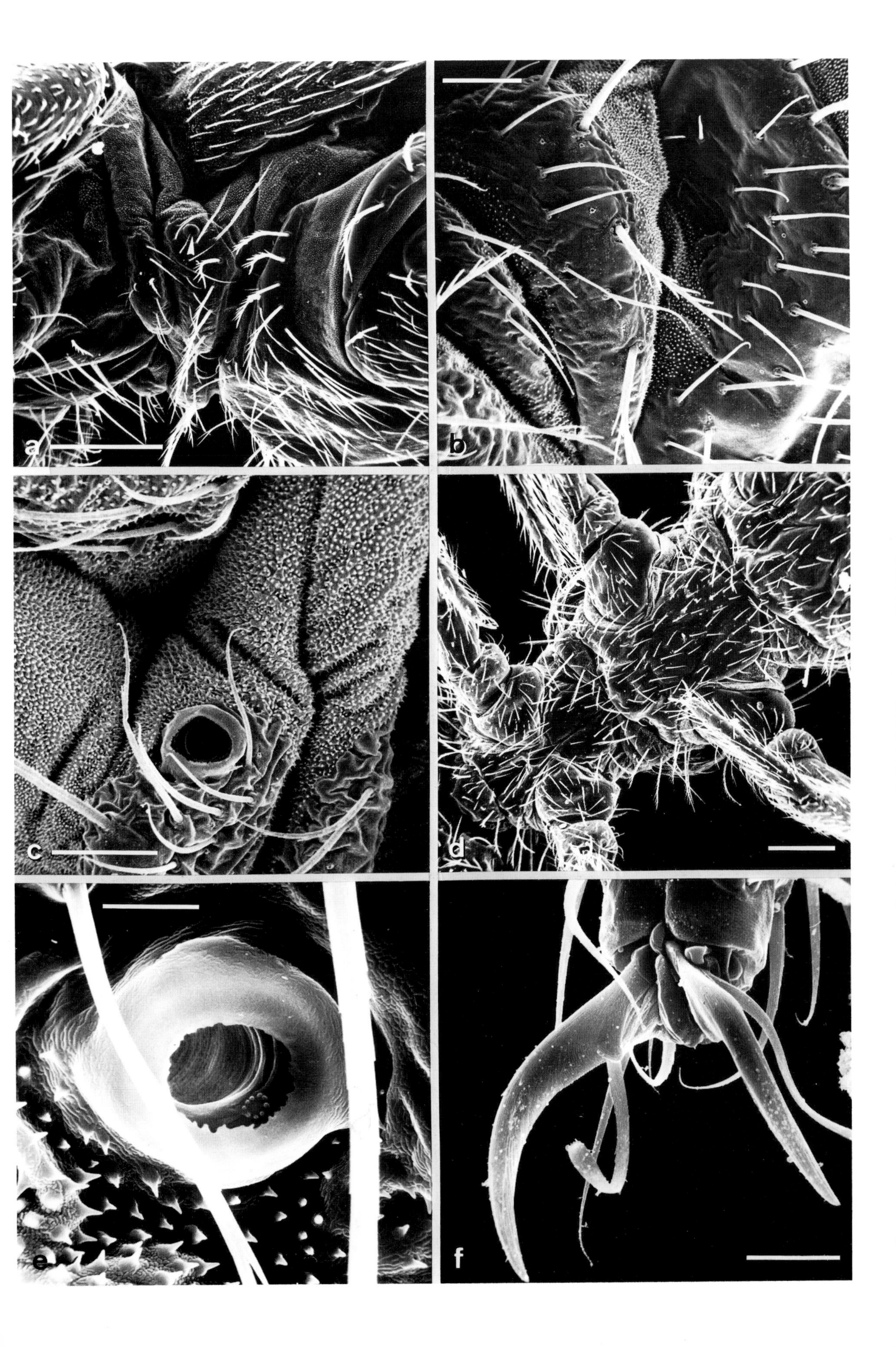

2.10.5 Osmoregulation in the Diplura

As in the Symphyla, Collembola, Protura and Thysanura, the Diplura also possess water-absorbing structures with which water can be taken up from the substrate. In this case they are delicate vesicles which are everted from the ventral posterior margin of the segments (Plate 84).

The ultrastructure of the tranport epithelium of the vesicles is, in principle, consistent with the structures found in the above-mentioned groups. However, a gradual modification of the epithelium is noteworthy. Eisenbeis (1976) described three regions on the epithelium of *Campodea staphylinus* (Fig. 111). Regions A and B resemble the usual scheme of transport cells, while region C at the base of the vesicle shows a transition to the normal epidermis. Weyda (1976, 1980) reported similar findings for *Campodea silvestrii* and *C. franzi*. The internal arrangement is almost identical to the external zones on the cuticle illustrated in Plate 84b.

The histochemical evidence of chloride ions in the transport epithelium also indicates a close relationship between water and ion transport (Eisenbeis, 1976). No quantitative measurements of the net absorption rates through the coxal vesicles, as carried out in Collembola and machilids, are at present available for this group.

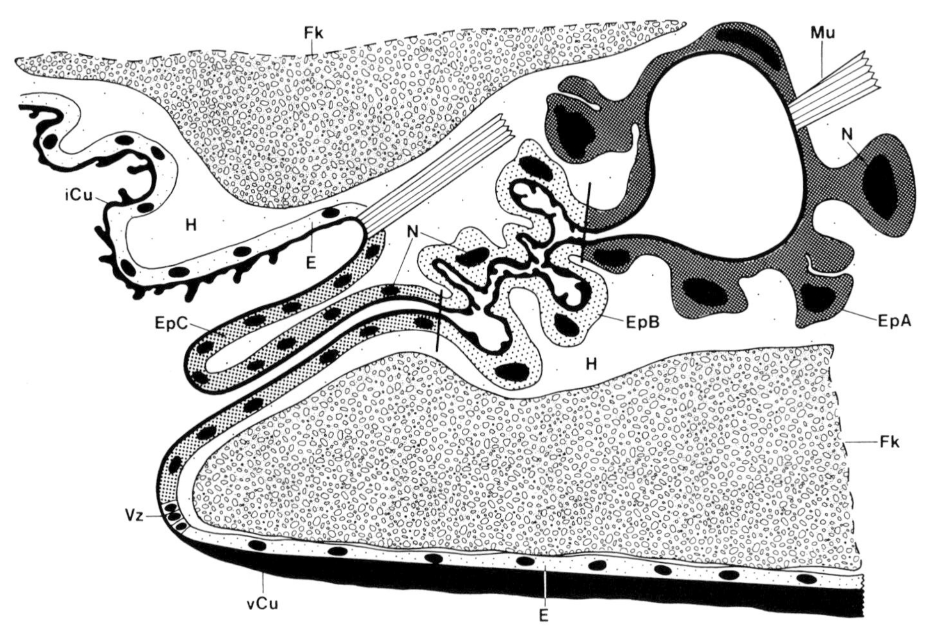

Fig. 111. Diagrammatic sagittal section through a retracted coxal vesicle in the body cavity (*H*) of *Campodea* (Eisenbeis, 1976). The vesicle is composed of three types of epithelium: *Ep A, B, C.* The cells of *Ep A* and *B* show the features of typical transport cells with the perikarya of *Ep A* floating freely in the haemolymph. Epithelium *C* forms a transition zone between the transport cells and the normal epidermis. *E* Epidermis; *Fk* fat body; *iCu* intersegmental cuticle; *vCu* ventral, sternal cuticle; *Mu* retractor muscle for the vesicle; *N* nucleus; *Vz* connective cells

---

Plate 84. *Campodea* sp. (Campodeidae).

a Abdomen, ventral, with everted coxal vesicles and styli. 100 μm.

b Coxal vesicle with differentiated cuticle. The transport epithelium is located in the distal region beneath the smooth and granular surface. 20 μm.

c Genital papilla of a male on the posterior margin of the eighth sternite. 25 μm.

d Cuticular structure of the intersegmental membrane. 5 μm.

e Lateral view of abdominal segments. The styli are inserted ventrally. 100 μm.

f Surface of sclerite with typical dermal gland outlet (cf. fine structure in Fig. 107). 1 μm

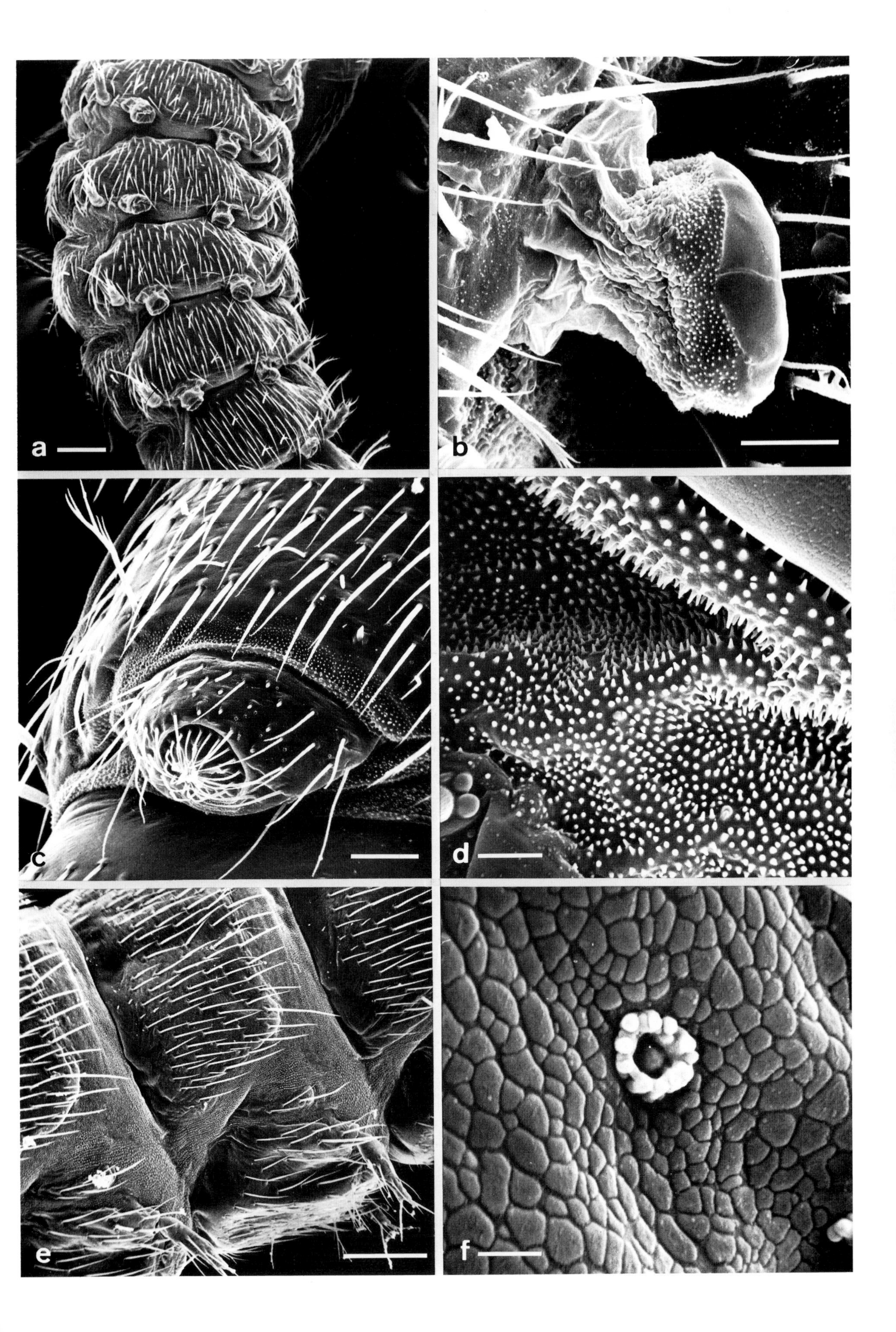

a
b
c
d
e
f

2.10.6 Adaptations of the Body of Euedaphic Diplura

Typical features of adaptation to their biotope, in particular, certain functional-morphological adaptations with respect to the abiotic factors of soil pore volume and air humidity, are recognizable in the body of the Diplura.

The worm-like, uniformly segmented appearance of these colourless animals, 3–10-mm long and less than 0.5-mm thick (Figs. 106 and 112; Plates 80 and 85), is characteristic. This shape enables the Diplura to penetrate 10–20 cm or deeper into the fine interstitial and pore system of the soil. The segmented extremities can be angled to lie close to the body if necessary. The long antennae, reminiscent of a string of beads, emphasize their well-adapted shape. Even the cerci of the campodeids are constructed like the antennae. They are covered with numerous sensilla trichodea (Plate 81e) and greatly extend the immediate radius of orientation of the eyeless animals. In the Japygidae they are modified to strong pincers but are still in keeping with the worm-like shape of the animals.

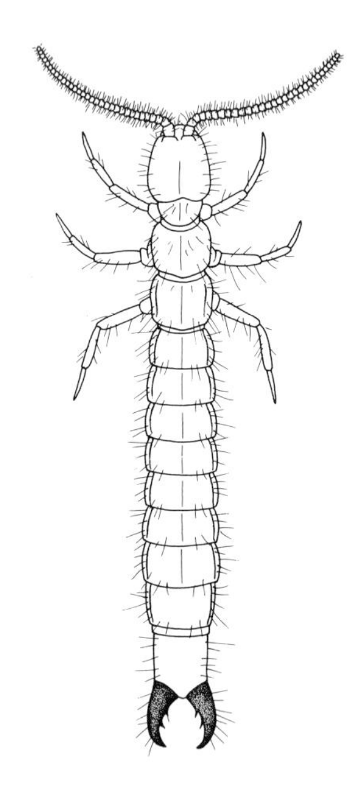

Fig. 112. Habitus of a japygid dipluran with segmented antennae and pincer-like cerci. (After Palissa, 1964)

Plate 85. Japygidae.
a Habitus, dorsal. 1 mm.
b Head and thorax, dorsal. 1 mm.
c Head capsule, lateral. 250 μm.
d Antennal bases, dorsal. Eyes are lacking. 250 μm.
e Head and prothorax, ventral. 0.5 mm.
f Region of the mouthparts with labium. 100 μm

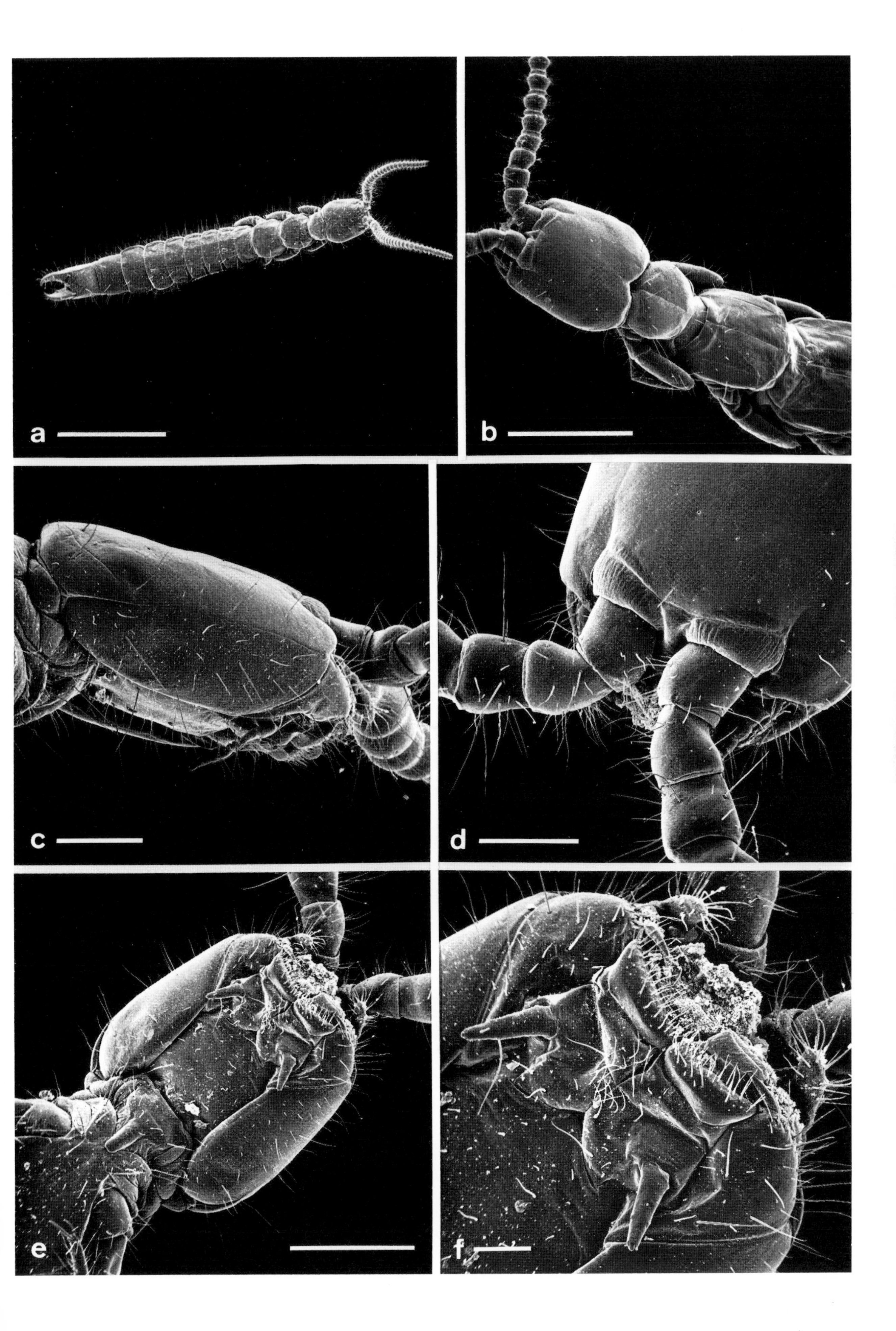

a
b
c
d
e
f

2.10.7 Predatory Life of the Diplura

Very few Entognatha are predators. But some of the Diplura, however, belong to the carnivores and feed on the smallest soil arthropods and Oligochaeta. *Campodea lankestri* has been observed to consume dipteran larvae (Marten, 1939), although campodeids are predominantly saprophytophagous and microphytophagous. The japygids, however, are decidedly predatory. Besides symphylans they mainly prey on Collembola (Simon, 1964) and even the caustic secretion from the pseudocelli of euedaphic Onychiuridae does not deter them. They also feed on campodeids (Kosaroff, 1935). The blind japygids perceive their prey by tactile stimulus of the sensory hairs on their long antennae. They crawl to and fro attempting to detect the body of their prey in the pore system of the soil. Then they grasp it rapidly in their mouthparts, bend their abdomen over their head and seize the prey in the powerful cerci which are modified to form pincers. In order to consume the prey the abdomen is once again bent over to the front presenting the food to the mouthparts (Fig. 113) (Schaller, 1949, 1962).

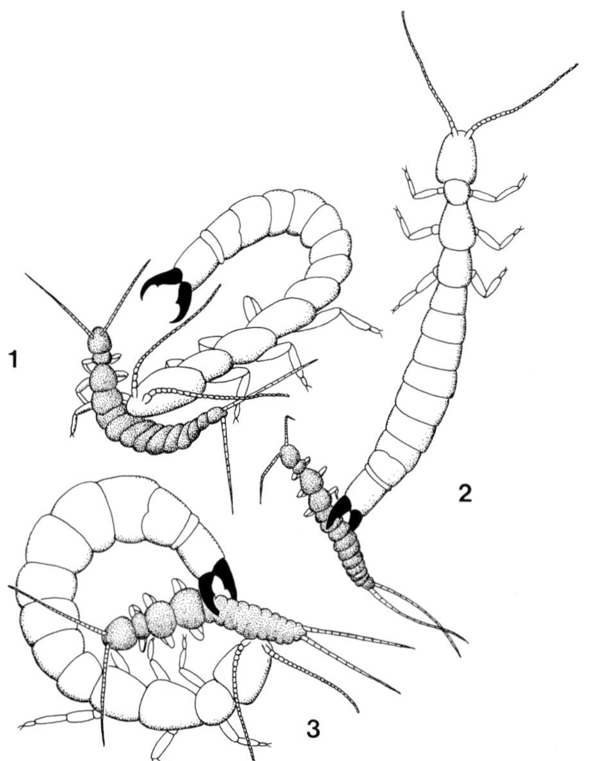

Fig. 113. Predatory behaviour of a japygid dipluran. (After Kosaroff, 1935).
1 Attack on a *Campodea* with raised abdomen. **2** Seizure of the prey with the pincers. **3** Transfer to the mouth and consumption of the prey

Plate 86. Japygidae.
a Abdomen, ventral, with coxal vesicles and styli. 0.5 mm.
b Abdomen, terminal, dorsal, with cerci. 1 mm.
c Stylus and coxal vesicle. 50 μm.
d Cerci (pincers) dorsal. 250 μm.
e Abdomen, caudal. 100 μm.
f Abdomen, caudal, with medial pincer base. Dorsal lies the anal operculum (*arrow*). 25 μm

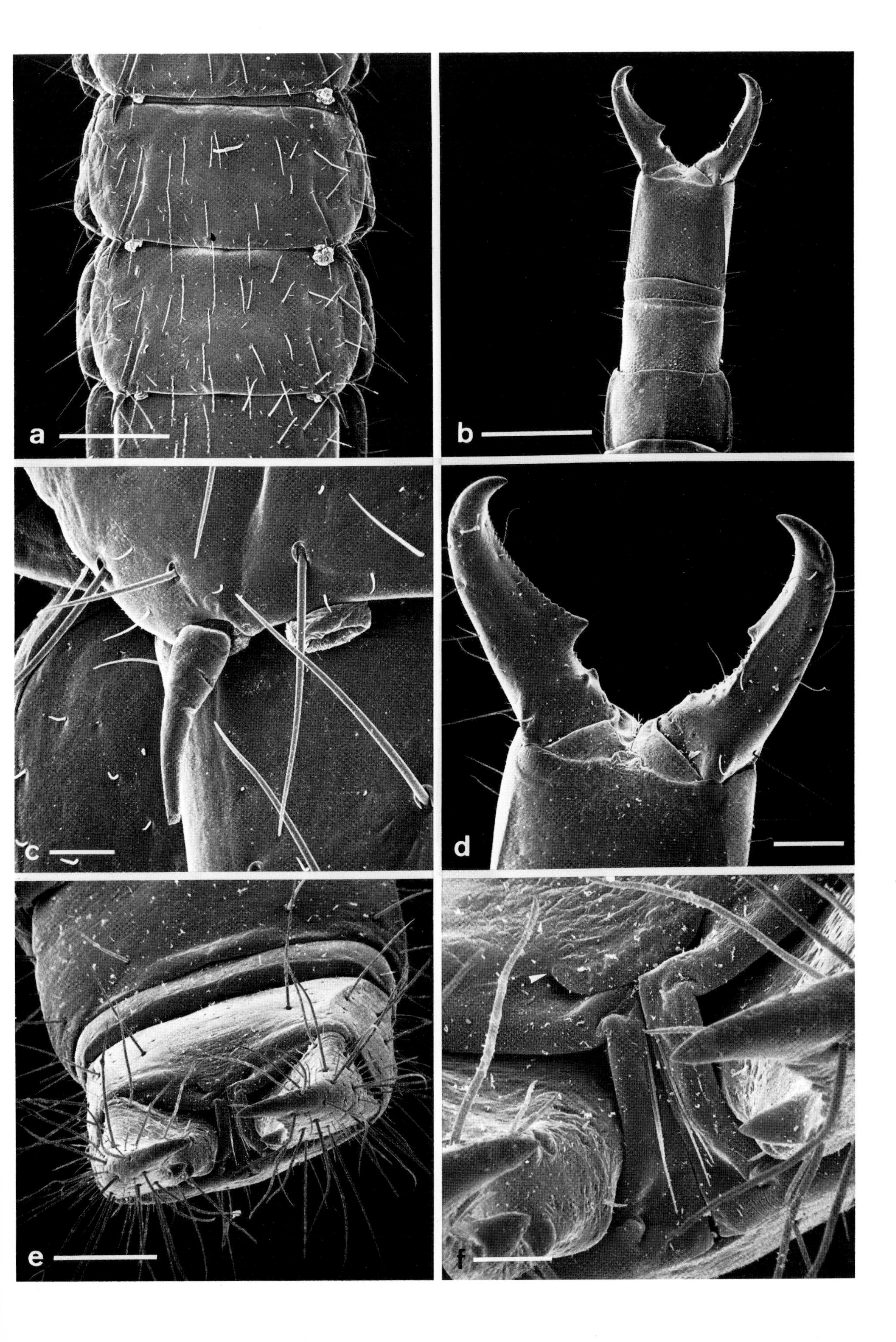
a
b
c
d
e
f

2.10.8 Biology of the Japygidae

We owe our knowledge of the life cycle of the japygids to the investigations of Pages (1967a, b, 1978), who rated them as xerophilic insects. A moisture preferendum of 85% r.h. was determined for *Dipljapyx humberti*, and this may not fall below 50%. A classification as mesophilic is more justified by these results. However, the japygids are definitely associated with relatively high temperatures. They occur predominantly in regions where the temperature at the soil surface exceeds 30 °C in summer and does not fall below 10 °C in the ground throughout the year. As a result they have been found only in places with suitable climatic conditions in central Europe (Simon, 1963).

The soil itself must be loose and well-ventilated, but still stable as the pre-existing cracks and crevices are further excavated to form tunnels and retreats. The japygids are distinctly territorial animals, constructing barriers to mark the boundaries of their territory. Their two main activities between moulting (6 months) are dormancy phases and feeding. A month before moulting they retreat to a chamber and gradually fall into a state of apathy. Females also show reduced activity before and after oviposition. The female remains isolated in a chamber where she deposits her eggs so that they hang freely from a stalk on the wall (Fig. 114). Thereafter she keeps them clean. She does not feed during this period. After the larvae have hatched they are constantly guarded and cared for.

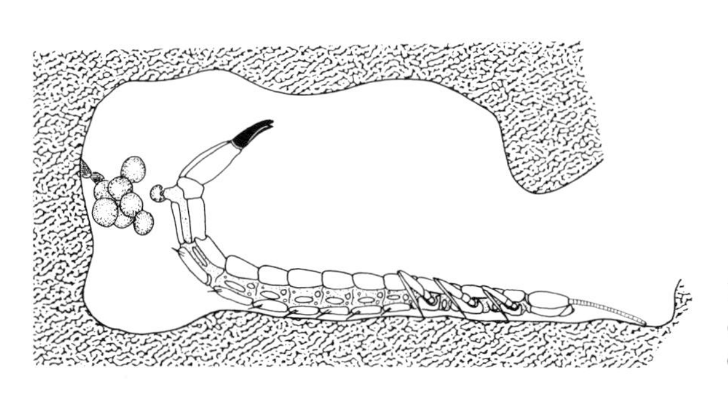

Fig. 114. Female of *Dipljapyx humberti* (Japygidae) during oviposition. The spherical eggs are firmly attached to the chamber wall by means of a secretion. (After Pages, 1967)

Plate 87. Japygidae.
a Base of the pincers, medial. 250 μm.
b Pincer claw, distal, with slightly sunken, conical sensilla. 25 μm.
c Base of pincer claw, medial, with toothed edge. 100 μm.
d Surface of pincer claw with sensillum. 3 μm.
e Toothed edge of a claw with sensilla. 20 μm.
f Surface of pincer claw with sensillum. The cuticle is highly perforated. 3 μm

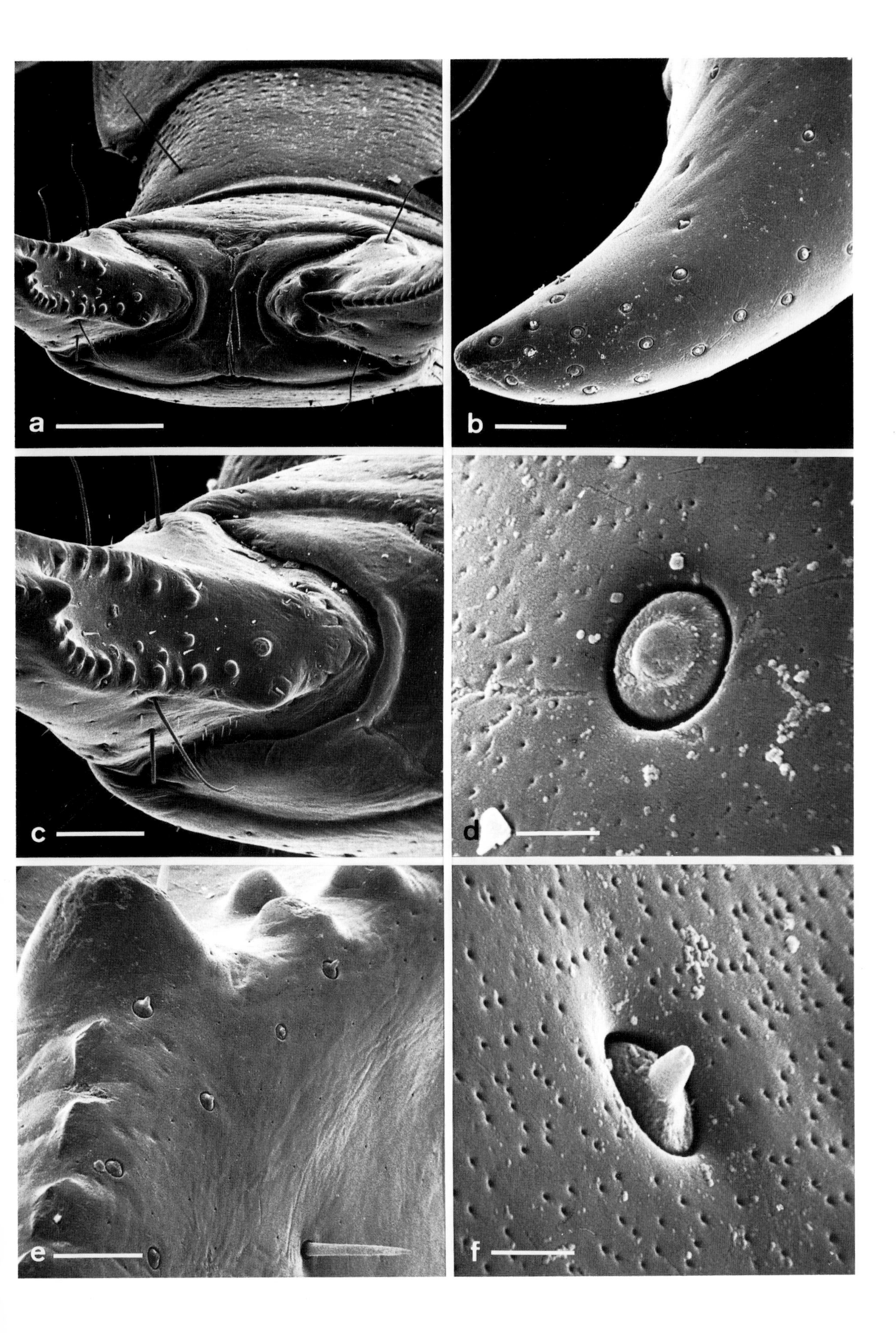
a
b
c
d
e
f

2.11 Order: Protura (Insecta)

General Literature: Tuxen, 1964; Janetschek, 1970; Nosek, 1973

2.11.1 Characteristics of the Protura

The proturans belong to the primary wingless insects. Adult animals are only 0.5–2.5-mm long. During postembryonal development the number of abdominal segments increases from 9 to 11.

The pear-shaped head lacks eyes and antennae (Plate 88). The mouthparts range from scratching-sucking to piercing-sucking (Plate 89). A pair of sensory organs, the pseudoculi (Fig. 105), are situated on the head capsule (Yin et al., 1986). Such organs were also described in the Pauropoda (Myriapoda) and they correspond to the Tömösváry organs of the myriapods (Bedini and Tongiorgi, 1971; Haupt, 1972). The pseudoculi and the postantennal organ in Collembola are homologous (Tuxen, 1931) and they are not to be confused with the pseudocelli of the Collembola (Francois, 1959; Haupt, 1979). The foremost of the three pairs of short legs, which bear a distinctly higher density of sensory hairs than the rest of the body, are held over head like feelers. Thus, unlike the mode of locomotion of other insects, the Protura use only their middle and hind legs for walking.

The abdomen bears certain structures which are of importance for life in the soil. On the three anterior abdominal segments there are rudimentary appendages with eversible vesicles (Fig. 116; Plate 89). The functional morphology of these organs is comparable to that of the coxal sacs of the machilids (Archaeognatha) and the Diplura or to the vesicles on the ventral tube of the Collembola, all of which participate in the water regime of the animals by means of water uptake. In addition, a pair of defensive glands with their openings on the eighth segment provide protection against attackers (Francois and Dallai, 1986).

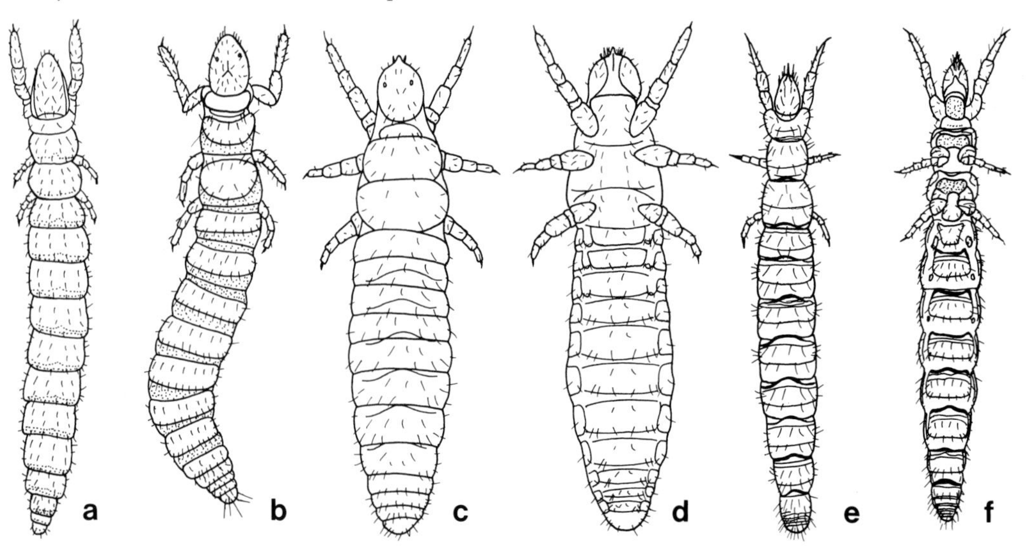

Fig. 115. Habitus of proturans. (After Janetschek, 1970). Eosentomoidea: **a** *Sinentomon erythranum* (Sinentomidae), dorsal; **b** *Eosentomon tuxeanum* (Eosentomidae), dorsal. Acerentomoidea: **c, d** *Fuijentomon primum* (Protentomidae), dorsal and ventral; **e, f** *Acerentomon maius* (Acerentomidae), dorsal and ventral

Plate 88. Acerentomoidea.
a Overview, lateral. 250 μm.
b Overview, ventral. 0.5 mm.
c Thorax and head with the pair of tactile legs. 100 μm.
d Head with tactile legs, ventral. 100 μm.
e Head in the prognathous position, with tactile legs, dorsal. 100 μm.
f Tactile leg, distal, with the terminal claw and long tactile hairs. 20 μm

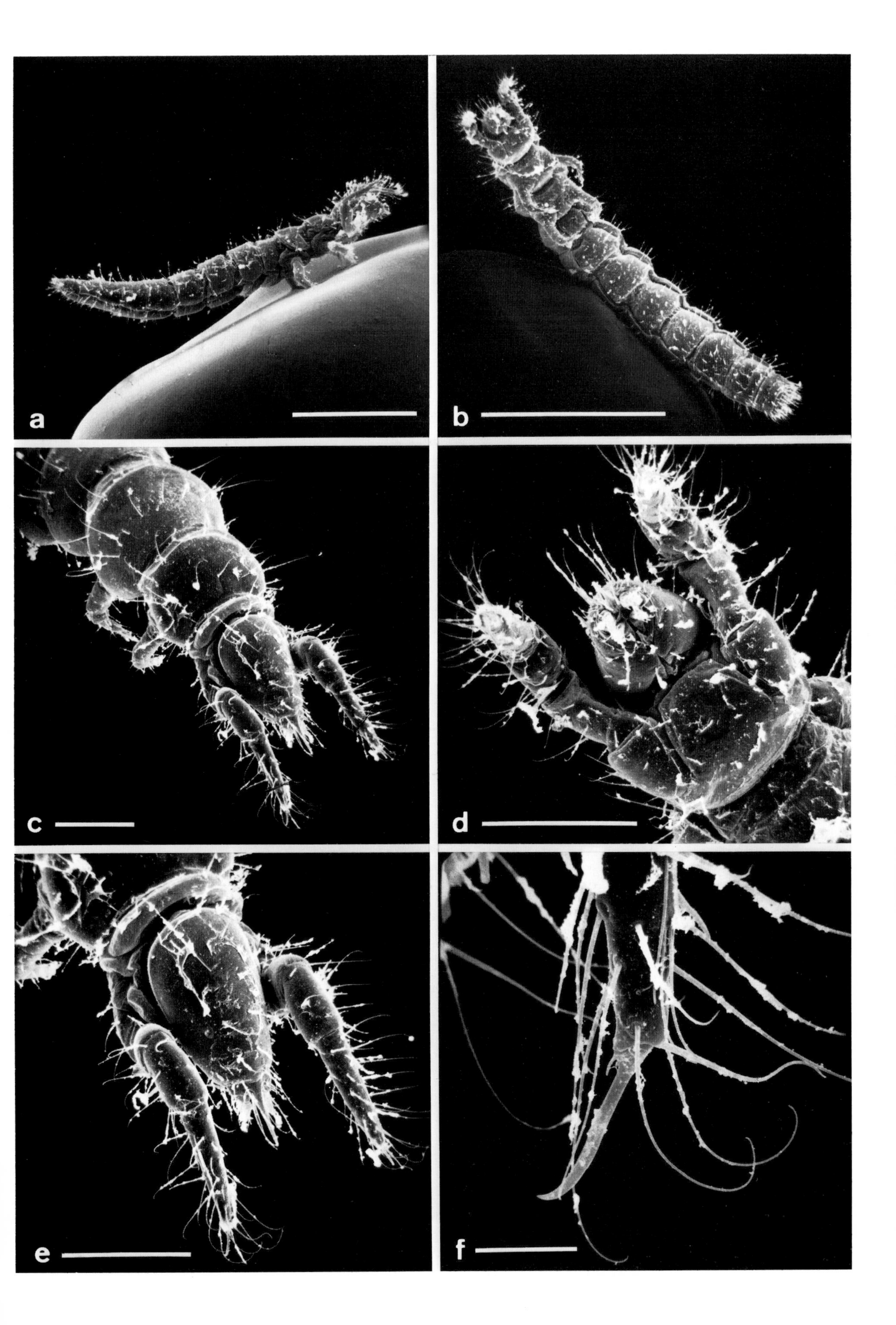
a
b
c
d
e
f

2.11.2 Soil Biological Aspects

The euedaphic proturans prefer the high air humidity and moderately warm temperature of the soil pore system. Their maximum density occurs at a depth of 3 – 10 cm. The abundance varies according to the biotope and reaches values of 2100 individuals/m^2 of soil surface in marshy meadows of East Holstein, West Germany (Strenzke, 1942), 3500 ind/m^2 in oak woodland of southwest Sweden (Gunnarsson, 1980), 4500 ind/m^2 in spruce woodland of the Lower Tatra (Nosek and Ambrož, 1964) and 9500 ind/m^2 in fir woodland in Denmark (Tuxen, 1931). Despite their high abundance, they are of no great significance to soil biology and decomposition of plant debris because of their low biomass. However, the catalytic effect of small, microphytophagous animals (consumers of fungal mycelia and bacteria) with regard to the microbial contamination of the soil should not be underestimated, as their activities influence the entire decomposition process.

The proturans exhibit striking adaptations to life in the soil, some of which they have in common with certain euedaphic Collembola. Their long body with its short legs, the lack of eyes, antennae and pigment, the tactile hairs on the body surface, the modified forelegs covered with sensory hairs which replace the missing antennae, the retractile and eversible abdominal vesicles and finally the defensive glands on the eighth abdominal segment are all adaptation mechanisms and prerequisites for life in the narrow, dark pores and channels of the soil biotope.

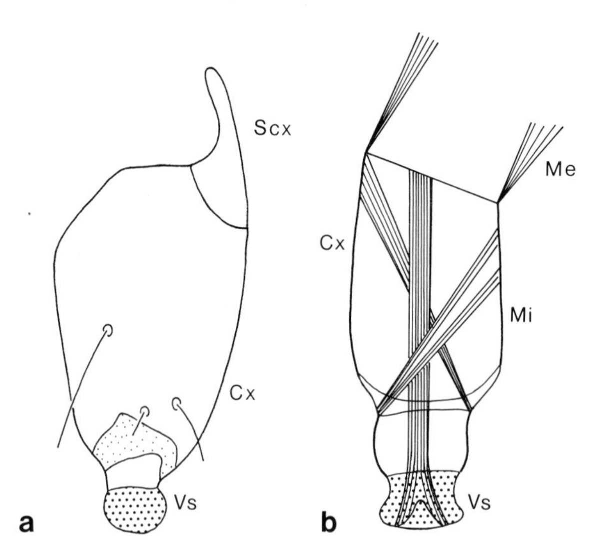

Fig. 116. Abdominal appendages of the protura. (After Snodgrass, 1935).
a Abdominal appendage of *Eosentomon germanicum*. *Cx* Coxa; *Scx* subcoxa; *Vs* vesicle (coxal sac); **b** Abdominal appendage of *Acerentomon doderoi* with internal and external musculature (*Mi, Me*). *Vs* Vesicle; *Cx* coxa

Plate 89. Acerentomoidea.
a Mouth cone with stylet, maxillary and labial palps and linea ventralis (*arrow*). 20 µm.
b Abdomen, ventral. 75 µm.
c Stylet (*arrow*), flanked by sensory bristles. The stylet is composed of the mandibles and the lobes of the maxilla, galea and lacinia. 5 µm.
d First abdominal segment, ventral, with double-segmented abdominal appendage. Distal there is an everted coxal vesicle (*arrow*). 10 µm.
e Mesothoracic leg showing segmentation. Medio-lateral sequence: subcoxa (*arrow*), coxa, trochanter, femur, tibia, tarsus and unguis. 25 µm.
f Abdominal segment 1 and 2, lateral, with abdominal appendages. 25 µm

a
b
c
d
e
f

2.12 Order: Collembola – Springtails (Insecta)

General Literature: Handschin, 1926; Stach, 1947–1960; Paclt, 1956; Gisin, 1960; Christiansen, 1964; Palissa, 1964; Schaller, 1970; Butcher et al., 1971; Christiansen and Bellinger, 1980; Joosse, 1983

2.12.1 Anatomy of the Collembola

The Collembola are primitive wingless insects. They are globally distributed, occurring even in the Antarctic and colonizing a variety of biotopes, often with high population densities. Despite their small size of only 0.2–9 mm, their abundance makes them important soil organisms, playing a significant role in decomposition processes.

The body of the springtails is clearly divided into the head, thorax and abdomen. The first of the six abdominal segments bears the ventral tube, which the springing organ or furca on the fourth and the retinaculum on the third segment. The genital aperture, lacking external genital appendages, is found on the fifth segment and the anus on the sixth. The thorax has three segments, each bearing a pair of legs divided into subcoxa, coxa, trochanter, tibiotarsus and pretarsus with claw. The four-segmented primary antennae are found on the anterior part of the head. Near their base lie simple compound eyes with a maximum of eight ommatidia. A few families have evolved a special sensory organ in this region, known as the postantennal organ. Collembola also possess striking entognathic mouthparts, which are united to form a mouth cone on the ventral side of the head.

The Collembola are divided into two groups, distinguishable by their body form: the elongate, cylindrical Arthopleona and the globular Symphypleona, the segments of which are largely fused together (Figs. 117 and 118). Research has increasingly focussed on the ecology of the Collembola since the work of Agrell (1941) and Gisin (1943).

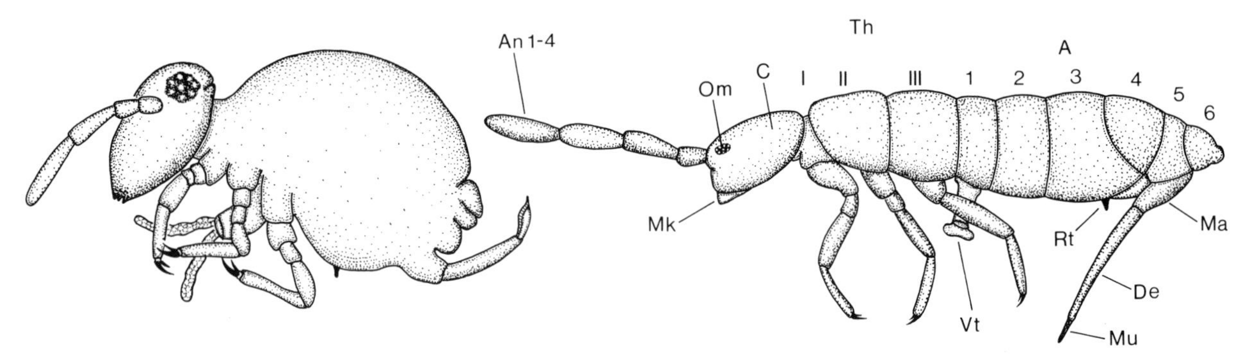

Fig. 117. Habitus of a symphypleonal Collembola. (Palissa, 1964)

Fig. 118. Segmental arrangement of an arthropleonal Collembola, *Isotoma viridis* (Isotomidae) (After Palissa, 1964). *A* Abdominal segments *1–6*; *An* antenna *1–4*; *C* head; *De* dens; *Ma* manubrium; *Mu* mucro; *Mk* mouth cone; *Om* eye (up to eight ommatidia); *Rt* retinaculum; *Th* thorax *I–III*; *Vt* ventral tube (coxal vesicles are everted)

Plate 90. *Tomocerus flavescens* (Tomoceridae).

a Habitus, lateral, with folded furca. 1 mm.

b Head, ventral, with mouth cone and ventral groove (*arrow*). 200 μm.

c Mouth cone, frontal. The *arrow* marks the beginning of the ventral groove. 100 μm.

d Mouthparts with labrum (*1*), mandible (*2*), maxilla (*3*), maxillary palp (*5*) and the hypopharynx (*4*). 50 μm.

e Ventral groove on the lower side of the thorax. It follows a discontinuous course as far as the ventral tube. It opens out in the region of tendinous plates. 20 μm.

f Abdomen, caudal, with the three flaps of the anal region, the genital papilla (*arrow*) and the furca base (manubrium, *arrowheads*). 200 μm

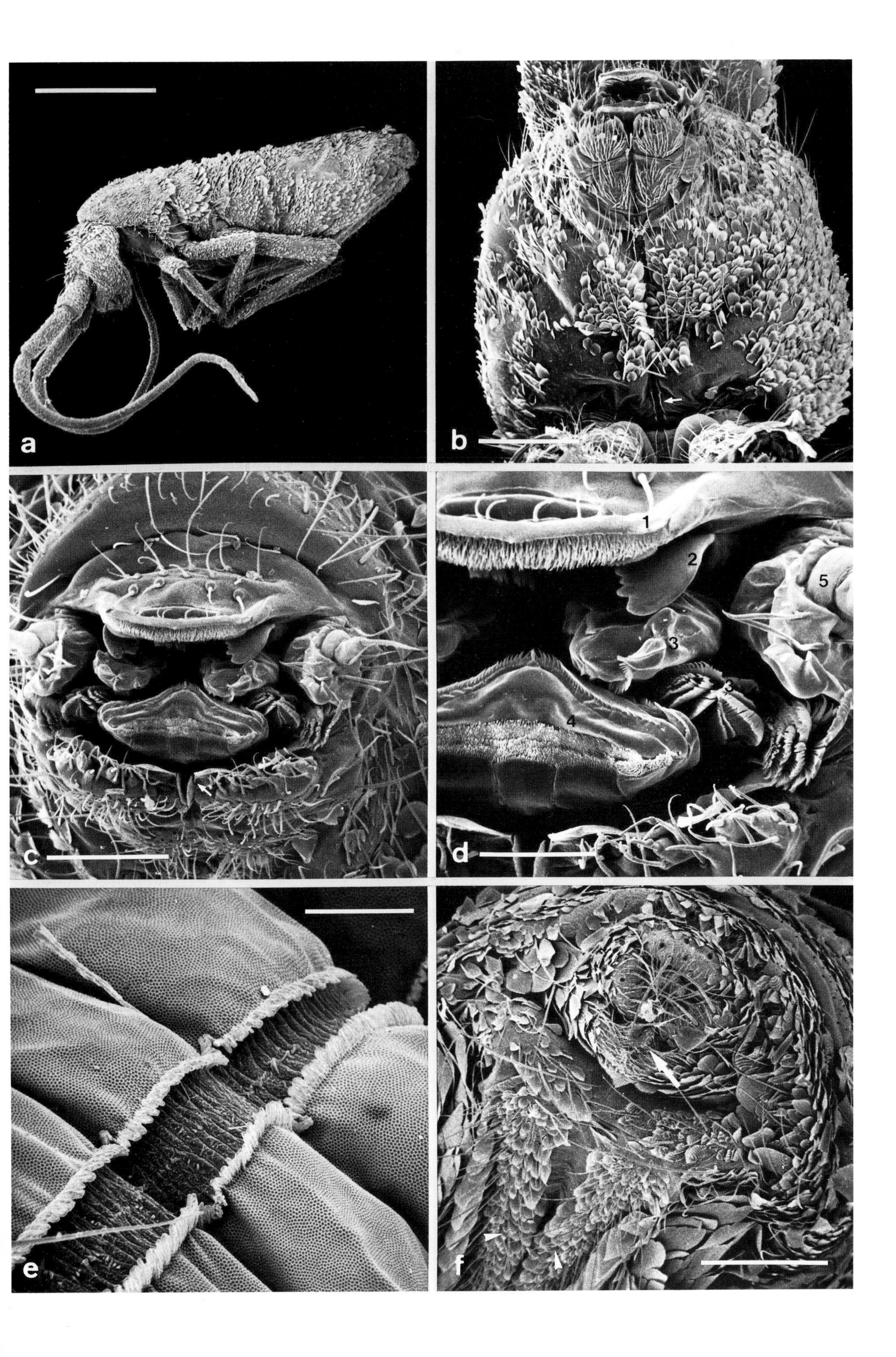
a
b
c
1
2
5
3
3
4
d
e
f

2.12.2 Life Forms of the Collembola

The Collembola are grouped into different life forms. Gisin (1943) distinguished between three types: atmobiotic, hemiedaphic and euedaphic life forms. However, this classification was later modified (Bockemühl, 1956; Christiansen, 1964). We suggest another scheme in which soil Collembola are divided into epedaphic and euedaphic life forms. Though not consistent with Gisin's system (1943), it conforms to our classification of soil arthropods as a whole (cf. Chap. 1.2).

The epedaphic species living on the ground and in the leaf litter layer are strikingly large compared to euedaphic Collembola. Their body surface has a richly pigmented pattern and often bears a dense covering of hairs or scales. The compound eyes are well-developed sensory organs, many species have eight ommatidia. Their four-segmented antennae are long and the fourth segment often shows secondary division. However, a postantennal organ is generally lacking. They possess a long, well-developed furca. The majority of the Entomobryomorpha and Sminthuridae are representatives of the epedaphic life form. These species are often encountered in atmobiotic habitats like the herbaceous and shrub layers and on tree trunks when the microclimatic conditions (e.g. humidity) permit a diurnal vertical migration (Gisin, 1943; Bauer, 1979).

Some epedaphic Collembola, especially certain species of the Isotomidae family, are progressively more adapted to life in the soil and exhibit adaptations resembling those of the euedaphic life forms. Though compound eyes are still present, they often have a reduced number of ommatidia. Their antennae are only moderately long and pigment is seldom distributed over the entire body surface. The postantennal organ, which is lacking in most epedaphic forms, is present, but often with a very simple arrangement (Fig. 128; Plate 103).

The euedaphic species inhabit the lower soil layers. Their elongated and cylindrical or worm-shaped body seldom exceeds 2 mm. Both legs and antennae are short. Their furca is frequently nonfunctional or has completely disappeared. In most cases pigment and hair are either absent or greatly reduced. Compound eyes are similarly lacking.

In contrast, the postantennal organ is complex (Figs. 122 and 128; Plate 96), the antennal organ on the third antennal segment (Plate 96) and the antennal sensory bristles are well developed. Finally, glandular pseudocelli, which have a protective function, are frequently distributed over the whole body. The Onychiuridae are typical representatives of the euedaphic life form.

Plate 91. *Tomocerus flavescens* (Tomoceridae).

a Antennae, proximal. 200 µm.

b Antenna, middle region, with secondary segmentation. 50 µm.

c Round scale. 10 µm.

d Antenna, distal. The secondary segments are covered with tubular hairs only. 20 µm.

e Hair base from the region of a coxal receptor organ (proprioreceptor). 3 µm.

f Surface structure of a ciliated macrochaeta on the mesonotum. 3 µm

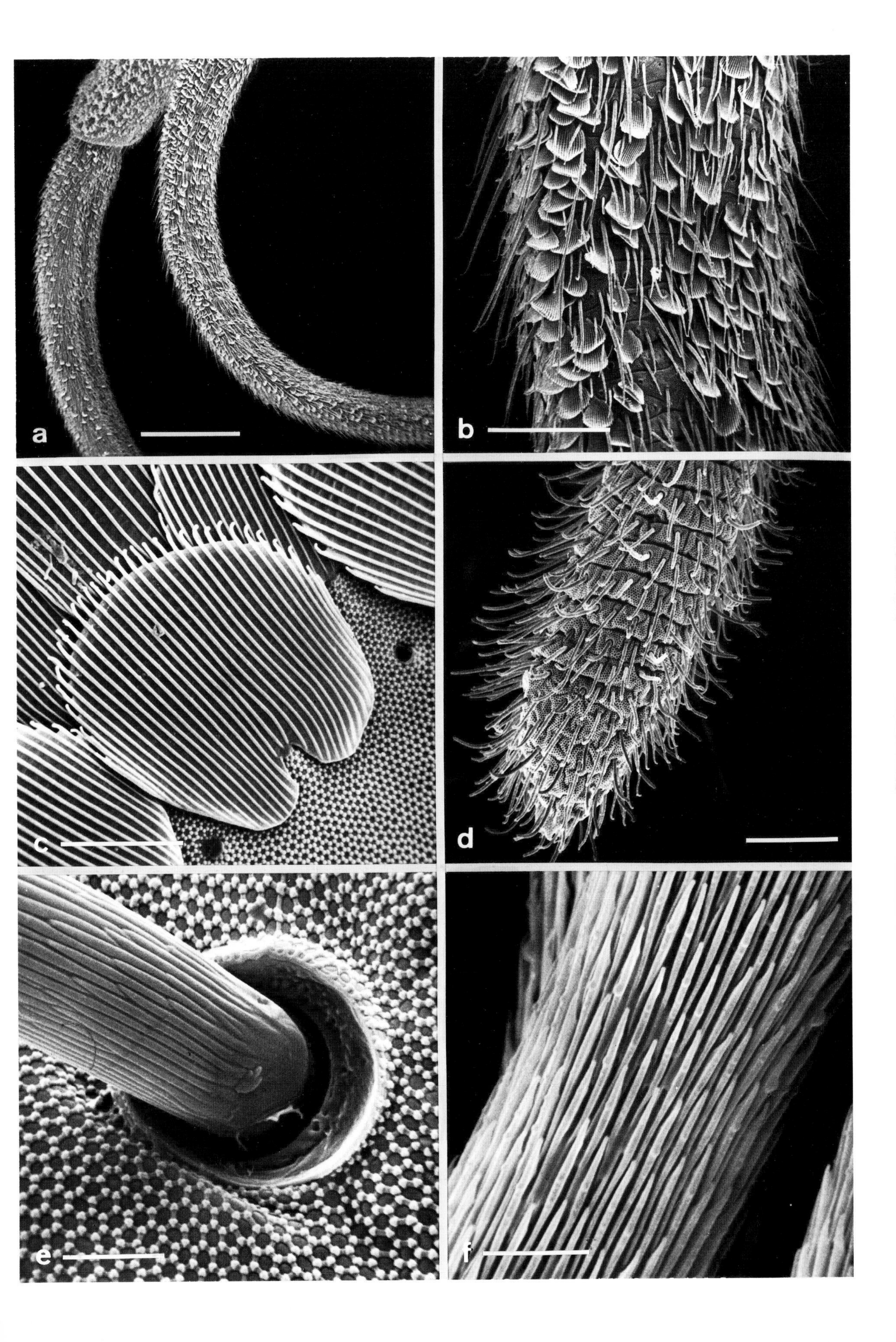
a
b
c
d
e
f

2.12.3 Compound Eye of the Collembola

A maximum of eight almost circular ommatidia form the functional eyes of the Collembola. Some species which live in caves or belong to the euedaphon lack ommatidia altogether (Barra, 1973; Strebel, 1963).

With few exceptions, the visual performance of the Collembola is restricted to distinction between dark and light. The resulting phototaxis is generally negative in most soil-dwelling species, while that of the atmobiotic animals and those living on the surface of water is positive.

Paulus (1972, 1971) investigated the fine structure of the compound eyes. He discovered that the collembolan ommatidia have one crystal cone (euconic eye type). Genuine main pigment cells were found in the Symphypleona, while most arthropleonal Collembola possess corneal cells without pigment. In *Tomocerus* the pigment coat round the crystal cone is formed by the neighbouring, peripheral epidermal cells. The pigment screen of the rhabdomeres is provided by a pigment coat within the retinula cells themselves.

With few exceptions the cuticular surface structure of the cornea is consistent with the normal cuticular structure of the body surface. A basically hexagonal pattern is formed by six circularly arranged, cuticular triangles (microtubercles or primary granules), which stand on small supporting ridges each connected to the neighbouring triangles. In *Tomocerus* this normal cuticular structure is only fully developed at the periphery of the ommatidia. The pattern is, however, smaller and flatter than on the neighbouring body surface. The top of the cornea is differently structured. Here, ridges are crowded together to form an entirely new pattern (Plate 92b, c).

The rhabdom of the ommatidia is constructed from eight retinula cells which do not bear eight rhabdomeres in each case. Unlike the other Collembola investigated, the rhabdom of *Tomocerus* is radially symmetrical, but the eighth retinula cell is reduced and lacks a rhabdomere.

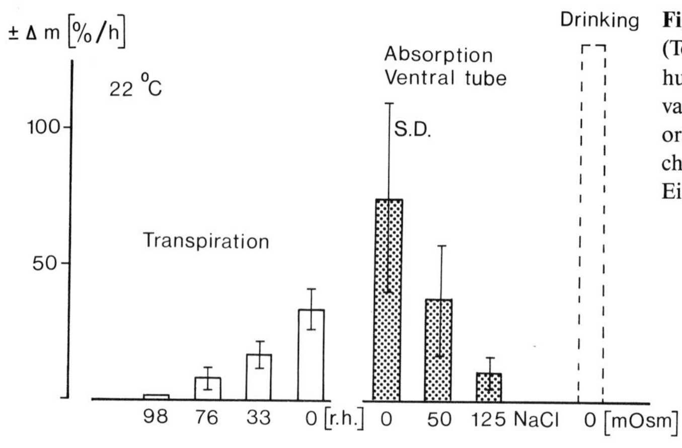

Fig. 119. On the water balance in *Tomocerus flavescens* (Tomoceridae). Comparison of transpiration rates at various humidities, absorption rates through the ventral tube at various saline concentrations (NaCl solution) and the rate of oral uptake of pure water. The rates refer to the percentage change in the total water mass m_0 of the animals. (After Eisenbeis, 1982) (cf. Chap. 2.12.4)

Plate 92. *Tomocerus flavescens* (Tomoceridae).

a Antennal base with simple compound eye consisting of six ommatidia. 100 μm.
b Ommatidium with cornea. 10 μm.
c Differentiated corneal surface. 2 μm.
d Basic pattern of the collembolan cuticle consisting of hexagonal tubercle rings (microtubercles, primary granules). It is assumed that these structures are primarily responsible for the hydrophobic properties of the cuticule. 1 μm.
e Tibiotarsus with pretarsus, distal, with long main and subsidiary claw (empodium) (*arrow*). Inserted dorsally there is a spatulate hair. 50 μm.
f Tarsal spatulate hair, distal. Secretion-conductive structures are to be found in the interior and the hair base is innervated by a mechanoreceptive sensory unit (cf. Blottner and Eisenbeis, 1984). 10 μm

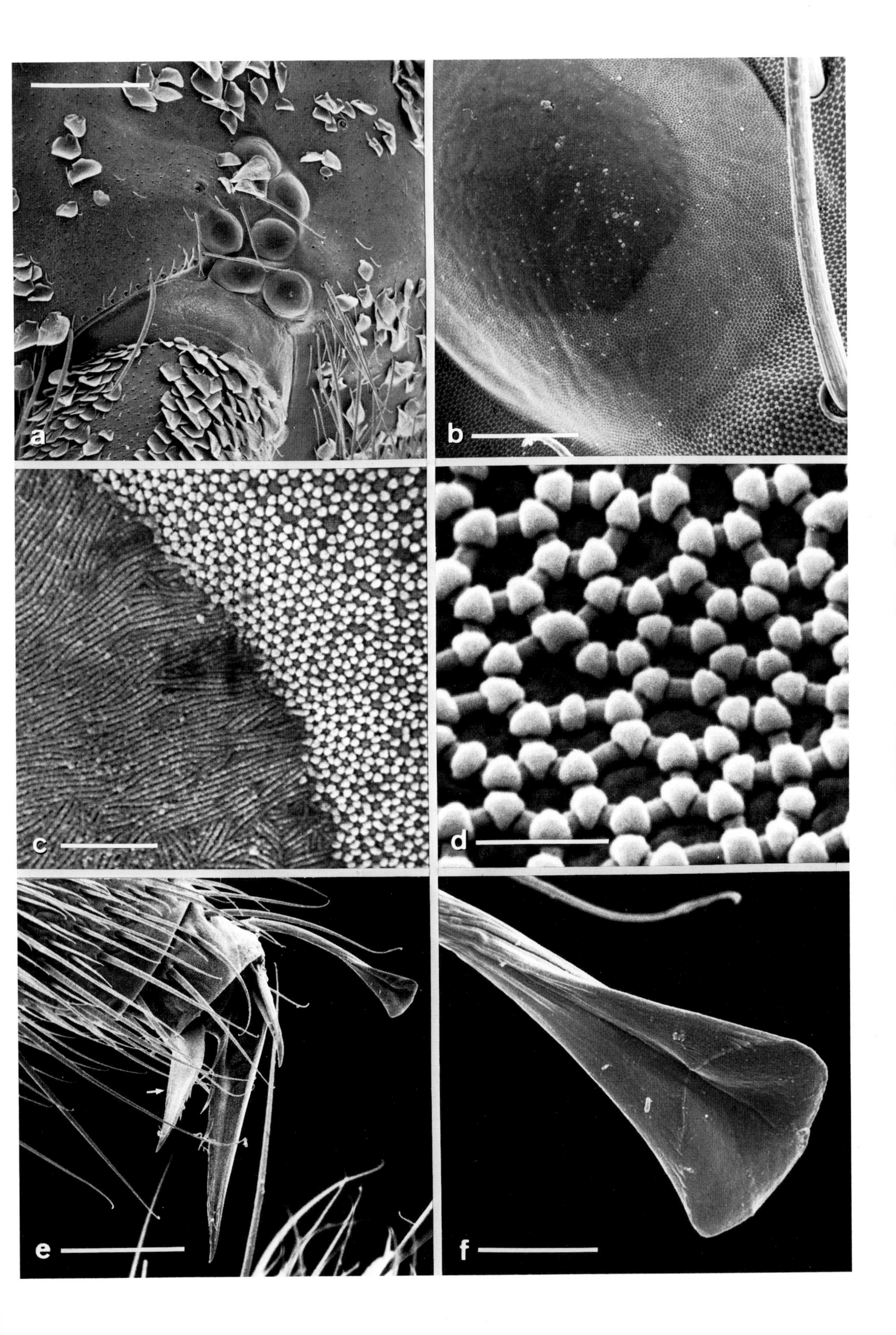
a
b
c
d
e
f

2.12.4 Ventral Tube of the Collembola

The ventral tube is inserted on the ventral side of the first abdominal segment and is composed of a basal plate, a cylinder, tube flaps and tube vesicles (coxal sacs). In *Tomocerus* the vesicles are everted by increasing internal haemolymphatic pressure and retracted into the cylinder by 12 retractor muscles (Plate 93) (Eisenbeis, 1976b, 1978). They are equipped with sensory organs (Fig. 124) which penetrate like sensors into the vesicle epithelium. These are interpreted as hygroreceptors or osmoreceptors (recently also as pH receptors), as the animals react to substrate moisture, increased salinity and lowered pH by eversion or retraction of their vesicles (Eisenbeis 1976a; Eisenbeis, 1982; Jaeger and Eisenbeis, 1984).

The individual parts of the ventral tube vary within the Collembola. In particular, the Symphypleona have developed long tubular vesicles, sometimes covered with wart-like papillae (Plate 102). The ventral tube is certainly a multi-functional organ. The main function must be regarded as transporting water and ions from the surroundings into the haemolymph, whereas respiration and adhesion onto the substrate surface are considered to be auxiliary functions.

The vesicle epithelium shows the typical features of transport epithelia, comparable to the rectal epithelium in insects (Eisenbeis, 1974; Eisenbeis and Wichard, 1975a,b 1977). Temperature, salinity and pH values within the substrate medium influence the absorption rate (Figs. 119 and 120). Desiccation causes the animals to evert their vesicles and to reduce the water deficit within a few minutes (Eisenbeis, 1982). This organ is presumably particularly significant for adaptation strategies when no water is available in drinkable form and only residual moisture can be taken up from the surface of leaves or from the ground.

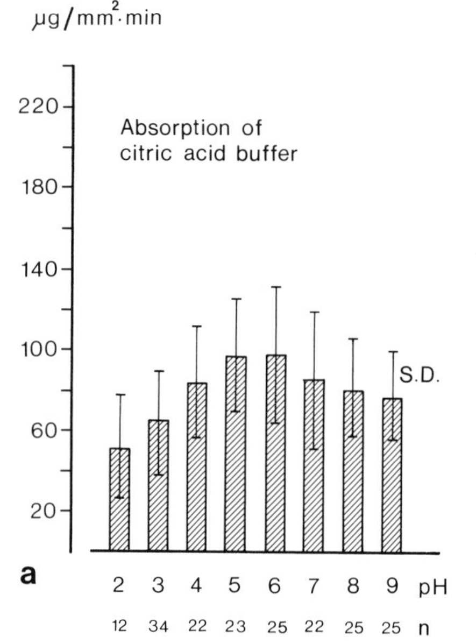

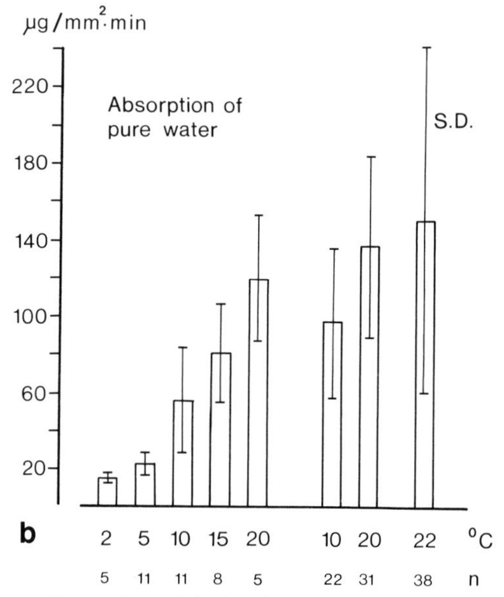

Fig. 120. Influence of various factors on the absorption rate by the ventral tube of *Tomocerus flavescens* (Tomoceridae). The rates are related to the effective vesicle surface of tube vesicles which makes contact with the substratum.

a Absorption of citric acid buffer (50 mOsm) at various pH values (cf. Jaeger and Eisenbeis, 1984). **b** Absorption of pure water at various temperatures (combined data from Bleicher, 1981; Eisenbeis, 1982 and Weissgerber, 1983)

Plate 93. *Tomocerus flavescens* (Tomoceridae).

a Ventral tube with tube cylinder, flaps and partly everted vesicles (coxal sacs). 100 µm.

b Everted vesicles, distal. The two vesicle hemispheres are separated by a medial furrow. A transport epithelium composed of 20 cells lies beneath the cuticle. 100 µm.

c Retinaculum, consisting of the corpus tenaculi (*arrow*) and the two rami which hold the furca in place. 50 µm.

d Cuticle of the tube vesicle. 3 µm.

e Paired hooks (rami) of the retinaculum, distal. 10 µm.

f Furca, distal, with the dentes (*arrowhead*) and toothed mucrones (*arrows*). 100 µm

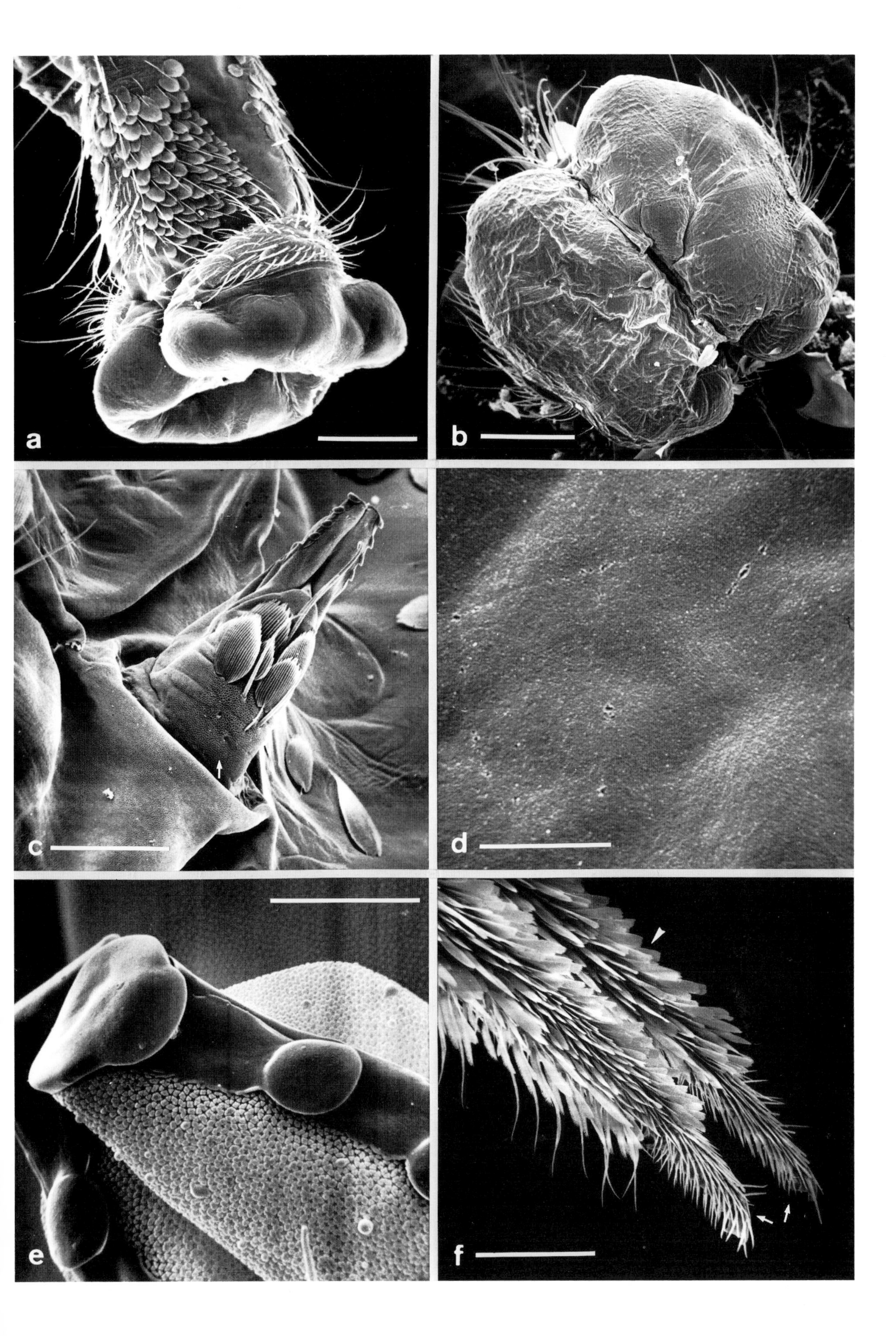

a
b
c
d
e
f

2.12.5 Jumping Apparatus of the Collembola

Many Collembola jump considerable distances spontaneously or when disturbed, using the furca as a springing organ and not the legs like other jumping insects; hence their common name, springtail. The furca is borne ventrally on the fourth abdominal segment and comprises an unpaired manubrium and the paired dentes and mucrones (Plates 93 and 94). The mucrones often show a variety of forms with teeth and lamellae. Those of *Sminthurides aquaticus*, which lives on the surface of water, are differentiated to form wing-like structures to improve their impact resistance (Plate 100).

In the resting position the furca is folded forwards and held in place by the retinaculum (Plate 93). The retinaculum is situated ventrally on the third abdominal segment. It is composed of an unpaired basic segment (corpus) and two, short-toothed branches (rami) to hold the furca more firmly between the paired dentes. Springtails can, however, also jump when the furca is not held by the retinaculum (Eisenbeis and Ulmer, 1978).

The animal is launched by thrusting the furca against the substratum with the help of both the powerful extensor system of the musculature and a hydraulic mechanism. Thus, appreciable distances, often of several centimetres (in *Allacma fusca* more than 20 cm high), can be sprung. The catapult-like jumps are executed within milliseconds. High-speed cinematography finally revealed that the animals turn a somersault in the air (Fig. 121) (Christian, 1978, 1979). Soil-dwelling animals show a rapidly decreasing tendency to jump. According to Bauer and Christian (1986) there is a correlation between the "flight behaviour" of springtails in the Entomobryidae and their habitat. The springing organ of euedaphic Collembola is largely reduced, in *Onychiurus* the furca is completely absent.

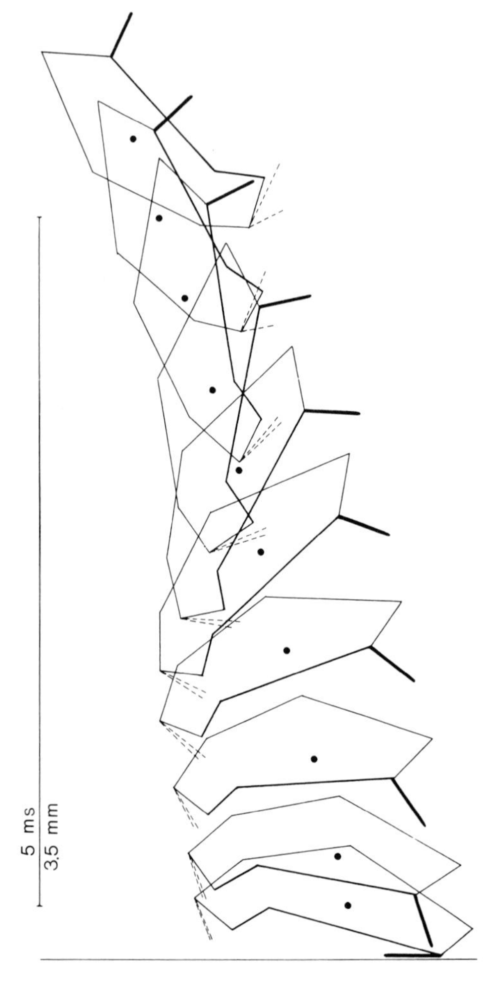

Fig. 121. Schematic representation of the jumping phases of a springtail showing a somersault directly after takeoff. The *dot* marks the position of the centre of gravity. (After Christian, 1978)

Plate 94. *Orchesella villosa* (Entomobryidae).
a Habitus, lateral. 1 mm.
b Springing organ, lateral. 0.5 mm.
c Base of the antenna with simple compound eye and some macrochaetae. 100 µm.
d Terminal segment of the furca: the mucro. 10 µm.
e Fine feathery antennal hairs immediately below the apex of the antenna. 10 µm.
f Macrochaetae, distal, with finely ciliated surface. 10 µm

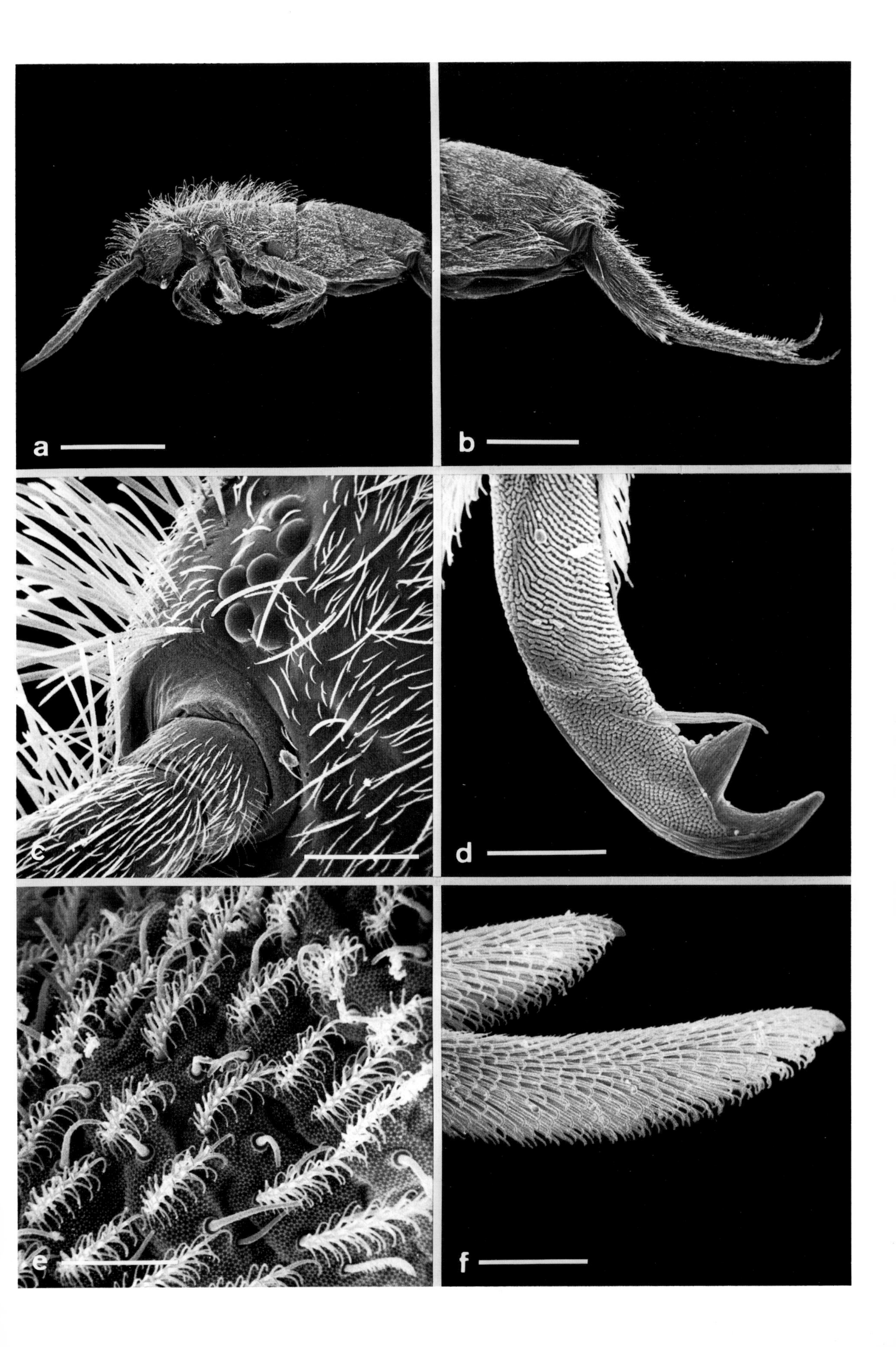
a
b
c
d
e
f

2.12.6 Euedaphic Collembola (Onychiuridae)

Onychiurids are regarded as characteristic members of the euedaphon, exhibiting adaptations appropriate to this way of life. Compound eyes are lacking, the antennae are short, the furca is reduced or no longer present and with certain exceptions (e.g. *Tetrodontophora*) they lack pigment and hairs are reduced in length and number. Instead they possess pseudocelli distributed over their body, a well-developed antennal organ on the third antennal segment and postantennal organs with numerous sensory subunits. They are usually less than 1 mm in length.

An important feature of structural adaptation to subterranean life is the reduction of the extremities. For instance, 70% of the total surface area of *Onychiurus* is concentrated in the trunk (thorax and abdomen), while more than 50% of the surface of the epedaphic *Tomocerus* is occupied by the head and extremities (Table 5). This more qualitative aspect is corroborated by quantitative evidence. The weight-related surface constant k_s for the surface formula $S = k_s \times w_0^{2/3}$ (S = surface area, w_0 = fresh body mass) is 8.84 for *Onychiurus* and 10.53 for *Tomocerus*. This means a reduction in the absolute body surface at the same weight level. Thus, the relatively high ratio of surface area/volume, caused by the diminutive size of *Onychiurus*, is decreased by the cylindrical shape of their body and the reduction of their extremities. This minimization of the body surface is meaningful as *Onychiurus* is an extremely hygric animal with a consistently high transpiration rate even in 100% humidity (Table 1). A similar minimization effect is realized in the globular-shaped Collembola (k_s for *Allacma fusca* 7.46, *Sminthurides aquaticus* 8.24; Table 5).

Table 5. Surface morphometry in Collembola showing the percentage proportion of the surface occupied by the various parts of the body and the surface constants k_s typical for each species

Species	Average proportion of body surface in %								
	Head	Ant.	Trunk	Legs	Furca	Ret.	Ventral tube	k_s	
Onychiurus sp. n = 40	10.38	4.05	70.56	14.97	–	–	–	8.84	Euedaphic
Tomocerus flavescens n = 30	7.46	10.59	46.50	25.93	7.17	–	2.31	10.53	Epedaphic-atmobiotic
Orchesella villosa n = 32	7.26	8.16	50.90	23.44	8.97	–	2.12	10.70	Epedaphic-atmobiotic
Allacma fusca n = 24	11.72	2.22	62.01	15.44	5.53	0.34	2.68	7.46	Epedaphic-atmobiotic
Sminthurides aquaticus n = 19	15.88	3.91	55.73	18.44	5.95	–	–	8.24	Atmobiotic (epineustic)

Plate 95. *Onychiurus* sp. (Onychiuridae).

a Habitus, lateral. The anal plate with spiny processes lies on the *left*. No springing organ is present. The postantennal organ is situated at the base of the antenna. The *arrow* marks the ventral tube. 200 µm.

b Anal plate with the triangular opening of the anus and two terminal spines. In front lies the transverse aperture of the female genital papillae. 50 µm

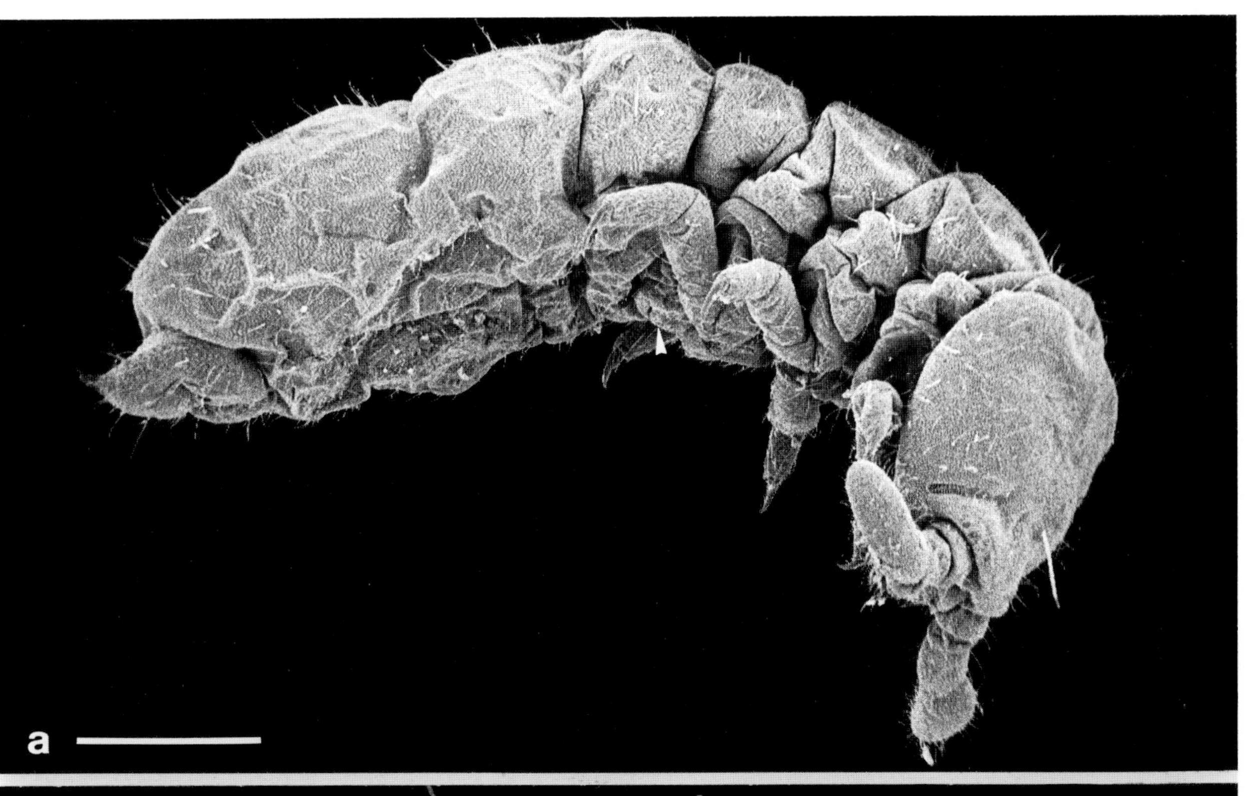
a

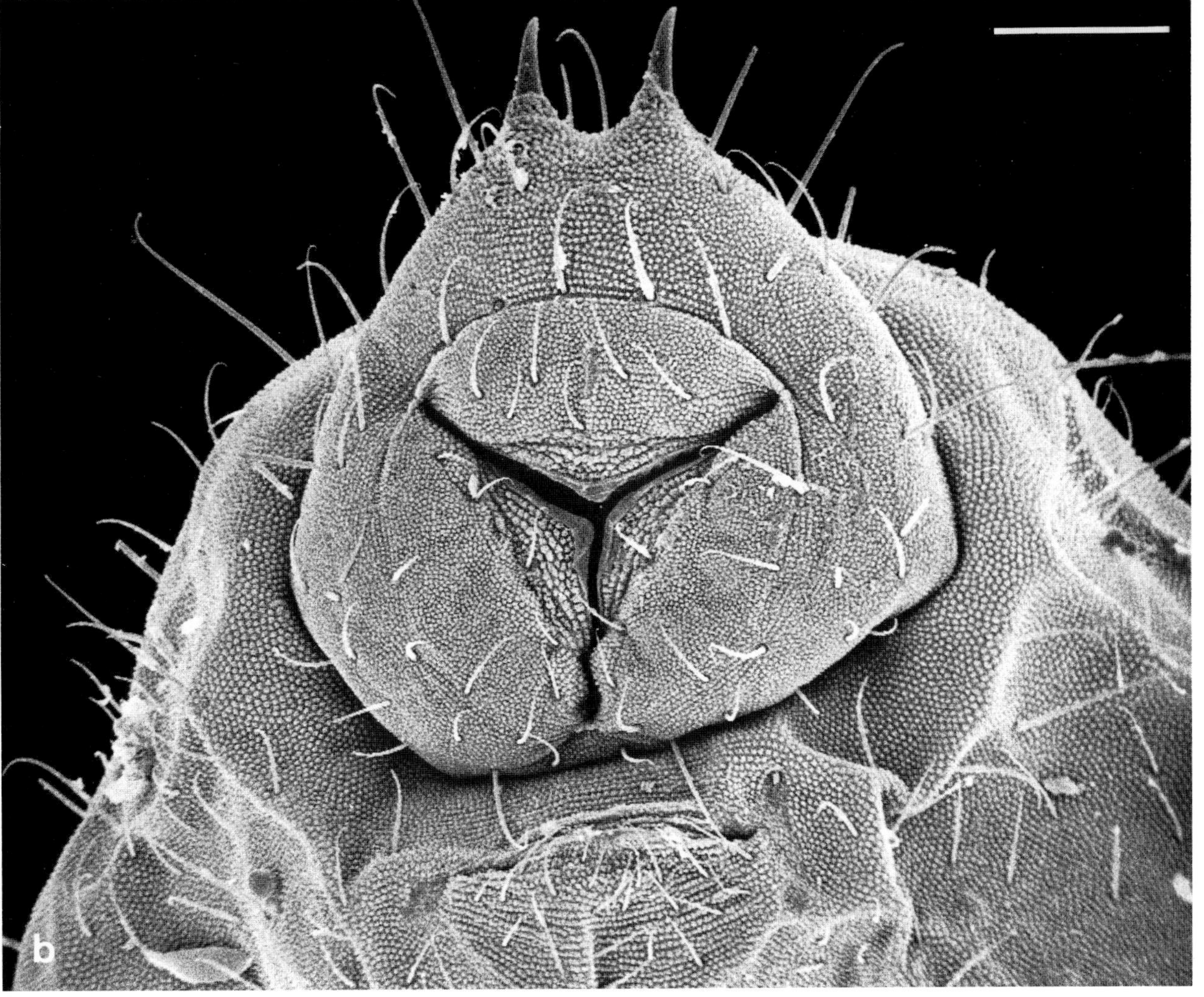
b

2.12.7 Sensory Organs of Euedaphic Collembola (Onychiuridae)

Blind, euedaphic Collembola possess mechanosensitive sensilla which assume the function of spatial orientation and perception. Sensory bristles, like those described in *Hypogastrura* (Altner and Ernst, 1974), are adaptations to life in the interstitial system of the soil.

Euedaphic forms also have well-developed, special sensory organs on the third antennal segment (antennal organs). In *Onychiurus* two types of sensilla with perforated walls, two basiconical and two globular sensilla, are situated in a hollow between five cuticular partitions (Plate 96). They are assumed to function as chemoreceptors. In front of the hollow lie four additional mechanoreceptors (Altner and Thies, 1972).

The postantennal organs (PAO), which lie near the base of the antennae set into the surface of the head capsule, do not occur in the epedaphic forms (Entomobryomorpha, Symphypleona), but are present in the euedaphic forms. In the onychiurids they are sunk into a cuticular groove and consist of mostly round to oval, finely perforated protuberances arranged in rows (Fig. 122; Plate 96). In the Isotomidae there is an unstructured, finely perforated lobe on the surface (Plate 103), with sensory cells connected to the nervous system beneath (Fig. 128) (Karuhize, 1971; Altner and Thies, 1976, 1978; Altner et al., 1970). No electrophysiological data are available as yet, but from the structural point of view it seems likely that PAO's are chemo-, hygro- or thermosensitive organs.

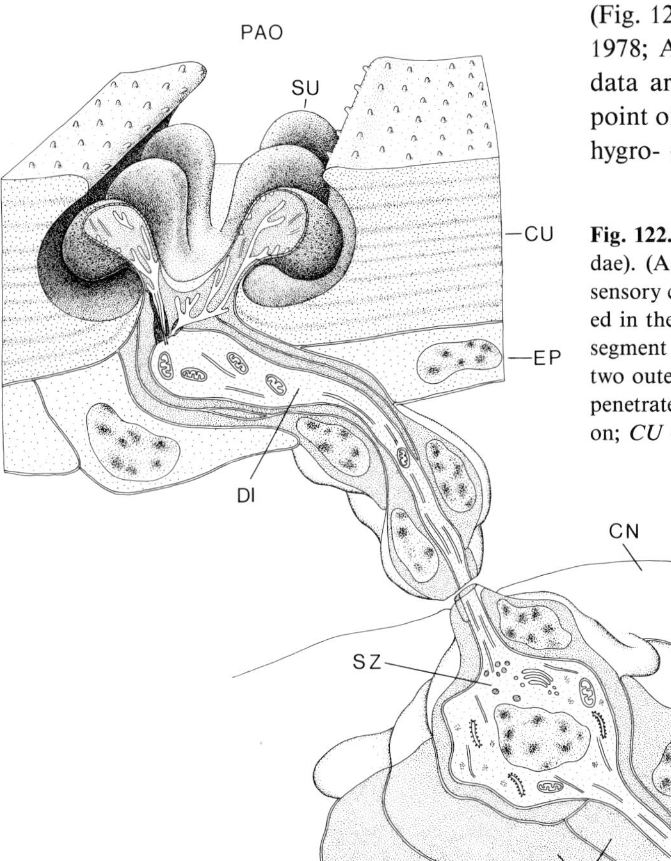

Fig. 122. Postantennal organ of *Onychiurus* sp. (Onychiuridae). (After Karuhize, 1971). The complex consisting of a sensory cell (*SZ*) and several envelope cells (*EC*) is embedded in the periphery of the brain (*CN*). The inner dendritic segment (*DI*) extends to the body wall where it divides into two outer dendritic segments (cilia), the branches of which penetrate into many bulbous sensory subunits (*SU*). *A* Axon; *CU* cuticle; *EP* epidermis; *PAO* postantennal organ

Plate 96. *Onychiurus* sp. (Onychiuridae).

a Head, frontal, with labrum, antennal base, pseudocelli and postantennal organ (*arrows*). 50 μm.

b Antennal base with pseudocelli and postantennal organ. 25 μm.

c Antenna, distal, with antennal organ. 25 μm.

d Sensory subunits of the postantennal organ. 1 μm.

e Antennal organ with comb-like subdivisions. Between the ribs there are modified sensilla (*arrows*). 4 μm.

f Pseudocellus. 3 μm

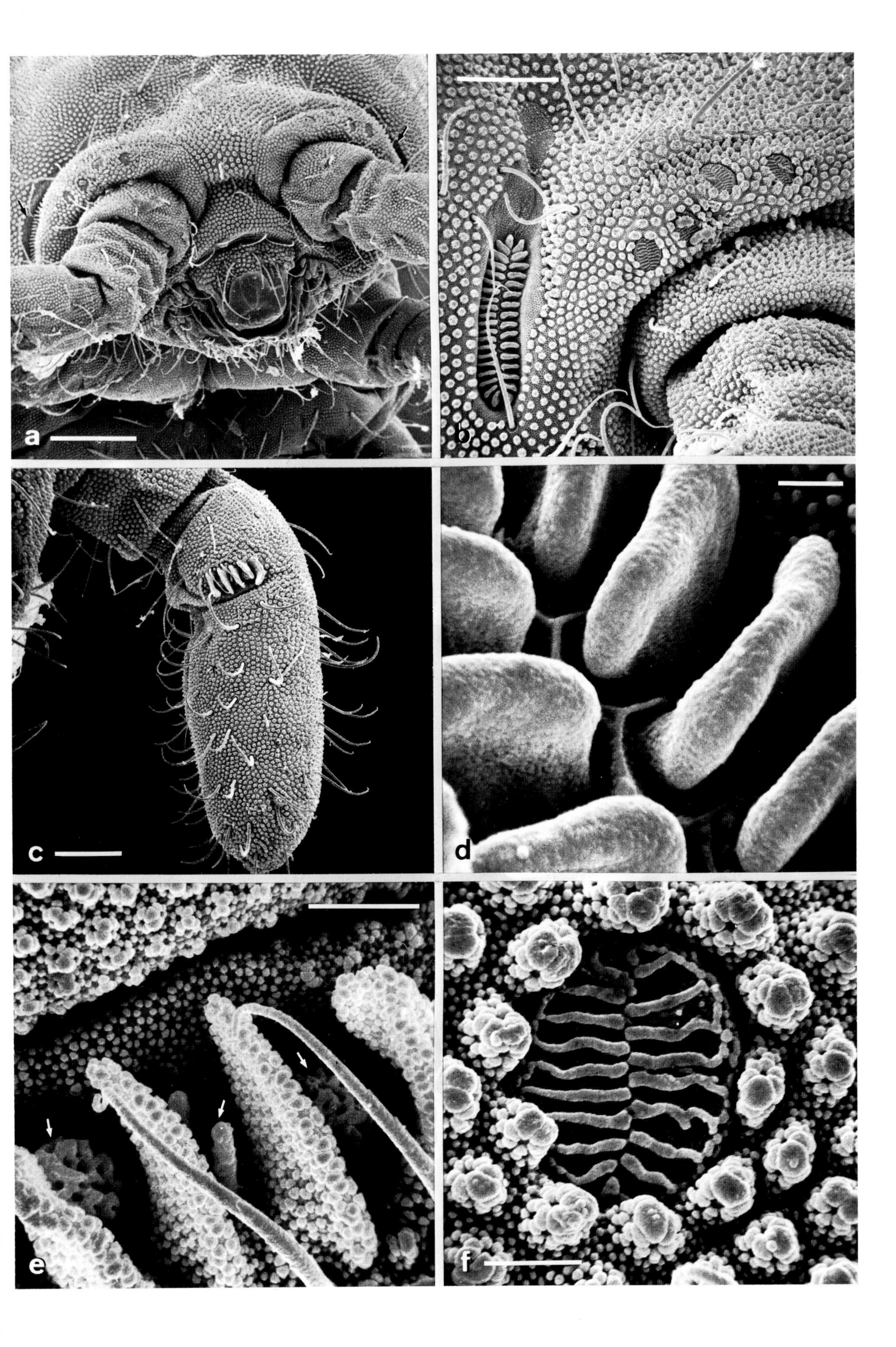
a
b
c
d
e
f

2.12.8 *Tetrodontophora bielanensis* (Onychiuridae)

Tetrodontophora bielanensis is an atypical member of the onychiurid family. With a maximum length of about 9 mm the adult animals are among the largest Collembola. Furthermore, they are well pigmented and blue-violet in colour. The furca is well-developed, but appears very small compared to the massive trunk (Plate 97). It is functional and the animals jump short distances when provoked. The animals live in the mountains under stones and in litter. Their highest population density occurs at an altitude of 1200–1400 m. Their main distribution areas are in the Carpathian and Sudeten Mountains (Stach, 1954). The bionomics of *Tetrodontophora bielanensis* were investigated by Dunger (1961), Vannier (1974, 1975), Jura and Krzystofowicz (1977) and Koledin et al. (1981).

Although *T. bielanensis* is 40–80 times heavier than the euedaphic *Onychiurus* species, the results of morphometric surface analysis show an identical surface constant (*Tetrodontophora bielanensis* 8.78, *Onychiurus* 8.84). This is to be expected according to the surface to weight ratio ($S = k_s \times w_0^{2/3}$) used for the calculation. The proportion of the surface occupied by the extremities is slightly higher in *T. bielanensis* due to the presence of a furca so that the trunk of *T. bielanensis* only accounts for 65% of the total surface (*Onychiurus* 70%, Table 5). The proportion of the surface occupied by the head, antennae and legs is almost identical in both species.

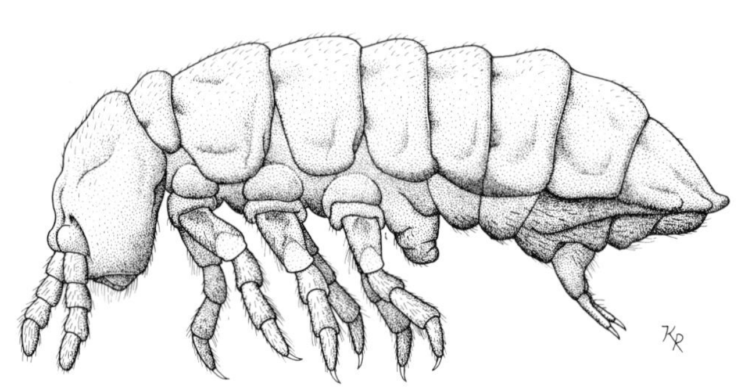

Fig. 123. *Tetrodontophora bielanensis* (Onychiuridae), lateral with everted furca. The blind animals are capable of executing short jumps as a means of escape. The ventral tube is inserted behind the legs

Plate 97. *Tetrodontophora bielanensis* (Onychiuridae).
a Overview, lateral. 1 mm.
b Overview, oblique caudal. 0.5 mm.
c Genital papilla, with the transverse genital aperture of a female. 100 µm.
d Abdomen, view of segments oblique dorsal, with pseudocelli at the posterior margin of the segments. In the hollows lie muscle insertions. 200 µm.
e Furca and retinaculum (*arrow*). The animals, weighing approximately 10 mg, use them for short jumps when disturbed. 200 µm

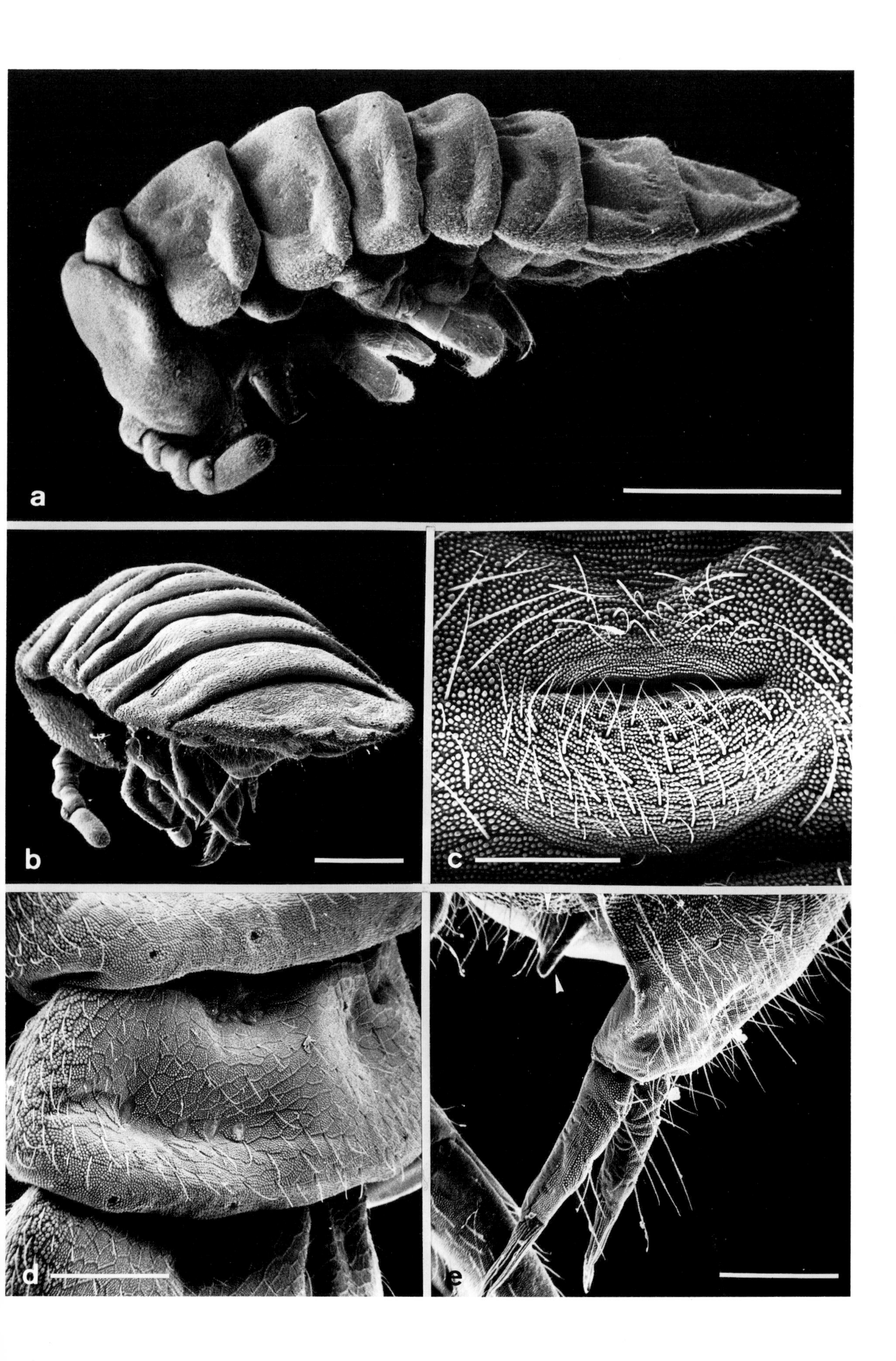
a
b
c
d
e

2.12.9 Pseudocelli of Euedaphic Collembola

Pseudocelli, situated in the integument, are organs typical of the onychiurid family (Plates 96–98). They are found on the head, on the thoracic and abdominal segments and occasionally on the legs. They are of round to oval shape with a diameter of about 4–8 μm. Unlike the surrounding, usually warty cuticular structure, the surface structure of their cuticle is almost smooth with two rows of fine ribs, separated by a central gap, providing stability for the thin plate. Beneath lie actively secreting cells, distinguishable from the surrounding epidermis cells, with their cell bases directly bordering the haemolymph cavity (Rusek and Weyda, 1981).

The first investigations on this subject were carried out on *Tetrodontophora bielanensis* (Konček, 1924). Liquid observed flowing out of the pseudocelli was believed to be blood, pressed out by increased haemolymphatic pressure (autohaemorrhage). The whitish-yellow droplets proved to be repellent to attackers as soon as they came in contact with the secretion. As most onychiurids are members of the euedaphon and are no longer capable of jumping to escape predators, these cuticular structures appear to be a necessary weapon. Mayer (1957) and Usher and Balogun (1966) regard the pseudocelli as defence glands.

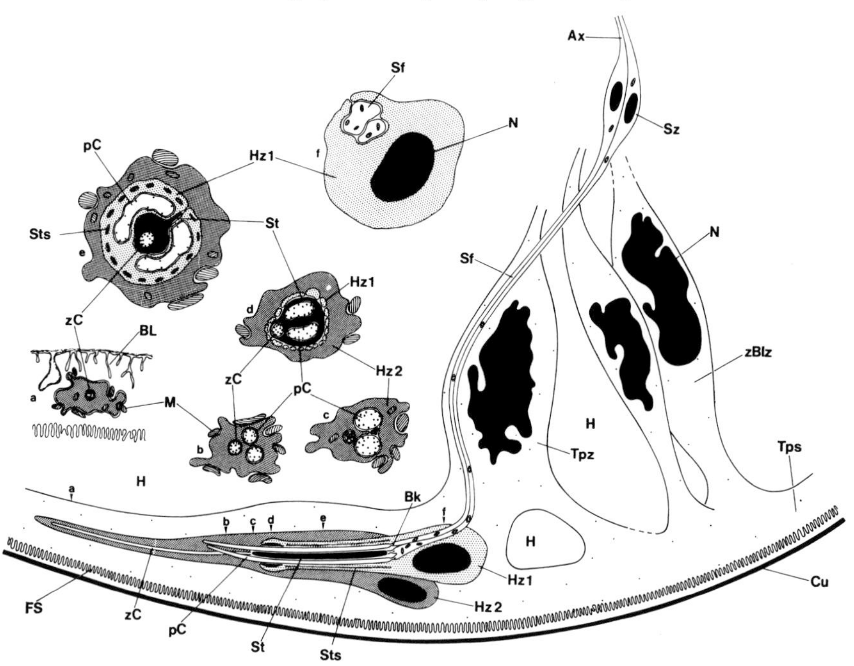

Fig. 124. Hygroreceptors or osmoreceptors in the transport epithelium of the ventral tube of *Tomocerus*, (After Eisenbeis, 1976a). Three dendrites (*Sf*), which terminate in distal cilia, penetrate into the transport epithelium (*Tpz, Tps*). The central portion of the cilia is enclosed and strengthened by an electron-dense sheath like a "scolops" as found in scolopidia of insects (*St*) (cross-sections *d, e*). In *Tomocerus* two different pairs of sensilla occur in the tube vesicles, in *Orchesella* both are identical, but differ in turn from those of *Tomocerus*. In this case the "scolops" is absent and they penetrate directly beneath the epicuticle. *Ax* Axon; *BK* basal body of the cilia; *Bl* basal labyrinth; *zBlz* central cells of the transport epithelium; *FS* microvilli border; *pC, zC* peripheral and central cilium; *H* haemolymphatic cavity; *Hz 1, 2* envelope cells; *M* mitochondria; *N* nuclei; *Sts* supporting rods. Cross-sections *a–e* are from the positions marked by *arrows* (cf. Chap. 2.12.4)

Plate 98. *Tetrodontophora bielanensis* (Onychiuridae).

a Head and thorax, oblique lateral. 0.5 mm.

b Antennal base with pseudocelli and the area of the postantennal organ. 200 μm.

c Abdomen, posterior margin of a segment with pseudocelli. 200 μm.

d Antenna, medial portion. 100 μm.

e Area of the postantennal organ. The inner structure of the two "spots" is unknown as yet. 25 μm.

f Pseudocellus on the antennal base. 10 μm

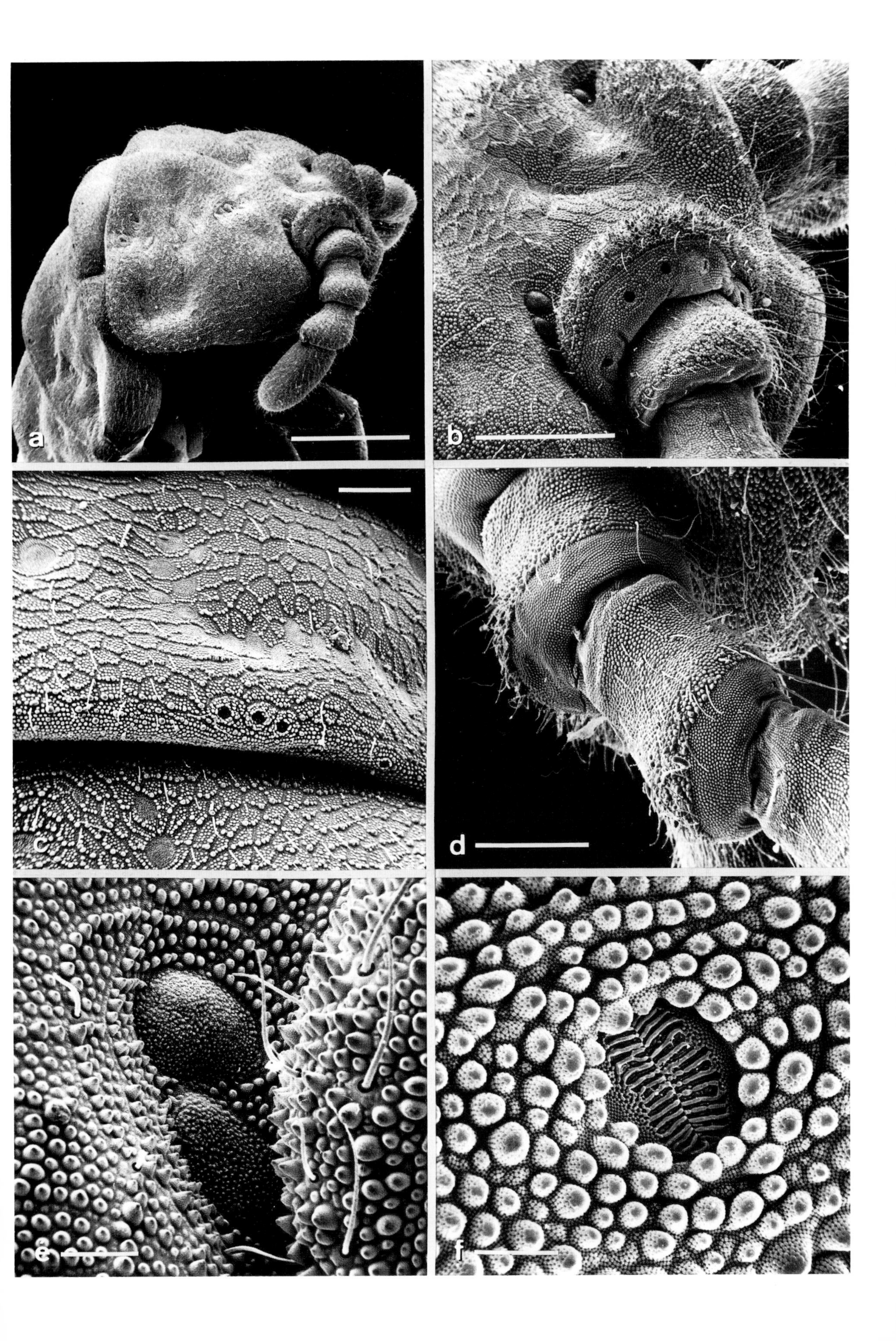
a
b
c
d
e
f

2.12.10 Surface Structure of the Collembola

The basic pattern of the collembolan cuticle is a hexagonal pattern which is schematically illustrated in Fig. 125 and has already been described above in *Tomocerus*. This honeycomb pattern undergoes specific and individual variations in the different Collembola, so that the basic arrangement often becomes unrecognizable (Massoud, 1969; Lawrence and Massoud, 1973; Dallai, 1977; Cassagnau and Lauga-Reyrel, 1985). It is frequently replaced by warty protuberances (macrotubercles) as in *Tetrodontophora*. They are of variable size and are either distributed uniformly and regularly or grouped into fields (Plates 98 and 99). In turn these protuberances are composed of the granular subunits (microtubercles, primary granules, "graines primaires") which often fuse together forming small, irregularly shaped plates towards the tip of the warty protuberances.

The surface structures must also be considered with regard to the animal's way of life. Euedaphic forms frequently have a relatively hairless cuticle (*Onychiurus*). Its hydrophobic properties are enhanced by the formation of macrotubercles (Hale and Smith, 1966; Plates 95 and 96). These animals are, however, unprotected against water loss and must be regarded as extremely hygric animals. Epedaphic and atmobiotic forms often show a higher density of microtubercles particularly on the freely exposed parts of their body. This is probably accompanied by a moderate reduction in the permeability of the cuticle (Lawrence and Massoud, 1973; Ghiradella and Radigan, 1974). As a hygric form *Tomocerus* is additionally protected by a dense coating of scales and hairs, others like *Orchesella villosa* only by a dense coat of hair. However, all these adaptations can, at best, only diminish transpiration. Only the deposition of further lipid and wax layers results in a distinct change in the surface structures and their properties to the extent that the basic structure completely disappears. This is convincingly exemplified by comparison of the surfaces and transpiration rates of *Allacma fusca* and *Sminthurides aquaticus* (Plates 100 and 101; Table 1).

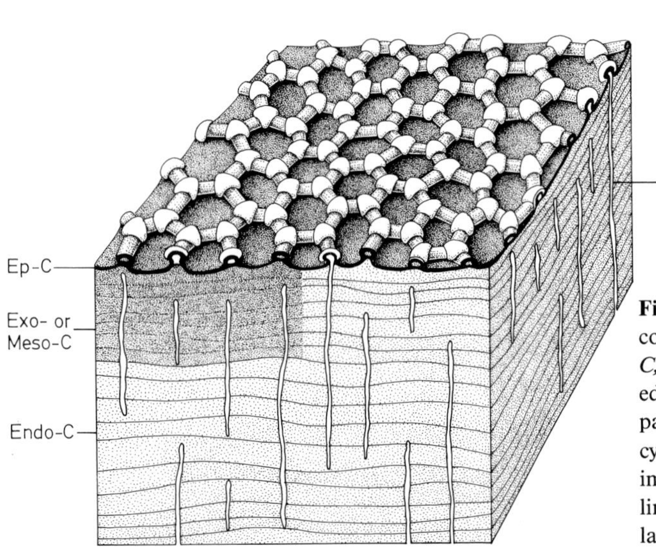

Fig. 125. Schematic representation of the basic pattern of collembolan cuticle; epi-, exo-, meso- and endocuticle (*Ep-C, Exo-C, Meso-C, Endo-C*). The cuticle is hardly sclerotized, the exocuticle or mesocuticle is often absent. The basic pattern of the epicuticle is formed from microtubercles and cylindrical links. It can be considerably varied by approximation and rearrangement of both the tubercles and the linking structures. The pore channels (*pc*) cross the lamellated cuticle, ending in the knobs of the microtubercles

---

Plate 99. *Tetrodontophora bielanensis* (Onychiuridae).

a Cuticle of the antennal base, composed of macro- and microtubercles. 5 μm.

b Abdomen, tergum, cuticle with macro- and microtubercles. 2 μm.

c Pattern of macrotubercles on the tergum of the abdomen. 50 μm.

d Abdomen, posterior margin of a segment. 50 μm.

e Muscle insertion on the tergum of the thorax. 20 μm.

f Variation of the microtubercles on a muscle insertion. 2 μm

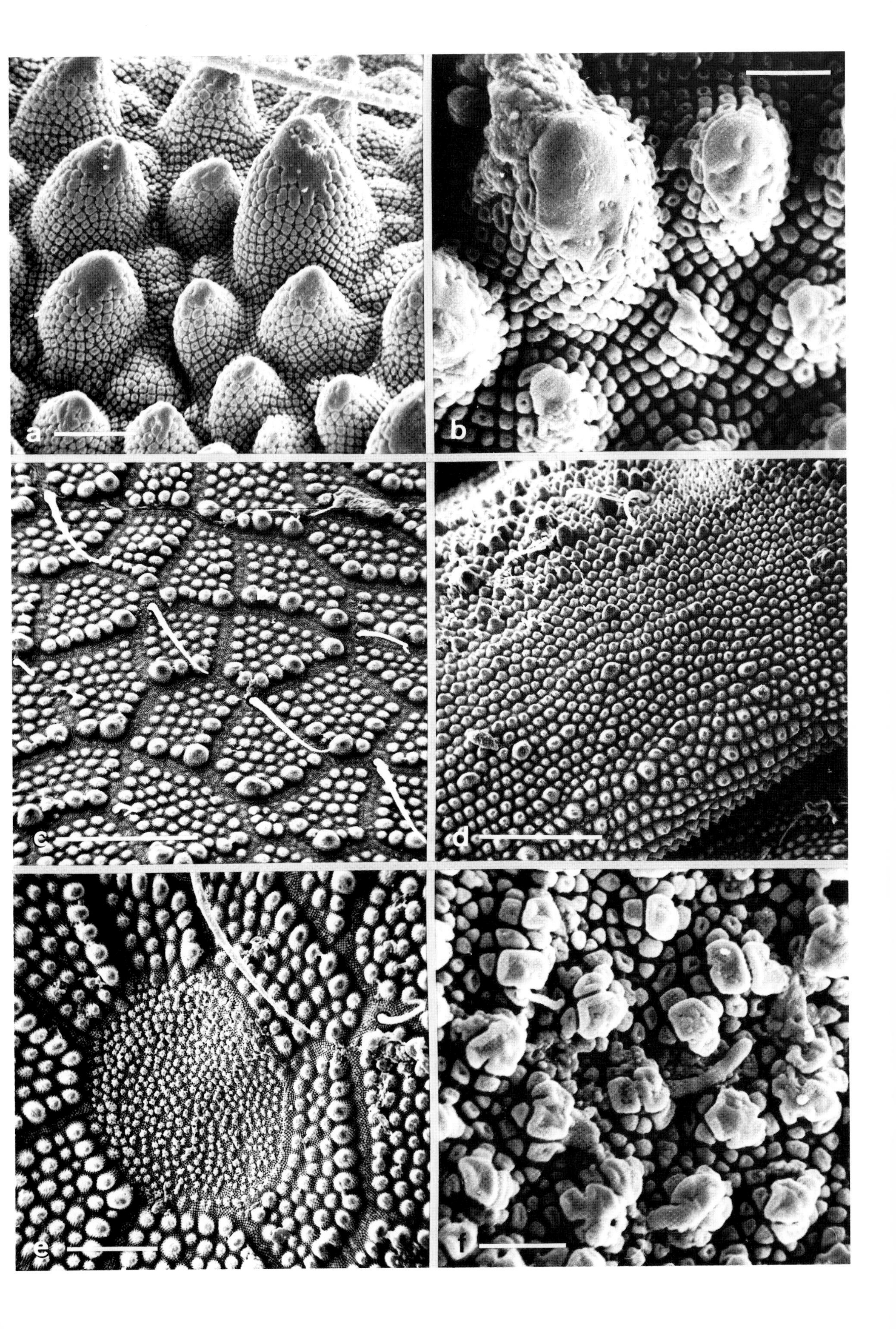
a
b
c
d
e
f

2.12.11 Mating Behaviour of the Collembola

Most arthropleonal Collembola never encounter their sexual partners. The males deposit sperm droplets on stalks (spermatophores) on the ground. The females take them into their genital aperture. This uptake of spermatophores is a chance occurrence dependent on the size and density of the spermatophore crop. The prevalent location for the transfer of sperm is the soil biotope because the uniformly high humidity of the interstitial spaces in the soil prevents desiccation of the sperm droplets before transfer is possible (Schaller, 1970; Döring, 1985).

In other species the males stroke willing females, deposit spermatophores near them and manoeuvre the females towards the spermatophores. Döring (1986) reported a definite contact behaviour between the sexes using their antennae in *Orchesella cincta*. Formation of mating pairs has mainly been reported in the Symphypleona (Mayer, 1957; Strebel, 1932) and is particularly marked in *Sminthurides aquaticus*. This globular Collembola shows a striking sexual dimorphism. Besides the primary sexual characters the partners show a great discrepancy in size. The distinctly smaller males have feelers differentiated to form clasping antennae, with which they attach themselves to the feelers of the large females. Thus, the males are carried around by their partners (Fig. 126; Plate 100), which activates them to deposit spermatophores and guide the females over the deposition site to enable the uptake of sperm. This mating behaviour of the enclasped partners often involves dance-like movements which are believed to facilitate sperm transfer, finding and uptake of spermatophores (Falkenhan, 1932; Schaller, 1970).

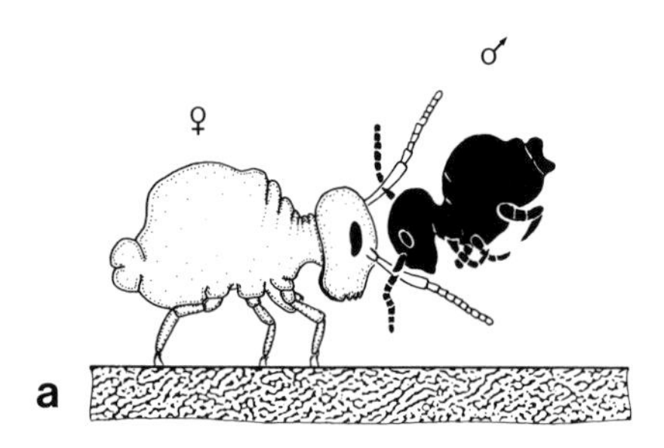

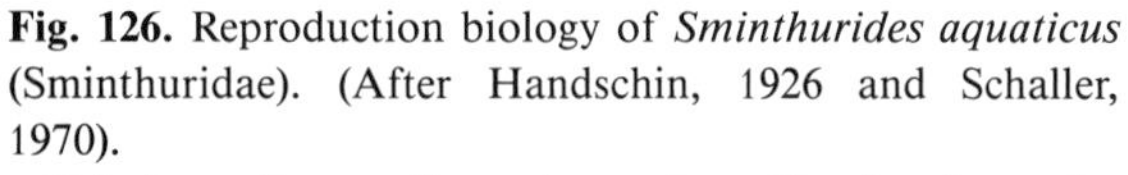

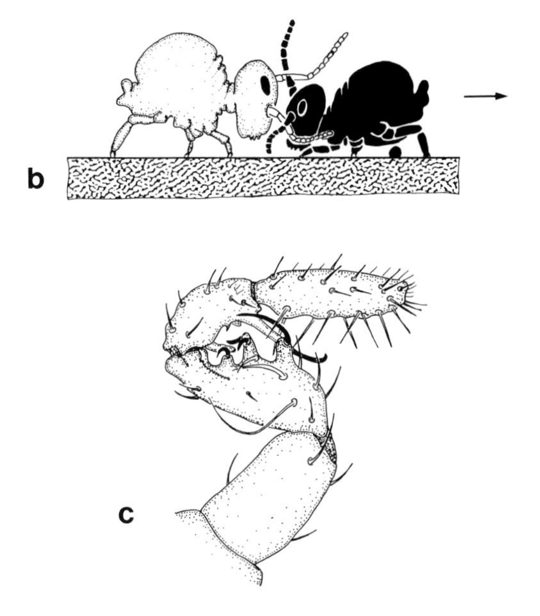

Fig. 126. Reproduction biology of *Sminthurides aquaticus* (Sminthuridae). (After Handschin, 1926 and Schaller, 1970).
a Pair formation on the water surface. The female carries the smaller male raised in the air. **b** A spermatophore has been deposited by the male. The female is subsequently guided over it to ensure its uptake. **c** Modified clasping antenna of a male

Plate 100. *Sminthurides aquaticus* (Sminthuridae).

- **a** Partners in clasping position, lateral. Large female (*left*), male (*right*). 250 µm.
- **b** Partners in clasping position, frontal. The small male clasps the antennae of the female with his modified antennae. 100 µm.
- **c** Partners in clasping position, lateral. 200 µm.
- **d** Clasping antenna of the male (*arrow*) in the clasped position. The third antennal segment is equipped with reinforced bristles to lock it in place. 25 µm.
- **e, f** Furca, lateral and caudal, with the wing-like broadened terminal segments, the mucrones, an adaptation to the epineustic way of life. 10 µm, 20 µm

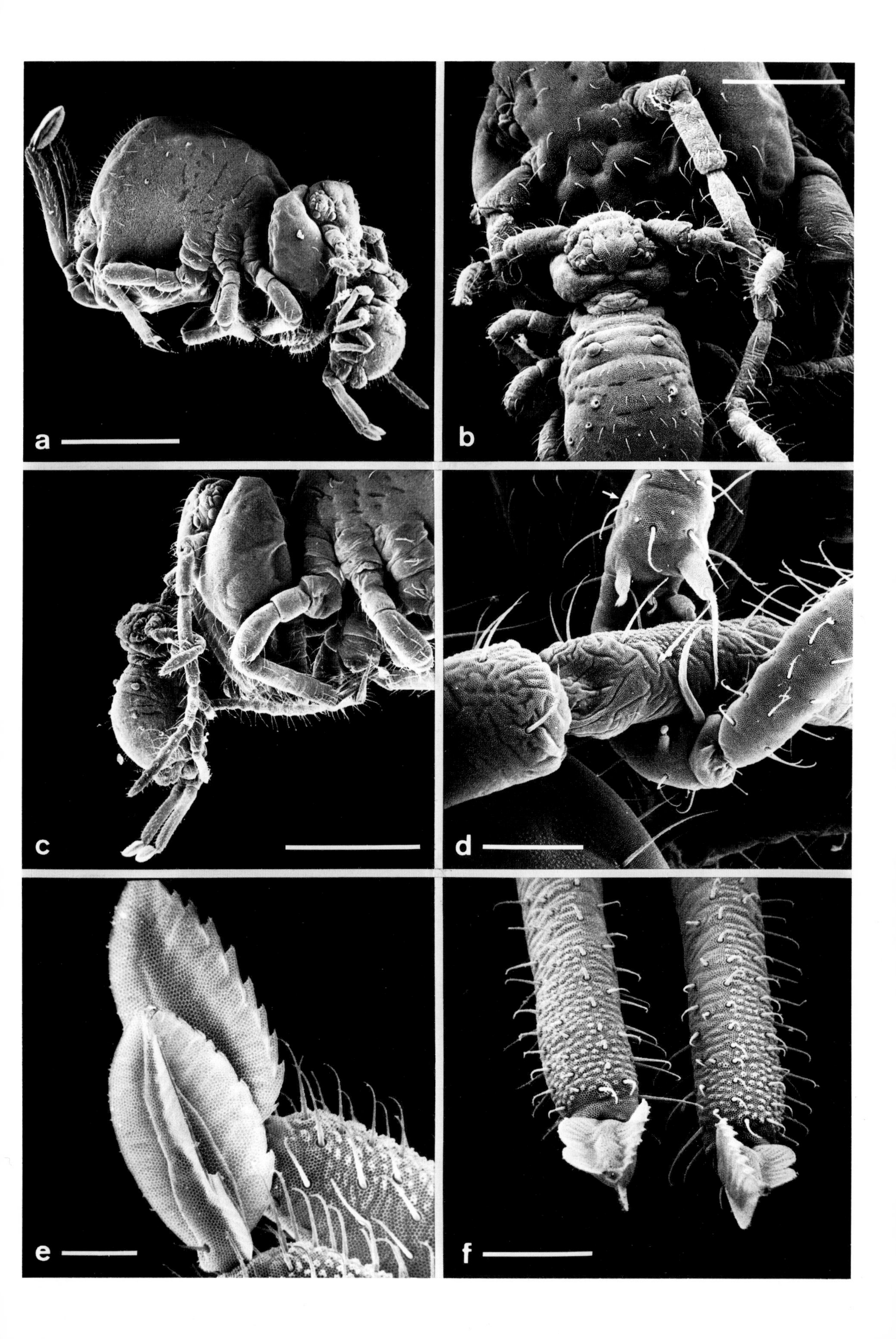
a
b
c
d
e
f

2.12.12 Vertical Distribution and Aggregation of the Collembola

The life forms (Gisin, 1943) are distributed vertically in the various horizons of the soil. The epedaphic Collembola prefer the litter layer (L-layer) and the euedaphic animals live permanently in deeper layers. When the humidity of the upper deposited horizons (L- and F-layers) is sufficient euedaphic forms, like *Onychiurus* also penetrate between the layers of leaves in the F-horizon during the day. The highest population densities are observed in the L- and F-layers. According to Wolters (1983) two-thirds of all Collembola are often to be found in the upper 3 cm of soil in a beechwood (Melico-Fagetum), where they feed microphytophagously on algae and fungi; skeletonization of plant material shows that Collembola also feed saprophytophagously on microbially decomposed plant fragments (Rusek, 1975; Hanlon and Anderson, 1979).

Fluctuations in populations and colony density are related to ecological factors and to rhythmic activities of the animals. Collembolan aggregations are primarily caused by their own activities and are evidently induced by aggregation pheromones (Verhoef et al., 1977; Mertens and Bourgoignie, 1977; Mertens et al., 1979). Ecological factors only play a secondary role in influencing them (Christiansen, 1970; Joosse, 1970, 1971, 1981; Peterson, 1980; Poole, 1964; Usher, 1969; Wolters, 1983). Thus, aggregation dynamics are a result of activity fluctuations which are determined, for instance, by moulting cycles and feeding phases and can therefore be designated as periodic aggregations (Joosse and Verhoef, 1974; Verhoef and Nagelkerke, 1977). Moisture and temperature are two of the effective ecological factors governing collembolan aggregations and inducing vertical migration (Milne, 1962; Usher, 1970; Takeda, 1978).

Plate 101. *Allacma fusca* (Sminthuridae).

a Overview, lateral. Next to the everted springing organ lies the small abdomen. 0.5 mm.

b Head, oblique frontal. 200 μm.

c Antenna. 100 μm.

d Antenna, distal, with secondary segmentation. Here the basic hexagonal pattern of the cuticle is recognizable. In the proximal region the microtubercles are overlaid by an additional coating. 20 μm.

e Antenna, distal, with tubular hairs and normal hexagonal cuticular structure. 5 μm.

f Antenna, proximal, with coated cuticle. 2 μm.

g Cuticle of the trunk, lateral, with overlaid and modified cuticular pattern. 5 μm

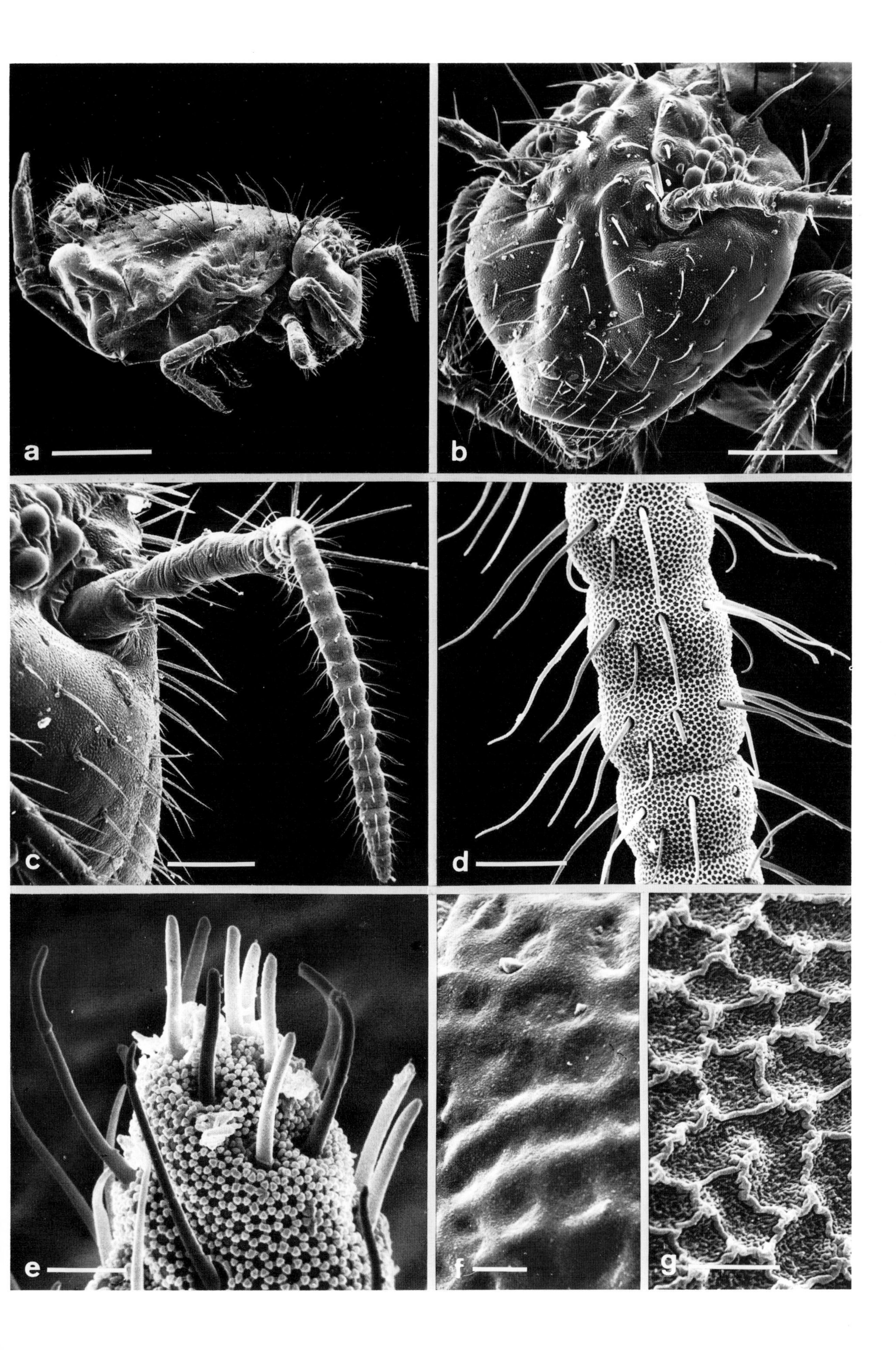
a
b
c
d
e
f
g

2.12.13 Respiration of the Collembola

Most Arthropleona breathe through the skin, their body cells being passively supplied with oxygen by means of diffusion. This mechanism can only function in tiny animals, with larger forms having a poorer oxygen supply than smaller ones. The 9-mm-long *Tetrodontophora bielanensis* is more sensitive to oxygen deficiency than the small euedaphic *Onychiurus*. Due to the long diffusion path, not all cells can be sufficiently supplied with oxygen when the oxygen partial pressure falls in the outer medium (Zinkler, 1966). This is also reflected in the water permeability of the skin. *Onychiurus* transpires about ten times faster than *Tetrodontophora*.

The small euedaphic forms, which have a distinctly higher oxygen consumption and CO_2 resistance than the epedaphic forms (Zinkler, 1966), would be endangered in the interstitial system of the soil if they were not especially adapted to temporary flooding of their biotope by rainwater. In addition to the basic pattern on the collembolan cuticle, the surface of these animals is differentiated so that it is strongly hydrophobic. If flooding occurs, the body surface becomes surrounded by a plastron, by means of which respiration can be maintained for several hours (Zinkler and Rüssbeck, 1986). Thus, it is functionally comparable to the gas gills of certain water insects.

The Symphypleona and the arthropleonal genus *Actaletes* possess a simple tracheal system. Oxygen is supplied to this system through a pair of spiracles situated laterally on the head near the neck region. These evidently cannot be closed. Tracheal branching is simple, without anastomoses. Its genesis in the course of early larval development was investigated in *Allacma fusca* (Fig. 127) by Betsch and Vannier (1977).

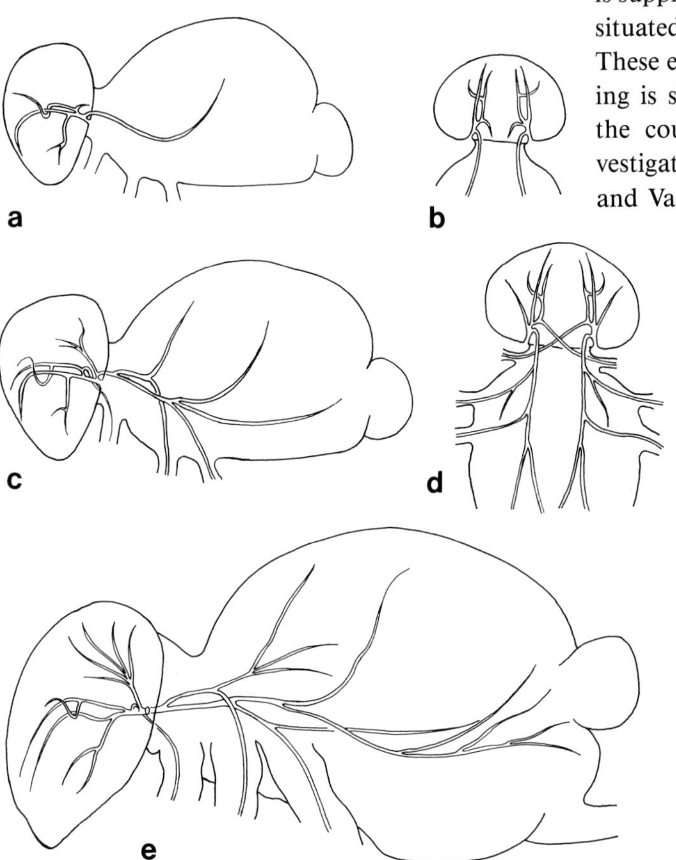

Fig. 127. Genesis of the tracheal system in juvenile stages I and II of the globular Collembola *Allacma fusca* (Sminthuridae). (After Betsch and Vannier, 1977).
a, b Juvenile stage I (1st intermoult stage); **c, d** Juvenile stage II (2nd intermoult stage); **e** Juvenile stage II (3rd intermoult stage).
In juvenile stage I the cuticle shows the normal collembolan pattern composed of microtubercles and accompanied by a high permeability. From the second stage onwards the structure changes, above all due to the addition of further overlying layers and variation of the tubercle pattern. There is a simultaneous growth in the tracheal system to compensate for the diminishing role played by respiration through the skin

Plate 102. *Allacma fusca* (Smithuridae).
a Large abdomen, caudal, with small abdomen (*arrow*) and springing organ (furca), lateral. 250 μm.
b Abdomen, caudal, with everted furca. 125 μm.
c Leg, distal. The claw is covered by a cap, the tunica. 50 μm.
d Retinaculum. 50 μm.
e Everted tubular vesicles of the ventral tube. 200 μm.
f Section of a tubular vesicle of the ventral tube with papilla-like surface. 25 μm

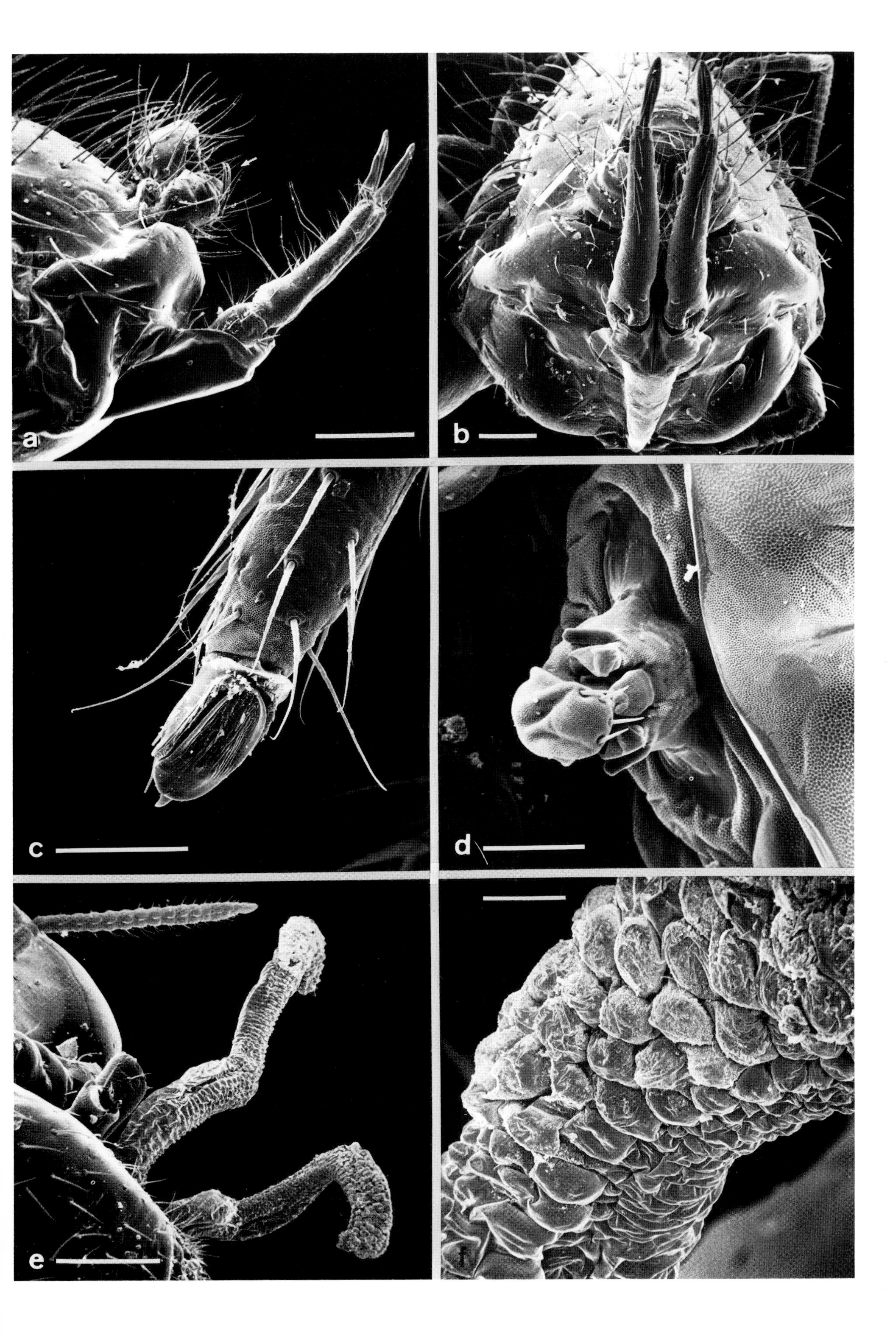

a
b
c
d
e
f

2.12.14 Dwellers on Ice and Snow: Glacier Fleas

The best known and most striking Collembola species living on the ice of the alpine mountain regions is *Isotoma saltans*, the common glacier flea (Steinböck, 1939; An der Lan, 1963; Schaller, 1960, 1963). Its habitat is the system of small crevices in the ice surface and within firn fields on the central part of glaciers. Mass concentrations are frequently found on the surface of firn fields and on new snow causing a sooty coating on the white surface. But Steinböck (1939) was able to discover glacier flea aggregations in ice as deep as 30 cm.

Their temperature preference is generally agreed to lie about 0 °C (approx. −4° to +5 °C). When disturbed, by air movement for example, they show a positive geotactic reaction. Although strong sunshine sometimes attracts them to the surface, they always return rapidly underground. Glacier Collembola feed on a dark-coloured mixture of organic and inorganic material known as "kryokonite" which is abundantly distributed over the ice surface. The inorganic, sharp-edged micaceous particles are enriched with algae, pollen grains, ciliates, rotifers, tardigrades and detritus.

Apart from the deep bluish-black *Isotoma saltans* there are several similar isotomids which belong to the glacial or nival fauna respectively so that careful determination is required. Recent investigations focus on the microclimate in the biotope of glacier fleas, their water balance and the ultrastructure of these interesting Collembola (Eisenbeis and Meyer, 1986; Eisenbeis, Meyer and Maixner, in prep.). Research on adaptation to cold and overwintering strategies in Collembola, mites and other soil arthropods was carried out by Aitchison (1979, 1983), Block (1983), Block and Zettel (1980), Brummer-Korvenkontio and Brummer-Korvenkontio (1980), Joosse, 1983, Leinaas (1983), Schenker (1983), Sømme (1976, 1976/77, 1979, 1982) and Zettel (1984).

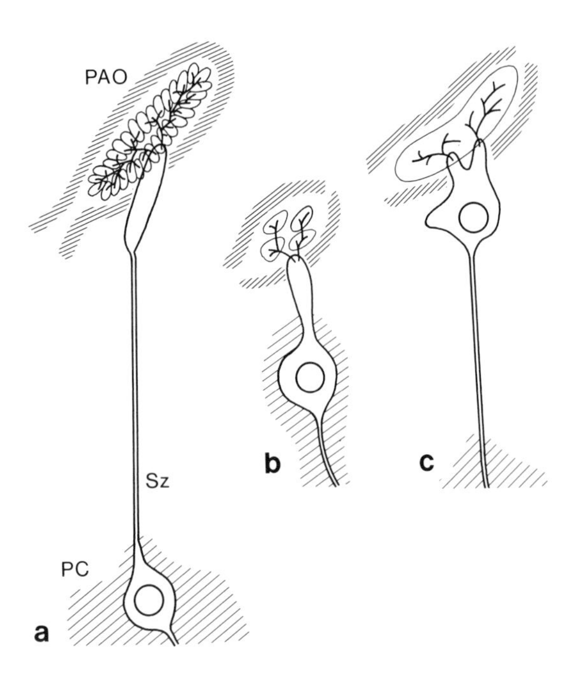

Fig. 128. Structural characteristics of the postantennal organs (*PAO*) of Collembola. (Altner and Thies, 1976). There are family-specific differences between the number of outer sensory units and between the position of the sensory cell (*Sz*) with relation to the protocerebrum (*PC*) (cf. Chaps. 2.12.2 and 2.12.7).
a *Onychiurus*; **b** *Hypogastrura;* **c** *Isotoma*

Plate 103. *Isotoma* "sp. G" (Isotomidae), glacier flea from glaciers in the Ötztal Alps (Austria) (cf. Eisenbeis and Meyer, 1986).
a Overview, lateral. 0.5 mm.
b Abdomen, oblique caudal, with triple flaps of the anal region and extended furca. 100 µm.
c Head, dorsal, with antennal bases, postantennal organs and ocular regions each consisting of six large and two reduced ommatidia. 100 µm.
d Abdomen, ventral, with retinaculum (*arrow*), ventral sclerites and manubrium (*arrow*). Rotation of the furca is facilitated by raising and lowering the large plates (cf. Eisenbeis and Ulmer, 1978). 50 µm.
e Ommatidia. 5 µm.
f Postantennal organ. 5 µm.
g Postantennal organ, perforated cuticle. 1 µm

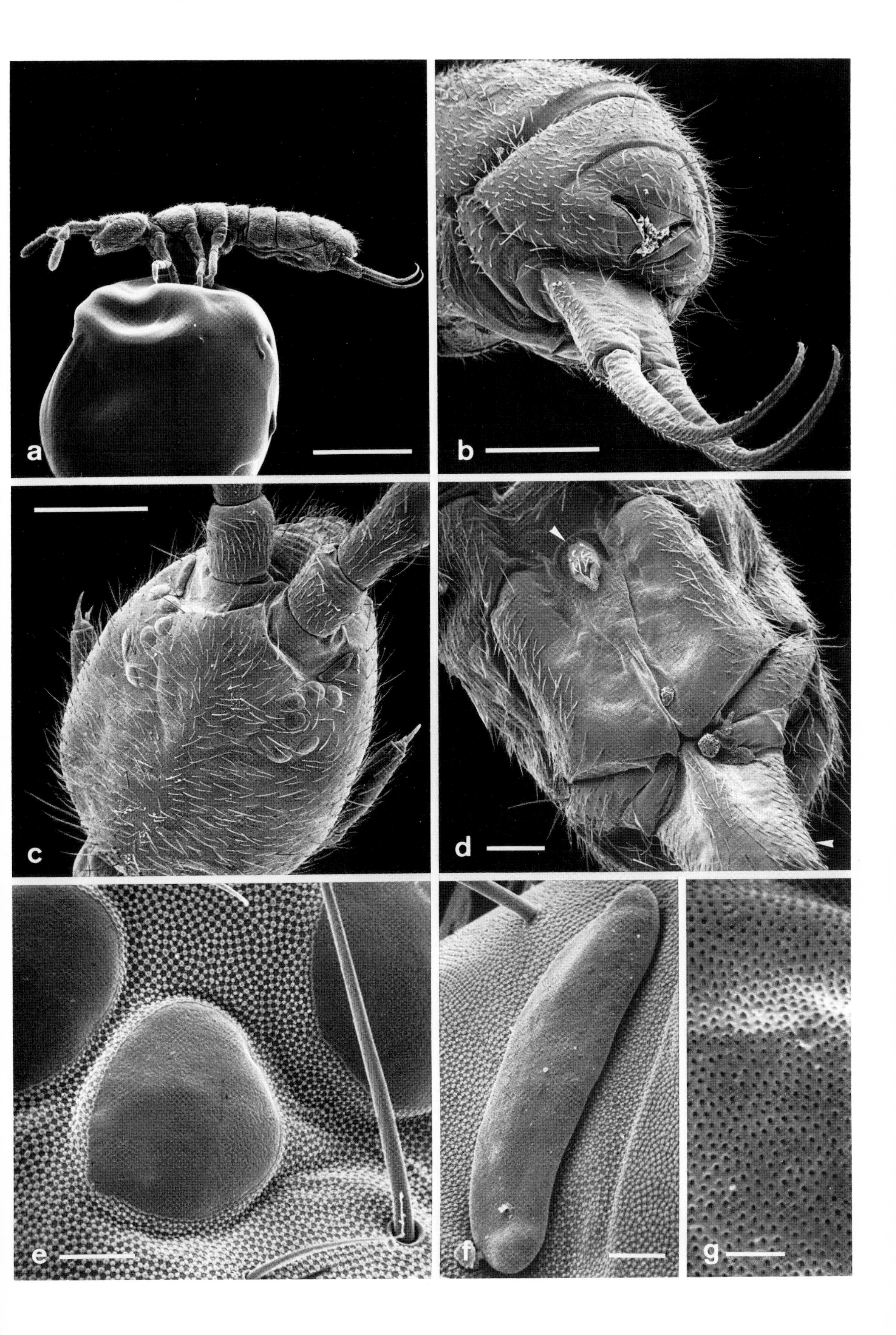

a
b
c
d
e
f
g

2.12.15 Dwellers on Ice and Snow: Firn Fleas

As far back as 1919 Handschin reported that Collembola were abundant in the nival fauna. Of the 90 collembolan species known in Switzerland at that time 27 could be regarded as belonging to this group. The firn fleas, *Isotoma nivalis* and *Isotomurus palliceps* are also encountered on glaciers, but more frequently in their bordering areas. A recent comparison of summer populations from the centre of a glacier with those of the bordering firn fields within the Ötztal valley near Obergurgl (Austria) showed that all the isotomids from the firn belonged to the *Isotomurus palliceps* species (Plate 104), while the purely glacial dwellers were an *Isotoma* species (Plate 103). Though superficial similarities between the two species are evident at first glance, careful examination reveals distinct differences (cf. Eisenbeis and Meyer, 1986).

The most striking difference between the firn flea and the deeply black pigmented glacier flea from the Ötztal Alps is that the legs, antennae and dentes of the former are white. Furthermore, they vary in the basic shape and proportion of their segments. A scanning electron microscope analysis revealed that the ornamentation on the labrum is an excellent determination character (Fig. 129).

Isotomurus palliceps disappears very rapidly into the snow when disturbed, especially when stones are lifted. Other individuals, possibly attracted to the surface by the light, appear suddenly from underground only to vanish immediately again.

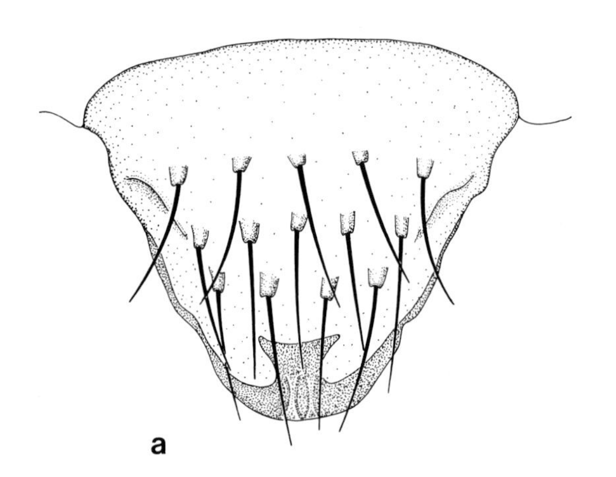

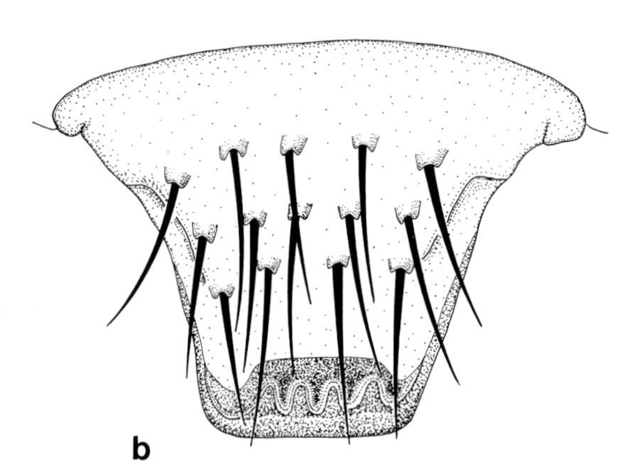

Fig. 129. Structure of the upper lip (labrum) of glacier dwellers from the Ötztal Alps (Austria). **a** Glacier flea, *Isotoma* "sp. G". **b** Firn flea, *Isotomurus palliceps*

Plate 104. *Isotomurus palliceps* (Isotomidae), firn flea from glaciers in the Ötztal Alps (Austria).

a Overview, oblique caudal, with extended furca. 100 µm.

b Overview, ventral, with folded furca. 250 µm.

c Head, with mouth cone, oblique ventral. 100 µm.

d Ocular area with eight well-developed ommatidia. The postantennal organ (*arrow*) is in the lower *left* corner of the picture. 25 µm.

e Upper lip (labrum), distal, with typical ornamentation. 10 µm.

f Ventral tube with ventral groove (*arrow*) and tube flaps. The retracted vesicles lie in the medial furrow. 25 µm

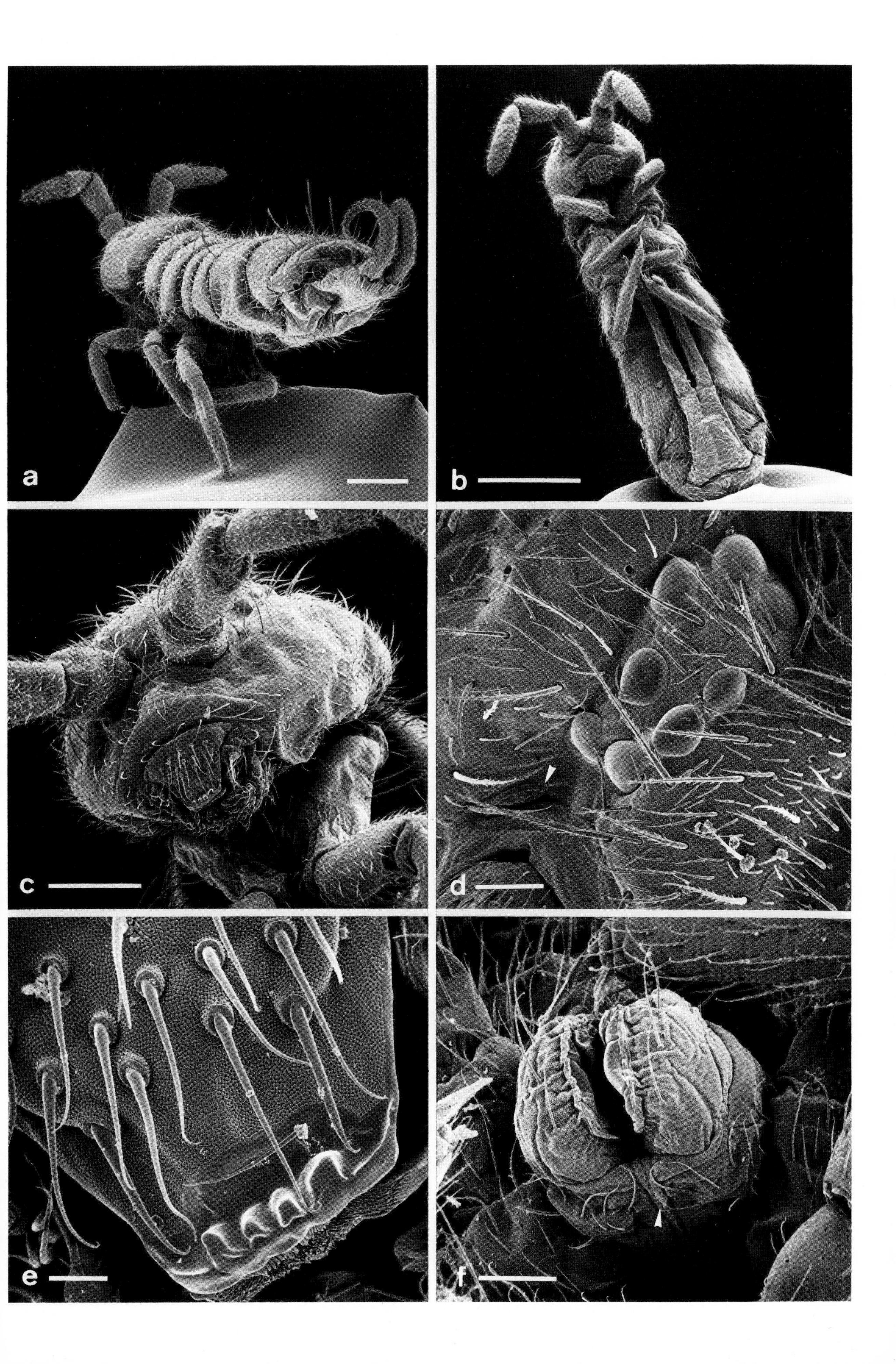
a
b
c
d
e
f

2.13 Order: Archaeognatha – Bristletails (Insecta)

GENERAL LITERATURE: Palissa, 1964; Handschin, 1929

2.13.1 Characteristics of the Archaeognatha

These bristletails belong to the order of the Archaeognatha and are represented in Europe by two families, Machilidae and Meinertellidae. The 10–15-mm-long animals have long feelers on their head and a central terminal filum which projects far beyond the paired cerci on the posterior of the abdomen (Fig. 130; Plates 105–108). The Machilidae have a dense coat of scales on the body surface, the extremities, antennae, cerci and terminal filum (Larink, 1976), whereas the Meinertellidae are without scales on the extremities and antennae (Sturm 1984). The bristletails often completely blend in with their surroundings by means of their body markings. They occur, on the one hand, in the rocky supralittoral zone at the sea (Delaney, 1959; Larink, 1968; Joosse, 1976), on the other, in rocky slopes and scree of warm highlands and mountains (Wygodzinsky, 1941; Janetschek, 1951; Sturm, 1980). They prefer light and warmth, tolerate low to moderate moisture and are, therefore, only superficially adapted to the soil which often has high humidity, little light and low temperatures. When threatened the bristletails jump several centimetres by thrusting the abdomen against the ground with the strong styli of the ninth abdominal segment and utilizing this force for the jump.

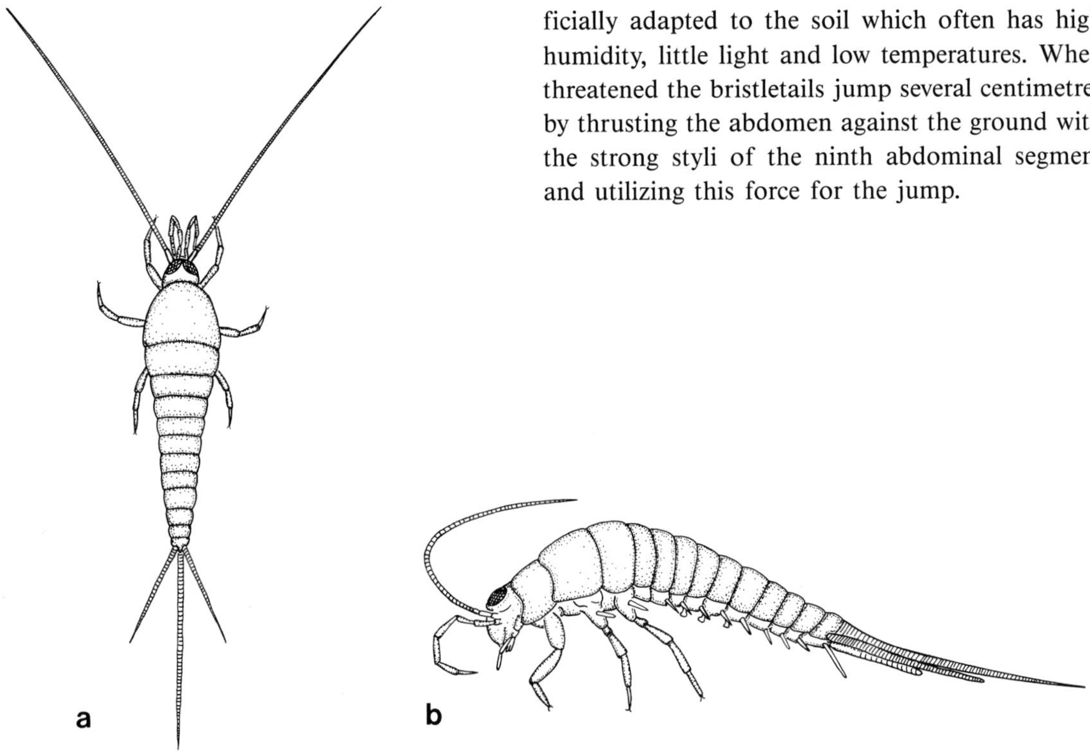

a b

Fig. 130. Habitus of bristletails (Machilidae). (After Handschin, 1929).
a *Machilis polypoda,* dorsal. **b** *Machilis* sp., lateral

Plate 105. *Trigoniophthalmus alternatus* (Machilidae).
a Overview, frontal, with the medially fused compound eyes, antennae, maxillary and labial palps and the legs. 250 μm.
b Compound eye, with sensory hairs and glandular outlets. 25 μm.
c Compound eye, surface of the ommatidia with sensory hair and glandular outlet. The surface of the cornea is composed of microtubercles (anti-reflection layer). 5 μm.
d Glandular outlet. 2 μm

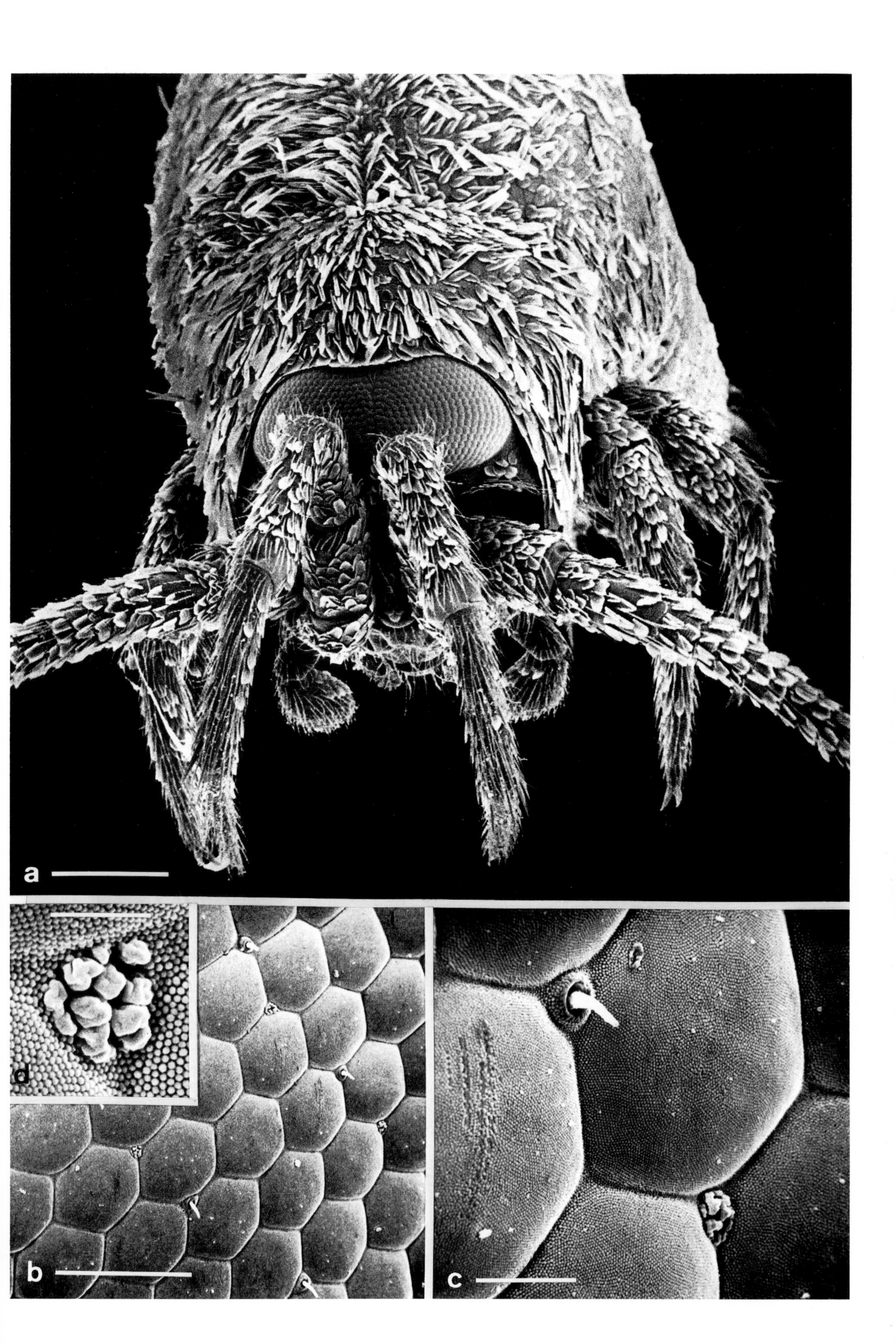
a
b
c
d

2.13.2 Sensory Organs on the Mouthparts of Bristletails (Machilidae)

Bristletails feed on algae and lichens. Their mouthparts (Larink, 1971) are highly sensitive to various stimuli. Thirteen different types of sensilla were found on them (Krüger, 1975). Some of the sensilla are distributed over all the mouthparts, others are situated in distinct fields on individual appendages.

On the distal tip of the labial palps of *Trigoniophthalmus alternatus* (Plate 106) there are 17–34 sensory complexes which cannot be fitted into any known classification of sensilla. They are probably not derived from genuine hairs, but emerge from the cuticle and the underlying epithelium. These protruding structures are rounded at the base, towards the tip they are obliquely flattened, with 4–17 small papillae on the plateau and trichomes or small genuine hairs on the flanks. Preliminary fine structural investigations suggest that they might be contact chemoreceptors. The complexes are innervated by about 100 dendrites with ciliary outer segments, whereby approximately ten sensory subunits, surrounded by their own cuticular sheath, are associated with one papilla.

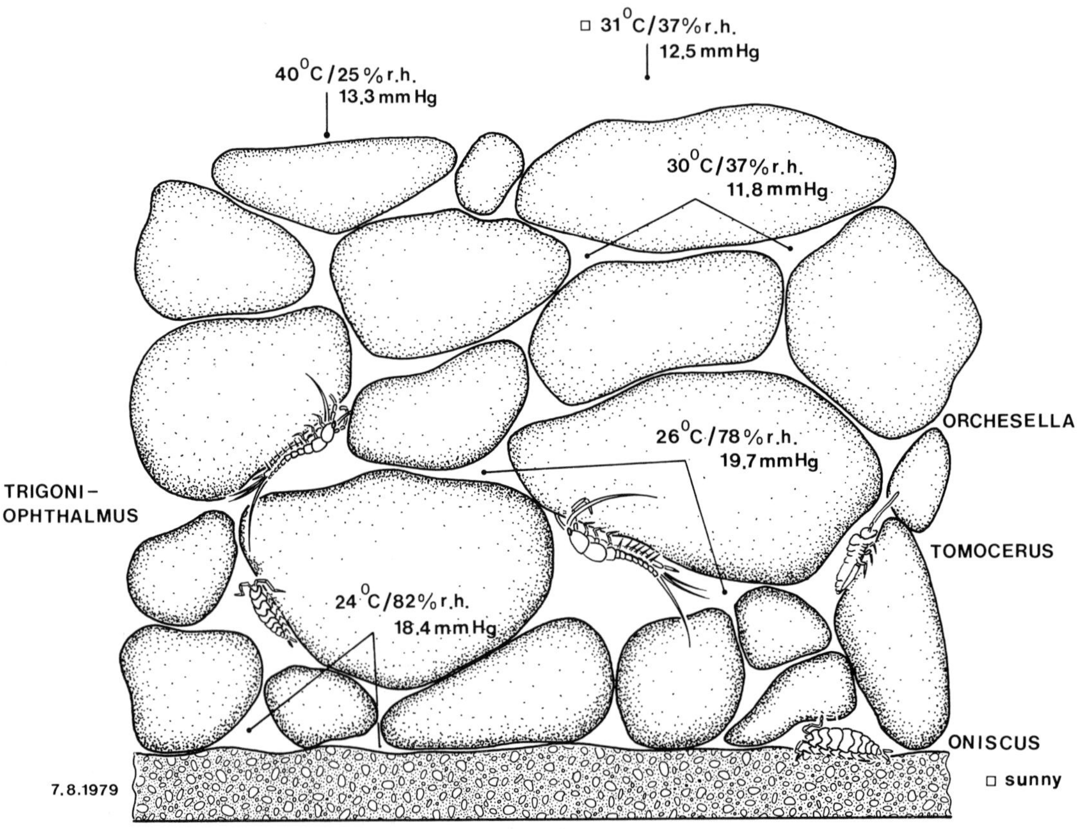

Fig. 131. Microclimate – temperature and moisture regime – in the habitat of the bristletail *Trigoniophthalmus alternatus* (Machilidae) in strong insolation. (After Eisenbeis, 1983). Under the conditions shown here *T. alternatus* prefers the third layer of stones (about 20 cm deep) where the ambient humidity is approximately 80%. When the surface of the stones is shaded, but conditions are otherwise constant, the animals migrate to the underside of the upper stone layer. Collembola and Isopoda were regularly encountered with the bristletails. The partial pressure of water vapour at the different sites is given in mmHg (1 mmHg ≈ 0.133 kPa)

Plate 106. *Trigoniophthalmus alternatus* (Machilidae).

a Anterior body, ventral, with antennae, maxillary and labial palps (*arrow*). 0.5 mm.

b Labial palp, distal. 25 μm.

c Sensory complexes (suprahairs) (*arrow*) on the apical labial palp surface. 20 μm.

d Sensory complexes (suprahairs) with small bristles or protuberances in the shaft region. 5 μm.

e, f Labial sensory complex (suprahair) apical with papillae. 2 μm, 1 μm

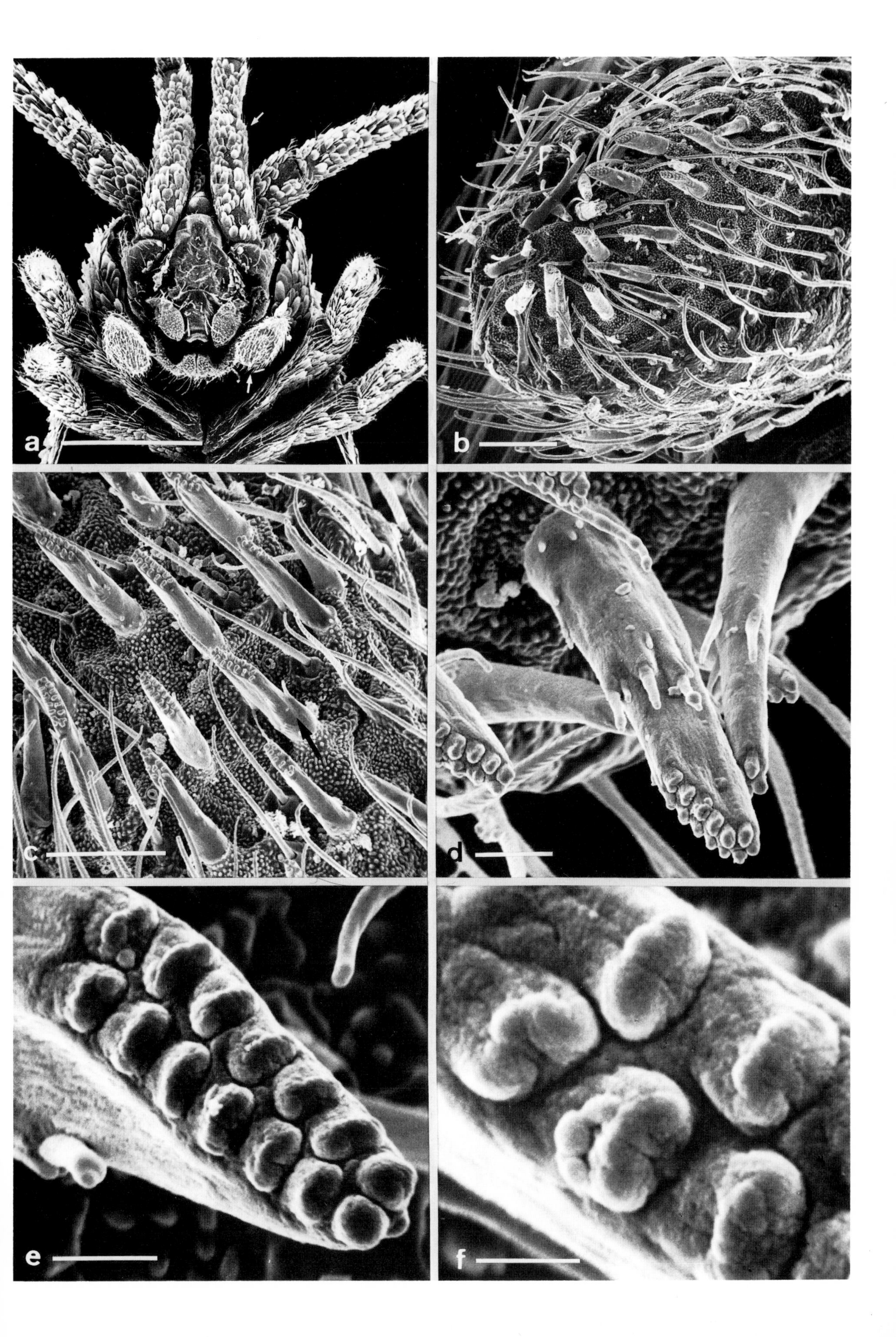
a
b
c
d
e
f

2.13.3 Water Balance of the Machilidae

The machilids are equipped with a variable number of coxal vesicles which are integrated into the body wall of the abdomen. Each segment can bear a maximum of four vesicles. They are everted by means of haemolymphatic pressure and retracted by muscles. The cuticle of the vesicle is pervaded by a fine capillary and pore system (Plate 107). According to Bitsch and Palevody (1973), Bitsch (1974) and Weyda (1974) the epithelium of the vesicles is a typical transport epithelium with basal and apical surface enlargement and numerous membrane-mitochondria complexes. The vesicles are employed in the uptake of water. A deficit of 20% of the normal water content in *Trigoniophthalmus alternatus* is compensated by the absorption of water through the vesicles in a few minutes (Fig. 132) (Eisenbeis, 1983). During absorption the number of vesicles activated varies, as they are everted and retracted independently of one another. Houlihan (1976) reported even higher rates of uptake for *Petrobius brevistylis.*

The coxal vesicles are regarded as accessory organs for water and ion uptake especially when the substrate (leaves, bark, stones) is only covered by a thin film of water. Their functional significance is: firstly as an adaptation strategy to enlarge the migration area, secondly they enable the animals to extend their stay in zones with unfavourable moisture conditions and thirdly they help to bridge the moisture-sensitive moulting phase. The organs are functional even during moulting, albeit at a reduced absorption rate.

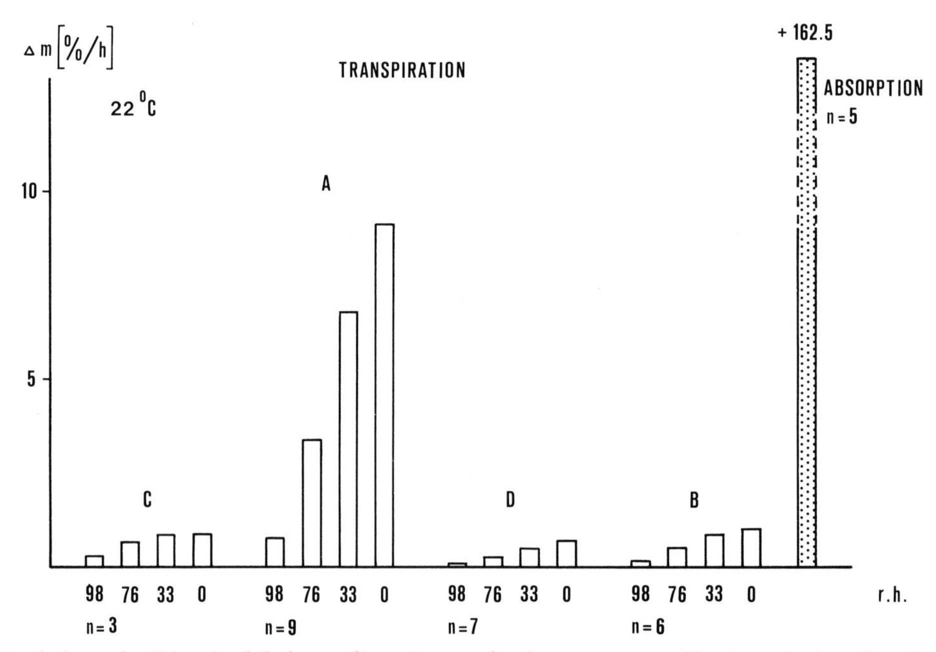

Fig. 132. Water balance in *Trigoniophthalmus alternatus* (Machilidae). (Eisenbeis, 1983). Transpiration at various ambient humidities and absorption of pure water through the coxal vesicles. The rates are expressed in the percentage change in the total water mass m_0. The newly hatched animals are more sensitive to water loss than the older instars. Generally, the capacity of water absorption exceeds the loss through transpiration by more than ten times. **A** Juveniles, 1/2–1 days after hatching. **B** Imagines. **C** Eggs. **D** Juveniles, 60 days after hatching

Plate 107. *Trigoniophthalmus alternatus* (Machilidae).
a Overview, lateral. 1 mm.
b Overview, ventral, with coxal vesicles. 1 mm.
c Abdomen, ventral, with everted coxal vesicles and styli. 0.5 mm.
d Pair of coxal vesicles. 100 µm.
e Cuticle of coxal vesicle, distal. 50 µm.
f Cuticle of coxal vesicle with microcapillaries. 2 µm

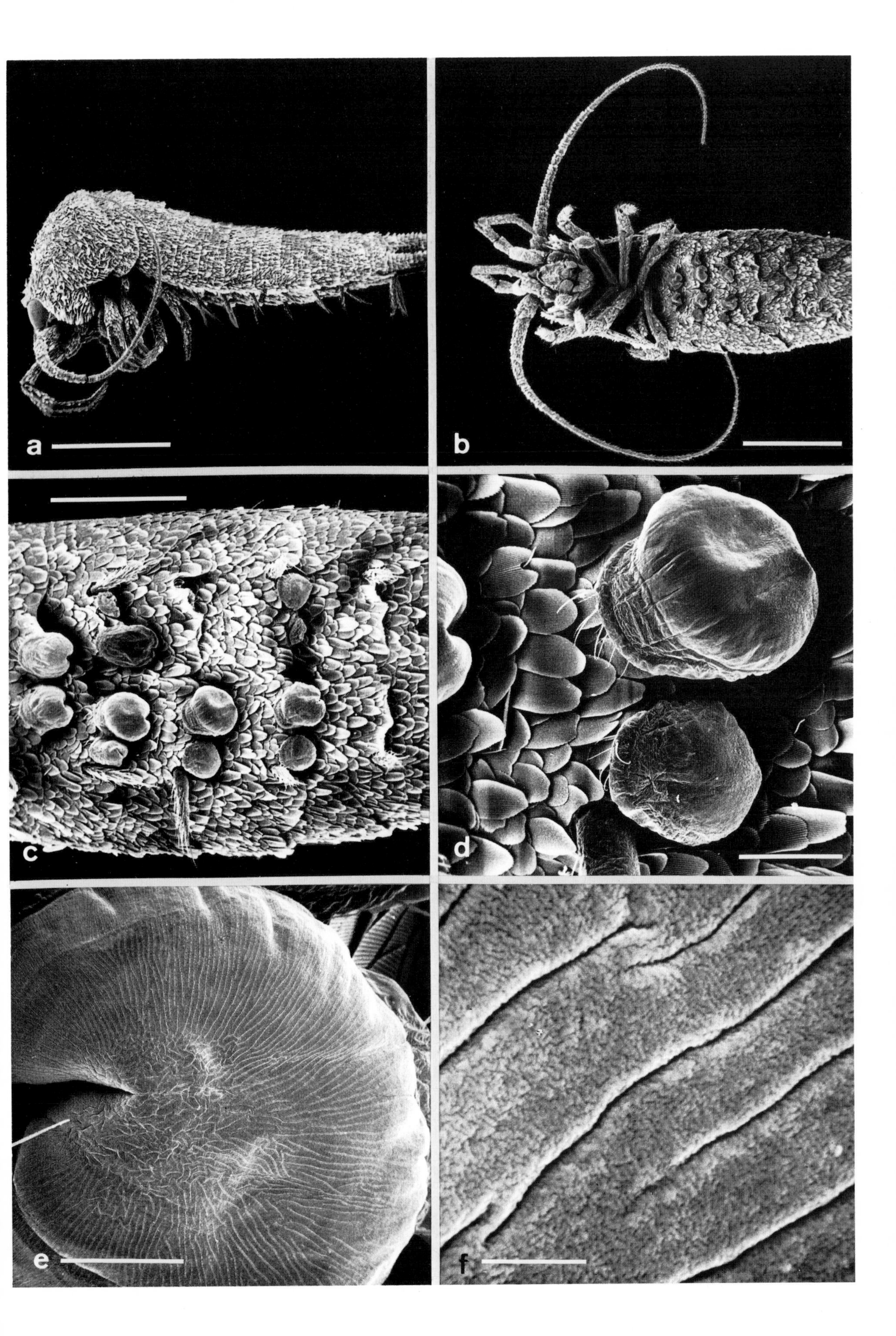

a
b
c
d
e
f

2.13.4 Outer Genital Appendages of the Archaeognatha

Knowledge of the outer genitalia of the bristletails is useful in determining species and also for better understanding of their reproductive biology.

On the posterior abdomen of the female of *Petrobius* there is an ovipositor, composed of four slender, segmented gonapophyses, two each on the eighth and ninth segments (Fig. 133). They are slotted together and adjustable with respect to each other. The gonapophyses of the female of many *Machilis* species form a short, strong ovipositor with thorn-like digging claws, which are thought to facilitate deposition of the eggs.

On the terminal segments of the male of *Petrobius* lies a penis consisting of two parts which form a cylinder broadening to a club-shape at the apex (Fig. 133). In addition, unsegmented moveable paramers bearing sensory bristles emerge from the ninth segment. In other bristletails the segmented paramers are formed by the eighth and ninth segments, but can also be lacking (Larink, 1970).

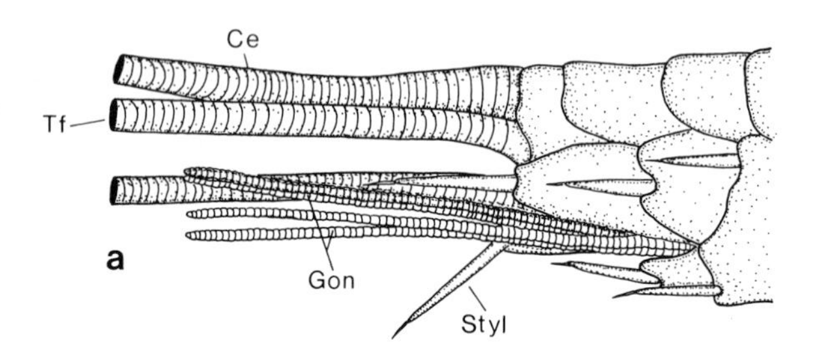

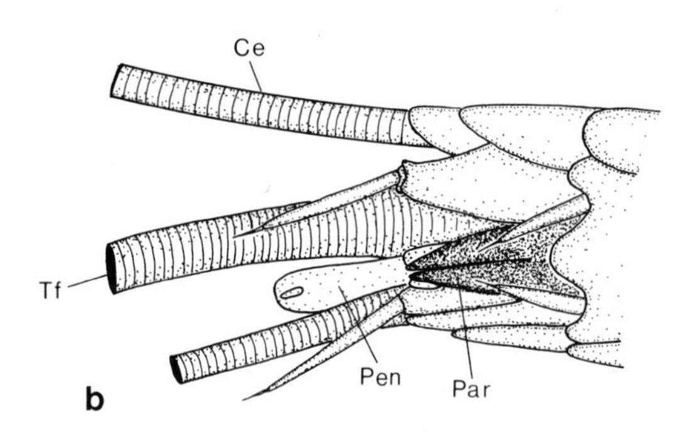

Fig. 133. Outer genitalia of *Petrobius brevistylis* (Machilidae). (After Larink, 1970).
a Female with segmented gonapophyses (*Gon*), which form the actual ovipositor. *Ce* Cercus; *Styl* stylus; *Tf* terminal filum. **b** Male with the outer genital apparatus consisting of the paired paramers (*Par*) and the bipartate penis (*Pen*)

Plate 108. *Trigoniophthalmus alternatus* (Machilidae).
a Antenna, central portion. 100 μm.
b Abdomen, caudal, lateral, with styli, cerci and terminal filum. 0.5 mm.
c Antenna, distal, without scales. 25 μm.
d Cerci and terminal filum. 200 μm.
e Apex of an antennal hair. 2 μm.
f Cercus, distal. 25 μm

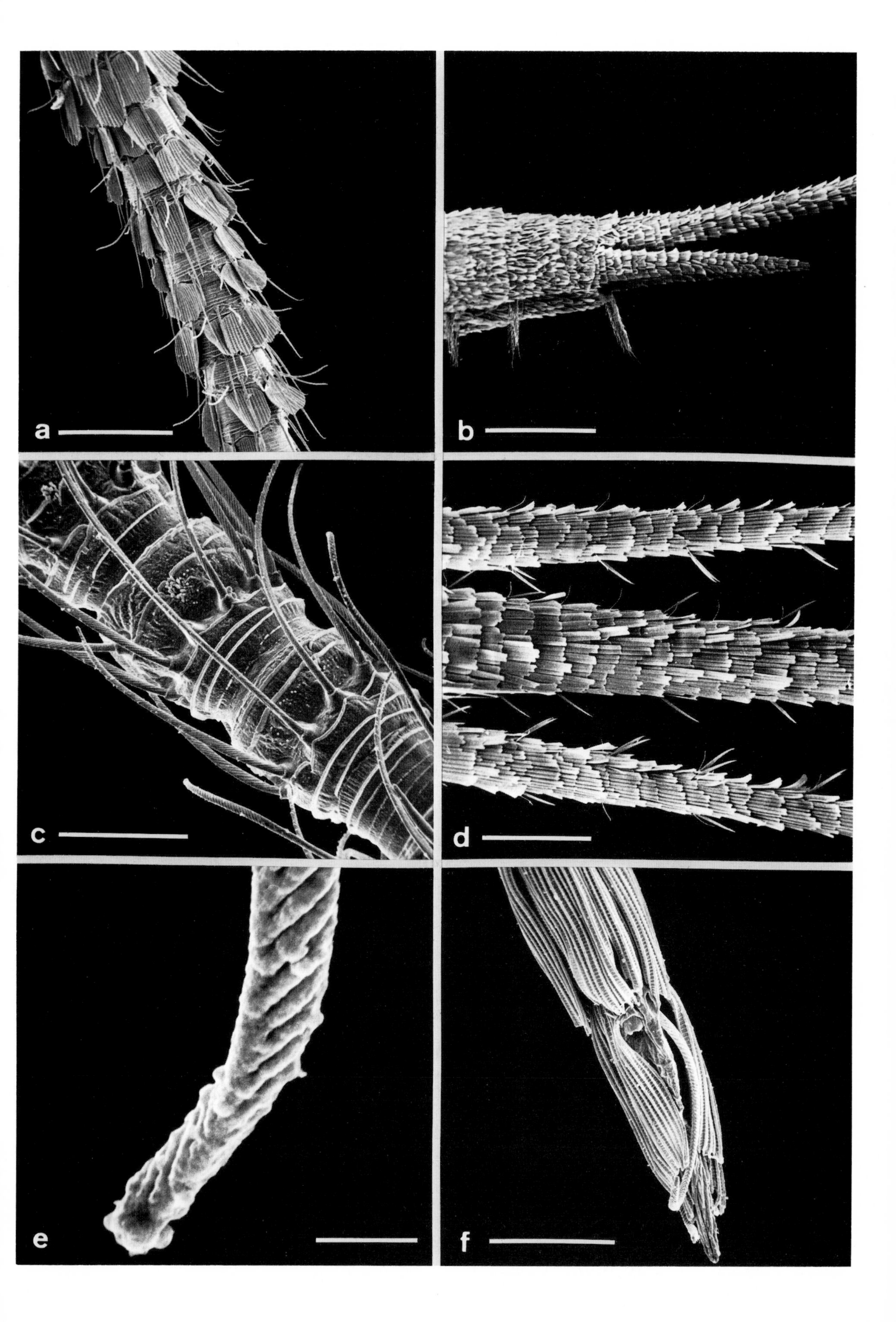
a
b
c
d
e
f

2.13.5 Mating Biology of Bristletails

As in many soil arthropods, mating and sperm transfer are accomplished indirectly without copulation in most of the bristletails. Their mating biology was described by Sturm (1952, 1955, 1960). At the beginning of courtship the partners touch each other with their feelers. If the female is willing to mate she raises her abdomen. The male manoeuvres the female from a position facing him until they are side by side and guides her to and fro with dancing movements. At the same time he presses his penis onto the soil, attaching a thread, which he stretches by moving forward and elevating his abdomen. This thread bears one to five sperm droplets. As a result of renewed fondling with the feelers and movements of the maxillary palps the male is able to manoeuvre the female in a semi-circle so that she is adjacent to the elevated thread.

The uptake of the sperm droplets through the ovipositor is stimulated by vibration of the male and, in *Machilis*, also accompanied by antennal movements (Fig. 134).

In the meantime three modes of sperm transfer have been determined in the bristletails. Besides the type involving the use of a sperm-bearing thread, which is observed in most of the Machilidae, "direct" transfer has been described in *Petrobius* (Sturm, 1978) and indirect transfer by means of spermatophores in the Meinertellidae family. In contrast to the mode of transfer observed in the Machilidae, the male deposits a stalked spermatophore on the ground and the sperm sac, covered by secretory products, is picked up by the female with the ovipositor (Sturm and Adis, 1984).

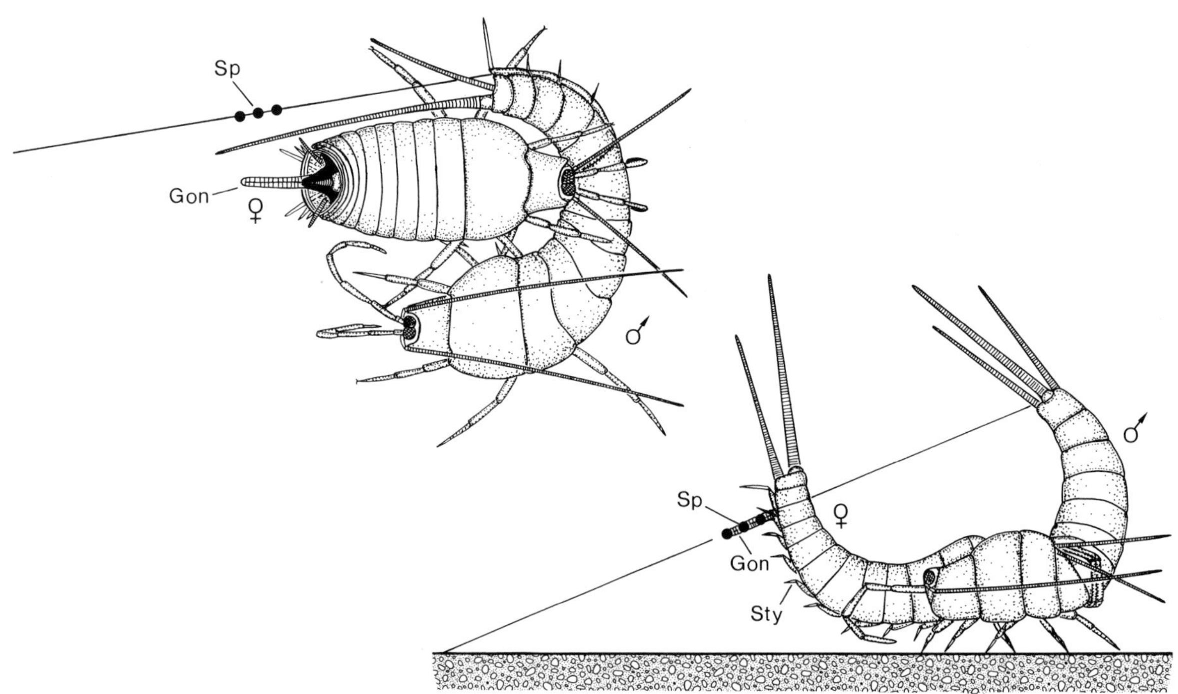

Fig. 134. Final stage of the courtship of the bristletails (Machilidae). (After Sturm, 1955). *Above*, the male guides the female by means of strong antennal movements and fondling with the maxillary palps towards the sperm droplets (*Sp*) deposited on a thread attached to the ground. When she is in the correct uptake position she takes the sperm droplets up with the gonapophyses of her ovipositor. (*Gon*); *Sty* styli

Plate 109. *Trigoniophthalmus alternatus* (Machilidae).

- **a** Abdominal segment, lateral, with external scale pattern. Most of the scales are missing, but their sockets are visible. 50 µm.
- **b** Abdominal cuticle with microridges. Between the scale sockets (dark holes) lie small bristles, which are possibly innervated like the scales. 5 µm.
- **c** Pattern of round scales. 10 µm.
- **d–f** Variations of the scale cuticle. 2 µm

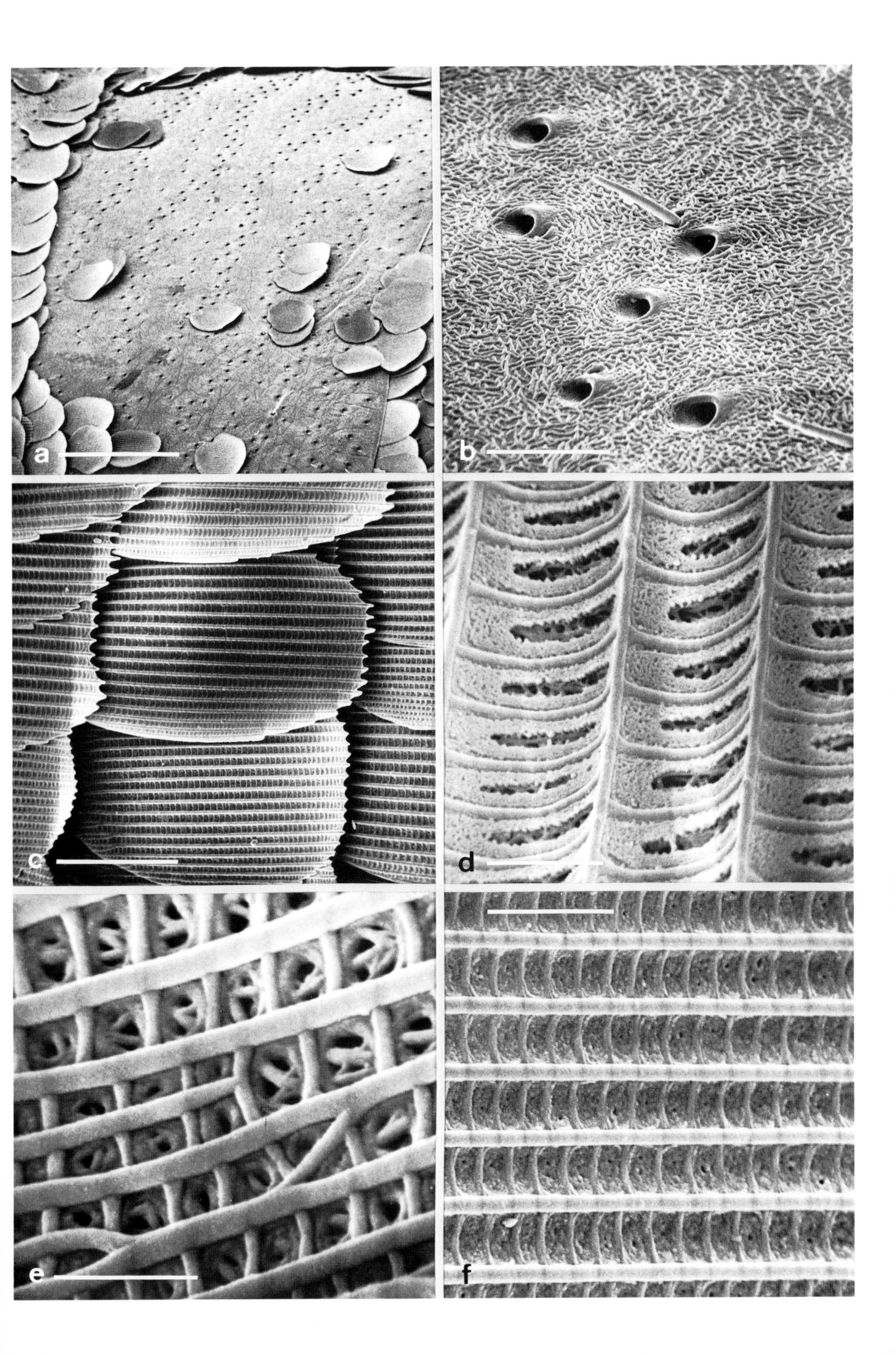
a
b
c
d
e
f

2.13.6 Egg Deposition in the Machilid *Trigoniophthalmus alternatus*

The eggs of *Trigoniophthalmus alternatus* are laid singly or in small groups in crevices on the surface of stones (Fig. 135). The animals have a lifespan of 2–4 years. As their embryonic development lasts longer than 400 days, the eggs must be adapted to the biotope during the long embryogenesis. Only a few days after oviposition a cuticle is secreted from the blastoderm as a lasting protection for the egg in addition to the chorion. The chorion splits a few days before hatching to expose the blastoderm cuticle of the outer egg surface.

The blastoderm cuticle has a bizarre, fine structure over the whole surface. But particularly in the centre of the flattened upper side, a finely branched, thread-like surface network is overlaid by a coarser polygonal pattern caused by convergence of the thread-like structures to form knotty concentrations at regular intervals (Plate 110). In the centre of the upper side the bizarre network is deeply sunk into the surface and held together by cuticular pillars and walls, which are connected to form a star-shaped, three-dimensional lattice (Larink, 1972, 1979).

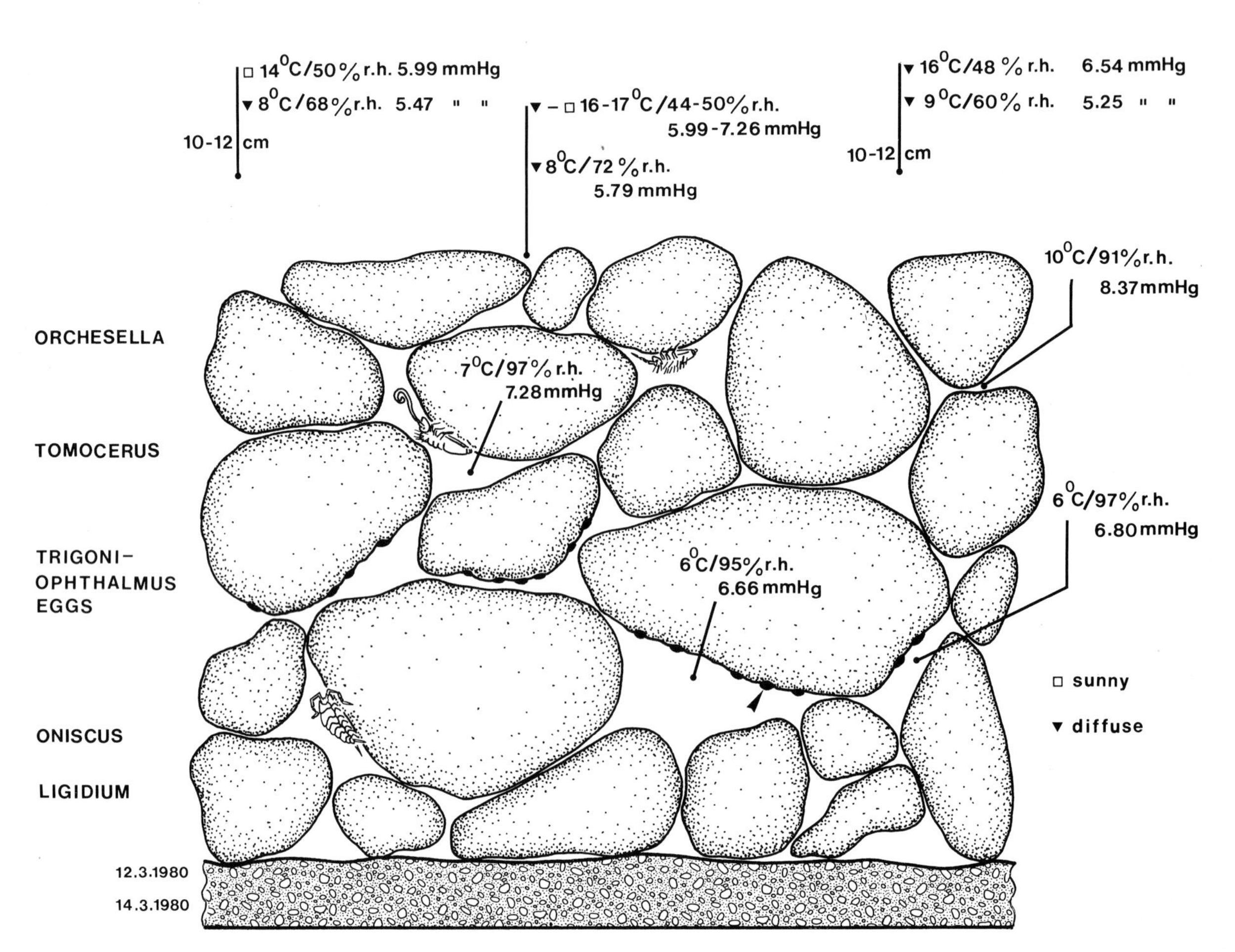

Fig. 135. Microclimate: temperature and moisture regime at the oviposition site of the bristletail *Trigoniophthalmus alternatus* (Machilidae). (After Eisenbeis, 1983). The eggs (*arrow*) are lentil-shaped, black structures which are attached with a secretion in the cavities of the stone surface, predominantly in the third stone layer (cf. Fig. 131). During the winter and spring the stones are covered by a thin film of water so that there is no danger of desiccation. The young bristletails emerge from about April/May onwards. The partial pressure of water vapour at the different sites is given in mmHg (1 mmHg ≈ 0.133 kPa)

Plate 110. *Trigoniophthalmus alternatus* (Machilidae).
a Upper side of an egg with attachment ring. 0.5 mm.
b Egg in profile. Its outer skin has burst open (upper side on the *right*). 0.5 mm.
c Upper side of the egg with open chorion. It bursts open some time before hatching to reveal the underlying blastoderm cuticle. 200 μm.
d Blastoderm cuticle with lattice structures. 3 μm

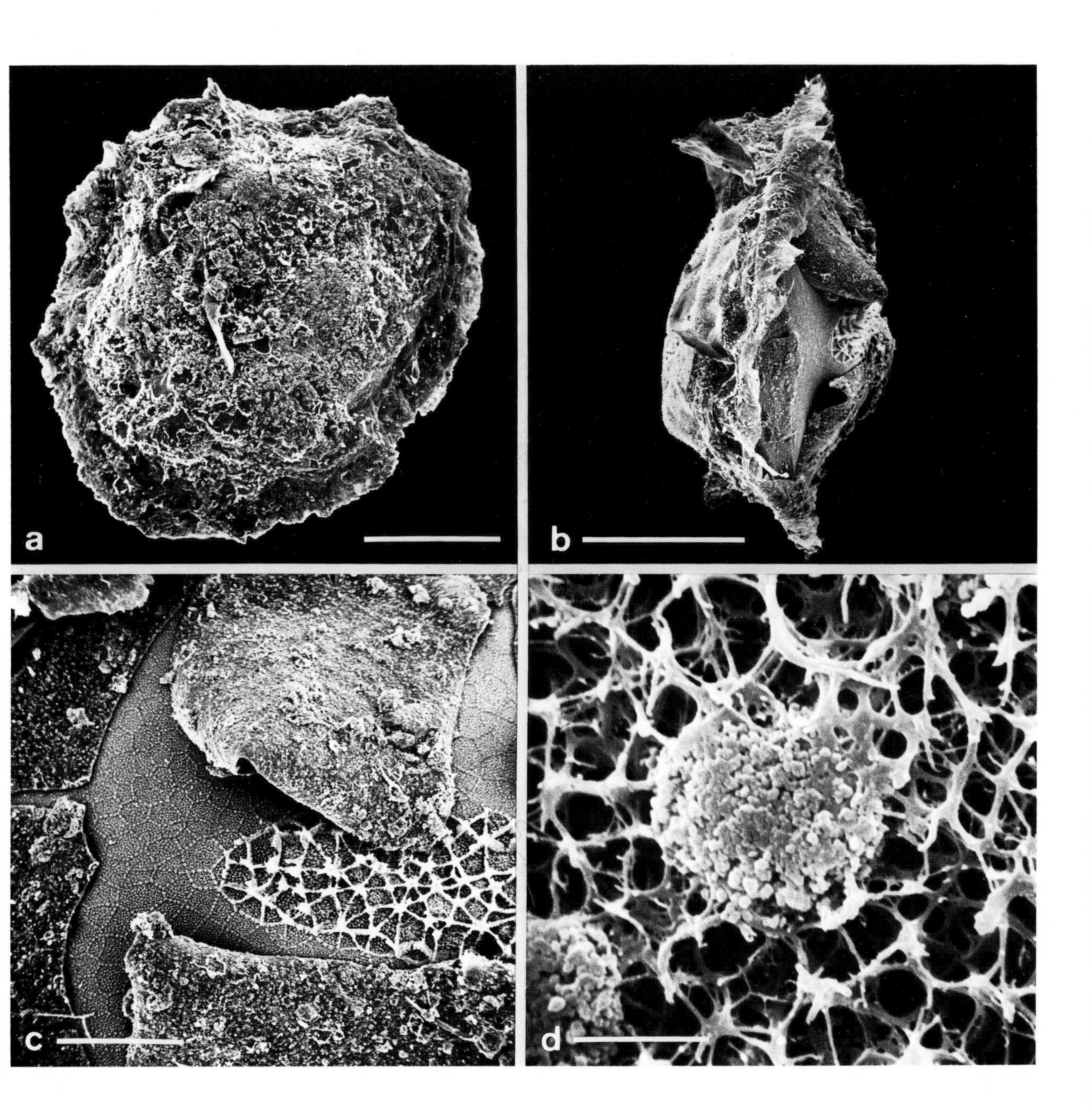
a
b
c
d

2.14 Order: Zygentoma (Insecta)

General Literature: Palissa, 1964; Handschin, 1929

2.14.1 Characteristics of the Zygentoma (Silverfish, Firebrats)

The silverfish or lepismatids were formerly grouped together with the bristletails or machilids into the Thysanura order. In accordance with the criteria of phylogenetic taxonomy the Archaeognatha order was created for the bristletails and the Zygentoma order for the silverfish. The Zygentoma, which are closely related to the Pterygota, comprise over 240 species. They originated in subtropical and tropical regions where the majority of their species live. In Europe the predominantly thermophilic forms, like *Lepisma saccharina*, avoid the soil biotope and are principally synanthrophic species. They often live in houses where they feed on various organic substances (polyphages) and where they occasionally occur in great numbers (Laibach, 1952).

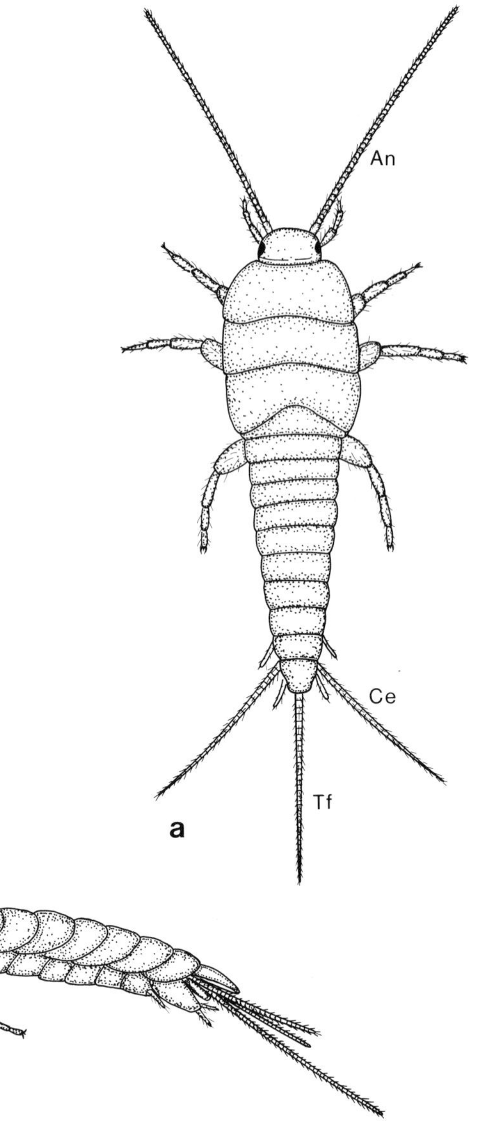

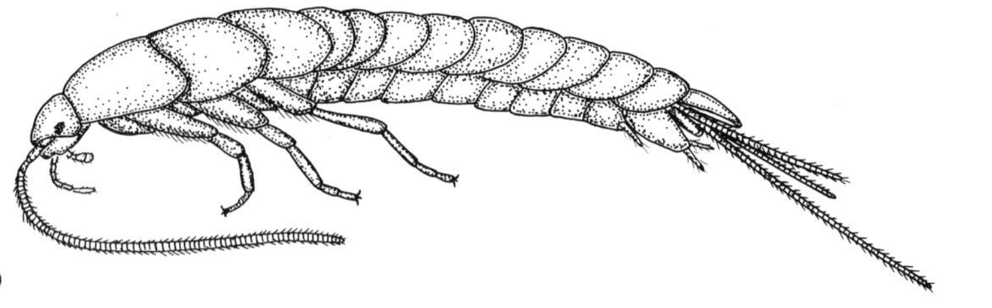

Fig. 136. Habitus of *Lepisma saccharina* (Lepismatidae). (After Handschin, 1929). **a** Dorsal view with antennae (*An*), cerci (*Ce*) and terminal filum (*Tf*); **b** Lateral view

Plate 111. *Lepisma saccharina* (Lepismatidae), silverfish, juvenile.

a Juvenile, newly hatched and lacking scales, lateral. At this stage the animals are white, segmentation of the body is clearly distinguishable. 0.5 mm.

b Head, oblique lateral, with compound eye and mandibular articulation (*arrow*). 100 µm.

c Abdominal segments, oblique lateral, with tergites and sternites. 50 µm.

d Compound eye and mandibular articulation (*arrow*). 20 µm.

e Abdominal segment, terminal margin. Though the cuticle has a scaly texture, its subsequent covering of genuine scales is not yet present. 10 µm.

f Section of the terminal filum. 25 µm

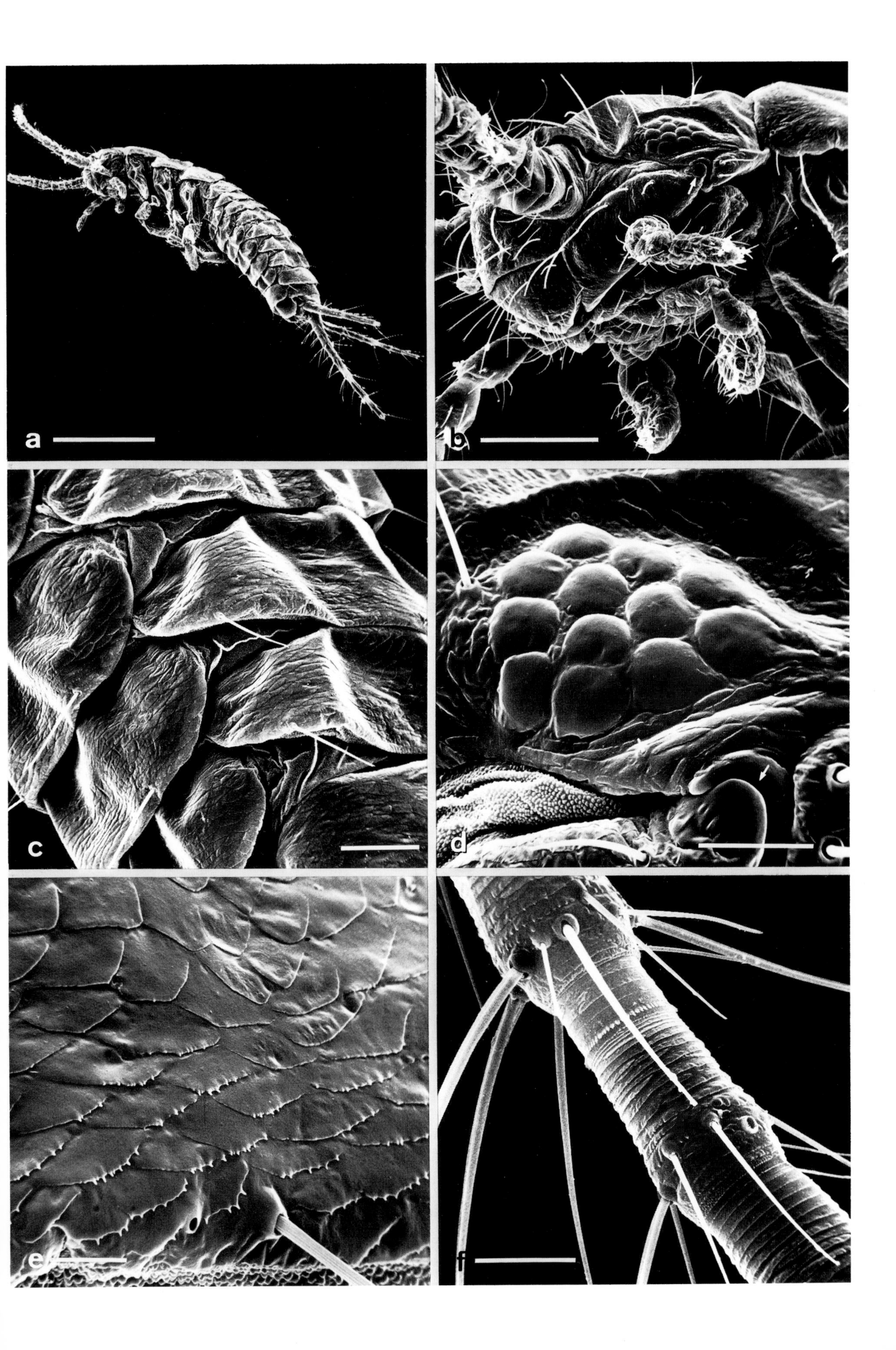

a
b
c
d
e
f

2.14.2 Structure and Function of the Scales

Like the bristletails, the silverfish *Lepisma saccharina* possesses a thick coating of scales (Plate 112). They are formed after the third larval moulting and are only absent on the intersegmental membranes, the ovipositor and the tips of the extremities. Adult animals have about 40000 scales. They lie in indistinct rows overlapping each other like roof tiles. The large scales covering the body surface are 100–250 µm long and 90–125 µm broad. Examination of the fine structure of the scales shows that they are composed of a continuous membrane with reinforcing longitudinal ribs running along the upper side of the scales parallel to the longitudinal axis. This solid construction of the Zygentoma scales distinctly differs from that of the Archaeognatha scales, in which an upper and lower lamella create a scale lumen connected to the exterior via pores. The lamellae are linked and stabilized by pillars and walls (Plate 109) (Larink, 1976).

The scales of *Lepisma saccharina* are formed by a trichogen cell. They are innervated and function as mechanoreceptive sensilla. In the final formative stage the trichogen cell completely withdraws from the scale to become one of the envelope cells within the sensillum (Fig. 137) (Larink, 1976).

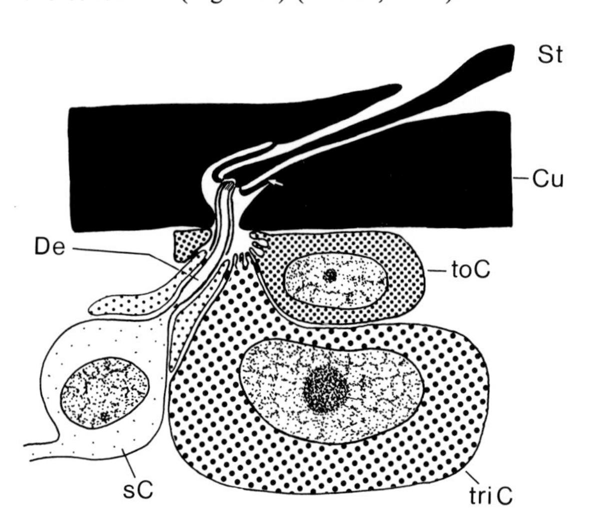

The life span of the silverfish with their attractive pattern of scales (Fig. 137; Plate 112) is 2–3 years. Initially the larvae lack scales (Plate 111). Their postembryonic development involves numerous moults which even continue at intervals of up to 100 days in the adults. The animals become sexually mature from about the tenth moult onwards. Data on the growth and development within the first five larval stages are provided by Sahrhage (1953) in Table 6.

Table 6. Growth and development of the silverfish, *Lepisma saccharina* (Lepismatidae), (Sahrhage, 1953)

Larval stage	Age in days	Body length (mm)	Weight (mg)
1	0–4	1.89	0.20
2	4–12	1.99	0.25
3	12–31	2.44	0.31
4	31–65	2.64	0.48
5	85	4.07	1.00

Fig. 137 *Lepisma saccharina* (Lepismatidae), micromorphology of a scale sensillum. (After Larink, 1976). The shaft of the scale (*St*) penetrates the cuticle through the channel-like scale socket and is connected to the cuticle by means of a delicate suspension (*arrow*). The dendrite (*De*) of a sensory cell (*sC*) is connected to the base of the scale. The tormogen cell (*toC*) and the trichogen cell (*triC*) as well as an additional envelope cell surround the sensory structure in its final state

Plate 112. *Lepisma saccharina* (Lepismatidae), silverfish.
a Habitus, lateral. 1 mm.
b Overview, frontal. 0.5 mm.
c Anterior of the body, ventral, with the typical flattened legs. 1 mm.
d Anterior of the body, lateral. 1 mm.
e Abdomen, pattern of scales. 50 µm.
f Thorax, pattern of scales. 50 µm

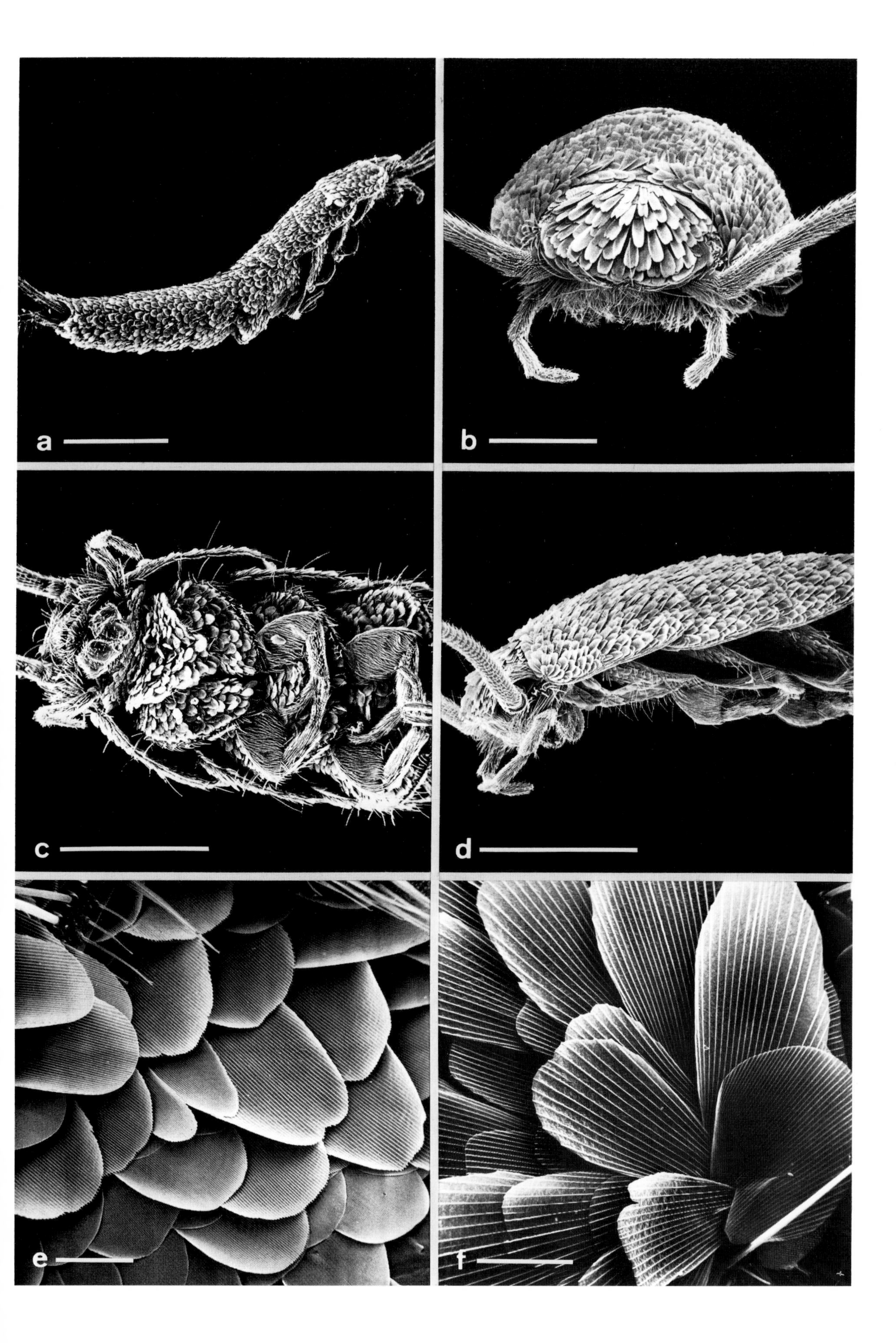
a
b
c
d
e
f

2.14.3 Labial Nephridium of the Silverfish

The excretory organ is situated within the labium and is known as the labial nephridium or gland. It is composed of three consecutive parts: the blind-ending sacculus, an almost straight, tubular labyrinth and the excretory canal (Fig. 138). The paired labial glands have a common, unpaired excretory duct which also serves as an outlet for the salivary glands and opens into the mouth cavity.

The sacculus is lined by podocytes (Fig. 138b) which enable the formation of primary urine in the sacculus lumen by means of ultrafiltration when a pressure gradient exists between the haemolymph and the sacculus lumen. The wall of the labyrinth is composed of typical transport cells which are characterized by a dense, apical border of microvilli and labyrinth-like infoldings of the basal cell membrane (Fig. 138b) which are closely associated with numerous mitochondria. They are assumed to have secretory and reabsorptive functions. In turn, the excretory canal is also lined by transport cells, indicating that reabsorption probably also occurs in this region. Thus, primary urine is produced by ultrafiltration in the sacculus and during its passage through the labyrinth and the excretory canal reabsorption of physiologically essential substances from the haemolymph takes place to give the final secondary urine (Haupt, 1965).

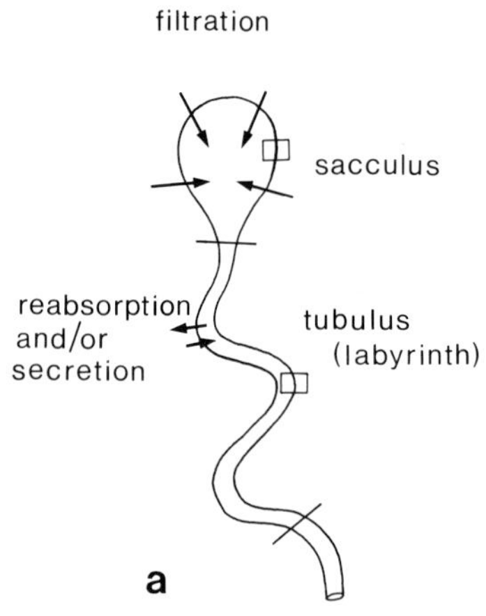

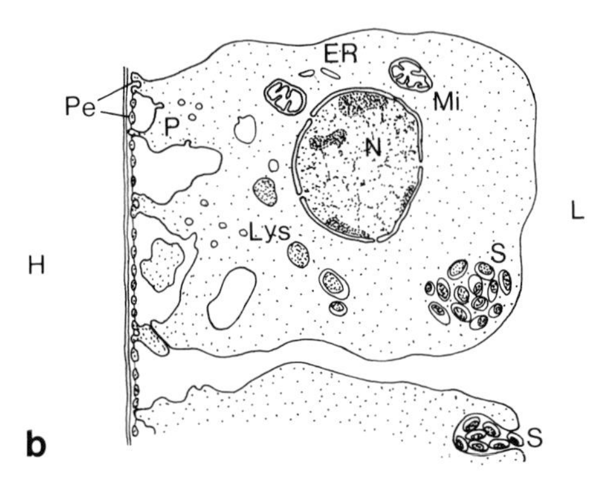

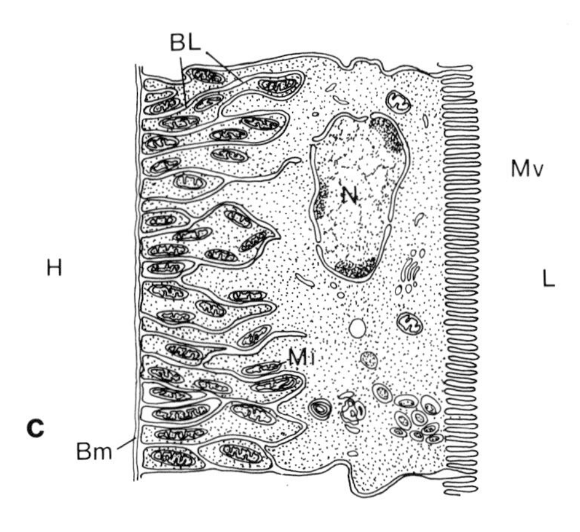

Fig. 138. Structure of the labial nephridium of a silverfish. (Berridge and Oschmann, 1972 and Haupt, 1965).
a Arrangement of the labial nephridium with sacculus, tubulus and excretory canal. The *rectangles* show the position of sections **b** and **c**. **b** Section from the sacculus epithelium. The cells situated on the luminal side of a basal membrane are called podocytes. They have fine processes which form a characteristic sieve structure attached to the basal membrane. *ER* Endoplasmic reticulum; *H* haemolymph cavity; *L* lumen; *Lys* lysosomes; *Mi* mitochondria; *N* nucleus; *Pe* pedicelli; *P* area of pinocytotic vesiculation; *S* accumulation and exocytosis of older lysosomes. **c** Tubulus – epithelium cell with typical transport structures. *Mv* Apical microvilli; *BL* basal labyrinth with mitochondria; *Bm* basal membrane (cf. **b** for the remaining abbreviations)

Plate 113. *Lepisma saccharina* (Lepismatidae), silverfish.
a Head, lateral, with antennae, maxillary and labial palps. 250 μm.
b Head, ventral, with upper lip (*arrowhead*), mandibles and palps. 250 μm.
c Antennal base (*arrow*) with compound eye. 50 μm.
d Pair of legs with flattened segments. 0.5 mm.
e Bristles near the eye. 10 μm.
f Tarsus, distal, with claws (ungues). 50 μm

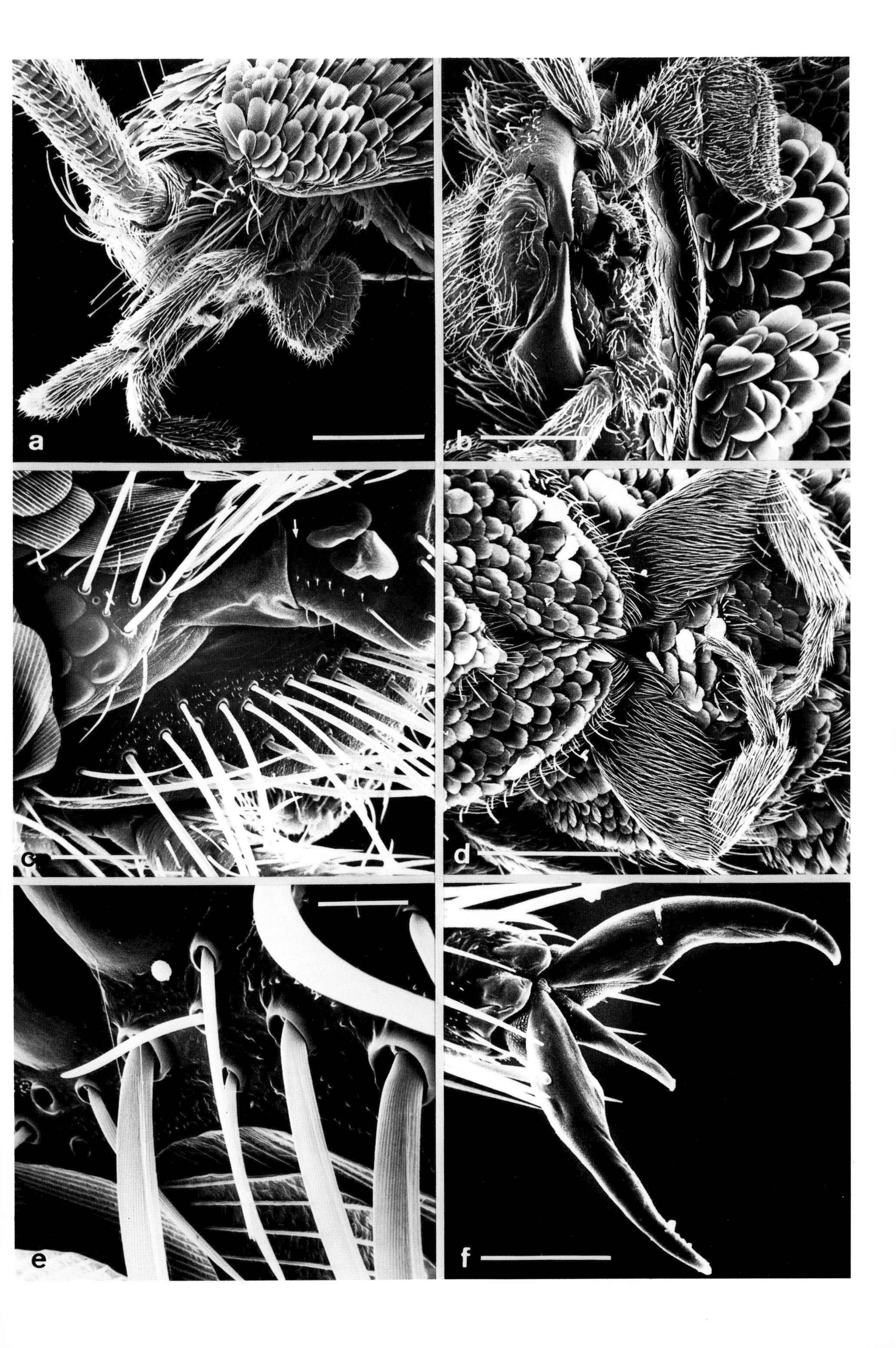
a
b
c
d
e
f

2.14.4 Outer Genital Appendages of the Zygentoma

The outer genitalia of the Archaeognatha and the Zygentoma are very similar. The female of *Lepisma saccharina* has a slightly curved, rod-like ovipositor formed by four gonapophyses (Plate 114). The gonapophyses are divided into 20 secondary segments without scales, but covered with bristles at the terminal end. The male penis is evidently non-segmented, but is split at the distal end. Its opening is ringed by short sensilla mounted on papillae. The paramers broaden to a club shape distally and are covered with bristles and possess glandular fields in the latero-ventral region where glandular papillae lie close together. They produce a secretion, presumably of a silky nature (Plate 114). The thread, which the male spins as a barrier to urge the female to take up the spermatophores in the course of the indirect sperm transfer possibly originates from these glands (Sturm, 1956a,b)

Plate 114. *Lepisma saccharina* (Lepismatidae), silverfish.
a Male, abdomen caudal, ventral, with cerci, terminal filum, styli and the paramers which belong to the genitalia. 0.5 mm.
b Glandular field on the ventral side of a paramer. 50 μm.
c Paramers. 100 μm.
d Part of the glandular field of the paramers. The white cone-like shapes on the glandular papillae are presumably secretions (silk). 10 μm.
e Female, ovipositor composed of four gonapophyses. 50 μm.
f Ovipositor, terminal. 25 μm

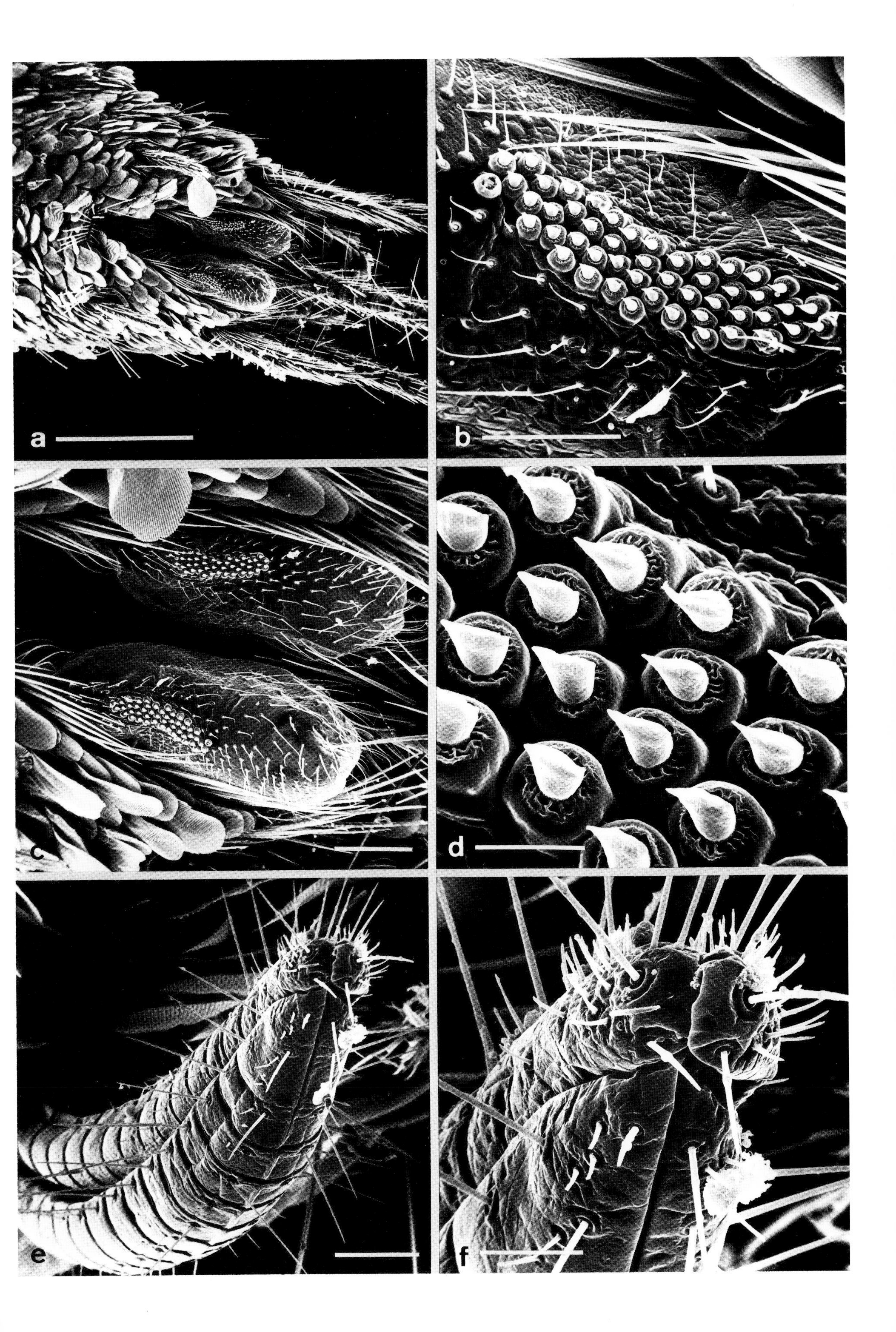
a
b
c
d
e
f

2.14.5 Sensilla on the Abdominal Appendages

Kränzler and Larink (1980) and Larink (1982) investigated the pattern of sensilla on the caudal appendages using the scanning electron microscope and described the occurrence and distribution of various types of sensilla there. Short, straight bristles emerge from the joint membranes of the cerci. Trichobothria or filamentous hairs are situated on the dorsal side of the terminal filum and isolated ones on the caudal margin of the cercal segments (Plate 115). On the ventral side of the styli, on the cerci and on the terminal filum there are short, hooked bristles about 12-µm-long. The smooth bristles without feathery edges are frequently found on the styli, but also on the cerci and the terminal filum. Supporting bristles are especially strong with longitudinal ribbing. They are situated in a double row on the ventral outer edge of the styli. Feathery bristles are long, laterally ciliated bristles with strong longitudinal ribbing and small cross ribs. They are inserted on the ventral side of the terminal filum and are mostly aligned parallel to the longitudinal axis of the body.

Even though no fine structural and electrophysiological investigations have yet been carried out, these sensilla can still be classified in terms of known sensilla forms: sensilla trichodea (short, straight bristles), sensilla chaetica (hooked bristles, smooth bristles, strong supporting bristles and feathery bristles). Trichobothria are certainly mechanoreceptors for the perception of sound and vibration.

On the flagella of the antenna of *Thermobia domestica* and *Lepisma saccharina* eight and nine types of sensilla are described by Adel (1984). The sensilla chaetica, sensilla trichobothria and sensilla campaniformia which appear on both species are most likely engaged in mechanoreception only, while the sensilla trichodea are assumed to be contact chemoreceptors. Coeloconic sensilla likewise occur on both of them. Three types of sensilla in *Thermobia domestica* and four types in *Lepisma saccharina* are characterized as basiconic sensilla. The analyses of the sensilla pattern show a highly specific distribution of sensilla on the annuli of the antenna. Moreover, there is often a distinct dorsoventral arrangement (Adel 1984).

Plate 115. *Lepisma saccharina* (Lepismatidae), silverfish.

a Abdomen, caudal, dorsal, with cerci and terminal filum. 0.5 mm.

b Terminal filum, central portion. 50 µm.

c Cercus, central portion. 100 µm.

d Terminal filum, proximal segments. 25 µm.

e,f Cercus and terminal filum with bristles, trichobothria and small glandular hollows. 20 µm, 20 µm

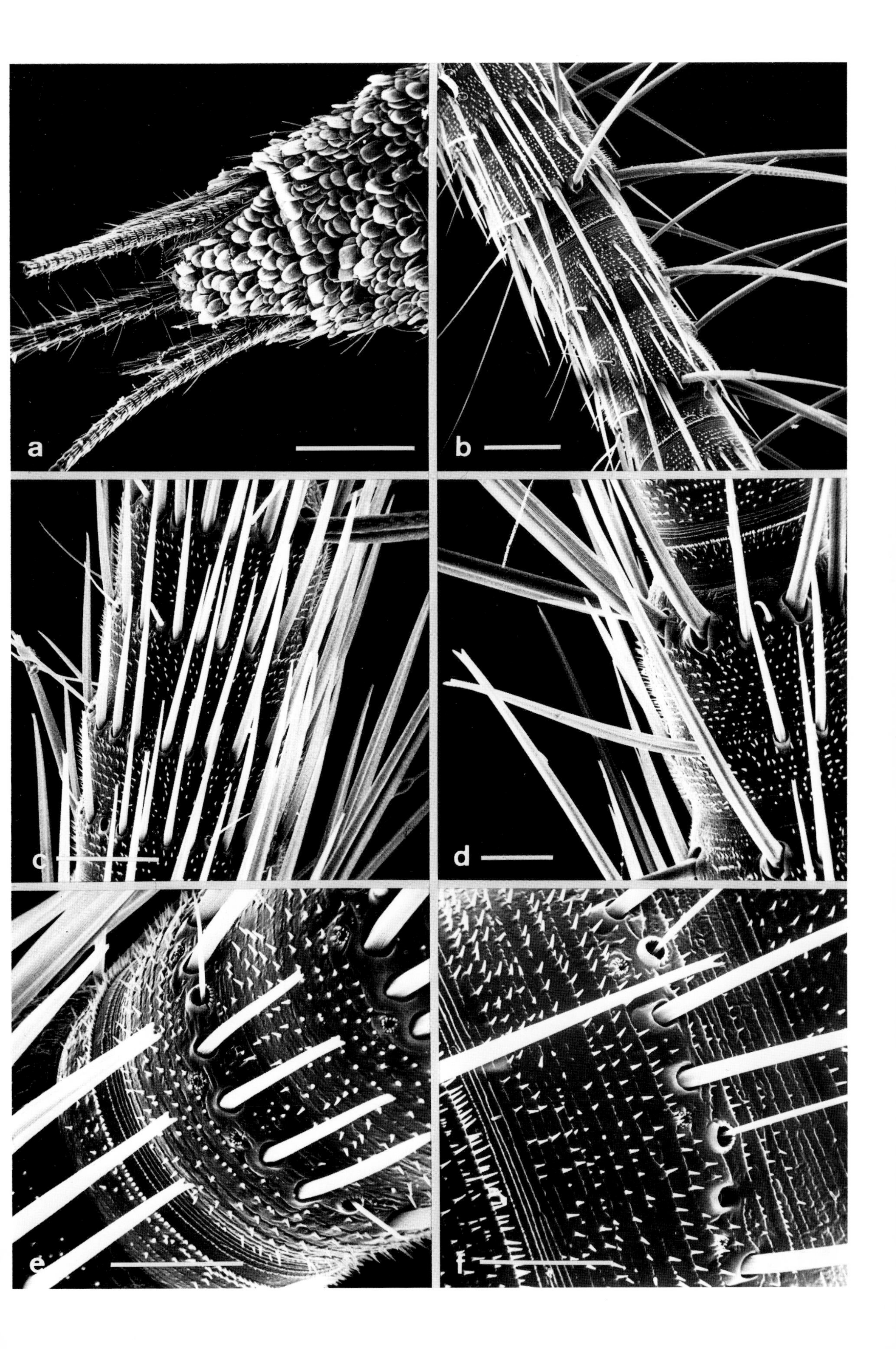
a
b
c
d
e
f

2.15 Order: Dermaptera – Earwigs (Insecta)

General Literature: Beier, 1953, 1959; Harz, 1957, 1960; Günther and Herter, 1974

2.15.1 Soil-Dwelling Earwigs

Earwigs, which belong to the Forficulidae family, are important to soil biology, as they often live in litter and feed on fresh plant fragments and fungal hyphae, but are also carnivorous. The globally distributed, 11–15-mm-long common earwig (*Forficula auricularia*) (Herter, 1965, 1967) and the wingless, 6–13-mm-long forest-dwelling *Chelidurella acanthopygia* (Fig. 139; Plates 116–118) (Franke 1985), which occurs in European deciduous woodland, are members of this group. They are nocturnal animals, hiding during the day in narrow crevices and cavities in which they have all-round bodily contact with the substratum (thigmotaxis).

The sand-dwelling earwig, *Labidura riparia* (Labiduridae) (Weidner, 1941; Herter, 1963), shows a preference for life in moist sand and penetrates deeper into the soil than other earwigs. They loosen the sand with their forelegs and transport it with their mandibles, thus excavating burrows 10–40-cm-deep, in which they evidently rest during the day. At night they hunt small insects and spiders (Dunger, 1974; Messner, 1963). Their burrow also serves as a winter quarter from which they re-emerge in May of the following year. In summer they excavate a brood chamber in the soil (Fig. 140). Their the female deposits her eggs and remains with them during the 9-day embryonic period (Caussanel, 1966, 1970).

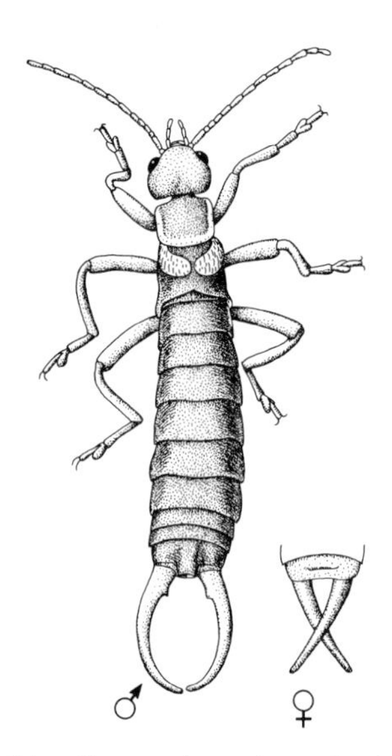

Fig. 139. *Chelidurella acanthopygia*, male, dorsal view. On the *right*: the cerci of a female. (After Brauns, 1964)

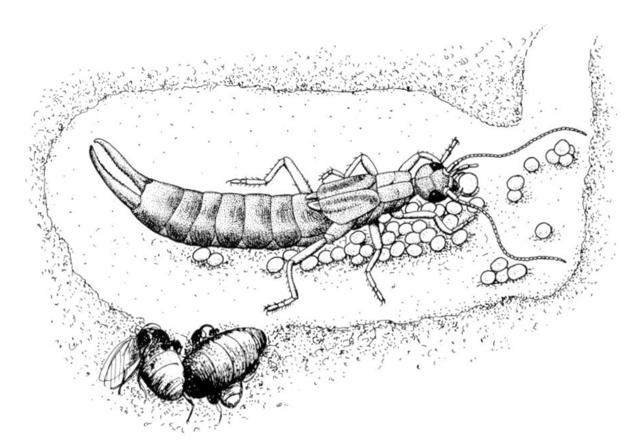

Fig. 140. *Labidura riparia*, female caring for her brood. Food is stored in the burrow: fly pupae and imagines. (After Caussanel, 1966 from Günther and Herter, 1974)

Plate 116. *Chelidurella acanthopygia* (Forficulidae), female **a, f** and male **b–e**.
a Head, lateral. 0.5 mm.
b Prothorax and head, ventral. 0.5 mm.
c Head, dorsal. 0.5 mm.
d Mouthparts, ventral. 100 µm.
e Surface of the ommatidia. 10 µm.
f Mesothorax, lateral. 0.5 mm

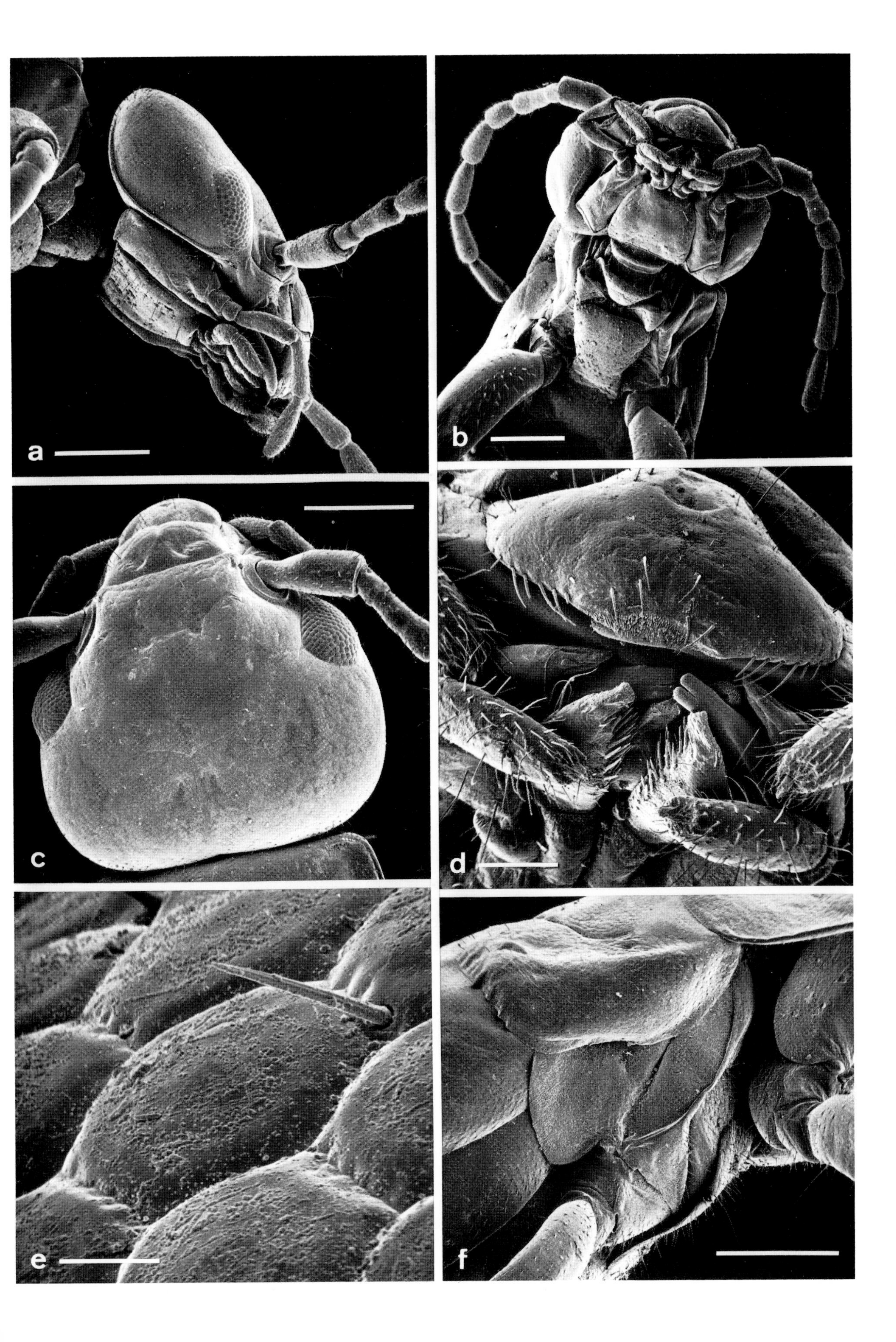
a
b
c
d
e
f

2.15.2 Sensory Perception and Thigmotaxis in Earwigs

Positive thigmotaxis is particularly striking in earwigs as soon as they retreat into narrow crevices, e.g. in bark, between leaves, etc., or in hollow spaces in the soil. The animals seek direct and almost all-round contact between as much of their body surface as possible, including their extended antennae, and the surrounding substrate. When disturbed they wedge themselves even more firmly into the protective soil crevice.

The mechanoreceptors and tactile sensory hairs, which are distributed with variable density over the body and antennae, are assumed to play an important role in this behaviour. Besides the tactile sensory organs, there are also chemoreceptors on the antennae for the orientation of the earwigs in their environment. Therefore, earwigs constantly explore the substrate in front of them with their incessantly waving antennae. Slifer (1967) has described the sensory organs on the antennae of *Forficula auricularia* (Fig. 141).

Fig. 141. Types of sensilla on the antenna of *Forficula auricularia*, the common earwig. (After Slifer, 1967 from Günther and Herter, 1974). *1* Tactile hair; *2* longer, thick-walled chemoreceptor; *3* shorter, thick-walled chemoreceptor; *4* thin-walled chemoreceptor; *5* coeloconic chemoreceptor

Plate 117. *Chelidurella acanthopygia* (Forficulidae), forest-dwelling earwig.
a Compound eye. 100 μm.
b Antenna, proximal, scapus-pedicellus joint. 50 μm.
c Tarsus with ventral brushes. 100 μm.
d Antennal segment from the middle region of the feeler. 25 μm.
e Tarsus, distal, with claws (ungues). 100 μm.
f Antenna, distal. 50 μm

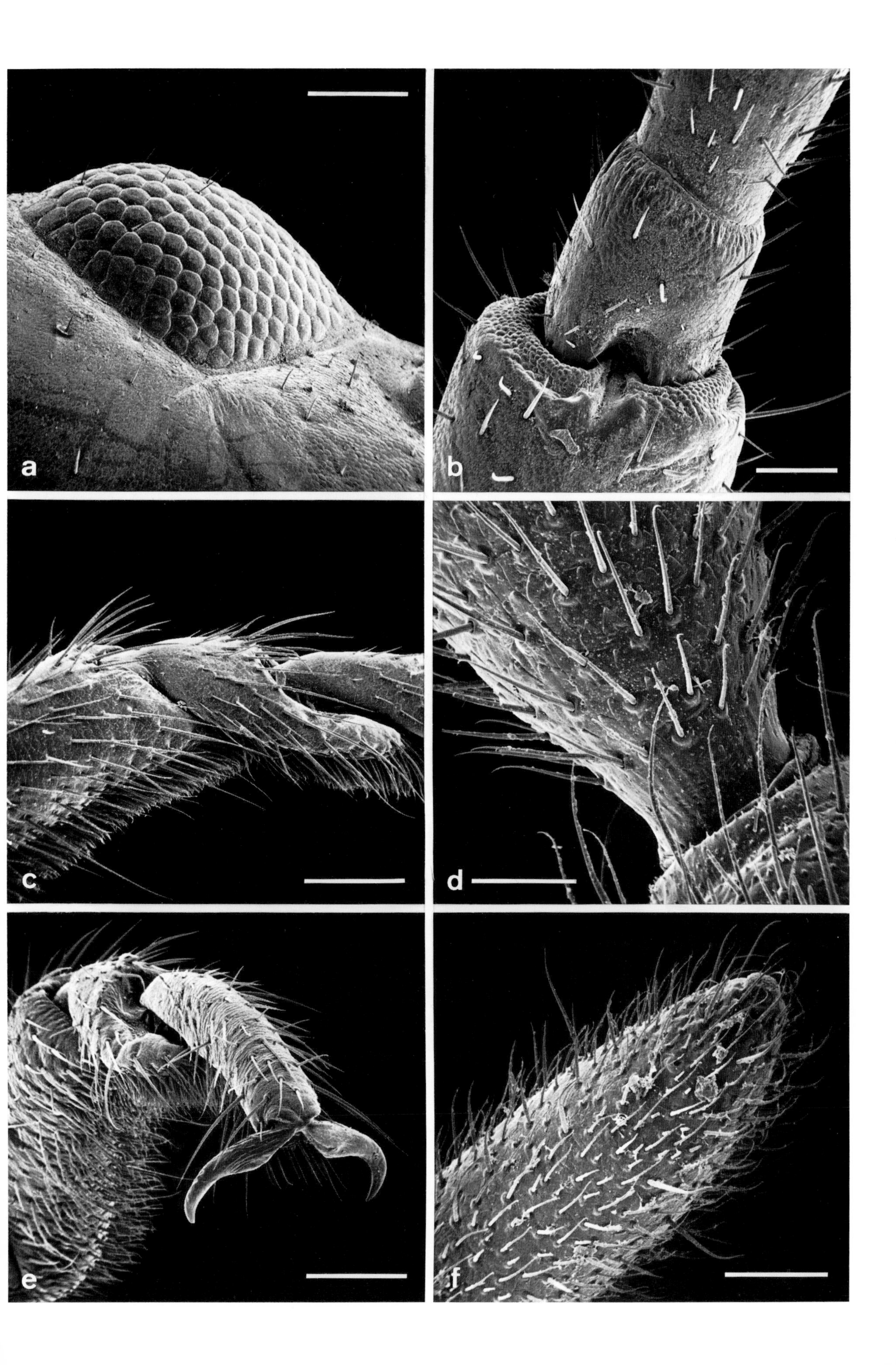
a
b
c
d
e
f

2.15.3 Feeding Habits and Defence of the Earwigs

Earwigs are both herbivorous and carnivorous. However, different families seem to exhibit typical dietary preferences associated with appropriately developed mouthparts (Popham, 1959). Thus, the Labiduridae prefer animal nutrition. *Labidura riparia* preys on small crabs, beetles, flies and other insects in the littoral zone, in drier biotopes it hunts millipedes, spiders and hairless caterpillars. In contrast, the Forficulidae are primarily herbivorous. *Forficula auricularia* feeds on fungal spores, green algae, lichen and moss, detritus of higher plants, but also on fresh parts of flowering plants. In addition, small insects (e.g. aphids), ant eggs and caterpillars are consumed. Examination showed that 30.5% of the stomach contents were of animal origin. *Chelidurella acanthopygia*, the forest-dwelling earwig, has similar feeding habits.

The pincer-shaped cerci (Plate 118) serve as a defence weapon with which the animals can hit out, tap threateningly on the ground, pinch and seize. But, despite the deployment of their mechanical and chemical weapons, earwigs are the victims of many predators, e.g. grasshoppers, crickets, predatory bugs, carabid, staphylinid and other beetles, predatory spiders and Chilopoda as well as lizards, birds and other insectivores (Günther and Herter, 1974).

Plate 118. *Chelidurella acanthopygia* (Forficulidae), female **a–e**, male **f**.

a, b Abdomen, terminal, dorsal (**a**) and lateral (**b**), with pincers. 1 mm, 1 mm.

c Female pincers, ventral, with pygidium (*arrow*). 0.5 mm.

d Abdominal segments, lateral, with overlapping sternites and tergites. 0.5 mm.

e Female pincers, proximal, lateral. 200 μm.

f Abdomen of a male, caudal, with the base of the pincers and pygidium (*arrow*). 0.5 mm

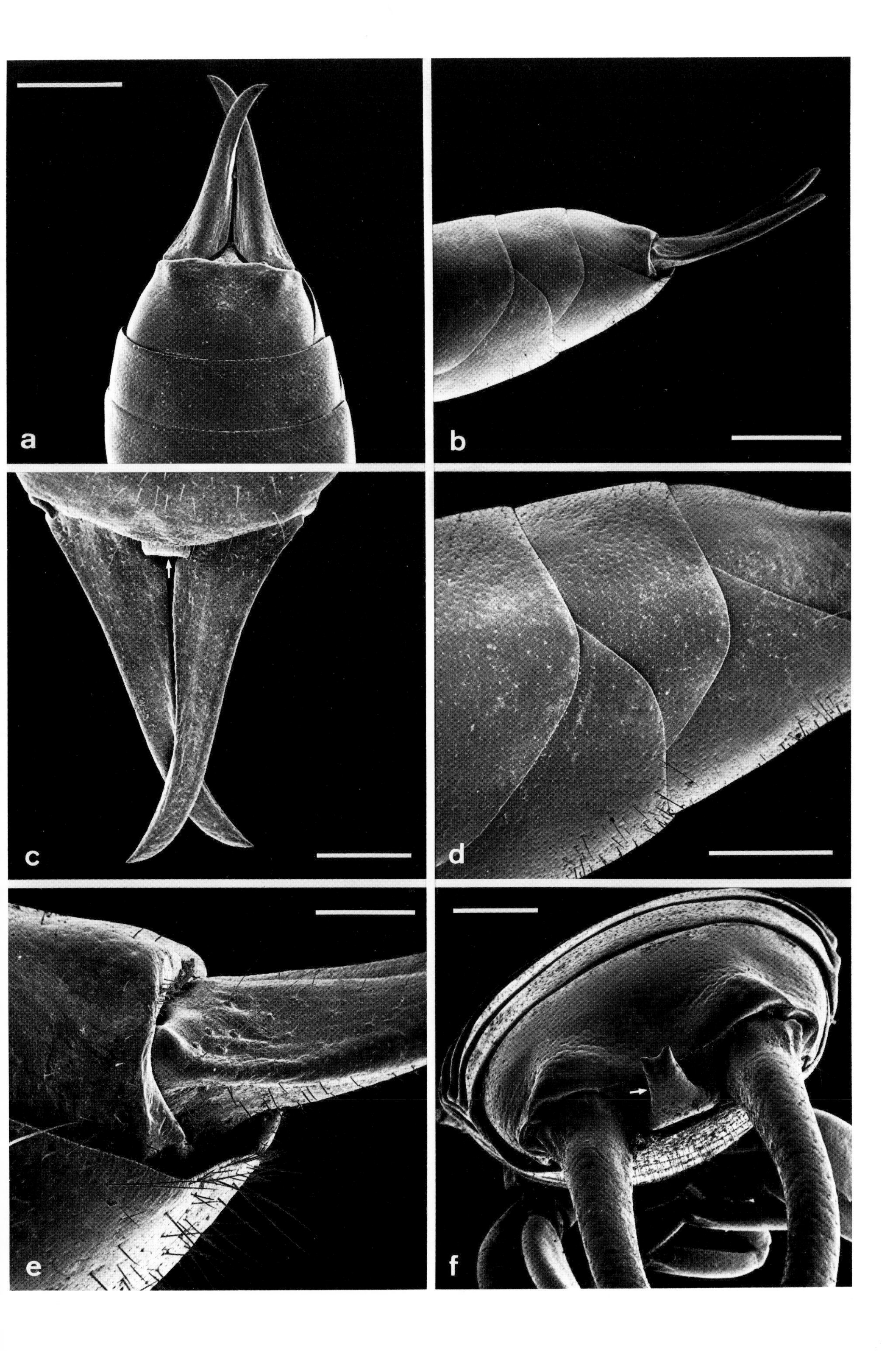
a
b
c
d
e
f

2.16 Order: Blattodea – Cockroaches (Insecta)

General Literature: Chopard, 1938; Harz, 1957, 1960; Princis, 1965; Beier, 1974

2.16.1 Characteristics of the Soil-Dwelling Cockroaches

The soil-dwelling Orthopteromorpha, to which the earwigs (Dermaptera), cockroaches (Blattodea) and crickets (Ensifera) belong, colonize the upper soil layers. Some Orthopteromorpha live hemiedaphically, but only temporarily. The epedaphic cockroaches belonging to the Ectobiidae family, represented in central Europe by the forest-dwelling species *Ectobius lapponicus, E. silvestris* and *E. panzeri* and several others (Brown, 1952; Roth and Willis, 1952; Harz, 1972), exhibit typical adaptations to this mode of life. Their bodies are dorsoventrally flattened and broadly oval. They have flat legs which can be tucked in close to the body and long sensitive antennae (Figs. 142 and 143; Plates 119 and 120). The cerci are equipped with many long filiform tactile hairs on the ventral side (Plate 121e).

The females are especially shy of light and hide in leaf litter during the day, whereas the more active males fly around by day. They feed on rotting plants. Due to their relatively low population density, they play only a minor role in soil biology.

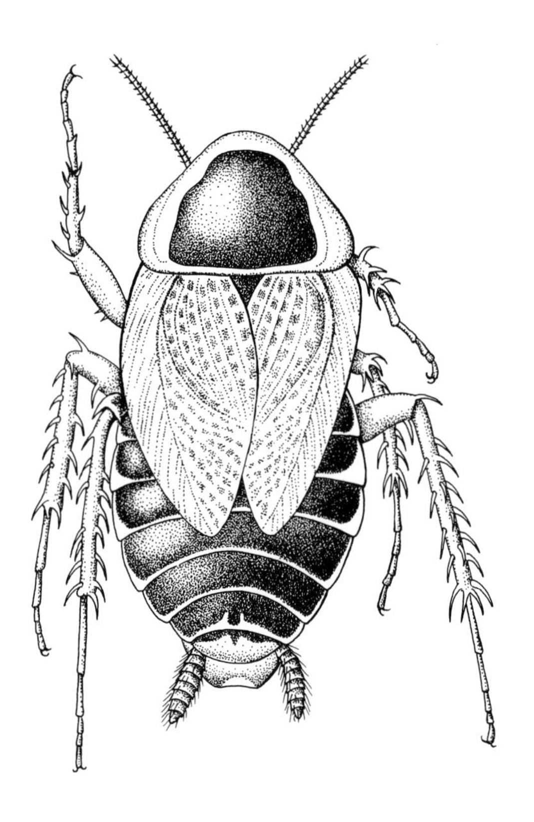

Fig. 142. Female of *Ectobius silvestris* (Ectobiidae), dorsal. (After Harz, 1960)

Plate 119. Larva of the Ectobiidae.

a, b Overview, lateral and ventral. Note the flattened coxae and femora of the legs. 1 mm, 1 mm.

c Head and thorax, ventral. 250 µm.

d Head and thorax, lateral. Note the hypognathous position of the head. 250 µm.

e Caudal view. 1 mm.

f "Cleaning brush" on the maxilla. 25 µm

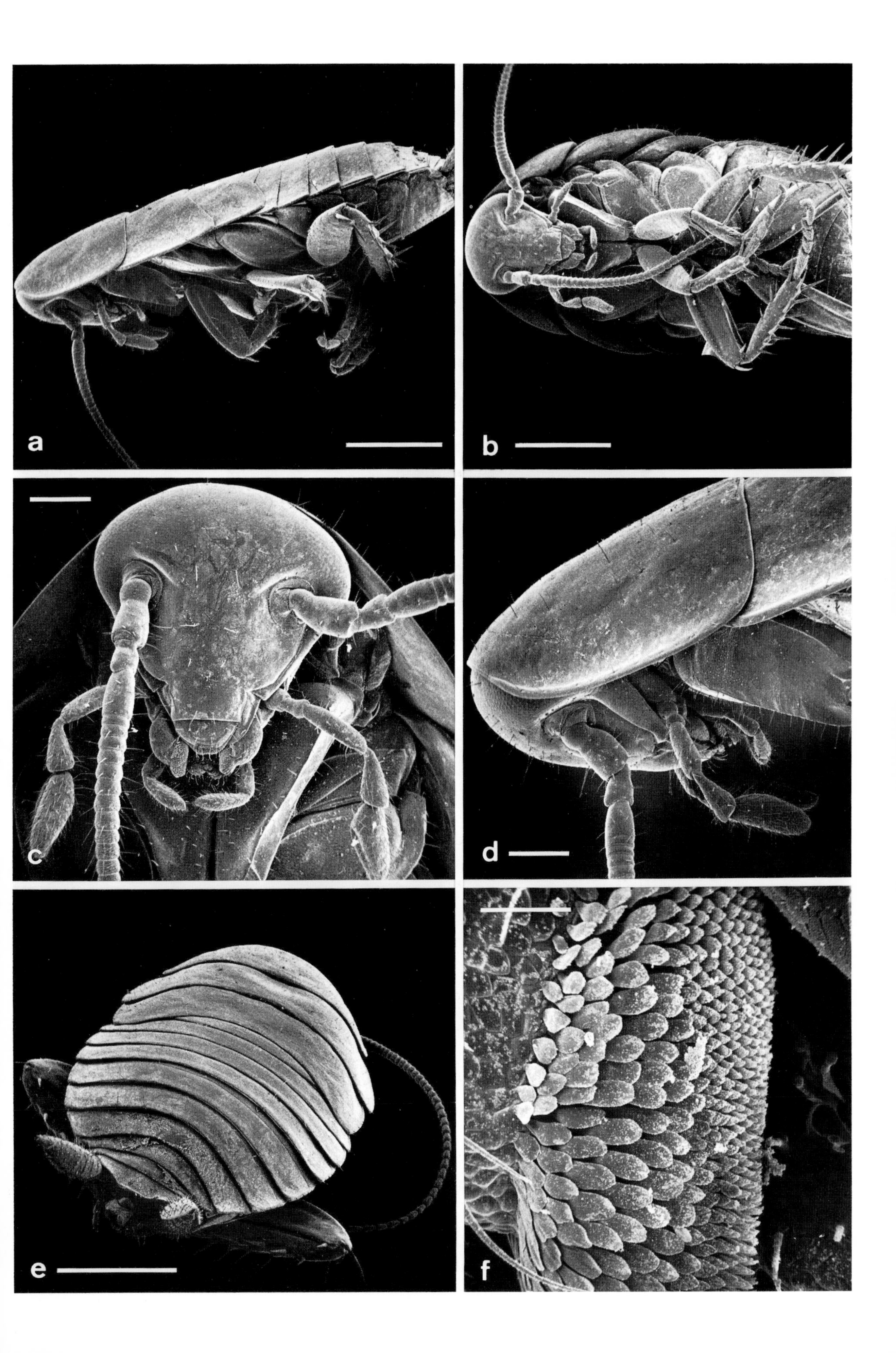
a
b
c
d
e
f

2.16.2 Antennae of the Epedaphic Cockroaches

Long, extended antennae are characteristic of epedaphic arthropods. Their sensory-physiological capabilities facilitate the spatial orientation of the animals. The antennae bear an abundance of sensory organs of variable structure and function, resulting in a wide range of sensory perceptions. Lengthening of the antennae accompanied by a high degree of flexibility allows a correspondingly greater area in front and in the immediate vicinity of the animal to be explored by continuous, mostly circular movements of the antennae.

The antennae of the epedaphic cockroaches are of this type. They are thread-like (Fig. 143), at least half as long, but normally as long as the body or longer. They are homonomously segmented. The long scapus and the somewhat short pedicellus are followed by multi-segmented antennal flagellum which gradually tapers towards the tip (Plate 120). Sensory organs such as tactile and olfactory receptors as well as those which perceive air humidity and temperature require further investigation (Beier, 1974; Eggers, 1924; Loftus, 1966, 1969; Slifer, 1968; Winston and Green, 1967).

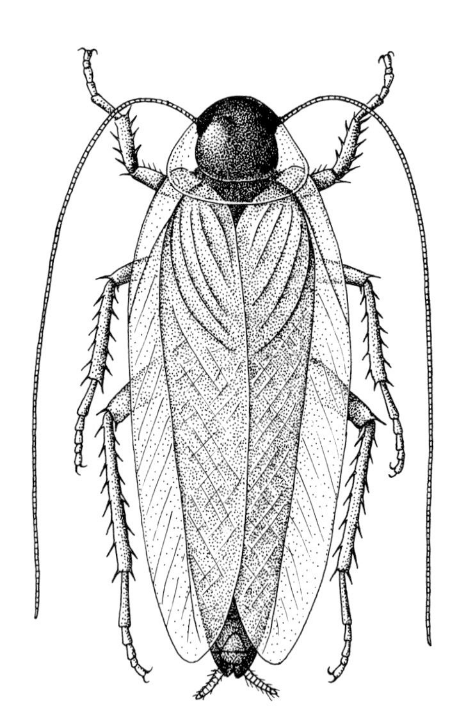

Fig. 143. Male of *Ectobius lapponicus* (Ectobiidae), dorsal. (After Brauns, 1964)

Plate 120. *Ectobius lapponicus* (Ectobiidae), male.
a Overview, frontal. The *arrow* marks the large coxae of the forelegs. 0.5 mm.
b Head and thorax, lateral. 1 mm.
c Pronotum, scutellum and proximal part of the elytra, dorsal. 0.5 mm.
d Surface of the ommatidia. 25 μm.
e Antenna, central portion of the flagellum. 25 μm.
f Antenna, proximal, with scapus, pedicellus and the base of the flagellum. 50 μm

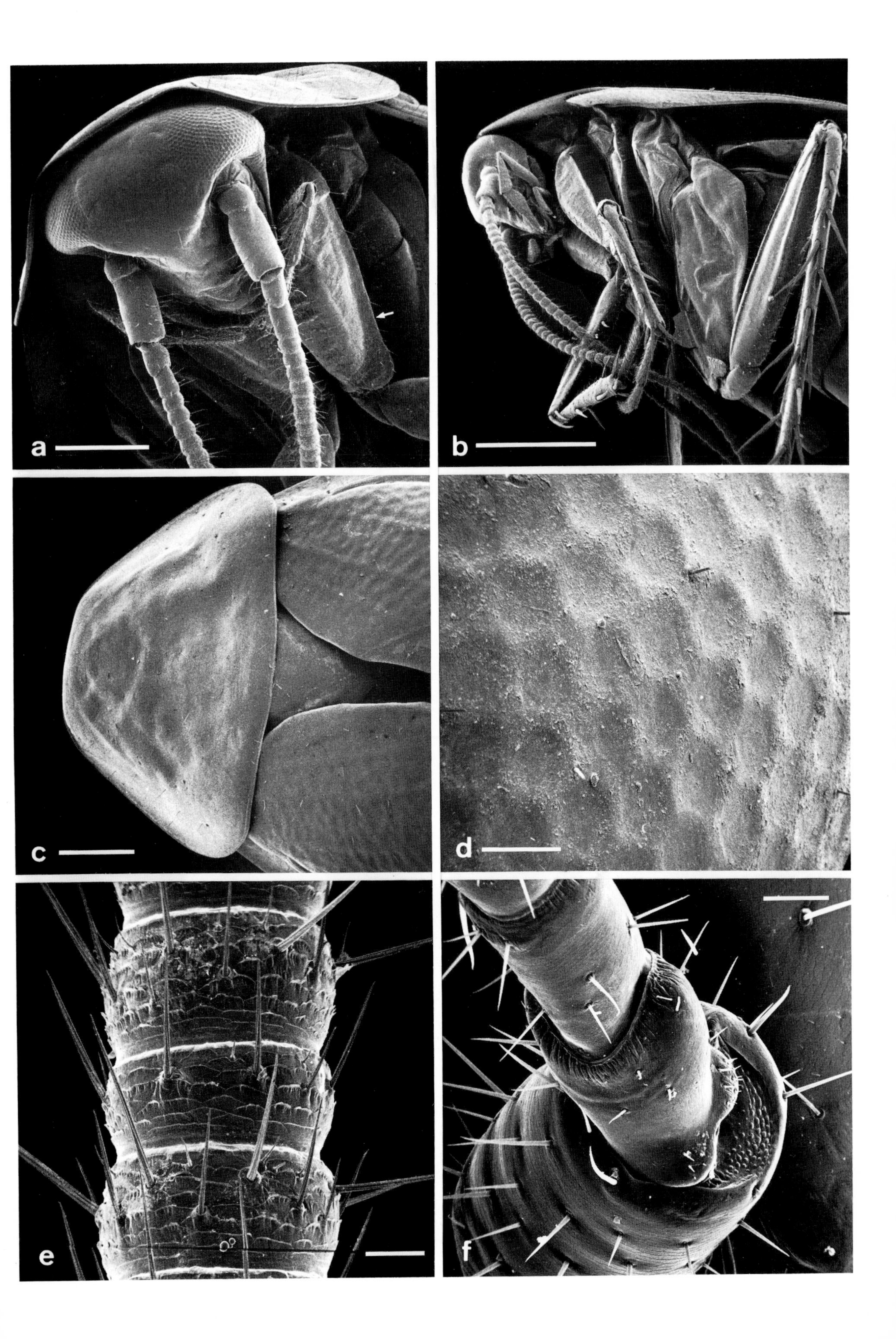
a
b
c
d
e
f

2.16.3 The Legs of Epedaphic Cockroaches

Cockroaches usually walk on the soil surface or within the interstices of the litter and like earwigs they are characterized by their thigmotactic behaviour. Thus, the coxae of the legs are flattened and can be folded closely against the body and the tibiae bear densely crowded thorns (Plate 121). The tarsi are composed of five segments. The first tarsal segment is covered by an elastic fold of skin at the tibio-tarsal joint which serves as a braking mechanism, the force being absorbed elastically by the upwards curving tarsus (Kupka, 1946; Beier, 1974). The fifth tarsal segment carries two claws for clinging to rough ground. Between them there is an attachment lobe (arolium) (Plate 121). The arolium and the small lobes on the tarsal segments, the so-called euplantulae, are set down on smooth surfaces (Fig. 144).

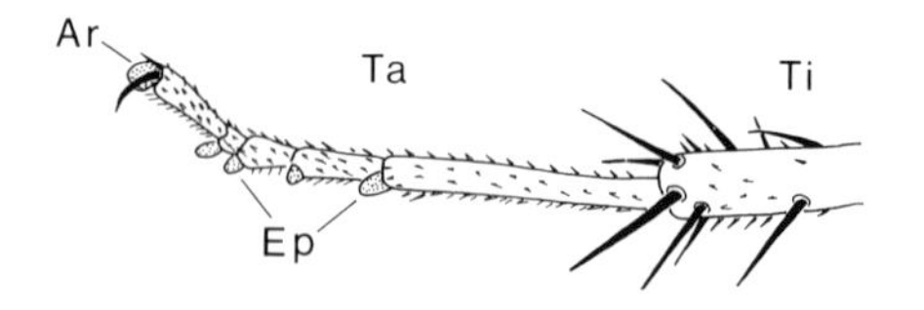

Fig. 144. Tarsus of *Blattella germanica* with attachment lobes. (After Beier, 1974). *Ar* Arolium; *Ep* euplantulae; *Ta* tarsus; *Ti* tibia

Plate 121. *Ectobius lapponicus* (Ectobiidae), male **a–d, f** and larva of the Ectobiidae **e.**

a Femur-tibia joint with spiny hairs bearing comb-like borders. 100 μm.
b Spiny hair, lateral, with marginal comb. 10 μm.
c Tarsus, distal, with claws (ungues) and attachment lobe (arolium). 50 μm.
d Abdomen, dorsal, with glandular hollow. 250 μm.
e Abdomen of a larva, caudal. Ventrally, the cerci bear long, tactile filiform hairs. 250 μm.
f Abdomen, glandular hollow with tuft of hairs. The latter enhances evaporation of the secretion. 50 μm

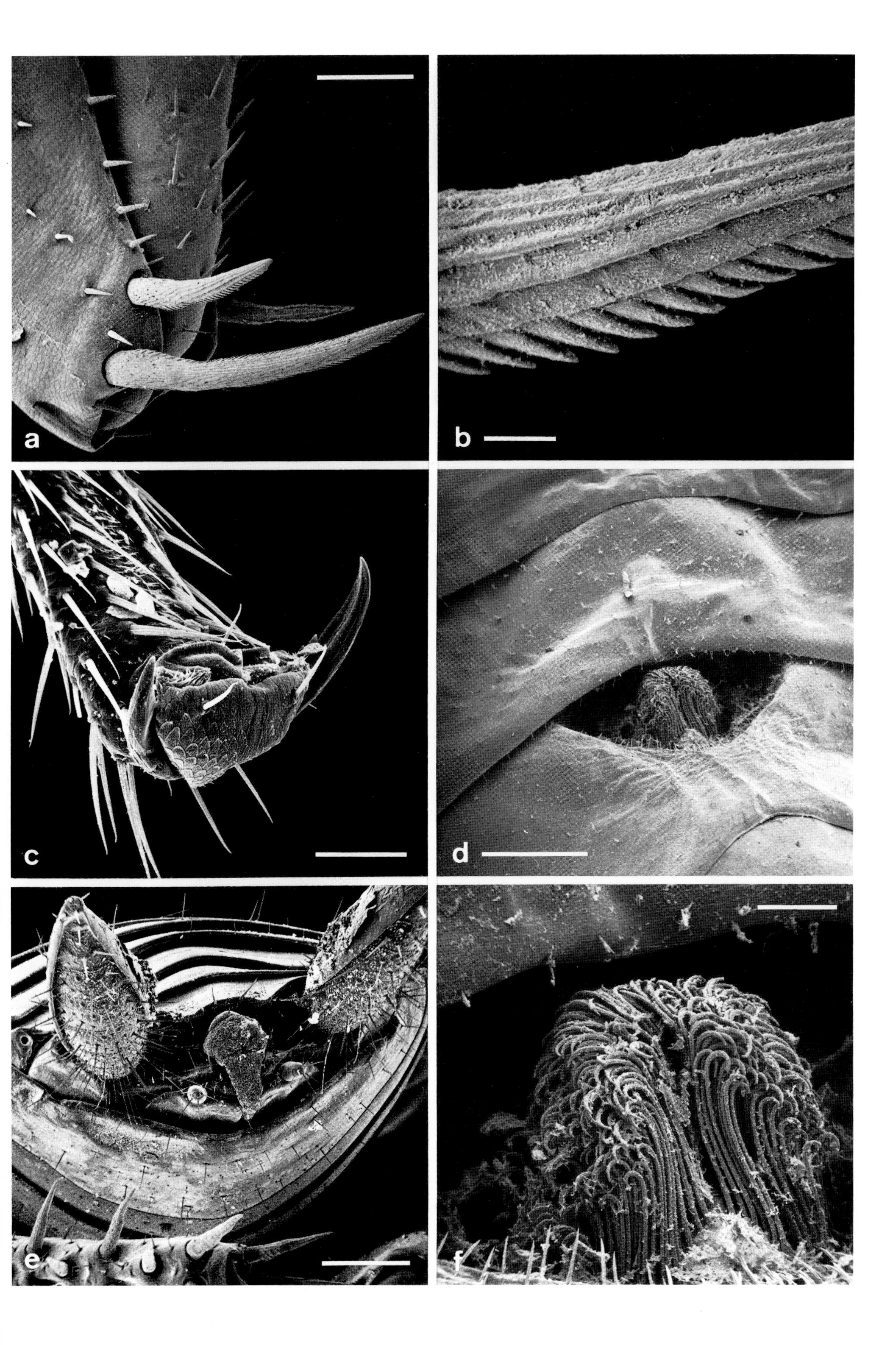
a
b
c
d
e
f

2.17 Order: Ensifera – Long-Horned Grasshoppers and Crickets (Insecta)

General Literature: Chopard, 1951; Beier, 1954, 1972; Harz, 1957, 1960

2.17.1 Forest- and Field-Dwelling Crickets

The dark brown forest cricket *Nemobius silvestris* (Plates 122–124), which is only 9–10-mm-long and the 20–26-mm-long field cricket, *Gryllus campestris*, belong to the Gryllidae, a family rich in species. Both species (Fig. 145) inhabit the soil (Röber, 1970).

The thermophilic field cricket, like the mole cricket *Gryllotalpa*, is at least temporarily hemiedaphic. The imagines of the field cricket dig burrows in dry meadows, sandy soils and open pine heath. They are predominantly phytophagous, dragging fresh plants and grasses into the burrow and consuming them as soon as they wilt (Harz, 1957).

The agile forest cricket, which is capable of running as fast as it can jump, leads an epedaphic life, primarily in the leaf litter of open woodland. Under the influence of the microclimate in various woodland areas, meadows and clearings, the forest cricket has proved to be a largely eurypotent species, tending only to avoid damp, cool and extremely shady areas (Brocksieper, 1978).

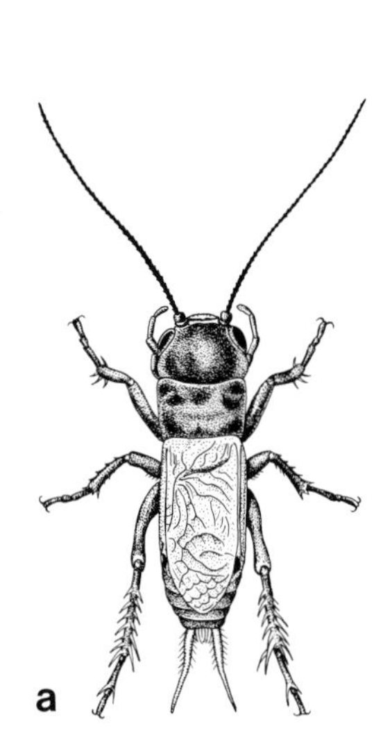

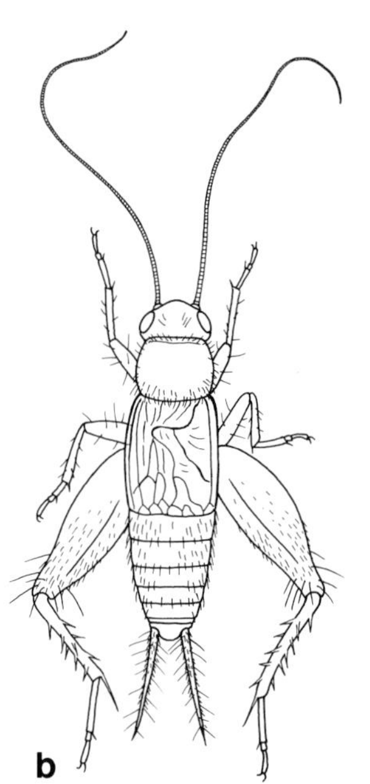

Fig. 145. Dorsal view of the field cricket (*Gryllus campestris*) (**a**) and the forest cricket (*Nemobius silvestris*) (**b**) of the Gryllidae family. (After Brauns, 1964 and Harz, 1957)

Plate 122. *Nemobius silvestris* Gryllidae (crickets), forest cricket, larva and imago.

a Larva, lateral, with its powerfully developed saltatorial legs. The caudal appendages are cerci (*arrow*). 1 mm.

b Head, lateral. Movement of the head is controlled by proprioreceptive hairs on the anterior margin of the pronotum. 0.5 mm.

c Mouthparts, ventral. 0.5 mm.

d Base of the antenna. On the second antennal segment, the pedicellus (*arrowhead*), there is a ring of scar-like pits (*arrows*). They indicate the position of the sensory organs in the interior, the scolopidia. 100 µm.

e Segments of the antennal flagellum with sensory hairs. 25 µm

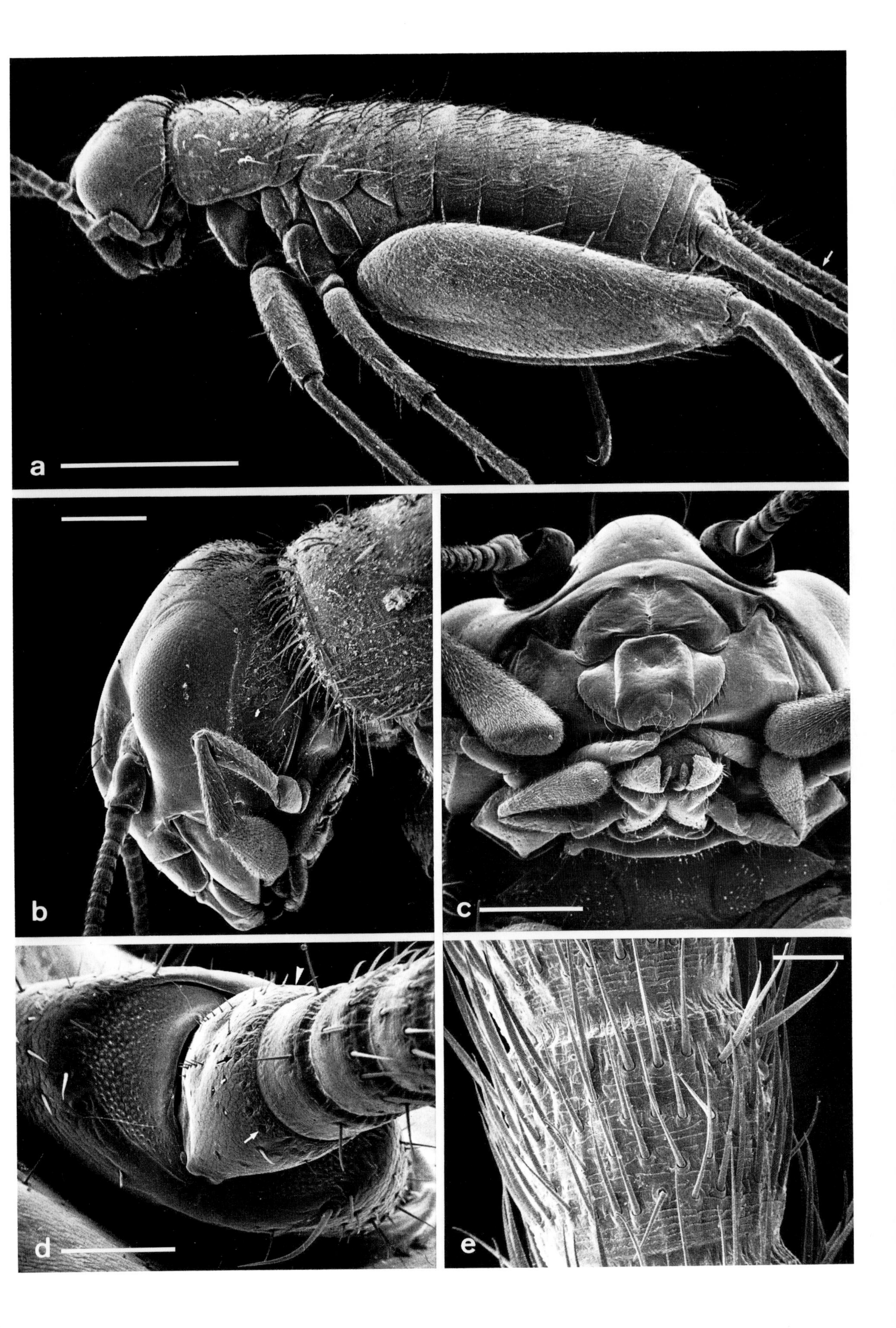
a
b
c
d
e

2.17.2 Hemiedaphic Life in Crickets

Among the hemiedaphic members of the Orthoptera, which dig and live in subterranean burrows, are the earwig (*Lapidura riparia*) – Dermaptera), the field cricket (*Gryllus campestris*) and the mole cricket (*Gryllotalpa gryllotalpa*).

The 50-mm-long, brownish mole cricket (Fig. 146) is particulary well adapted to the hemiedaphic way of life. Its head is covered with fine hair and its antennae are comparatively short. The powerful, broad front legs are differentiated to form shovel-like fossorial legs by shortening of the leg segments, modification of the tibia with four claw-like thorns and with the addition of two distinct processes on the laterally inserted tarsus.

The finger-thick tunnels, in which the animals seek food (insects and earthworms) and are normally encountered, have an irregular course through the soil. At mating time, however, they abandon their tunnel dwellings and run with surprising agility on the surface of the ground. They can also be observed in flight during this time (Halm, 1958; Godan, 1961).

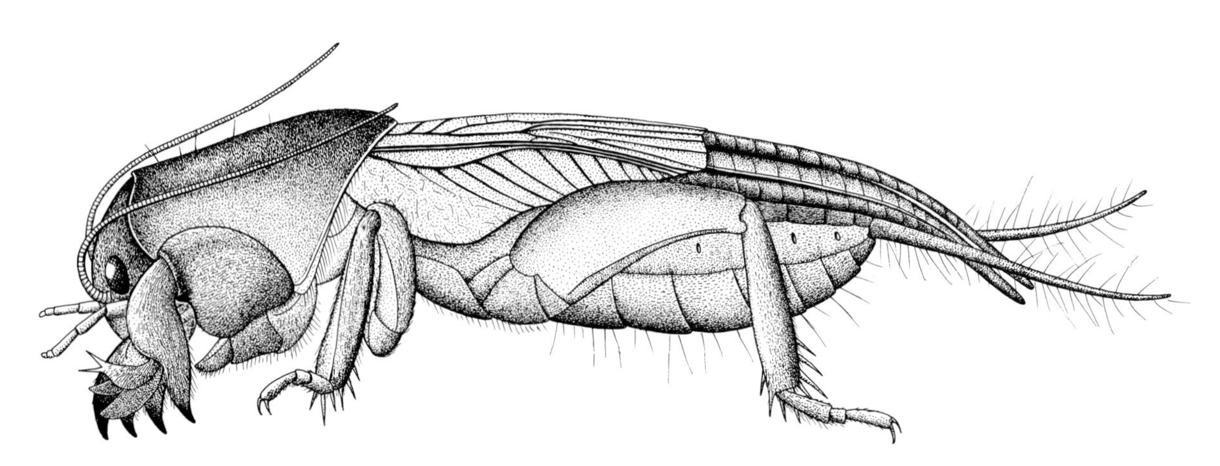

Fig. 146. *Gryllotalpa gryllotalpa* (Gryllotalpidae), mole cricket. Lateral view showing spade-like front legs. The tips of the hind wings extend far beyond the abdomen. The animals burrow into sandy soil within seconds, whereby the streamlined unit, consisting of the head and pronotum, acts like a plough

Plate 123. *Nemobius silvestris* Gryllidae (crickets), forest cricket, adult female.

a Head and thorax, dorsal. 1 mm.

b Metathorax with stunted wings. 0.5 mm.

c Femur-tibia joint of the front leg, frontal view. The *arrow* marks the tympanic membrane of the tibial auditory organ. 250 µm.

d Femur-tibia joint of the front leg, lateral. 200 µm.

e Tympanic membrane. 50 µm.

f Tibio-tarsal joint of the front leg. 100 µm

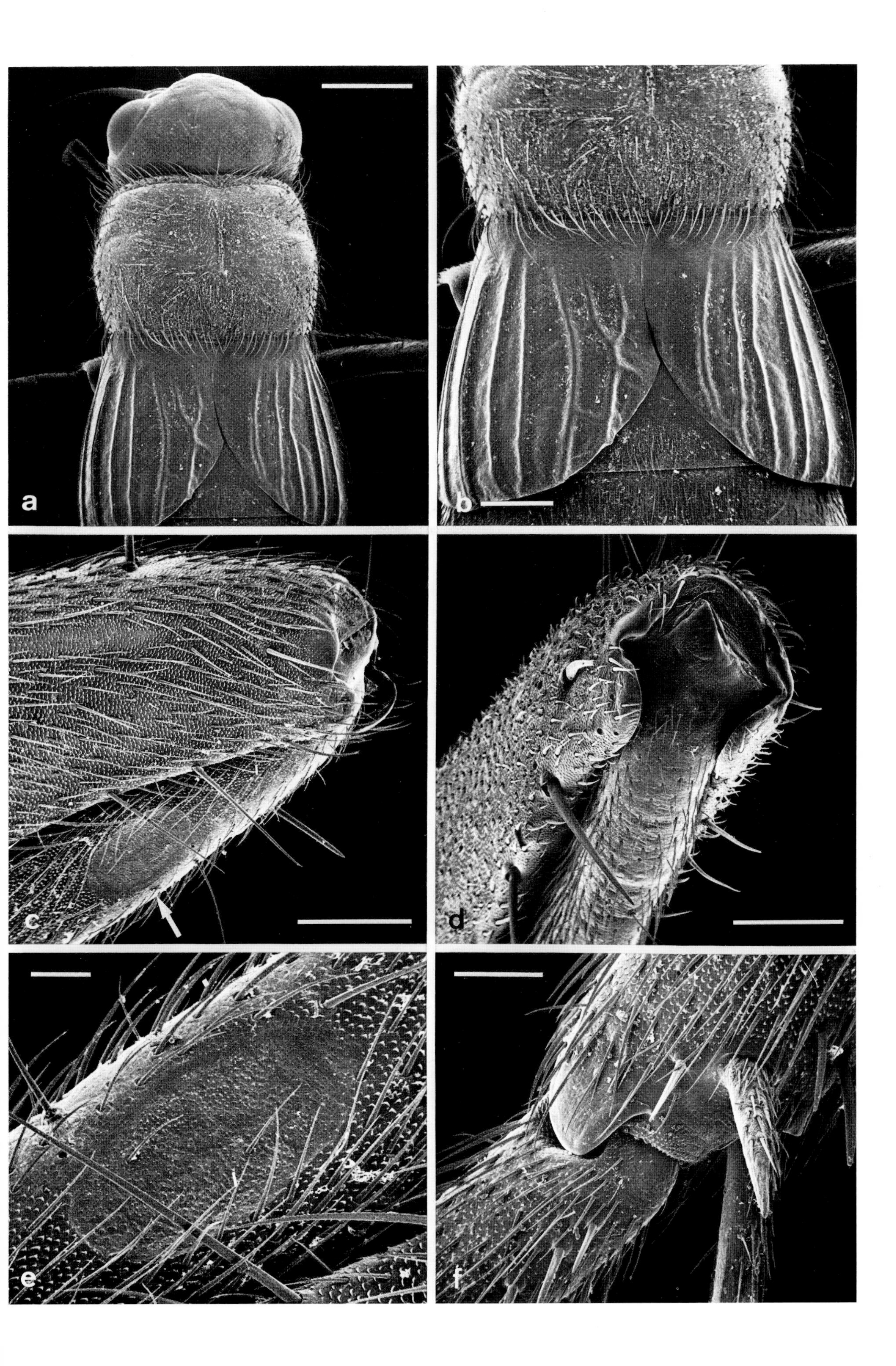

a
b
c
d
e
f

2.17.3 Sensory Hairs on the Cerci of Crickets

Several different types of sensory hairs are found on the cerci of the Mediterranean cricket, *Gryllus bimaculatus*: long filiform hairs, club-shaped hairs and short bristles (Gnatzy and Schmidt, 1971, 1972a, b; Schmidt and Gnatzy, 1971, 1972) which are also present in other crickets, e.g. the field cricket *Gryllus campestris* (Sihler, 1924) and the forest cricket *Nemobius silvestris* (Plate 124).

The filiform and club-shaped hairs have a similar internal ultrastructure and are equipped with campaniform sensilla (1–5 in the former and 1–2 in the latter) under the cuticle on the outer rim of their cupulate socket. Both hairs are mechanoreceptors. The long filiform hairs perceive the slightest air disturbances. They probably vibrate towards the cup wall facing the moulting canal when stimulated (Fig. 147). The vibrational direction of the hair is perceived when deformation of the cupulate socket is conveyed to the campaniform sensilla (Gnatzy and Schmidt, 1971; Gnatzy and Tautz, 1980). As their structure suggests, the club-shaped hairs function in a similar manner and are interpreted as gravity receptors (Nicklaus, 1969). The short bristles have a pore at the tip through which the outer dendritic structure has contact with the exterior. Thus, they are thought to be contact chemoreceptors which can simultaneously perceive mechanical stimuli (Schmidt and Gnatzy, 1972).

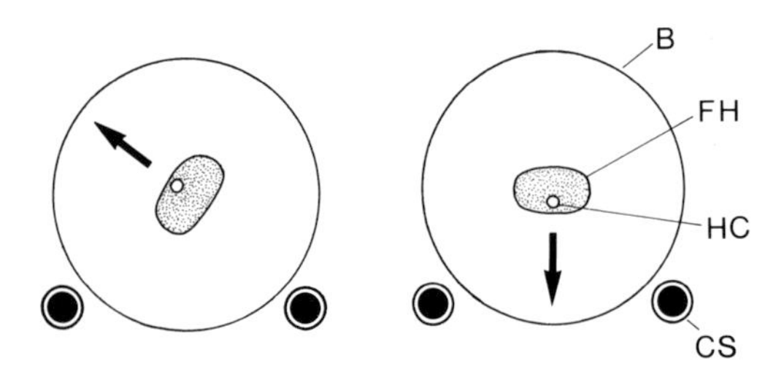

Fig. 147. *Gryllus bimaculatus* (Gryllidae). Presumed vibrational direction of two filiform hairs (*FH*) (*arrows*) on the cercus according to the position of the moulting canal (*HC*) and the campaniform sensilla (*CS*). The latter are thought to perceive deformation of the cupulate hair socket (*B*). (After Gnatzy and Schmidt, 1972)

Plate 124. *Nemobius silvestris* – Gryllidae (crickets), forest cricket, larva (**a**), and adult female (**b–f**).

a Abdomen of a female larva, terminal, ventral, with cerci and the ovipositor in its early developmental stage. 1 mm.

b Abdomen of an adult female, terminal, lateral with cerci and the base of the ovipositor. 0.5 mm.

c Cercus, proximal, with filiform and club-shaped hairs. 250 μm.

d Central portion of the cercus with short bristles and long filiform hairs. The latter are inserted into cupulate sockets and are stimulated in accordance with a directional characteristic (cf. Fig. 147 and text). 100 μm.

e Abdomen, skin on the flank with spiracle. 100 μm.

f Cercus, distal. 10 μm.

g Club-shaped hair on the base of the cercus. 10 μm

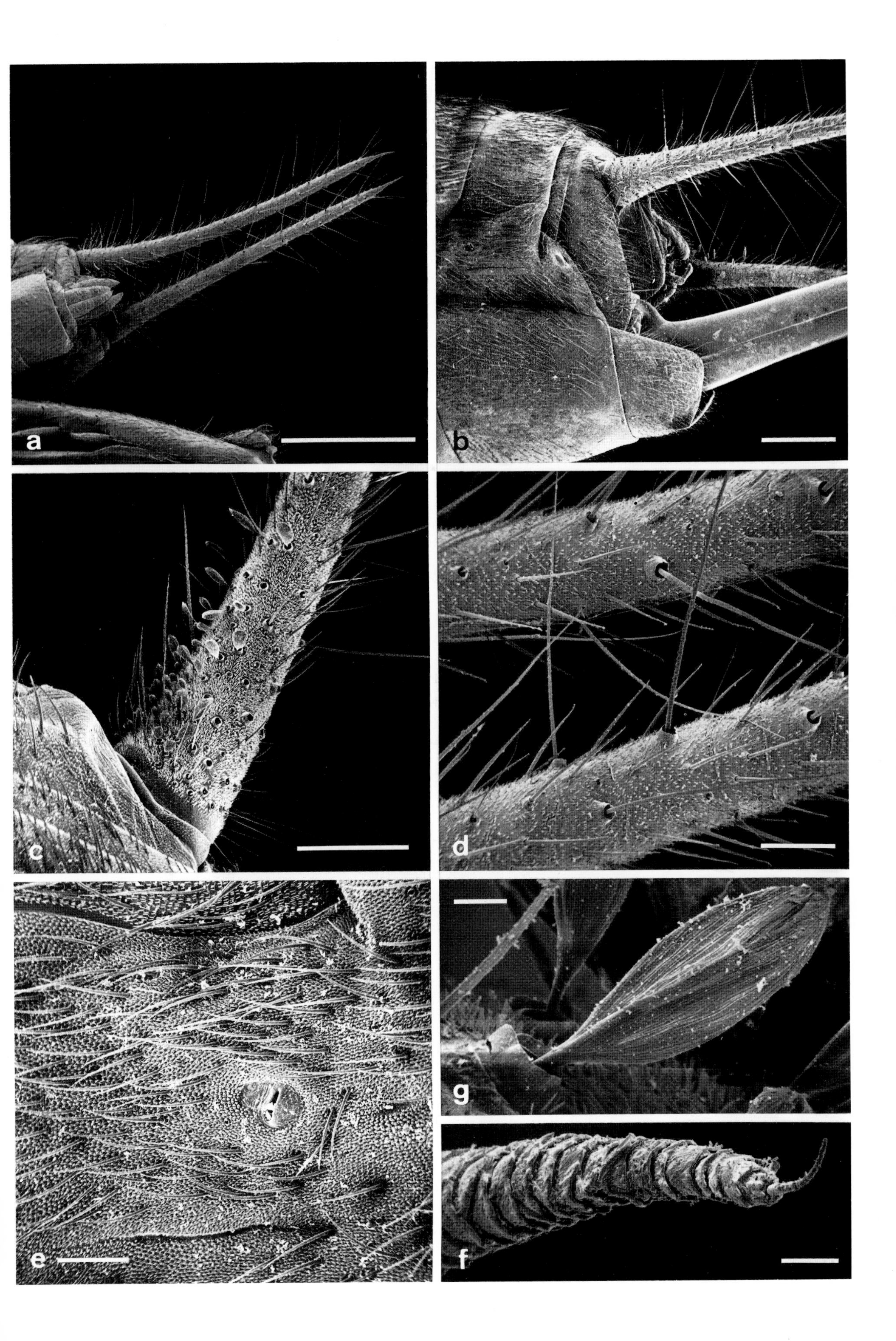
a
b
c
d
e
g
f

2.18 Order: Hemiptera (Insecta)

General Literature: Weber, 1930; Wagner, 1966; Jordan, 1962, 1972

2.18.1 Burrower Bugs (Cydnidae)

Only a few species within the hemipteran suborders, Heteroptera and Homoptera (e.g. some species of the Cicadina, Aphidina and Coccina), are true soil-dwellers. In some species only the larvae are restricted to the soil. Of the bugs, the burrower bugs or Cydnidae are most representative of the edaphic way of life. Their bodies are mostly dark, oval and more or less convex. Many species burrow in the upper soil layers, but species of the Mediterranean genus *Brysinus* reach a depth of 40 cm, and feed on juices from plant roots. *Cydnus aterrimus* is the largest species occurring in central Europe with a length of 8–12 mm (Fig. 148; Plates 125–127). It is distributed throughout the Palaearctic region with the exception of northern Europe and northern Asia (Wagner, 1966).

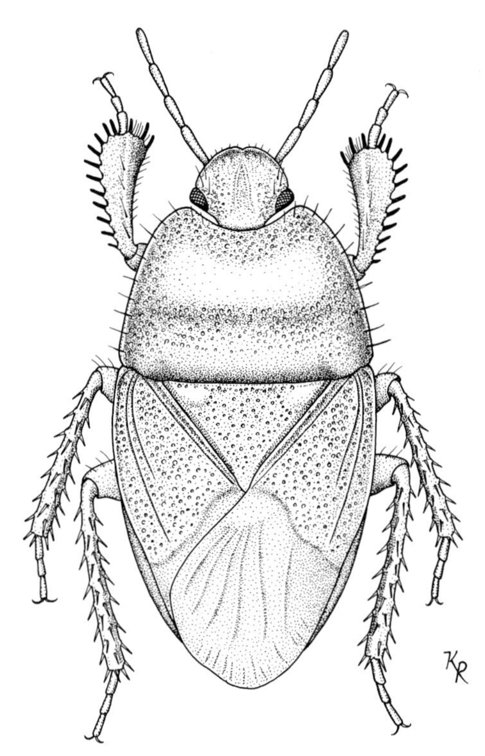

Fig. 148. *Cydnus aterrimus* (Cydnidae), burrower bugs, habitus, dorsal

Plate 125. *Cydnus aterrimus* (Cydnidae), burrower bugs.
a Habitus, dorsal, with head, scutum, scutellum and elytra divided into a leathery portion (corium) and the membrane. 1 mm.
b Caudal view of the wing membrane, corium and scutellum. 0.5 mm.
c Head and scutum, dorsal. Smaller ocelli (*arrow*) lie between the compound eyes. 1 mm.
d Compound eye, dorsal. 100 µm.
e Surface of the compound eye with ommatidia. 20 µm

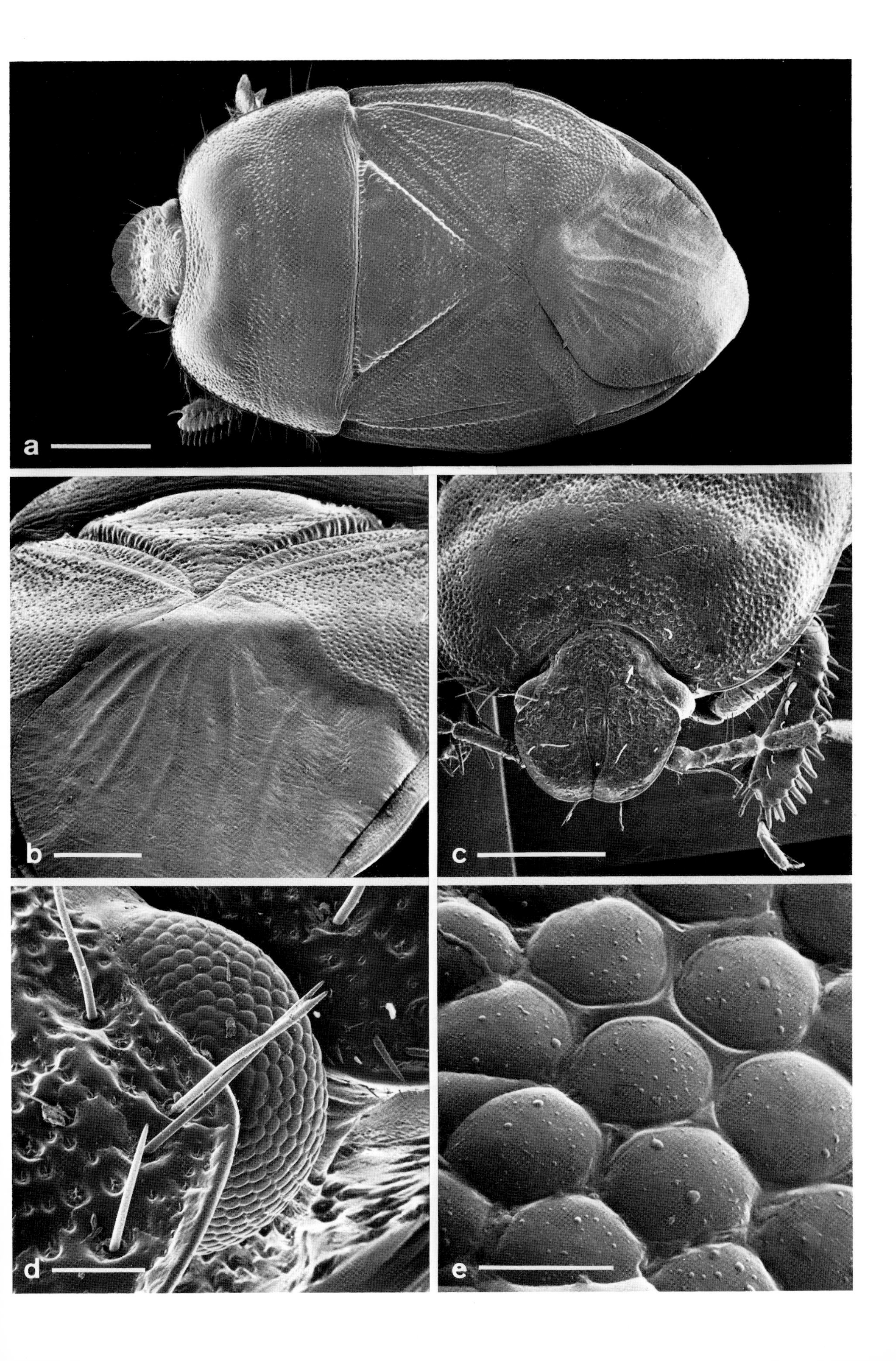
a
b
c
d
e

2.18.2 Fossorial Life of the Hemiptera

Bugs frequently obtain nourishment from plants using their usually piercing-sucking mouthparts with the mandibles and maxillae forming stylets. Thus, apart from several predatory bugs, which occur in the litter layer, the Hemiptera are generally restricted to the atmobiotic biotope. This is especially true of the Coccina and the Aphidina, but even within the Sternorrhyncha there are some root-boring species (Coccina, Margarodidae), which have evolved spade-like front extremities for excavation (Fig. 149b). The larvae of the cicadas (Cicadidae) are similarly adapted to soil life, an exception within the Auchenorrhyncha. Their larval development takes place almost exclusively underground where they burrow through the soil with well-developed fossorial legs (Fig. 149a), feeding on the roots of plants. Adult burrower bugs (Cydnidae) suck the roots of *Euphorbia* plants. They also use their flattened front extremities, equipped with rows of strong bristles, as fossorial legs for their edaphic way of life (Fig. 149c) (Weber, 1930; Jordan, 1972).

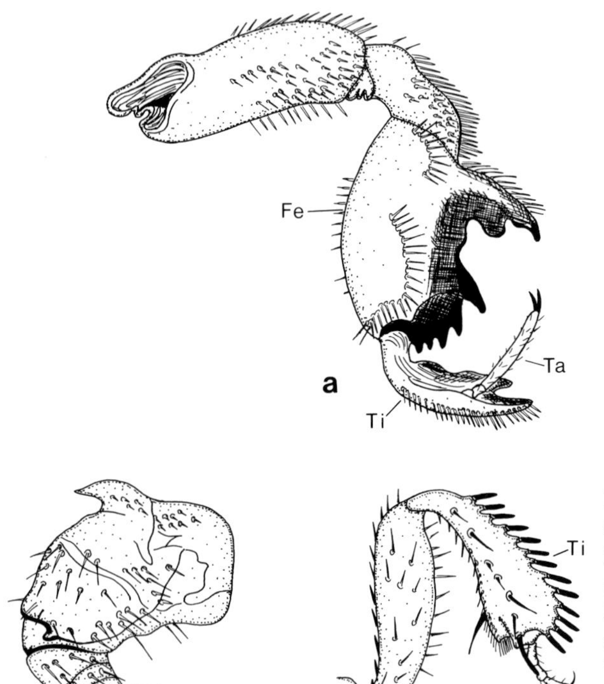

Fig. 149. Types of fossorial legs found in the Hemiptera. **a** Fossorial leg of a cicada, *Tibicen septendecim*, in the sixth instar. The major part of the excavators are formed by the femur (*Fe*). *Ta* Tarsus; *Ti* tibia. (After Weber, 1930). **b** Fossorial leg of a coccinean louse, *Margarodes meridionalis*. The whole leg functions as an excavator. (After Weber, 1930). **c** Fossorial leg of a burrower bug, *Cydnus aterrimus*. The tibia is broadened to form an excavator and is reinforced by strong spiny hairs. (After Schorr, 1957)

Plate 126. *Cydnus aterrimus* (Cydnidae), burrower bugs.

a Head and thorax, ventral, with a sucking stylet (rostrum) and the front legs, modified to fossorial legs. 1 mm.
b Fossorial leg, dorsal, with tibia and tarsus. 0.5 mm.
c Base of the sucking stylet. 250 μm.
d Surface of a fossorial leg (tibia), ventral. 100 μm.
e Stylet tip; the bundle of piercing bristles is ensheathed by the lower lip (labium). 200 μm.
f Tarsus of fossorial leg with three segments. 250 μm

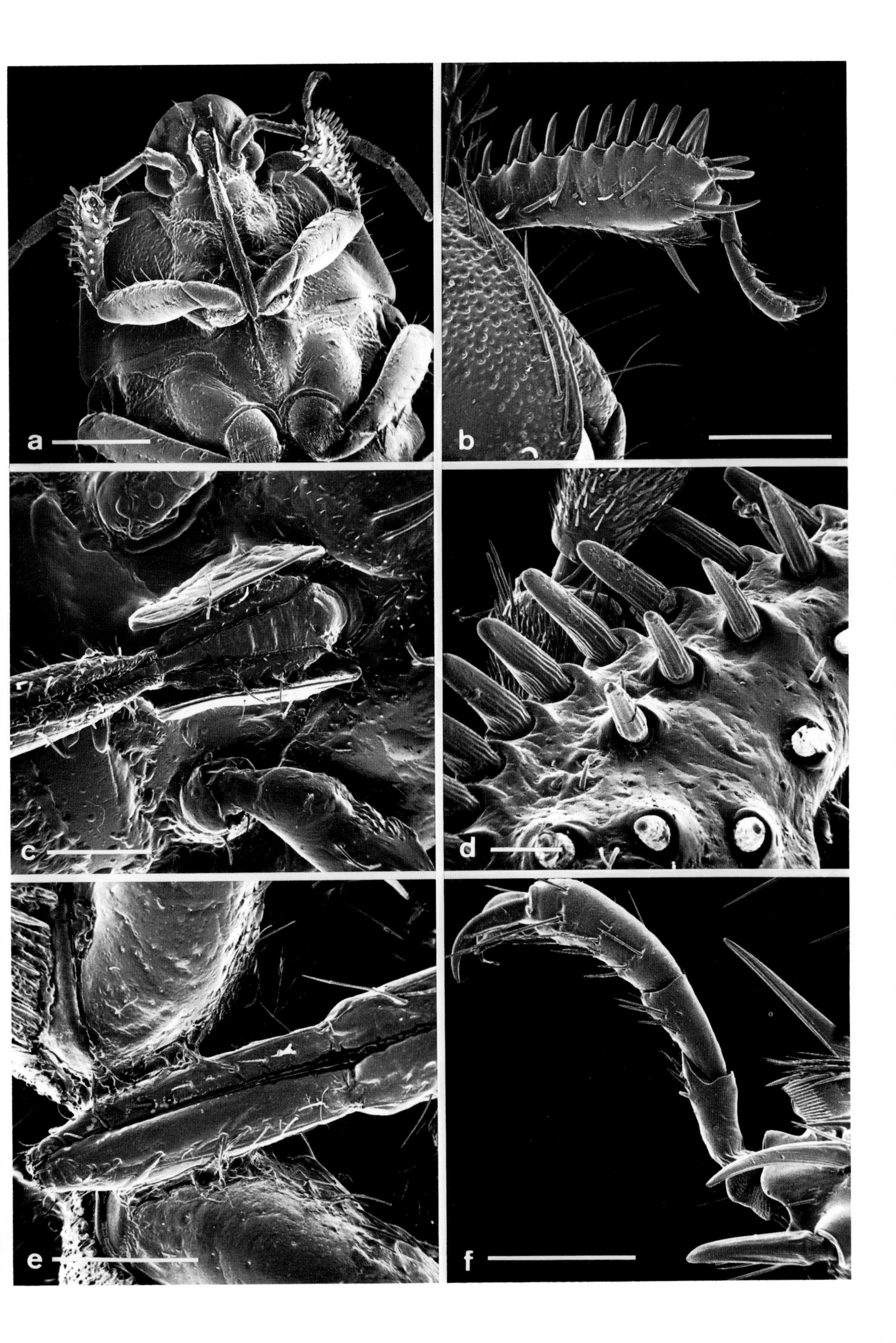

a
b
c
d
e
f

2.18.3 Defensive Stink Glands in Bugs

Bugs possess stink glands which produce a contact poison. Their enemies are immobilized when the poison penetrates the skin and through the spiracles of predatory insects, if they have not already been deterred by the repugnant odour (Remold, 1963). In adult bugs the stink glands are usually paired and open onto the metapleurites on the ventral thorax near the coxa of the third pair of legs. Around the outlet there is often a field of bizarrely differentiated cuticle, which spreads the outflowing secretion over a larger area. In comparison to the glandular outlet, this enlarged area allows rapid evaporation of the secretion, which intensifies the effect of the repellant odour. Thus, this area surrounding the glandular outlet is called the evaporation field (Plate 127). The dorsal abdominal stink glands of larval and some adult bugs are similarly constructed. The inner structures are shown in Fig. 150 for the red bug *Pyrrhocoris apterus*.

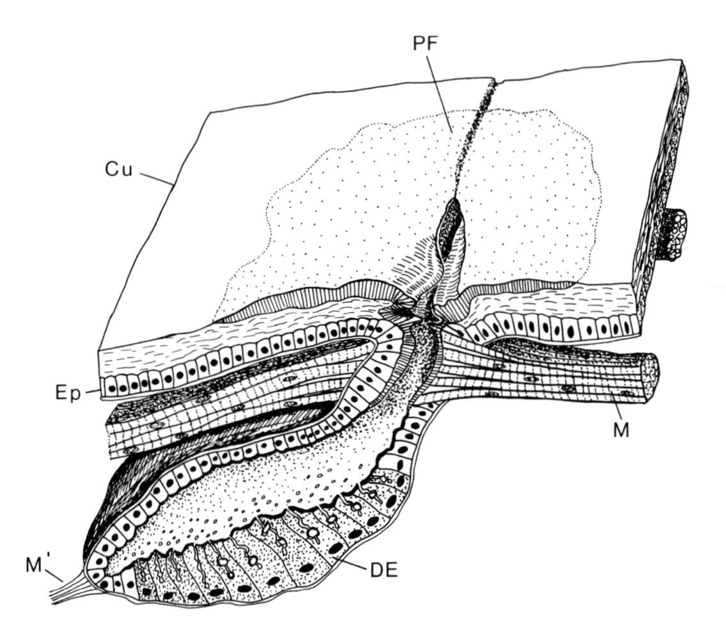

Fig. 150. *Pyrrhocoris apterus* (Pyrrhocoridae), red bugs, anatomy of a dorsal stink gland. (After Weber, 1930). The glandular epithelium (*DE*) is invaginated in the dorsal surface and covered by a thin cuticle (intima). The secretion produced by the stink gland is extruded through the outlet by a muscular occlusion and pressure mechanism (*M,M'*). The pigmented field (*PF*) is comparable to the evaporation field of the thoracic stink glands. *Ep* Epidermis; *Cu* cuticle

Plate 127. *Cydnus aterrimus* (Cydnidae), burrower bugs.
a Stink gland in the metapleural thoracic wall. 0.5 mm.
b Microstructures on the evaporation field with bristle hair. 6 µm.
c Stink gland with ear-shaped outlet and evaporation field. 250 µm.
d Microstructures on the evaporation field. 6 µm.
e Glandular opening. 25 µm.
f Transition area between the evaporation field and the normal cuticle. 10 µm

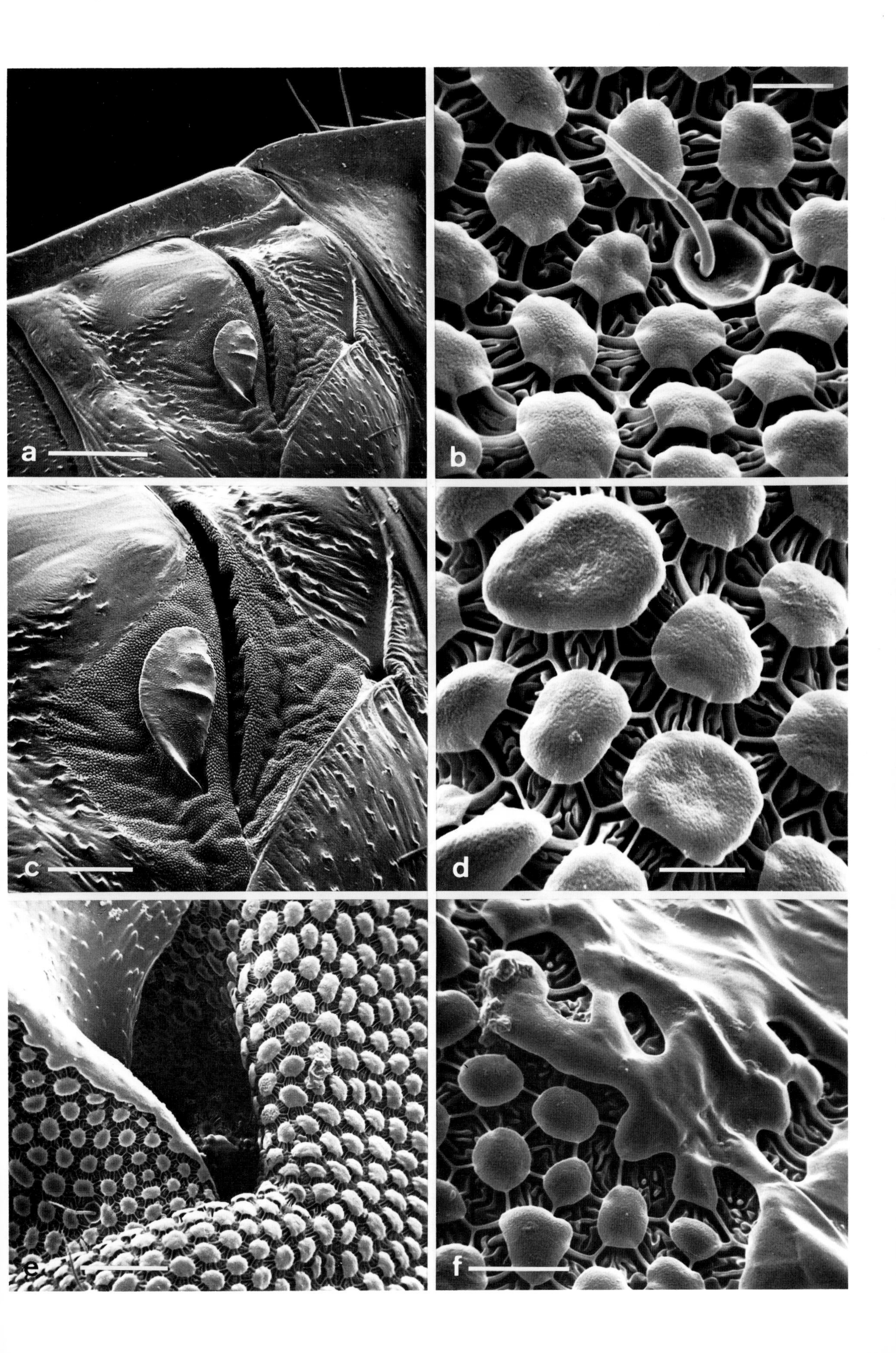

a
b
c
d
e
f

2.18.4 Care of the Brood and Larval Development

Brood care by the female of *Cydnus aterrimus*, described by Schorr (1957), begins when she deposits about 30–50 eggs on the soil in the vicinity of the food plant *Euphorbium* and continues in her defence of the clutch from possible enemies. After hatching the larvae remain 8–9 days with their mother. During this time they suck droplets excreted form their mother's anus. With this initial nutrition the larvae also ingest symbiotic intestinal bacteria, without which digestion of their food would be impossible.

The larvae undergo paurometabolic development, passing through five larval instars. The gradual but continuous development of the larva to the form of the imago is recognizable even in the first instar (Schorr, 1957). Initially the thoracic segments are all identically shaped. Later the mesonotum grows and the elytra and wing sheaths develop. In the last instar the scutellum and the elytra "anlagen" are clearly formed and extend caudally over the abdomen.

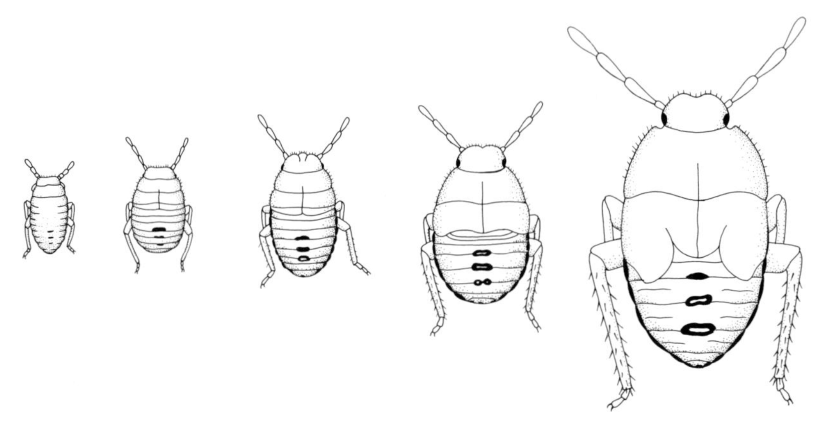

Fig. 151. *Cydnus aterrimus* (Cydnidae), burrower bugs. Example of a stepwise larval development (paurometabolism) accompanied by moults and with successive formation of the wing "anlagen". (After Schorr, 1957)

Plate 128. Larva of a bug from the soil litter.

a Habitus, lateral. The sucking stylet (rostrum) extends from the head along the ventral side of the thorax. 0.5 mm.

b Dorsal view. 250 μm.

c Head with stylet base, oblique dorsal. 250 μm.

d Abdomen, caudal, with anus. 100 μm.

e Caudal view of a compound eye. 50 μm.

f Ommatidia of the compound eye. 10 μm

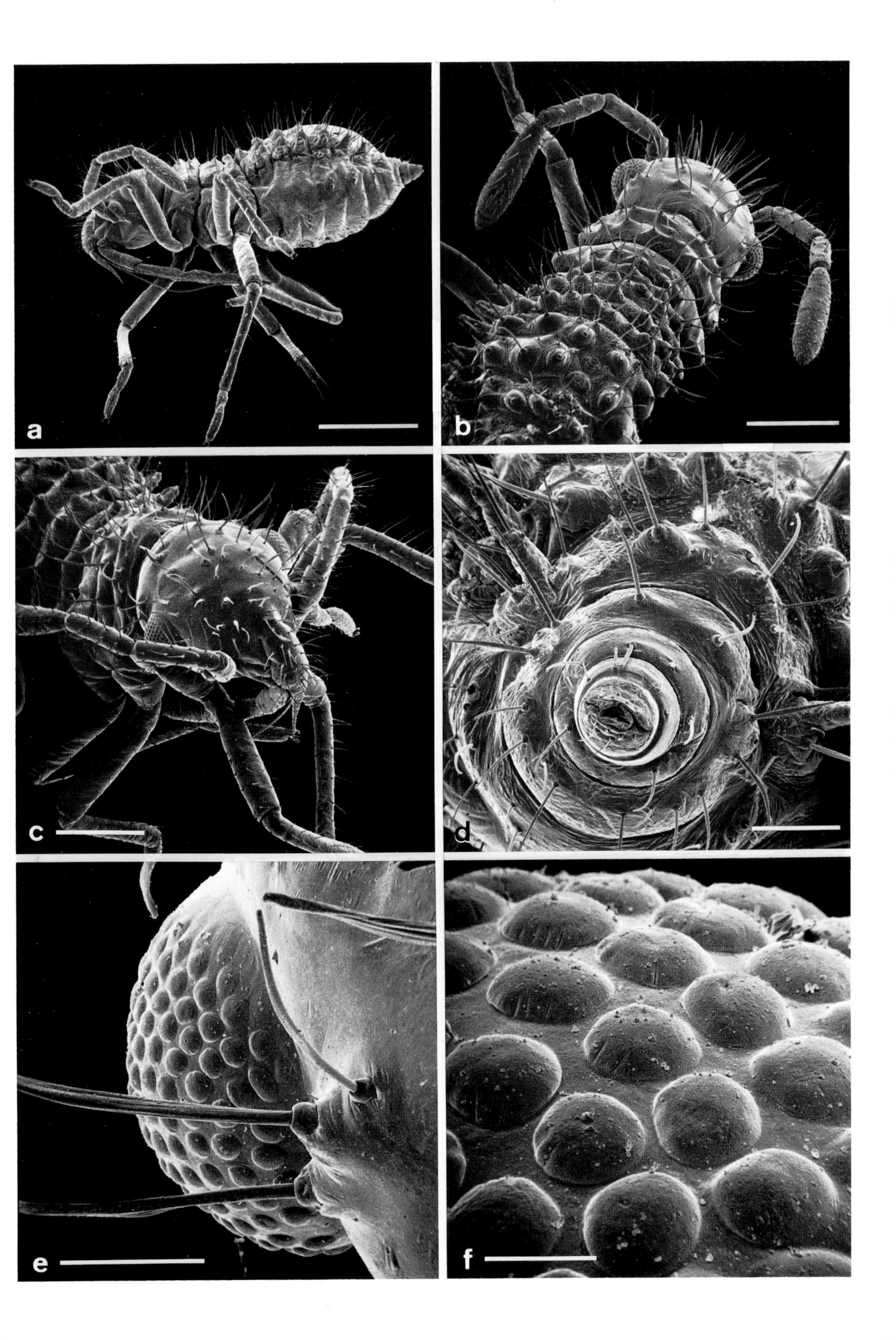
a
b
c
d
e
f

2.19 Order: Planipennia (Insecta)

General Literature: Aspöck et al., 1980

2.19.1 Ant Lions (Myrmeleonidae)

The larvae of the Myrmeleonidae family, known as ant lions, have a remarkable way of life. Like all Planipennia larvae, they are predaceous and usually live hidden in surface sand, where they trap arthropods, often ants, walking above. However, relatively few larvae, among them those of the genera *Myrmeleon* (Figs. 152 and 153; Plates 129–131), *Euroleon, Solter* and *Cueta* in Europe (Aspöck et al., 1980), construct the well-known pitfall trap in loose sand. These larvae sit at the bottom of the sand funnel, buried up to their head in sand, and lie in wait with their powerful mouthparts outstretched ready to seize their prey (Fig. 154). When an insect reaches the rim of the funnel, sand begins to slide down the slope of the pit, the insect's hold is loosened and it slides into the funnel where it is immediately seized by the pincers of the ant lion, immobilized by a secretion, predigested and finally sucked out (Doflein, 1916; Eglin, 1939; Neboer, 1960; Plett, 1964; Steffan, 1975; Matthes, 1982a, b).

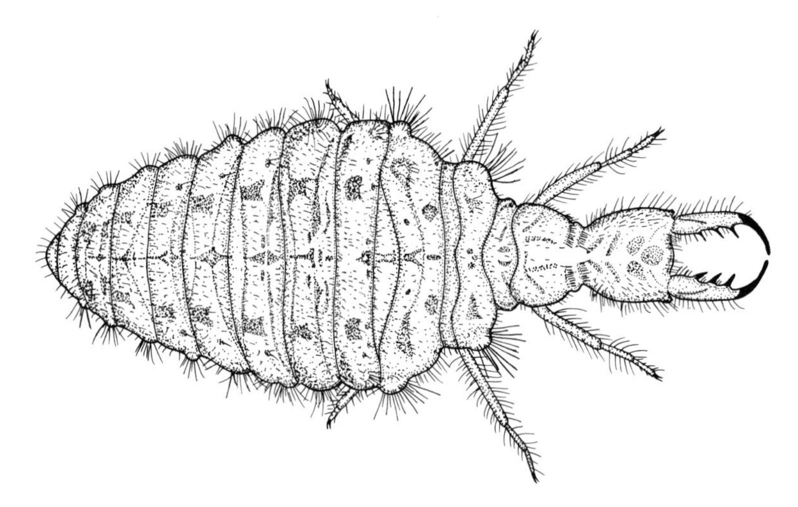

Fig. 152. Larva of *Myrmeleon inconspicuus* (Myrmeleonidae), dorsal. (After Principi, 1943 from Aspöck et al. 1980)

Plate 129. *Myrmeleon formicarius* (Myrmeleonidae), ant lion.

a Head and thorax, oblique frontal. 1 mm.
b Head with pincers, dorsal. 1 mm.
c Pincers, ventral. The stylet-shaped maxillae are fitted into a groove in the mandibles. 0.5 mm.
d Pincers, dorsal. At their base lie the eye tubercles and the antennae. 0.5 mm.
e Pincer finger, ventral. 150 μm.
f Pincer finger, distal, with conical sensilla in shallow depressions. *Inset*: Pincer sensillum. 25 μm, 5 μm

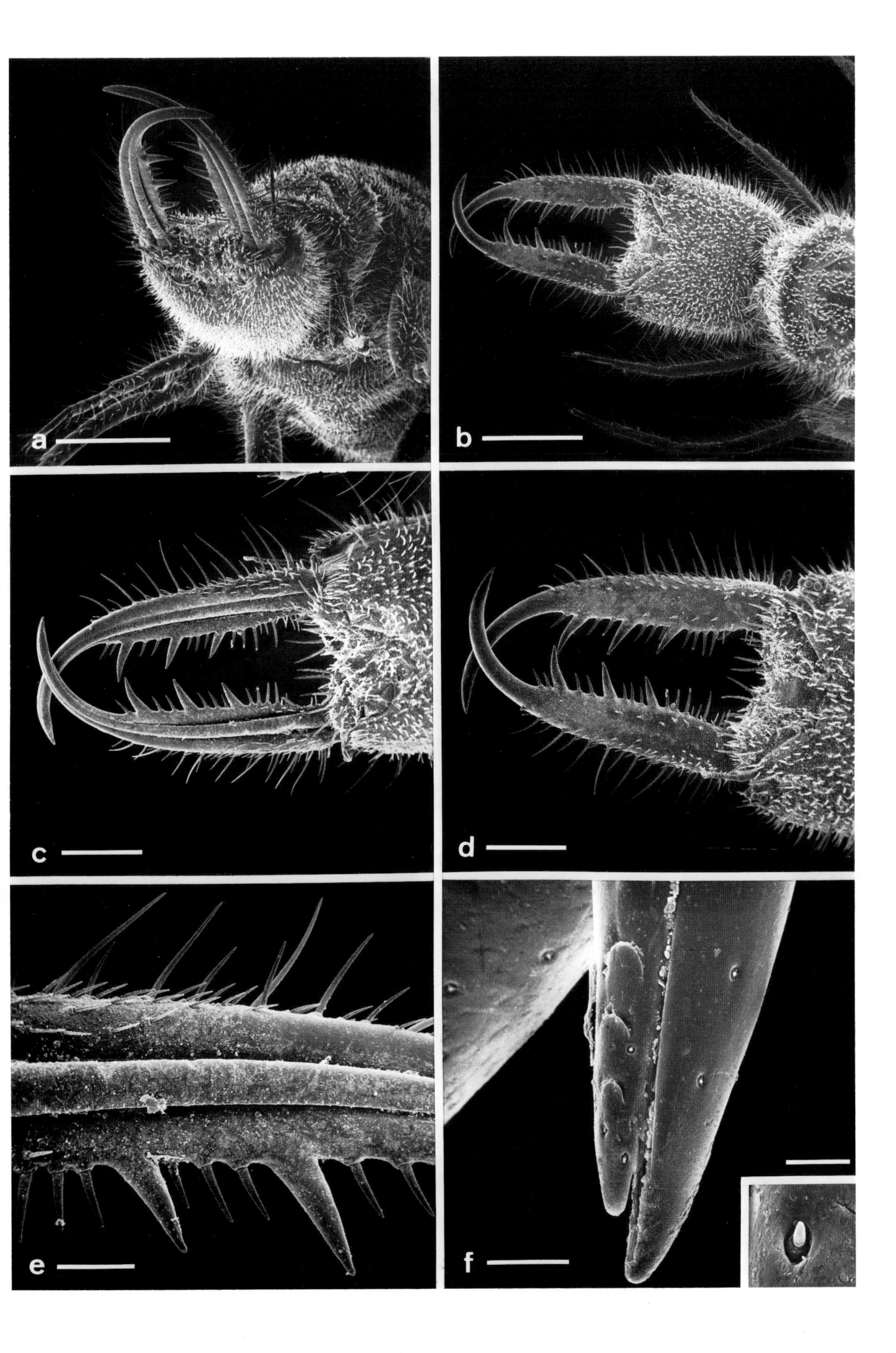
a
b
c
d
e
f

2.19.2 Sensory Perception in the Ant Lion

The larva lurking at the bottom of its sand funnel is sensitive to external stimuli, especially vibration. Although the antennae and eyes seem small and relatively insignificant compared to the mighty mouthparts (Fig. 153; Plates 129 and 130) the sensory organs as a whole, and the mechanoreceptors in particular, are of importance in perceiving prey, as are the mouthparts in seizing the prey.

The eyes of the ant lion are situated in groups on a tubercle at each side of the head (Plate 130). Six ocelli form a functional complex at an angle of nearly 30° to the longitudinal axis of the larval body. On each side, separate from the ocelli complex, a small, stunted seventh eye, is situated at the base of the eye stalk (Plate 130c). Whereas the rudimentary eye looks downwards, the ocelli of the eye stalk covers the area of activity of the lurking ant lion. (Jokusch, 1967).

Fig. 153. Head of the larva of *Myrmeleon formicarius* (ant lion), Myrmeleonidae, ventral view. (After Doflein, 1916). The stylet-shaped maxillae (*Mx*) are folded out of their groove in the mandibles (*Md*). *A* Eye tubercle; *An* antenna; *Kf* maxillary groove; *Lb* labium with palps; *Md* mandible; *Mx* maxilla; *Tr* comb of finely feathered bristles (trichomes, palisade hairs) on the base of the pincers

Plate 130. *Myrmeleon formicarius* (Myrmeleonidae), ant lion.

- **a** Eye tubercle and antenna at the base of the pincer finger, dorsal. 250 µm.
- **b** Part of the lower lip (labium) with labial palp (feeler) at the base of the pincer finger, ventral. Under the feeler lies a palisade of trichomes, which are stimulated by the movement of the pincers. 100 µm.
- **c** Eye tubercle with antennal base and palisade hairs at the base of the pincers, latero-ventral. 100 µm.
- **d** Palisade hairs (trichomes). 25 µm.
- **e** Surface of a single eye. 5 µm.
- **f** Eye tubercle. 50 µm

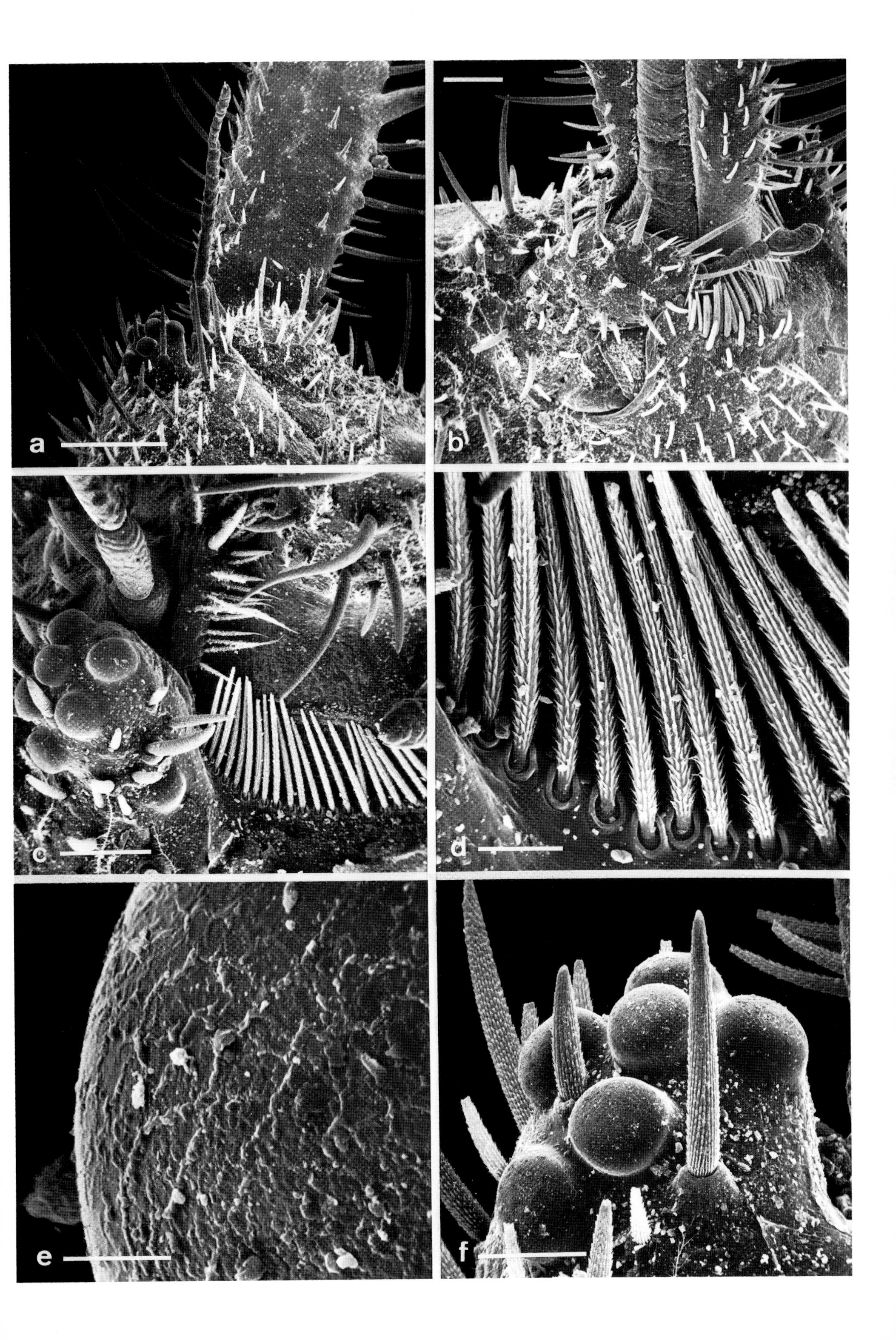

a
b
c
d
e
f

2.19.3 The Effect of Temperature on Ant Lions

Even before prey is sighted at the rim of the funnel, the lurking ant lion is alerted by vibrational stimuli. The eyes play a subordinate role in the capture of prey. However, Jokusch (1967) believes the eyes have an important task in perceiving solar radiation shining into the funnel and relates this perception to the ant lion's sensitivity to heat.

As their abdomen is buried in the sand at the bottom of the funnel, the animals are protected from intense heat. But on cloudless days the slopes of the sand funnel are warmed by the sun, which influences the hunting behaviour of the ant lion. The temperature begins to increase at sunrise on the western slope, reaches maximum values on the northern slope when the sun is at its meridian and decreases on the eastern slope of the funnel in the afternoon. As the temperature increase follows its course, the ant lion changes its position accordingly. In the morning the animal is often concealed in the sand of the eastern slope at the bottom of the funnel, towards noon it moves round via the southeast to the south. In the afternoon the southwestern and later the western slopes are preferred. Midday temperatures of over 40 °C at the bottom of the funnel cause the larvae to cease all hunting activity and retreat into cooler sand layers. Only starving larvae can still be observed lying in wait with their mouthparts outspread, ready to capture prey (Geiler, 1966).

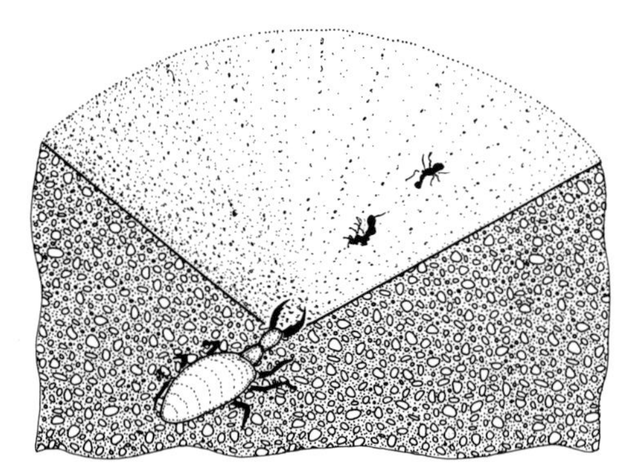

Fig. 154. Larva of *Myrmeleon formicarius* (ant lion), Myrmeleonidae, at the bottom of its funnel trap. The front of the head and the pincers are exposed and await their prey, falling ants. (After Jacobs and Renner, 1974)

Plate 131. *Myrmeleon formicarius* (Myrmeleonidae), ant lion.

a Abdomen, caudal, with rows of long spiny and filiform bristles. 0.5 mm.

b Abdomen, ventral, with lateral tufts of filiform bristles. 1 mm.

c Anal region with different types of bristles (sensilla). 200 µm.

d Abdomen. lateral, tuft of filiform bristles. 100 µm.

e Row of long, spiny bristles in the anal region, used for anchoring the abdomen in sand. 50 µm.

f Thick, bulbous sensilla in the anal region. 25 µm

a
b
c
d
e
f

2.20 Order: Coleoptera – Beetles (Insecta)

General Literature: Böving and Craighead, 1931; Burmeister, 1939; Crowson, 1981; Evans, 1975; Freude et al., 1965 ff.; Klausnitzer, 1982

2.20.1 Life Forms of the Soil-Dwelling Beetles

The Coleoptera order has a great diversity of forms; many of its species are soil-dwelling, belonging to the euedaphic, hemiedaphic and epedaphic life forms.

Euedaphic species are characterized by a slender, elongated form, as illustrated by the staphylinid beetles, and by shortening of the extremities and reduction of the tarsal segments. In addition, shortening of the elytra and reduction of the membranous hind wings can be observed. This consistent transformation, which is sometimes accompanied by a reduction of eyes and pigment, is only evident in the typical euedaphic beetles, e.g. a subfamily of the Staphylinidae: Leptotyphlinae (Fig. 172) (Coiffait, 1958, 1959). The majority of soil-dwelling beetles, especially in central Europe, are epedaphic to hemiedaphic. They are predators on the soil surface, but penetrate the soil to a greater or lesser extent by burrowing. Only the larvae of certain beetles fulfil the morphological conditions for euedaphic life in the soil.

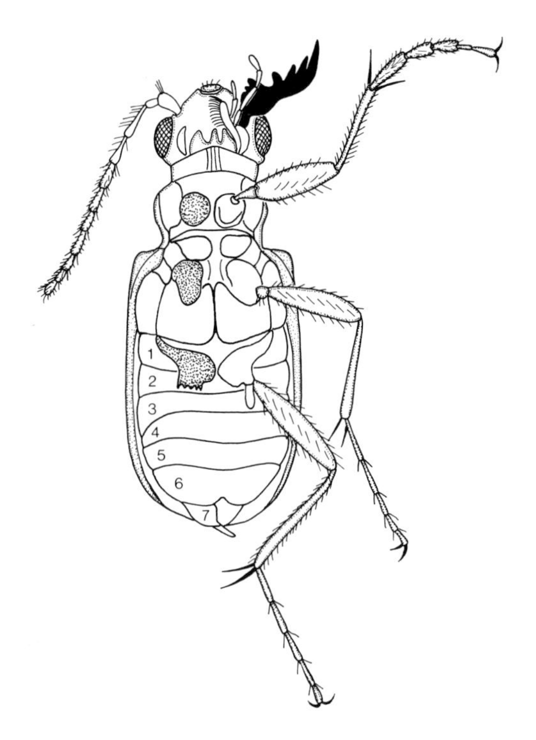

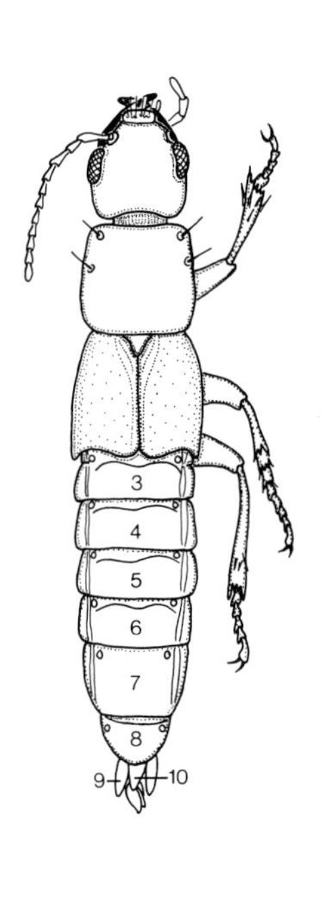

Figs. 155, 156. Schematic diagram showing two forms of beetles. (After Klausnitzer, 1982).

Fig. 155. Cicindelid beetle, ventral. The mouthparts are extremely simplified (mandibles are *black*). The legs with their coxal segments have been removed from one-half of the body. The *numbers* denote the sternites of the abdominal segments.

Fig. 156. Staphylinid beetle, dorsal. Beneath the shortened elytra lie the folded, functional hind wings. The *numbers* denote the tergites of the abdominal segments

Plate 132. Larva of *Cicindela* sp. (Cicindelidae), tiger beetle.
a Thorax and head, oblique dorsal. 0.5 mm.
b Thorax and head, lateral. 1 mm.
c Head, ventral. 1 mm

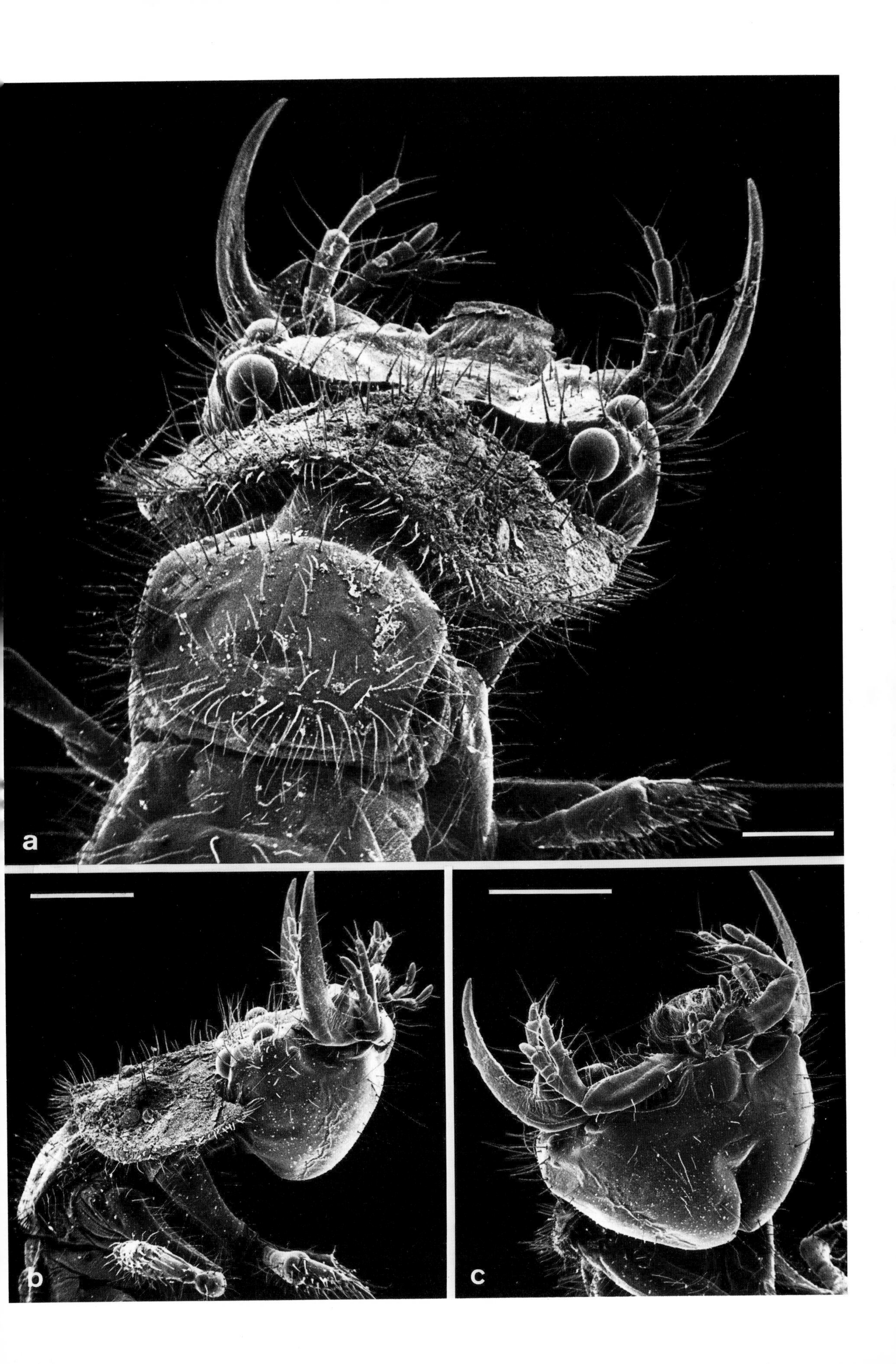
a
b
c

2.20.2 Tiger Beetle Larvae (Cicindelidae)

Tiger beetle larvae (Fig. 158; Plates 132–134) lead a hemiedaphic life; they dig shafts of up to 40 cm depth, depending on their age and size, in loose sandy soil. The first instars of *Cicindela hybrida* and *C. campestris* burrow 15–20 cm deep, the third instars live in vertical shafts about 40 cm deep, lying in wait at the entrance. The circular opening of the burrow is sealed by the broad head, which is bent forward ventrally, together with the wide pronotum (Fig. 157). The upper sides of the head and pronotum are flattened. Furthermore, the animals camouflage themselves by attaching sand, earth and detritus particles to the pronotum so that they completely blend in with their surroundings (Plates 132 and 133).

The outspread, dorsally directed, powerful mouthparts as well as two larger pairs of individual eyes (stemmata) protrude clearly from the upper side of the head. Four pairs of eyes, two of which are inserted laterally, are very small or rudimentary. The larvae lurk in this position waiting for prey which they perceive not only by means of vibrational stimuli but also, due to their wide field of vision, by sight.

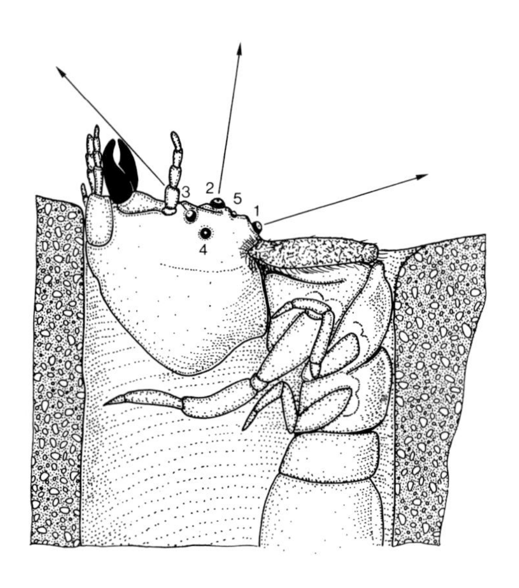

Fig. 157. Larva of *Cicindela* (Cicindelidae), tiger beetles, lying in wait. The *arrows* show the optical axes of the important stemmata. (After Weber, 1933 and Faasch, 1968)

Plate 133. *Cicindela* sp. (Cicindelidae), tiger beetle.

a Head and pronotum, dorsal. 1 mm.

b Two large stemmata and the smaller stemma on the *left*. 250 μm.

c Mouthparts, lateral, with the large mandibles as grasping organs, the antennae (*arrowhead*) and the labial (*arrow*) and maxillary palps. 0.5 mm.

d Mouthparts, ventral. The *arrowhead* marks the upper lip (labrum). 250 μm.

e Maxillary palps, lateral. 250 μm.

f Surface of the pronotum covered by particles of earth and sand interspersed with long setae. 100 μm

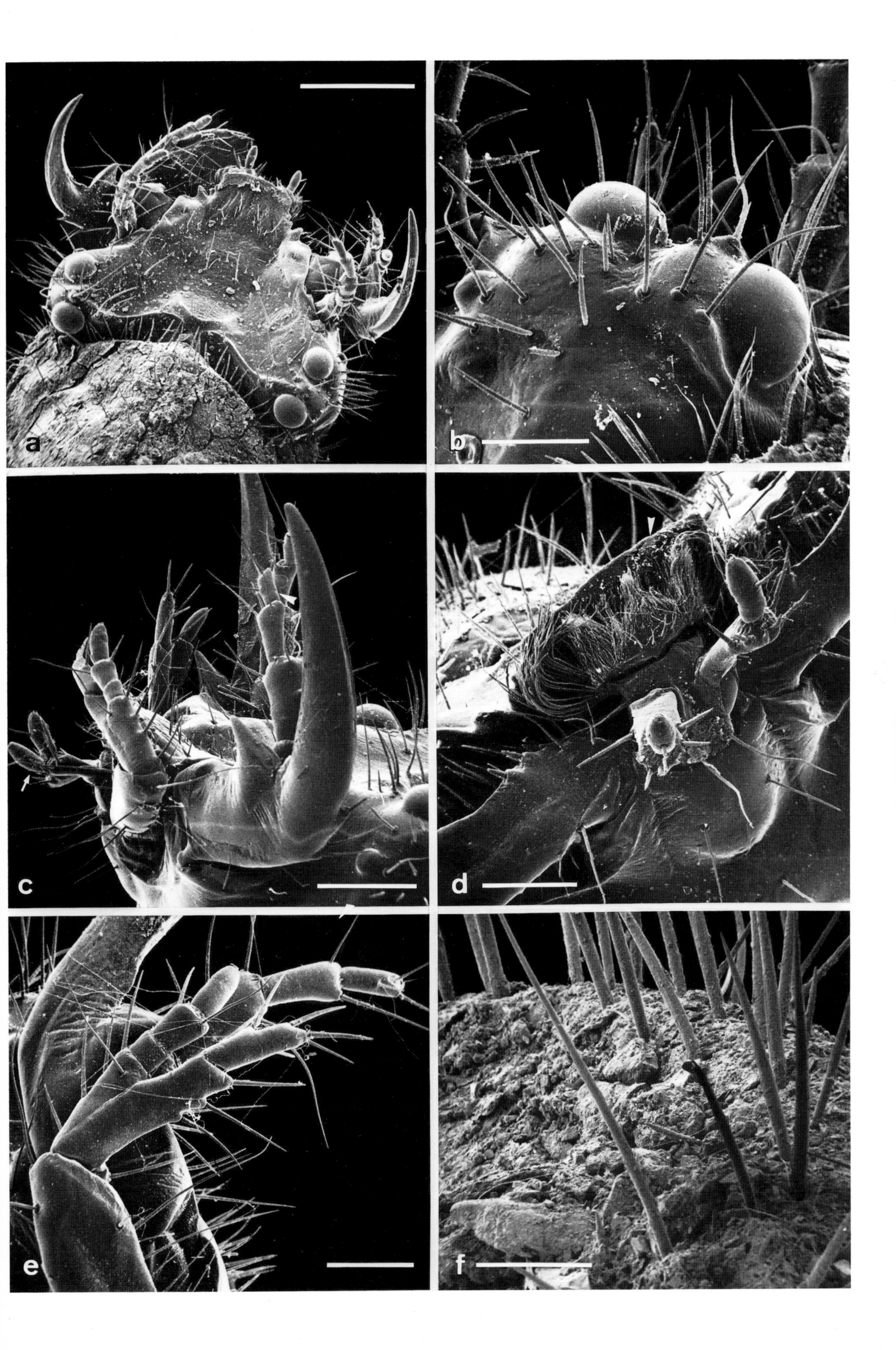

a
b
c
d
e
f

2.20.3 Predatory Life of Tiger Beetle Larvae

While lying in wait for prey the larva seals the opening of its burrow with its head and pronotum (Fig. 158). It anchors itself firmly in the burrow with the terminal end of its abdomen and a pair of hooks on the anchor organ situated on the fifth abdominal segment. When the larva observes an approaching insect it prepares to seize the prey by carefully detaching its abdomen from the burrow wall. It remains momentarily anchored to the burrow only by its anchor organ. If the insect then approaches to within a few millimetres, the tiger beetle immediately presses its freely moveable abdomen against the burrow wall again, this time about the height of the anchor organ. It simultaneously loosens its hold on the wall with the hooked dorsal hump by stretching swiftly forwards, at the same time thrusting itself out of the burrow and seizing the prey in its outstretched mandibles. The captured insect is subsequently dragged into the burrow and devoured at the bottom (Faasch, 1968; Evans, 1965).

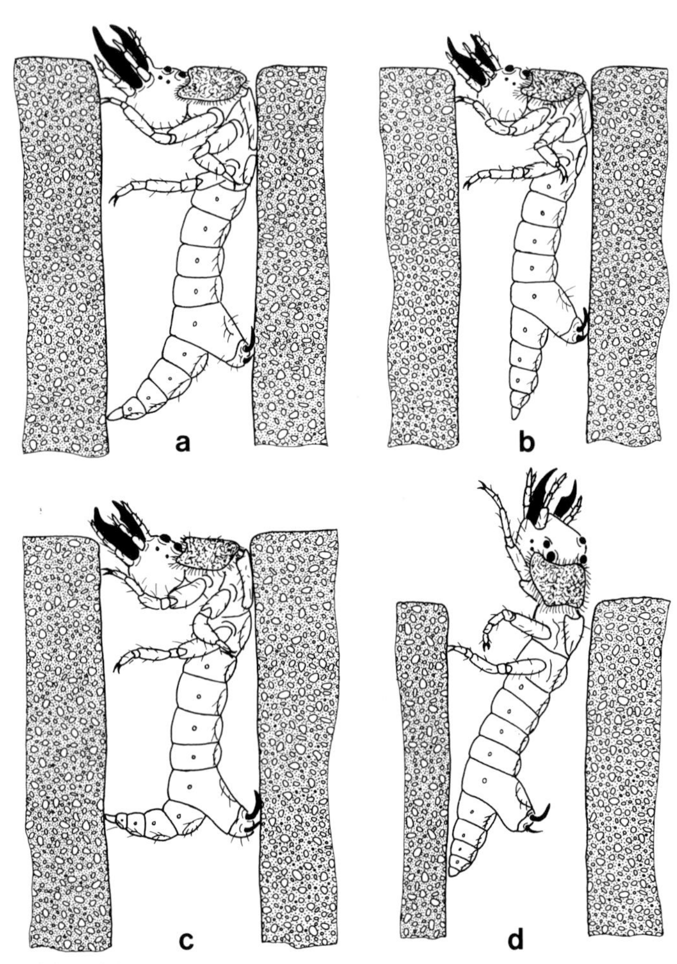

Fig. 158. Larva of *Cicindela* (Cicindelidae), tiger beetles capturing prey. (After Faasch, 1968).
a Larva lying in wait with its abdomen firmly anchored into the burrow wall.
b The larva has sighted an object and stretches its abdomen.
c The larva braces itself firmly against the wall, ready to spring.
d The larva thrusts its body out of the burrow and seizes the prey

---

Plate 134. Larva of *Cicindela* sp. (Cicindelidae), tiger beetles.
a Abdomen with anchor organ, lateral. 1 mm.
b Anchor organ, lateral. 0.5 mm.
c Anchor organ, caudal. 0.5 mm.
d Pair of hooks on the anchor organ. 0.5 mm.
e, f Hook apparatus on the anchor organ. 250 μm, 250 μm

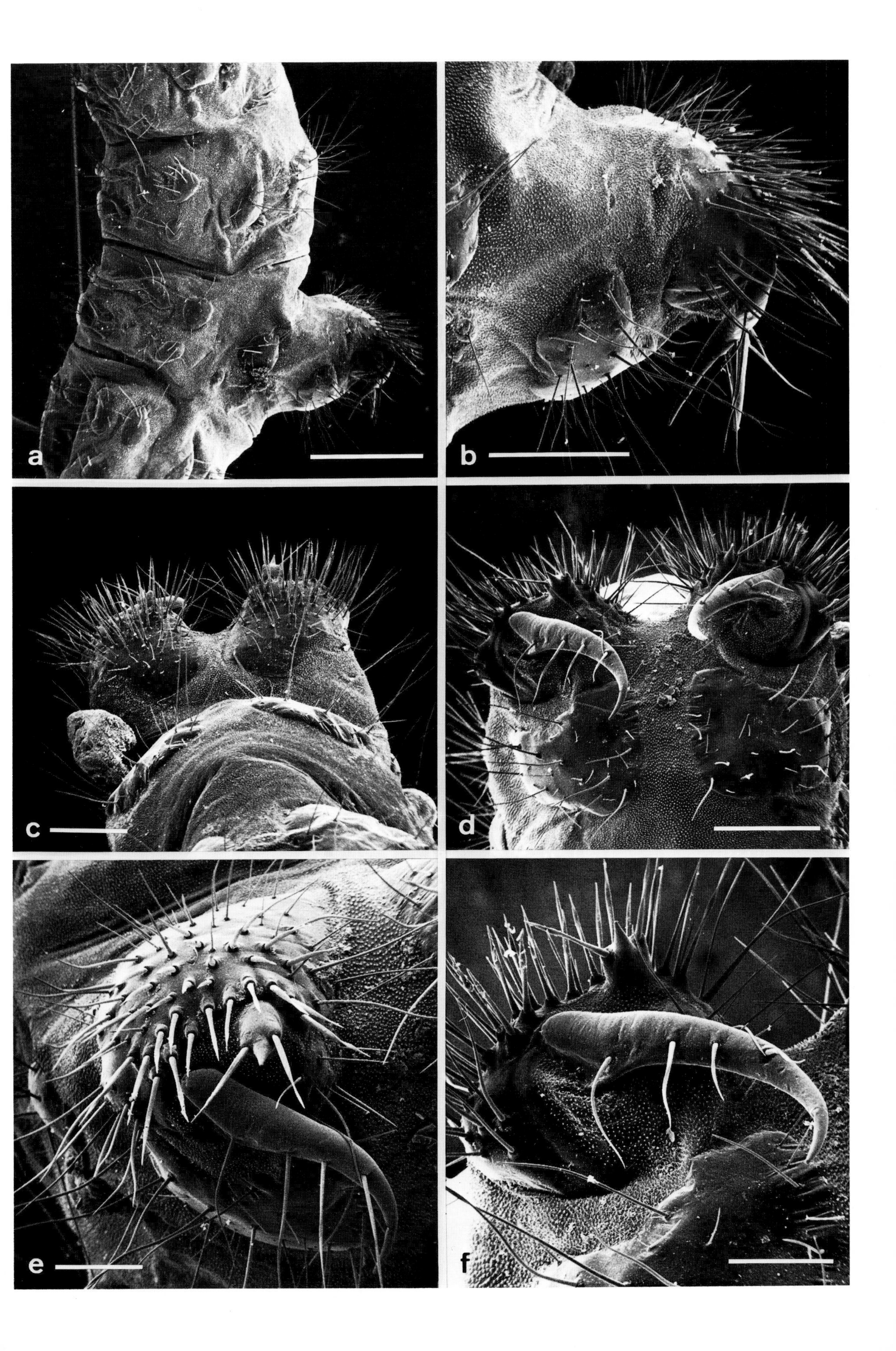
a
b
c
d
e
f

2.20.4 Nutrition of the Epedaphic Ground Beetles (Carabidae)

Carabid beetles (Plates 135–142) are primarily predatory soil animals (Loreau, 1982), which are restricted to life on the ground by the loss or reduction of their hind wings, and also by their long legs. During the day they hide in leaf litter or under stones, but at night they hunt other small litter-dwellers such as insects, snails and earthworms. Prey is usually attacked from behind or the side. The animal is seized swiftly in the powerful, toothed mandibles, crushed by striking the mandibles together, covered with digestive juices (midgut secretion), predigested extraintestinally and finally ingested as chyme. Some carabids have evolved a successful visual hunting strategy for capturing fast-moving prey like Collembola (Bauer, 1981, 1985a). Other species like *Loricera pilicornis* and two species of the genus *Leistus* use setal traps to enclose a springtail during an attack. These traps are constructed from enlarged setae on the antennae and on the ventral surfaces of the head (Bauer, 1982b, 1985b; Hintzpeter and Bauer, 1986). The daily food intake of carabids exceeds their own body weight. *Carabus auratus* can eat up to two and a half times its own weight. However, the average daily food consumption is 0.875 g for a body weight of 0.640 g (Scherney, 1959, 1961).

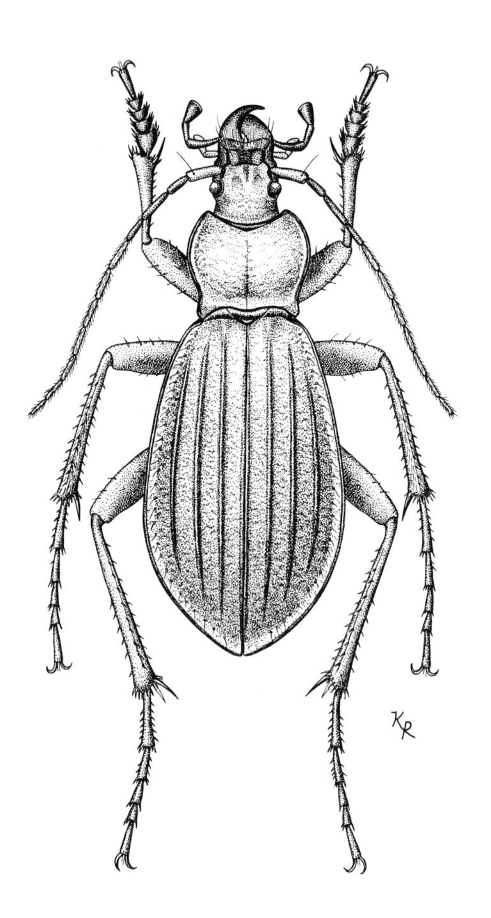

Fig. 159. *Carabus auronitens* (Carabidae), ground beetles, habitus, dorsal view

Plate 135. *Carabus auronitens* (Carabidae), ground beetles.
a Head, lateral. 1 mm.
b Clypeolabrum (*arrow*), mandibles, antennal bases and palps, dorsal. 1 mm.
c Maxillary palp, distal, with fields of sensilla. 100 μm.
d Head, dorsal. 1 mm.
e Caudal view of an elytron. 1 mm.
f Proprioreceptive hairs between the head and pronotum. 50 μm

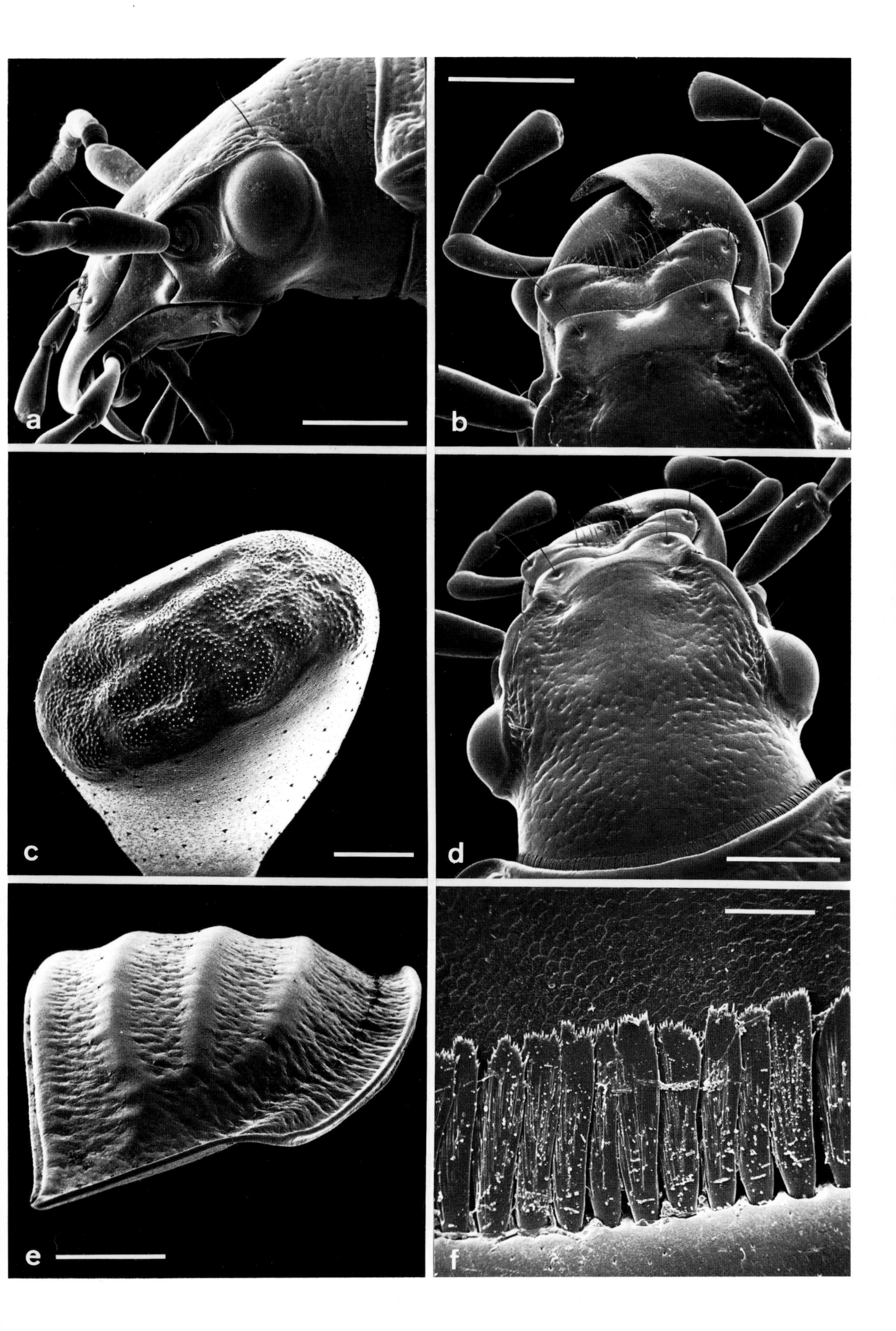
a
b
c
d
e
f

2.20.5 The Carabid Fauna of Central European Forests

A summary of numerous faunistic investigations (Thiele, 1977) carried out in a variety of habitats shows a significant association of carabid species and a close relationship between these associations and plant communities of central European forests. The association of forest carabids were assigned by Thiele (1977) to the following plant communities (associations) comprising two plant sociological orders (Table 7):

Order: Fagetalia Silvaticae (beech and deciduous forests)

Associations:

1. Fagetum – mountain beech forest
2. Querco-Carpinetum – oak-hornbeam forest of the mountain regions
3. Querco-Carpinetum – oak-hornbeam forest of the lowlands
4. Fraxino-Ulmetum – ash-elm of the water-meadow forest

Order: Quercetalia Robori-Petraeae (oak-birch forests)

Associations:

1. Querco-Betuletum – oak-birch forest
2. Fago-Quercetum – beech-sessile oak forest

Table 7. The occurrence of important carabid species in some forest associations of central Europe (Thiele, 1977)

	Fagetalia				Quercetalia	
	Fagetum (mountains)	Querco-Carpinetum (mountains)	Querco-Carpinetum (lowland)	Water-meadow forests (usually Fraxino-Ulmetum, lowland)	Fago-Quercetum (mountains) and Querco-Betuletum (mountains)	Querco-Betuletum (lowland)
Number of stands investigated	7	9	5	6	9	6
Peak in Fagetum						
Molops elatus	+ + +	○	–	–	○	–
Carabus auronitens	+ +	–	○	–	+	○
Pterostichus metallicus	+ +	○	–	–	+	–
Harpalus latus	+ +	○	+	+	–	–
Peak in mountain and/or lowland Fagetalia						
Pterostichus vulgaris	+ + +	+ +	+ + +	+ + +	+	○
Nebria brevicollis	+ +	+ +	+ +	+ +	+	○
Molops piceus	+ + +	+ + +	+ +	–	○	–
The same, but also in mountain Quercetalia (although in much reduced abundance)						
Abax parallelus	+ + +	+ + +	+ +	+	+ +	○
Abax ovalis	+ + +	+ +	–	–	+ +	+
Trichotichnus laevicollis (incl. *T. nitens*)	+ + +	+ +	–	–	+ +	–
Pterostichus cristatus	+	+	–	–	+ +	–
Peak in certain societies of the Fagetalia						
Pterostichus strenuus	+ + +	○	+ +	+ + +	+	+
Pterostichus madidus	+	+ +	+ +	○	○	+
Cychrus caraboides	+	+	+ + +	–	○	○
Peak in water-meadow forests						
Agonum assimile	+	+	+ +	+ + +	+	○
Carabus granulatus	○	○	+	+ + +	+	○
Leistus ferrugineus	–	○	–	+ + +	–	+

Plate 136. *Notiophilus biguttatus* (Carabidae), ground beetles.
a Overview, oblique lateral. 1 mm.
b Head and prothorax, lateral. 0.5 mm.
c Surface of an elytron. 100 μm.
d Head, oblique frontal. 250 μm.
e Proprioreceptive hairs between the head and prothorax. 25 μm.
f Mouthparts, frontal. 100 μm

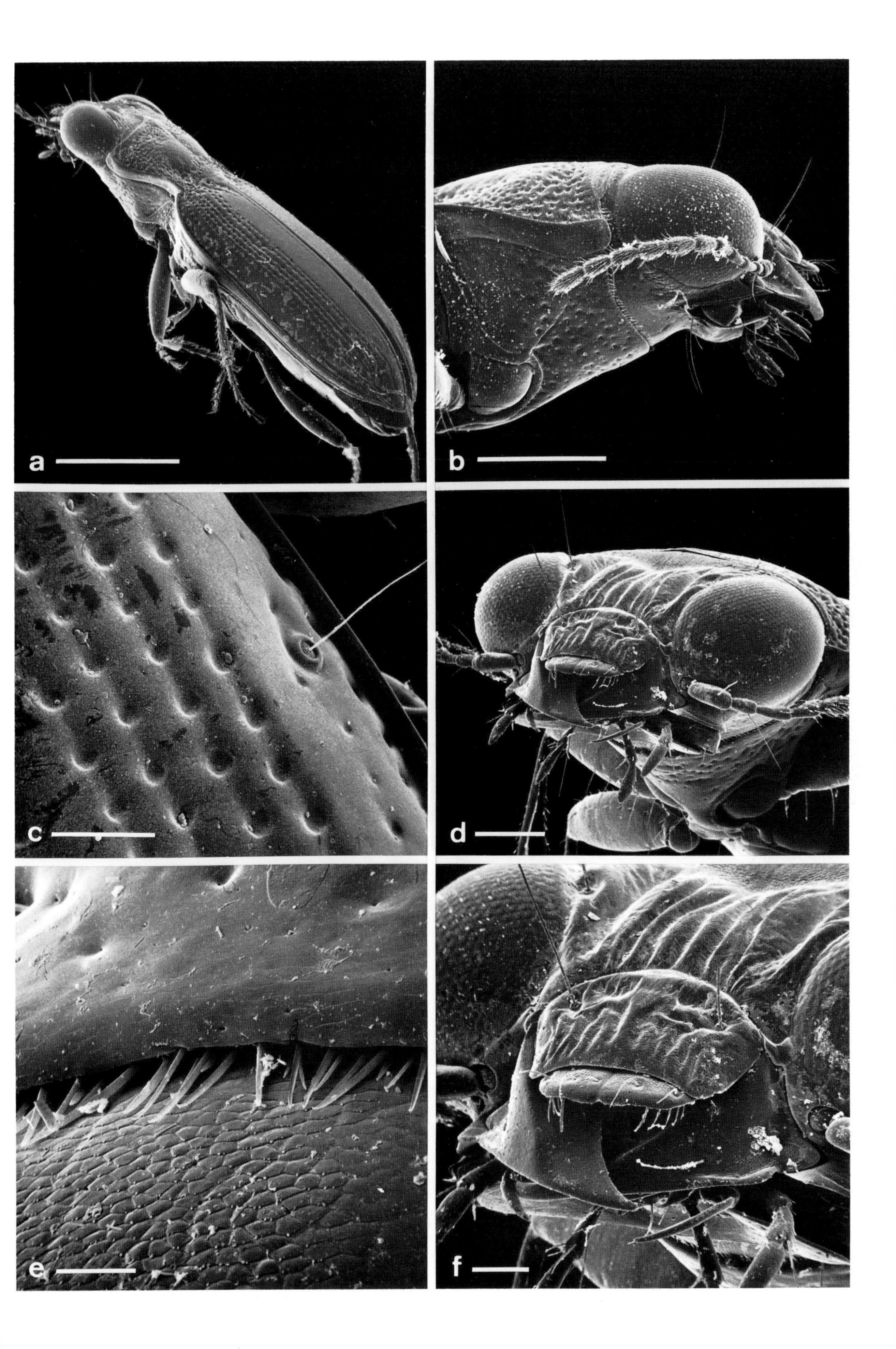
a
b
c
d
e
f

Table 7 (continued)

	Fagetalia				Quercetalia	
Number of stands investigated	Fagetum (mountains) 7	Querco-Carpinetum (mountains) 9	Querco-Carpinetum (lowland) 5	Water-meadow forests (usually Fraxino-Ulmetum, lowland) 6	Fago-Quercetum (mountains) and Querco-Betuletum (mountains) 9	Querco-Betuletum (lowland) 6
Clivina fossor	–	–	–	+ + +	–	–
Pterostichus anthracinus	–	–	–	+ + +	–	–
Agonum micans	–	–	–	+ + +	–	–
Pterostichus cupreus	○	○	○	+ +	–	–
Patrobus atrorufus	○	○	+	+ +	○	–
Trechus secalis	–	○	○	+ +	○	–
Asaphidion flavipes	○	–	○	+ +	–	–
Pterostichus vernalis	–	○	○	+ +	○	–
Agonum viduum (incl. *A. moestum*)	–	–	○	+ +	–	+
Bembidion tetracolum	–	–	–	+ +	–	–
Agonum fuliginosum	–	–	–	+ +	–	+
Agonum obscurum	–	–	–	+ +	–	○
Peak in mountain Quercetalia						
Cychrus attenuatus	+	–	–	–	+ +	–
Peak in lowland Quercetalia						
Notiophilus rufipes	○	+	○	–	○	+ +
Notiophilus palustris	–	○	+	+	○	+ +
Calathus micropterus	○	–	○	○	–	+ +
Peak independent of forest type in mountains						
Carabus problematicus	+ +	+ +	+	–	+ + +	+ +
Carabus coriaceus	+ +	+ + +	+	–	+ +	+
No strict habitat affinity						
Abax ater	+ + +	+ + +	+ + +	+ +	+ + +	+ +
Carabus nemoralis	+ + +	+ + +	+ + +	+ +	+ + +	+ +
Pterostichus oblongopunctatus	+ + +	+ + +	+ + +	+	+ + +	+ + +
Pterostichus niger	+ +	+ +	+ + +	+ + +	+ +	+ +
Trechus quadristriatus	+	+ +	–	+ +	+ + +	○
Notiophilus biguttatus	+ +	○	○	+ + +	+	+ + +
Loricera pilicornis	+	+	○	+ +	○	+ +
Pterostichus nigrita	+ +	–	+	+ +	○	+ +
Carabus violaceus or *C. purpurascens*	+ +	○	○	○	+	+ +
Number of species	33	32	30	31	32	28
Number of species with a presence of ≥50%	20	14	12	25	12	12

(The frequency with which the species occurred in the different plant societies is indicated; – absent; ○ up to 24%; +25 to 49%; + + 50 to 74%; + + + 75 to 100%. Only species which attained a frequency of 50% in at least one of the six habitat complexes investigated were included).

Plate 137. *Notiophilus biguttatus* (Carabidae), ground beetles.

a Metathorax and abdomen, ventral, with the bases of the hind legs. 250 μm.

b Epipleura of the elytron. 100 μm.

c Tibio-tarsus joint on the middle leg. 25 μm.

d Tarsus, distal, with claws (ungues). 25 μm.

e Abdomen of a female, caudal. 250 μm.

f External genital organs of a female with the vaginal palps (*arrows*) and the supragenital area. 50 μm

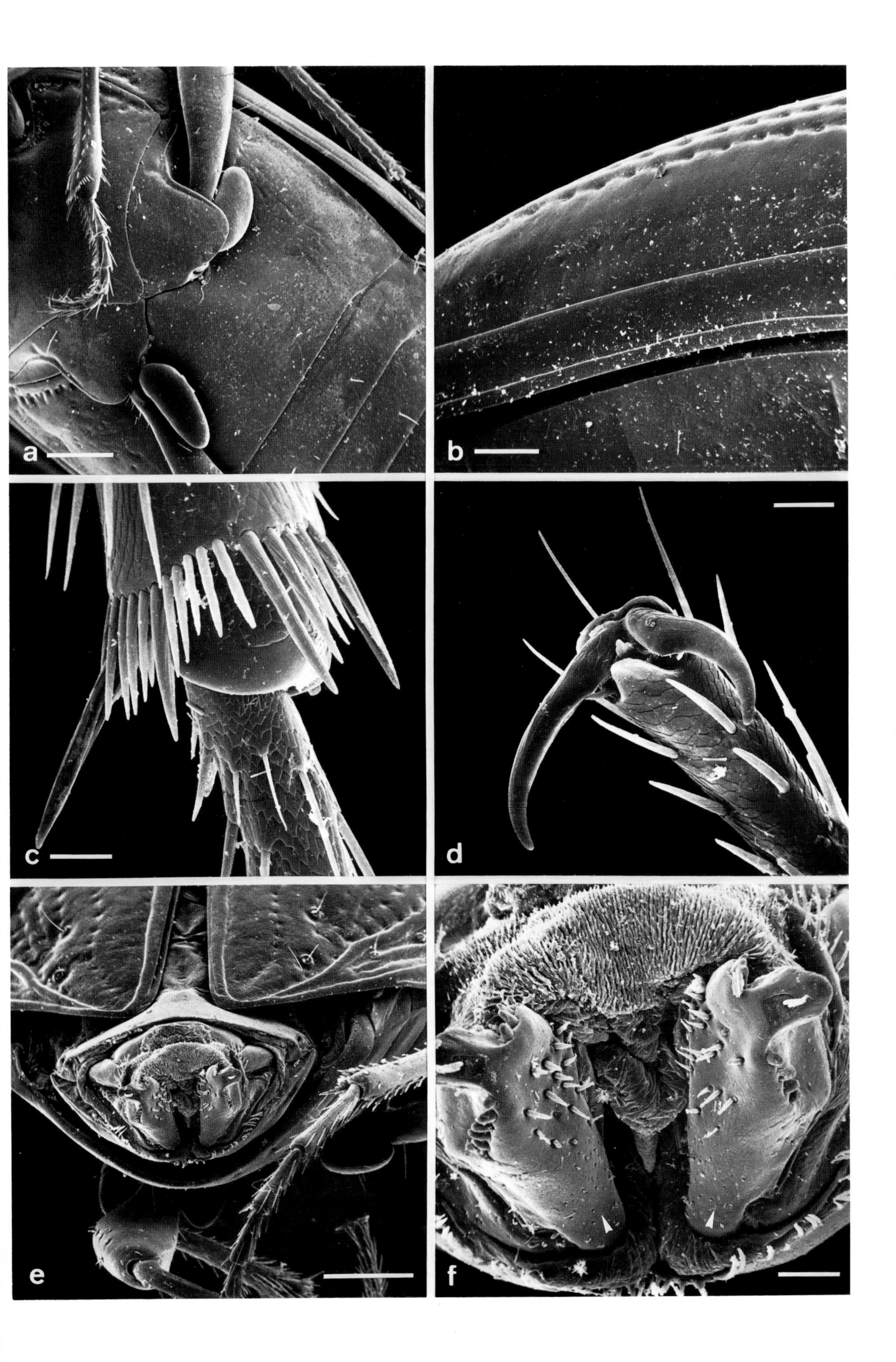
a
b
c
d
e
f

2.20.6 Habitat Affinity of Carabid Beetles

Habitat affinity shown by the carabids and the affinity of communities for certain plant associations depend on a variety of biotic and abiotic ecological factors, among which climatic factors such as temperature, moisture and light are important. Compared to the eurytopic *Pterostichus* species, which lives under widely varying ecological conditions, *Agonum assimile* is a stenotopic carabid which lives in moist, cool beech and deciduous forest (Fagetalia Silvaticae) and especially in water-meadow woodland (Fraxino-Ulmetum) and in oak-hornbeam woodland (Querco-Carpinetum) (Gersdorf, 1937; Thiele, 1956; Wilms, 1961; Paarmann, 1966; Thiele, 1977). The preference of this species for reduced light intensity and moderate temperature conditions illustrates the habitat affinity of this stenotopic species (Fig. 160) (Thiele, 1964, 1967; Neudecker, 1974; Neudecker and Thiele, 1974; Wasner, 1977). Habitat selection in the Carabidae is influenced by the perception of habitat cues like olfactory stimuli associated with soil microorganisms. The microclimatic conditions within habitats mainly serve to maintain favourable body temperatures and to regulate the timing of various activities (Evans, 1983).

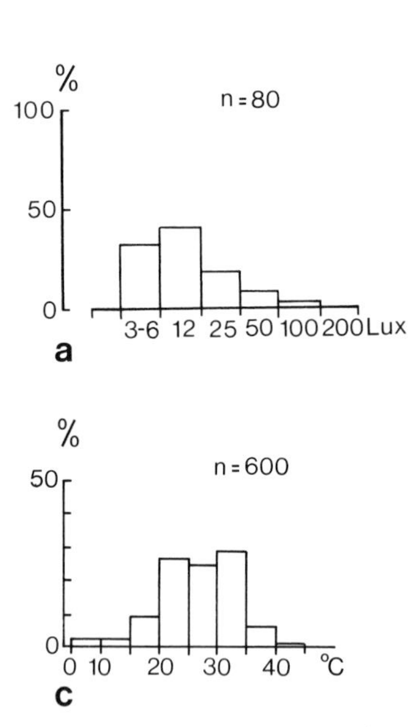

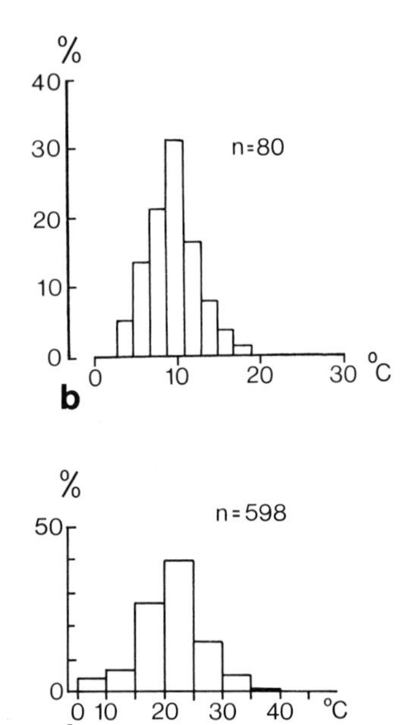

Fig. 160. Distribution of *Agonum assimile* in **(a)** light **(b)** temperature preference experiments. (After Neudecker, 1974); distribution of two *Pterostichus* species in a combined temperature-light intensity gradient. The gradient extended from 5 °C at 1750 Lux to 45 °C at 2 Lux; **(c)** *P. angustatus* **(d)** *P. oblongopunctatus* (Paarmann 1966)

Plate 138. *Agonum marginatum* (Carabidae), ground beetles.
a Anterior of the body, dorsal. 1 mm.
b Head, oblique frontal. 0.5 mm.
c Head, ventral, with maxillary and labial palps. 250 μm.
d Mouthparts, frontal. 250 μm.
e Pronotum, scutellum and elytra. 0.5 mm.
f Cuticular structure of the elytra. 50 μm

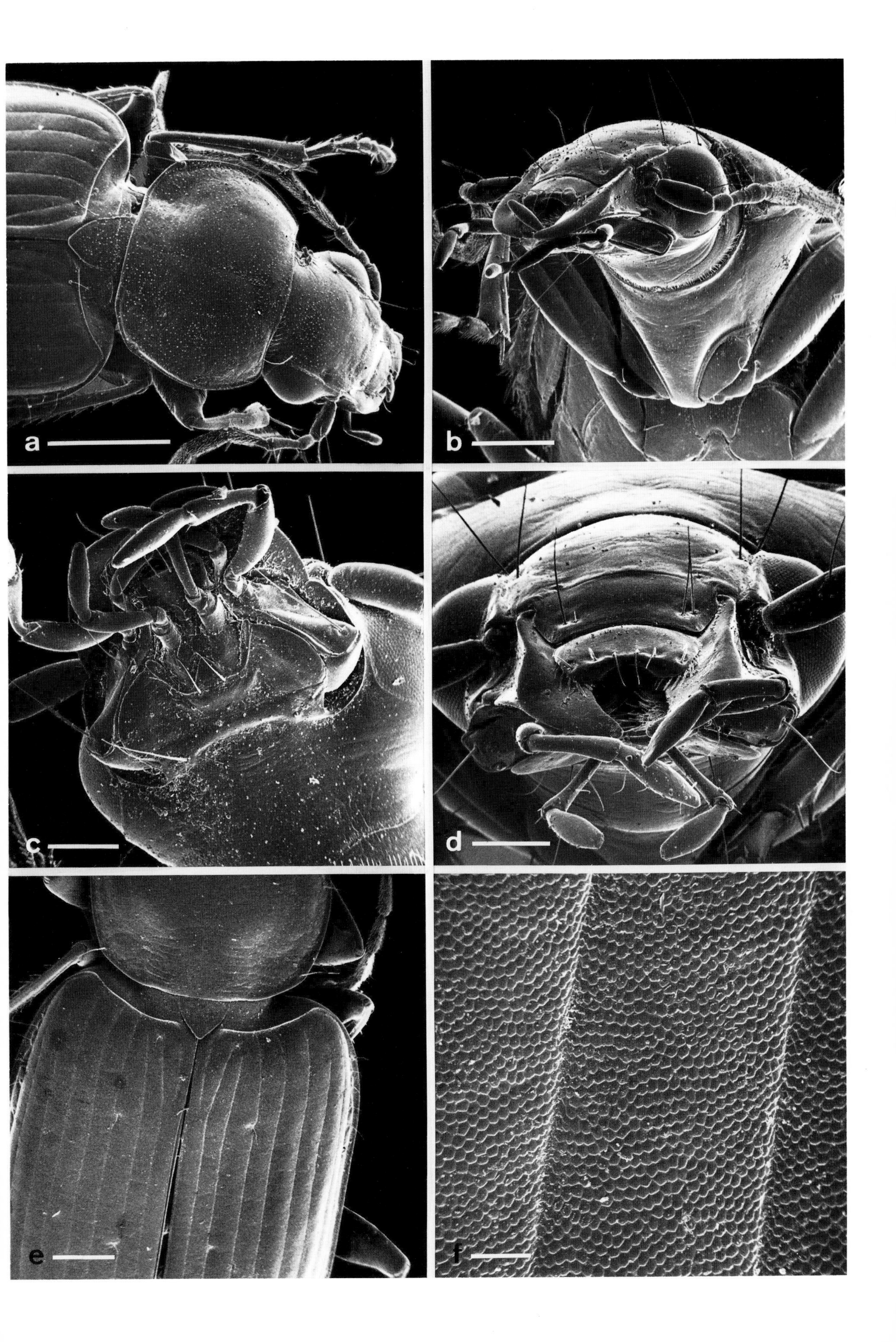
a
b
c
d
e
f

2.20.7 Nutrition of Carabid Larvae

Most carabid larvae (Fig. 161; Plates 139 and 140) are soil-dwellers and, like the imagines, are primarily predators with extraintestinal digestion. Small soil arthropods are among the prey of the carnivorous larvae (Bauer, 1982a). In addition, some prefer snails and earthworms. Among the larvae which eat earthworms are *Abax ater* and *A. parallelus* (Löser, 1972; Lampe, 1975; Thiele, 1977). Certain species of the genus *Carabus* and *Cychrus* are specialized snail-eaters (Sturani, 1962). The larvae enter the shell of their prey from the side and make their way between the snail and its shell, crawling with their ventral side towards the inner wall of the shell in order to avoid having their spiracles and extremities covered by the increased amount of slime exuded by the snail (Fig. 162).

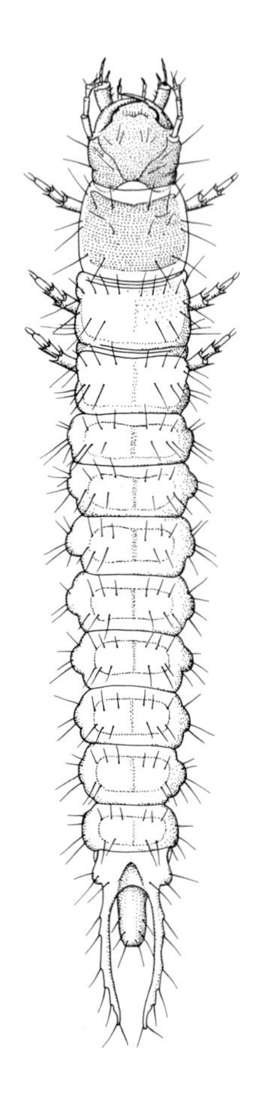

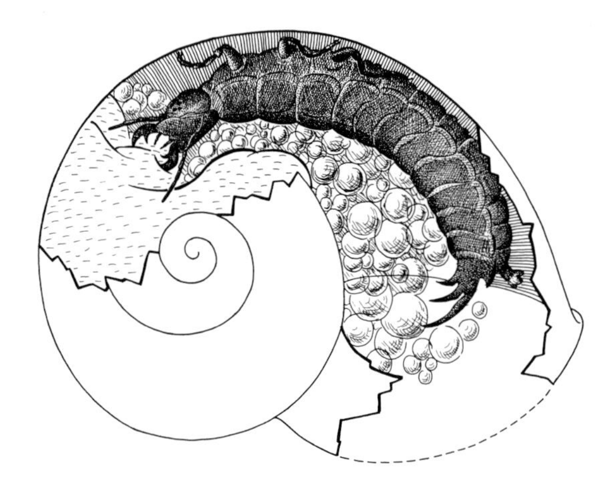

Fig. 161. Habitus of a carabid larva, dorsal. (After Böving and Craighead, 1931)

Fig. 162. Larva of *Carabus* devouring a snail. It presses its ventral side close to the shell to minimize contact with the slime mass. (After Sturani, 1962)

Plate 139. Larva of the Carabidae (ground beetles).
a Head and pronotum, dorsal. 0.5 mm.
b Head and prothorax, lateral. 1 mm.
c Head, dorsal. 0.5 mm.
d Eye spot with stemmata. 100 µm.
e Labial palp, distal. Several digitiform sensilla are situated subapically (*arrowhead*). 50 µm.
f Labial palp, distal, with sensilla. 10 µm

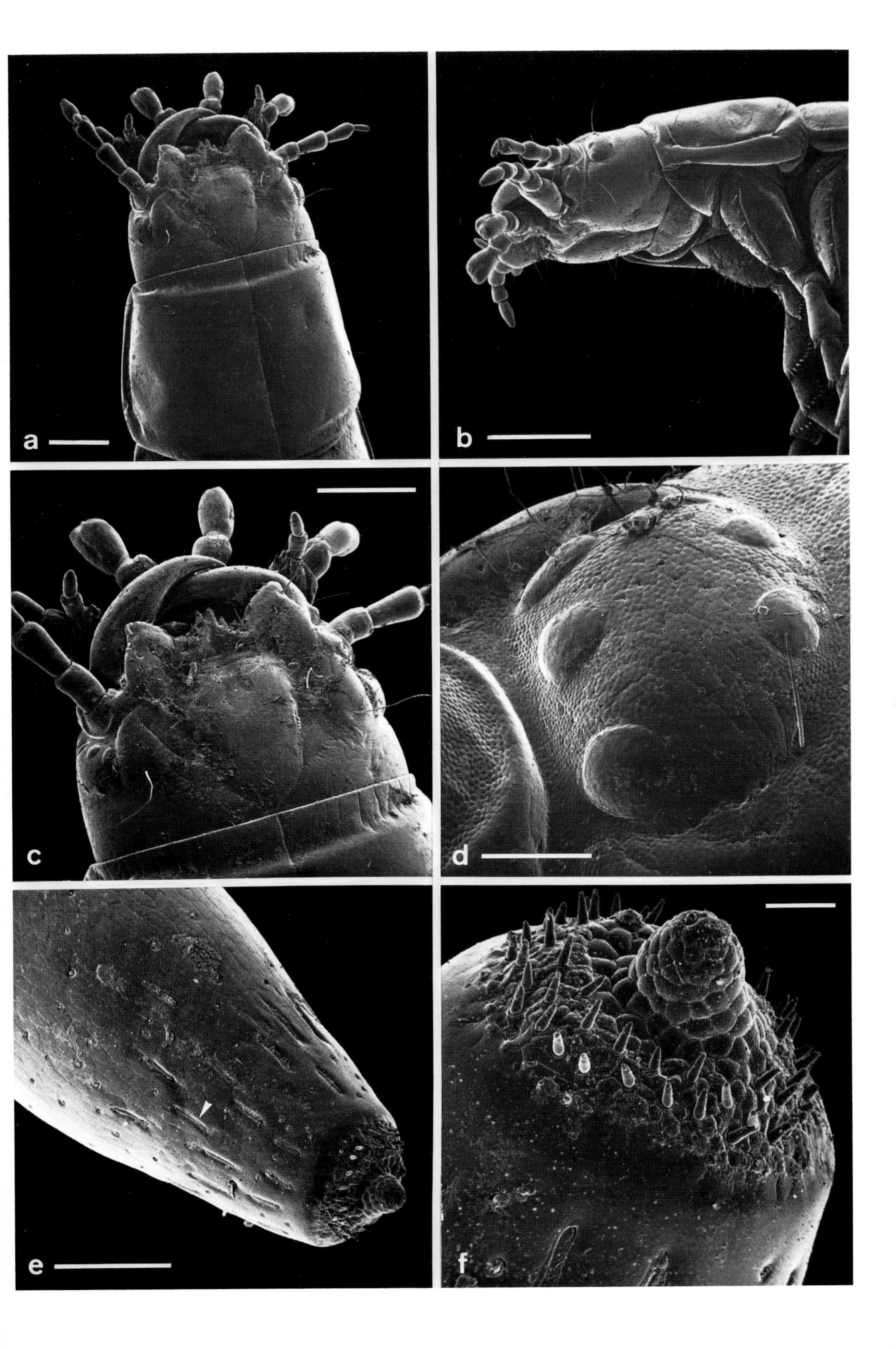
a
b
c
d
e
f

2.20.8 Life Forms of Carabid Larvae (Carabidae)

The mode of life and the body form of many carabid larvae permit characterization into euedaphic, hemiedaphic and epedaphic life-form types. Sarova (1960) already distinguished between nine different morpho-ecological types which reflect adaptations to life in the soil (Dunger, 1983).

Larvae of the genera *Trechus, Bembidion, Pterostichus, Nebria, Agonum* and *Abax* belong to the predatory, epedaphic larvae which hunt on the soil surface and in the litter layer, only occasionally hiding in pre-existing burrows.

Many larvae of the genera *Carabus* and *Cychrus* live almost completely hemiedaphic lives. Although the larvae hunt above ground they excavate their own burrows which serve as retreats and pupal chambers (Fig. 163).

The predatory larval stages of the subfamilies Elaphrinae and Omophroninae are permanent soil-dwellers. As appropriate for their euedaphic mode of life they have a more or less wedge-shaped head and fossorial legs. The eyes are reduced, whereas tactile and olfactory organs are obviously well developed (Dunger, 1983; Raynaud, 1974).

Fig. 163. Pupal cell of a ground beetle (Carabidae). The pupa is supported by stilt-like tufts of dorsal bristles. At the side of the chamber lies the discarded larval exuvia. (After Klausnitzer, 1982)

Plate 140. Larva of the Carabidae (ground beetles).
a Palps situated beneath the mandibles. 20 μm.
b Surface of the mandibles with conical sensilla in shallow pits. 20 μm.
c Abdomen, dorsal, with two spiny cerci. 1 mm.
d Abdomen, ventral, with two spiny appendages (cerci, urogomphi) and anal tube. 1 mm.
e Tarsus, distal, with claws (ungues). 100 μm.
f Abdomen, caudal, with cerci and anal tube. 250 μm

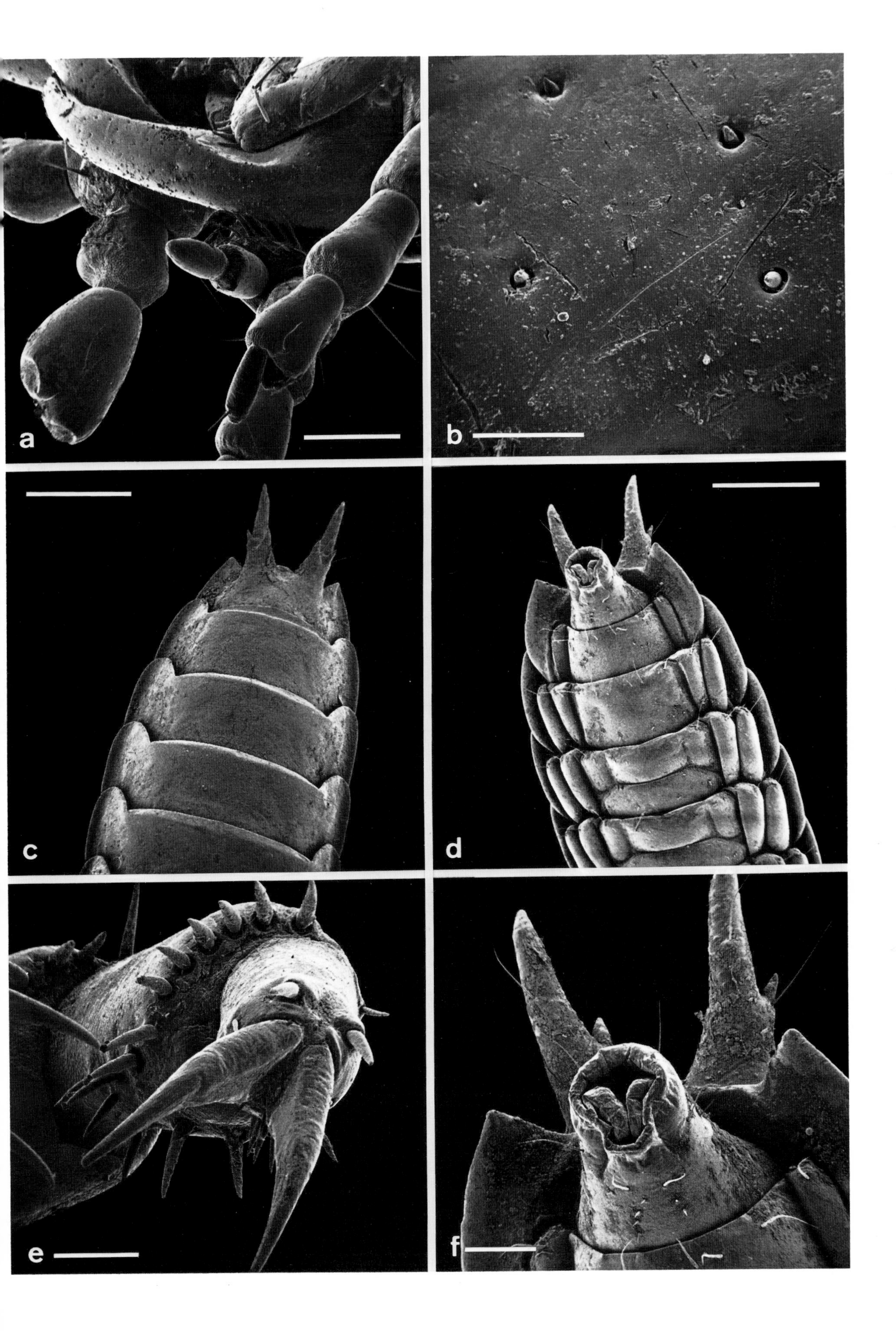
a
b
c
d
e
f

2.20.9 The Bombardier Beetle: *Brachinus crepitans* (Carabidae)

Four members of the genus *Brachinus* are found in central Europe (Plate 141). Together with the genus *Aptinus*, represented by one species, they form the Brachininae subfamily. These small epedaphic beetles possess a unique defensive weapon. When threatened, a gas from the paired pygidial glands on the anus is released and explodes with an audible pop. The complex structure of this gland is schematically illustrated in Fig. 164.

The cells of the pygidial gland secrete a mixture of hydrogen peroxide, hydroquinone and toluhydroquinone which is stored in a collecting bladder. An enzymatic secretion, consisting of catalases and peroxidases, is excreted into the lumen of the neighbouring firing chamber by enzyme-producing glands in its wall. These enzymes trigger off the explosion mechanism as soon as the bladder orifice is opened and the mixture is forced under pressure from the collecting bladder into the firing chamber. The catalases split the hydrogen peroxide into water and oxygen, while the peroxidases oxidize the hydroquinones to yellow-violet benzoquinone and toluquinone. The gas is explosively expelled at temperatures of up to 100 °C, which has a repellant effect on small enemies of the beetle (Schildknecht et al., 1968; Schnepf et al. 1969).

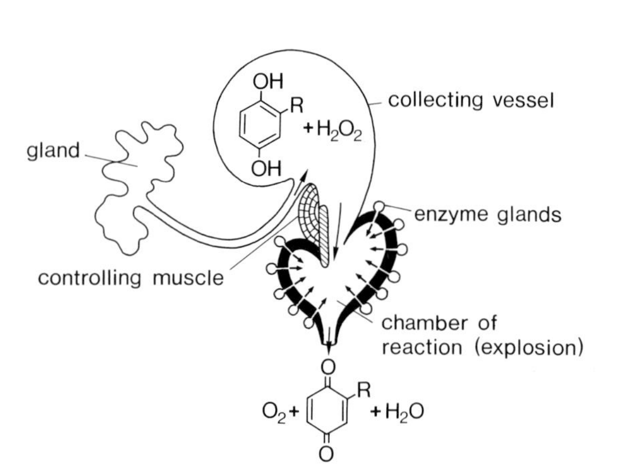

Fig. 164. Schematic diagram of the pygidial apparatus in the bombardier beetle; *R* represents H or CH_3. (Schildknecht et al., 1968)

Plate 141. *Brachinus crepitans* – Carabidae (ground beetles), bombardier beetle.
a Habitus, dorsal. 1 mm.
b Head and anterior pronotum, dorsal. 0.5 mm.
c Abdomen, oblique caudal. 0.5 mm.
d Maxillary palp, distal, with sensilla. 20 μm.
e Pygidial region. 200 μm

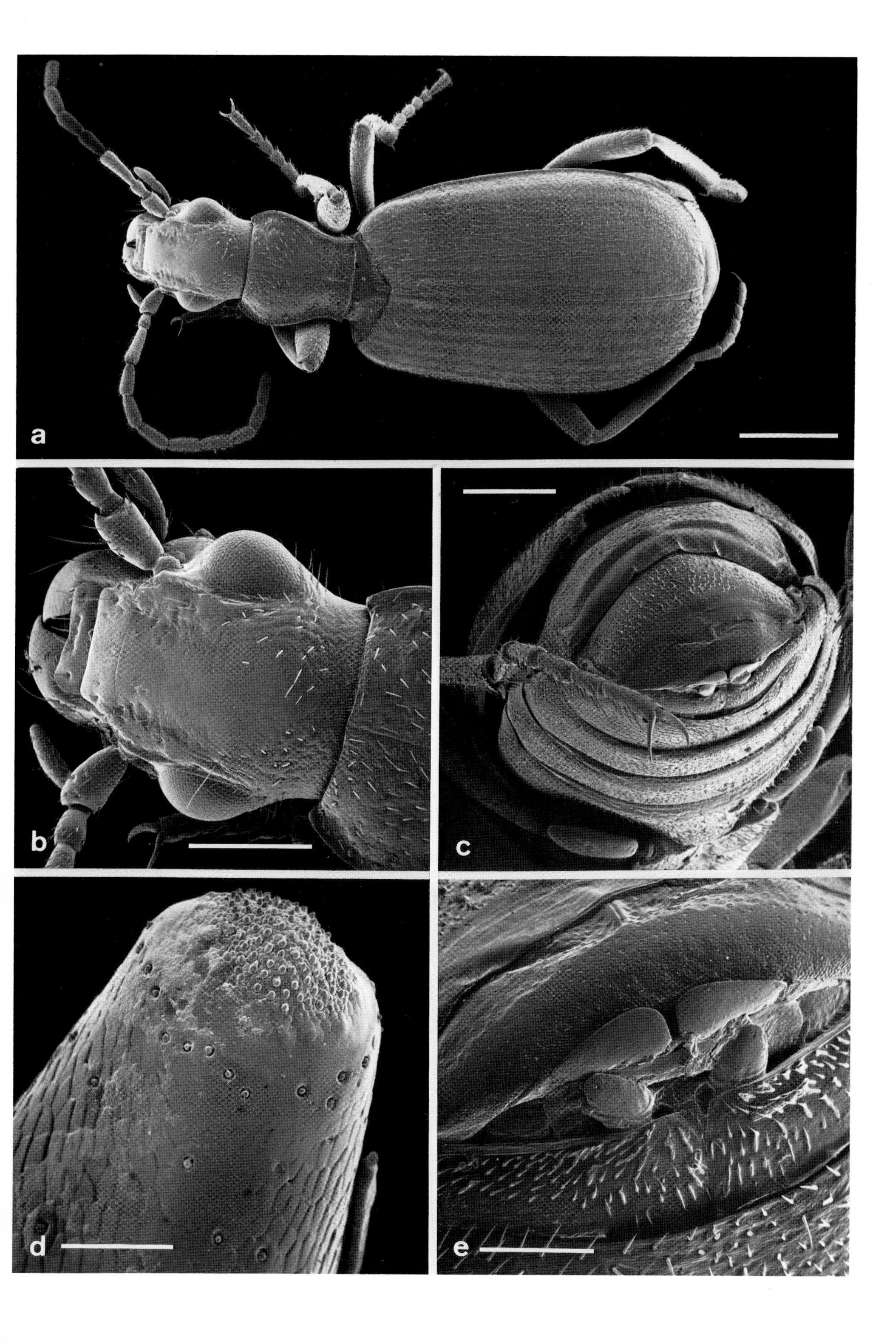
a
b
c
d
e

2.20.10. The Burrowing Ground Beetle: *Dyschirius thoracicus* (Carabidae)

In general, ground beetles belong to the epedaphic soil arthropods. Their long legs enable them to skillfully waylay prey in the loose litter layer. However, some ground beetles belonging to the Scaritinae subfamily have forelegs modified to form fossorial legs with broad, anterio-laterally extended tibiae which are often equipped with thorns and a cleaning apparatus (Plate 142). These burrowers have adapted to a hemiedaphic way of life and can also pursue their prey in deeper layers. Species of the genus *Dyschirius* (Figs. 165 and 166) are frequently encountered with hemiedaphic staphylinids and heterocerids which also possess fossorial legs (*Bledius, Heterocerus*) and live gregariously in self-excavated burrows. The carnivorous *Dyschirius* species specialize in hunting these beetles and burrow in search of them using their mouthparts and fossorial legs.

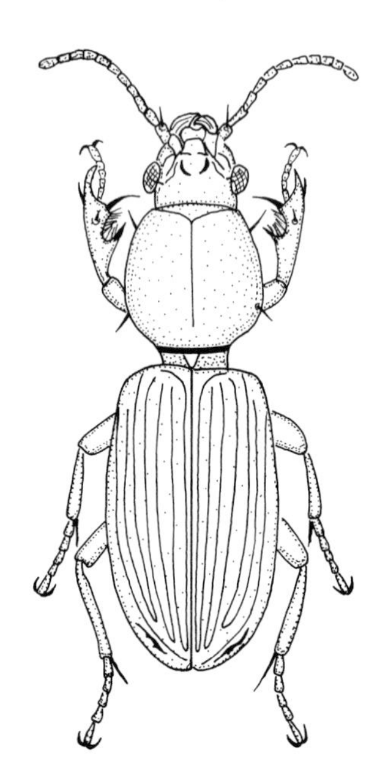

Fig. 165. Habitus of *Dyschirius*, dorsal. (After Brunne, 1976)

Fig. 166. Habitus of the larva of *Dyschirius thoracicus*, dorsal. (Reitter, 1908)

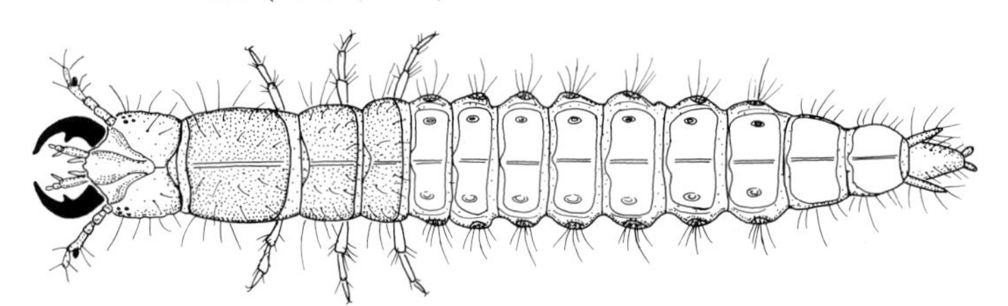

Plate 142. *Dyschirius thoracicus* (Carabidae), ground beetles.
a Overview, lateral. 1 mm.
b Overview, oblique frontal, with fossorial legs. 0.5 mm.
c Prothorax and mesothorax, ventral, with leg bases. 250 µm.
d Fossorial leg, with cleaning comb on the broadened and elongated tibia. 100 µm.
e Cleaning comb on the tibia of the foreleg. 50 µm

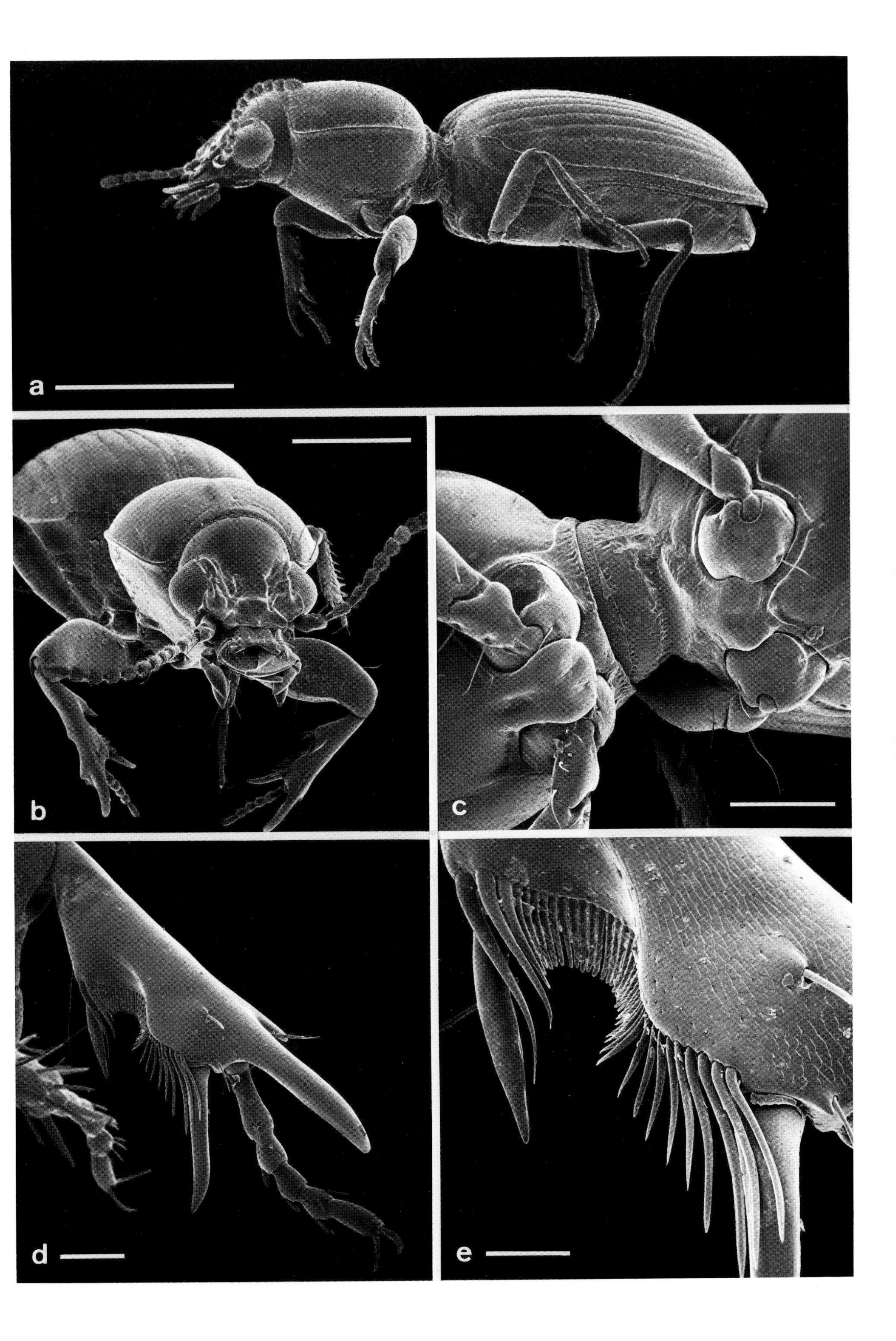
a
b
c
d
e

2.20.11 Necrophagous Beetles (Silphidae)

Soil arthropods specializing in the consumption of excrement and carrion (coprophages and necrophages) actively participate in the decomposition of animal debris. It is predominantly the silphid (*Silpha, Necrophorus, Oeceoptoma,* etc.) and the staphylinid beetles (*Omalium, Lathrimaeum, Proteinus, Atheta,* etc.) which feed on carrion and are engaged in its successive degradation (Tischler, 1976; Lundt, 1964; Topp et al., 1982). Only a few silphids are predatory rather than necrophagous, e.g. the snail-eating *Phosphuga atrata* (Plate 143).

The "carrion burier", *Necrophorus vespillo* (Fig. 167), buries the bodies of small, dead animals. Several hours are required for the work, in which both females and males participate. The carrion is gradually pushed into a spherical shape and laid into an excavated cavity (Fig. 168), the crypta, above a sloping tunnel. From the crypta the female digs a second tunnel (secondary tunnel) and deposits the eggs in its wall. After a short embryonic period the emerging larvae, attracted to the carrion by its odour, are initially fed with predigested carrion by the female. Within a period of 7 days the larvae have already passed through three instars and have grown from 0.5 to 2.8 mm. Pupation takes place in the soil near the crypta (Pukowski, 1933; v. Lengerken, 1954).

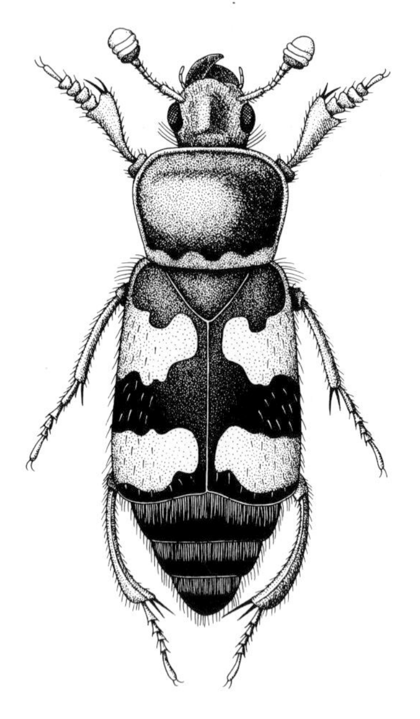

Fig. 167. Habitus of *Necrophorus vespillo* (Silphidae), carrion beetles, dorsal. (After Bechyně, 1954)

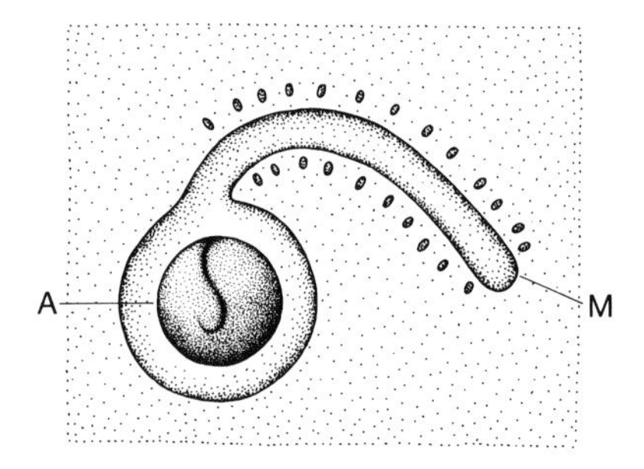

Fig. 168. Horizontal section through the crypta of *Necrophorus vespillo* (Silphidae). (After v. Lengerken, 1954). The egg chambers are established along the secondary tunnel (*M*). A ball of carrion (*A*) lies ready for the larvae in the neighbouring cryptal chamber

Plate 143. *Phosphuga atrata* (Silphidae), carrion beetles.
a Pronotum and elytra, oblique dorsal. 2 mm.
b Head, oblique frontal. 250 µm

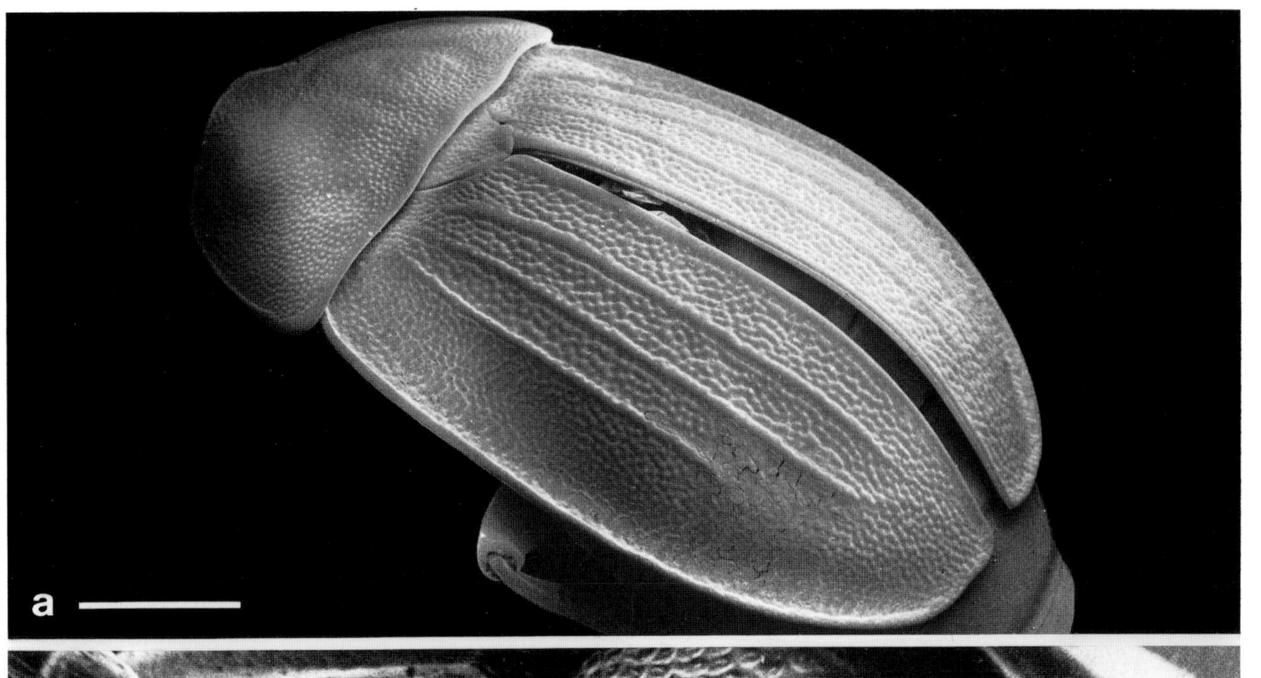
a

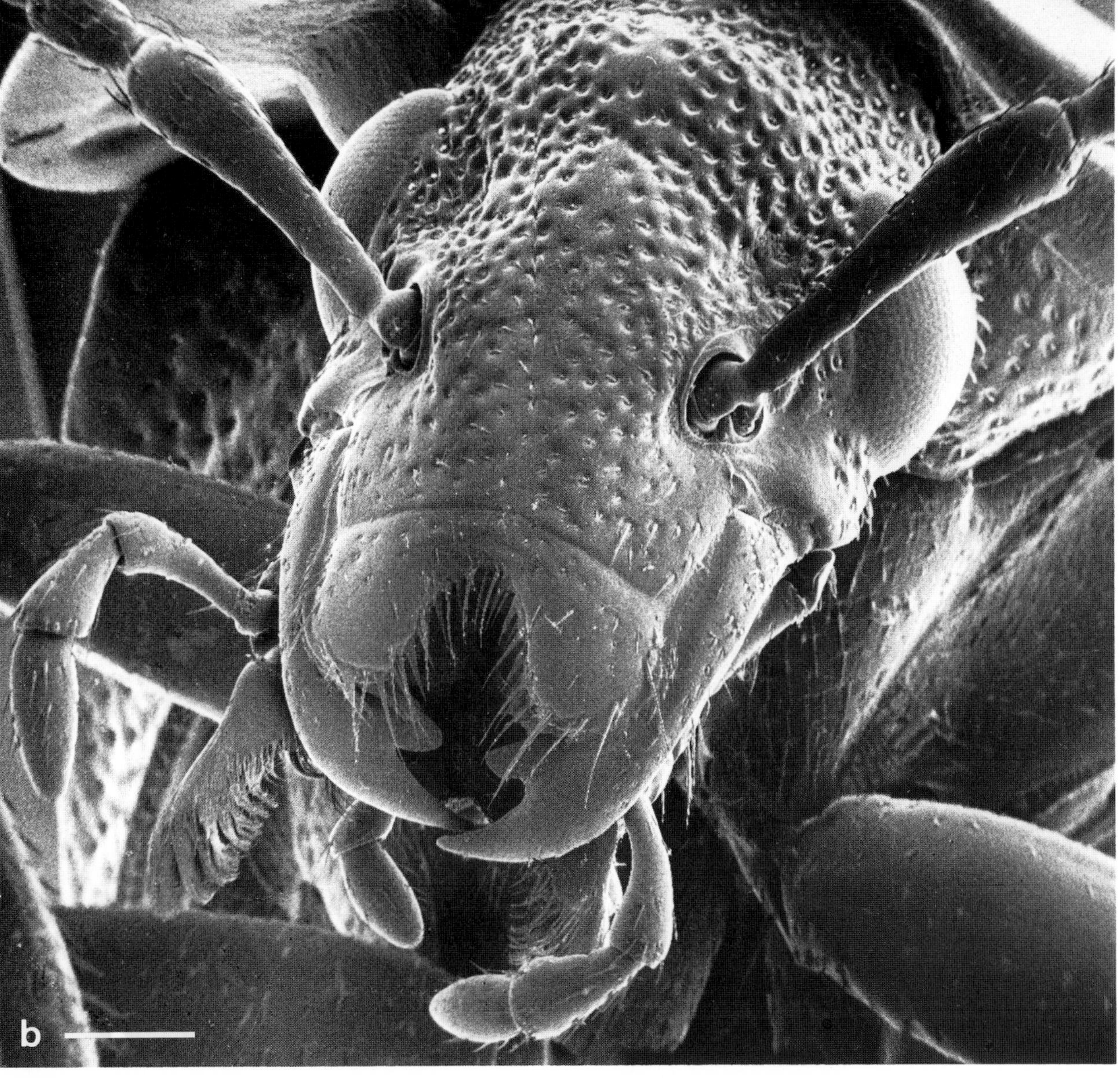
b

2.20.12 Larvae of the Scydmaenidae (Ant Beetles)

The little-known larvae of the Scydmaenidae (Fig. 169; Plates 144 and 145) are, like the imagines, predominantly epedaphic. Their dorso-ventrally flattened body resembles that of a woodlouse. It permits them to inhabit the flat hollows between mouldy leaves in the litter layer or to hide under stones, bark and in moss cushions. Both the larvae and imagines are carnivorous, feeding on mites, especially oribatids (Schuster, 1966a, b). The larva of *Cephennium thoracicum* curls itself round the captured mite, kills it by biting it with the stiletto-like mandibles and injects it with digestive juices. The contents of the body are subsequently sucked out. In *C. austriacum* Schuster (1966b) found that the larvae pick up the mite from the ground using their mandibles. They then walk around carrying their prey, allowing it no contact with the soil. Sometimes they also curl round the captured mite and kill it as described above. The sucking process can take some hours.

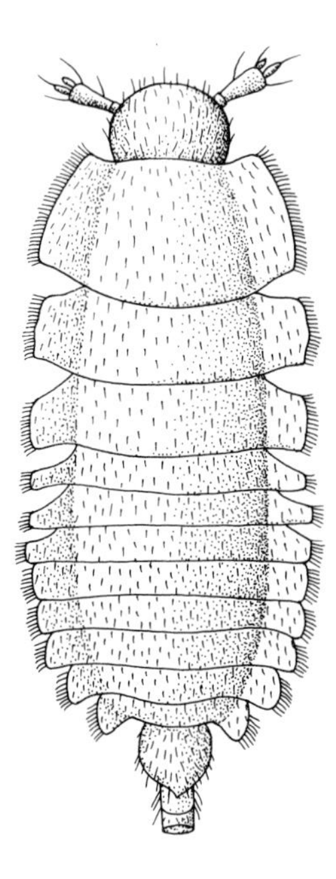

Fig. 169. Larva of *Scydmaenus tarsatus* (Scydmaenidae), habitus from the dorsal side. (After Klausnitzer, 1978)

Plate 144. Larva of the Scydmaenidae.
a Overview, lateral. 1 mm.
b Anterior of the body, dorsal. 0.5 mm
c Head and part of the prothorax, lateral. 200 μm.
d Head, dorsal, with antennae and mouthparts. 100 μm.
e Mouthparts, lateral. The *arrowhead* marks an antenna. 50 μm.
f Cuticle of the head. 10 μm

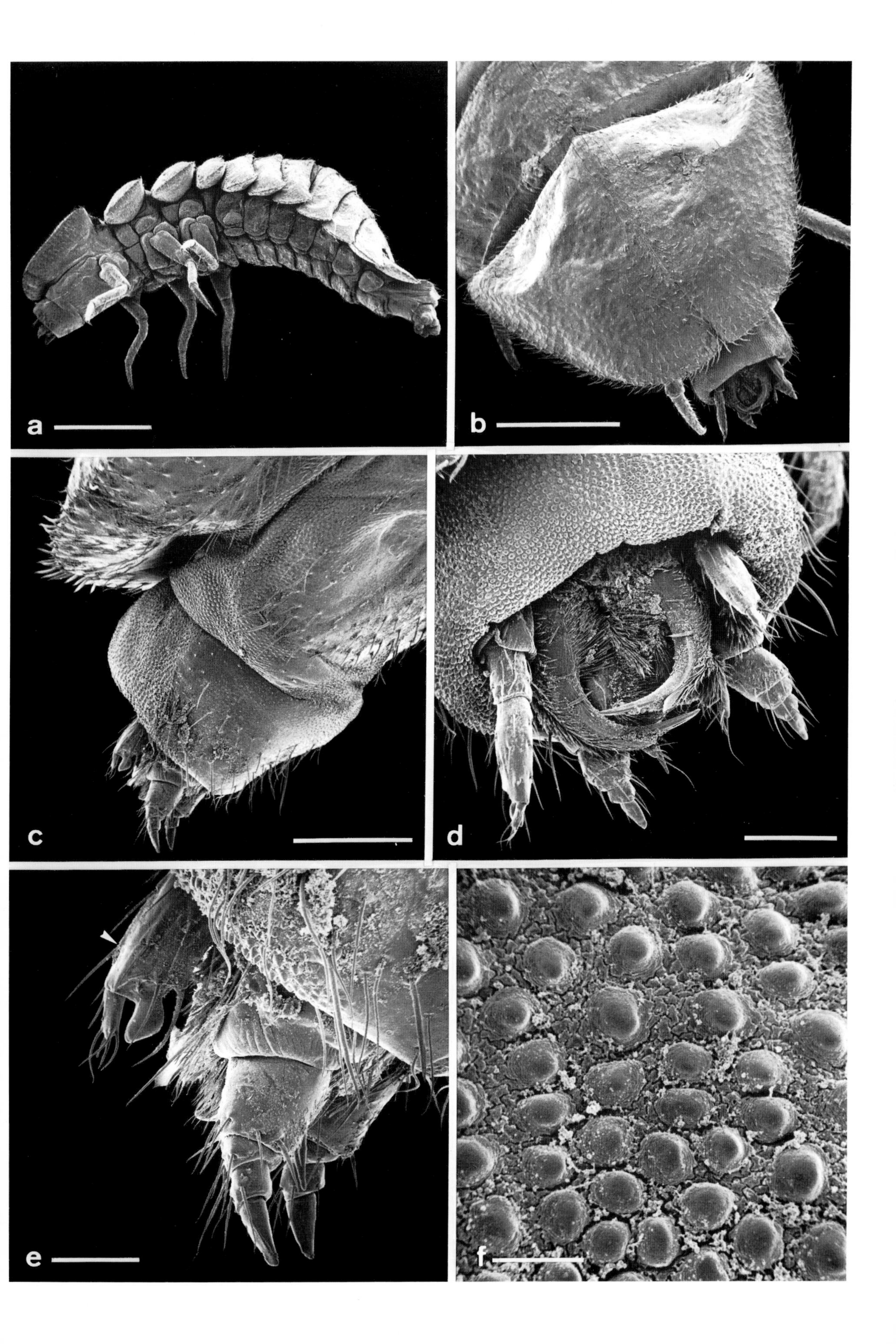

a
b
c
d
e
f

2.20.13 Anal Papillae of Beetle Larvae

Many soil-dwelling Coleoptera and Diptera larvae possess more or less conspicuous lobes in the anal region which can often be everted and enlarged by means of haemolymphatic pressure (Plate 145). They resemble the anal papillae of many aquatic and semi-aquatic insect larvae which use these organs for osmotic and ionic regulation. Among the aquatic Coleoptera the larvae of Helodidae possess five contractile anal papillae with typical transporting epithelia (Wichard and Komnick, 1974). The epithelium is fine-structurally characterized by dense infoldings of the apical plasma membrane, by deep infoldings of the basal plasma membrane and by a high content of mitochondria, which possess densely packed cristae. Histochemically detectable amounts of chloride ions are localized in the epicuticle and the zone of apical plasma membrane infoldings. These fine-structural and histochemical results suggest that the anal papillae of some aquatic Coleoptera larvae also participate in osmoregulation by the absorption of salt.

Analogous structures (ventral tube, coxal sacs) for the regulation of the water regime are to be found in the soil-dwelling, primitive, wingless insects (Apterygota) and other arthropods. It is, therefore, conceivable that some anal lobes of soil-dwelling Coleoptera larvae are, like the anal papillae, eversible extensions of the rectal transport epithelium for the absorption of water and ions (cf. Plate 158a, b).

Plate 145. Larva of the Scydmaenidae.
a Overview, caudal. 0.5 mm.
b Abdomen, terminal, lateral, with everted anal papillae. 0.5 mm.
c Abdomen, dorsal. 100 µm.
d Anal region, lateral, with everted anal papillae. 250 µm.
e Anal papillae. 100 µm.
f Cuticle of the anal papillae. 10 µm

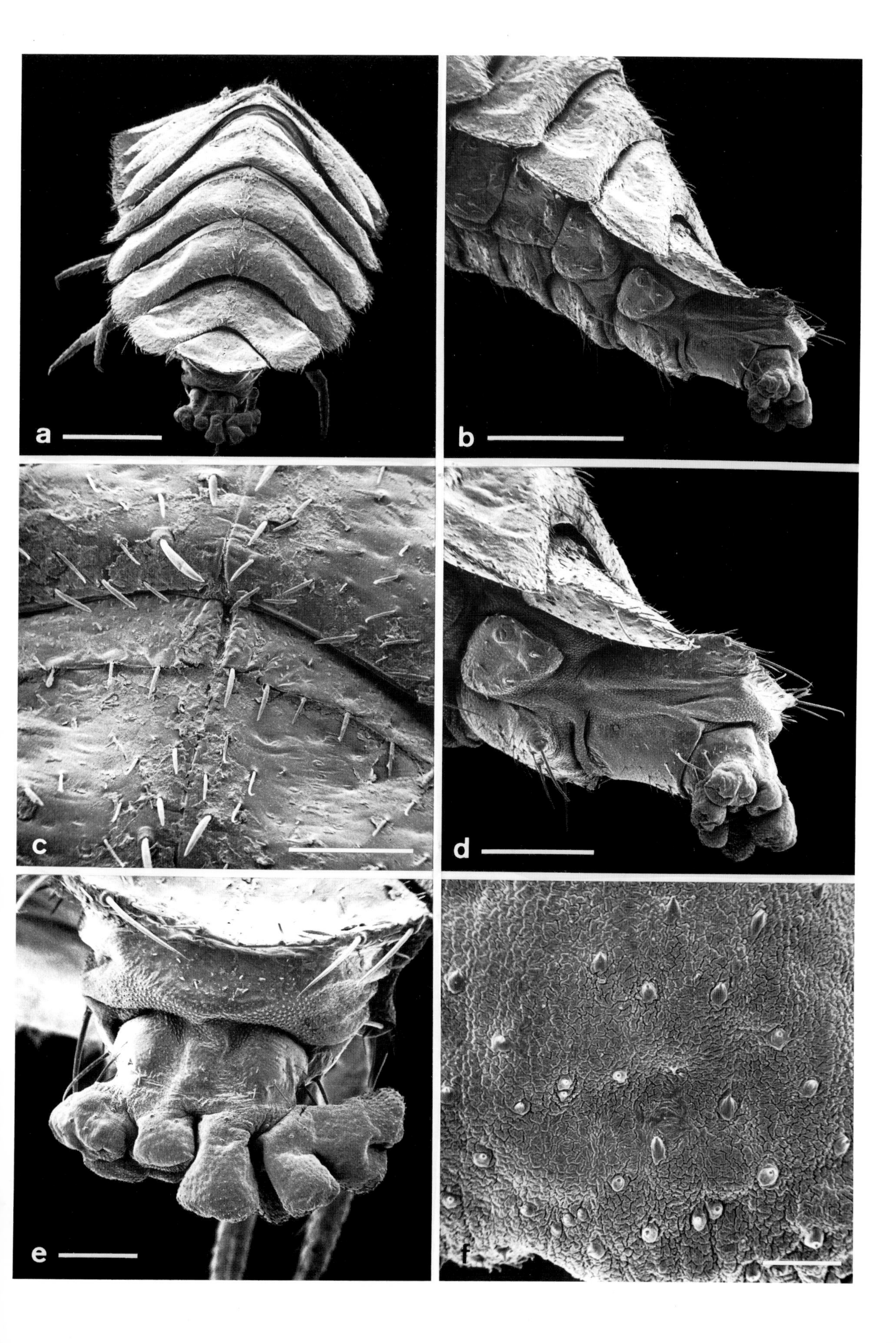
a
b
c
d
e
f

2.20.14 Feather-Winged Beetles (Ptiliidae)

The Ptiliidae family includes very small beetles (Fig. 170; Plate 146), many of which are less than 1 mm in length. A member of this family, the North American beetle *Nanosella fungi*, is the smallest known beetle with a length of only 0.25 mm. The larvae and imagines occur in rotting wood and decaying plant fragments and leaves. They feed on fungal spores.

Obviously these beetles have no notable importance to bioproduction, as they are not only characterized by their diminutive size, but also by their small populations. For instance, single specimens of six different species were collected using various methods in a beech forest. Only *Acrotrichis intermedia* occurred regularly and accounted for just over 3% of the total density of soil beetles in the photo-eclectors (Friebe, 1983). The isolated occurrence of aggregations of this species in the leaf litter is attributed to the effect of pheromones (Tips et al., 1978a, b, c).

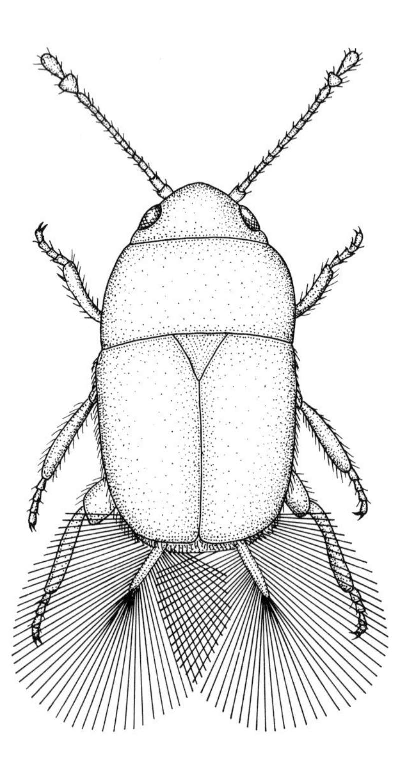

Fig. 170. *Acrotrichis sericans* (Ptiliidae), feather-winged beetles. (After Jacobs and Renner, 1974)

Plate 146. *Acrotrichis* sp. (Ptiliidae), feather-winged beetles.
a Habitus, dorsal. 250 µm.
b Frontal view with unfolded elytra. 250 µm.
c "Flight position", lateral. 0.5 mm

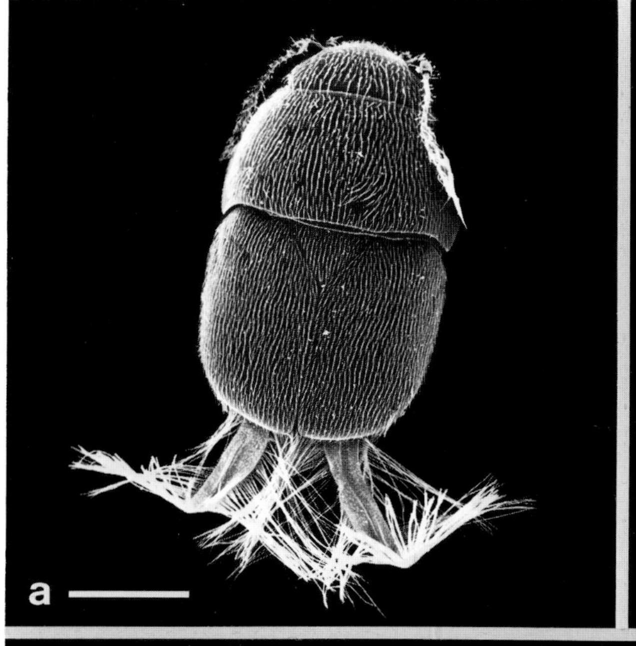
a
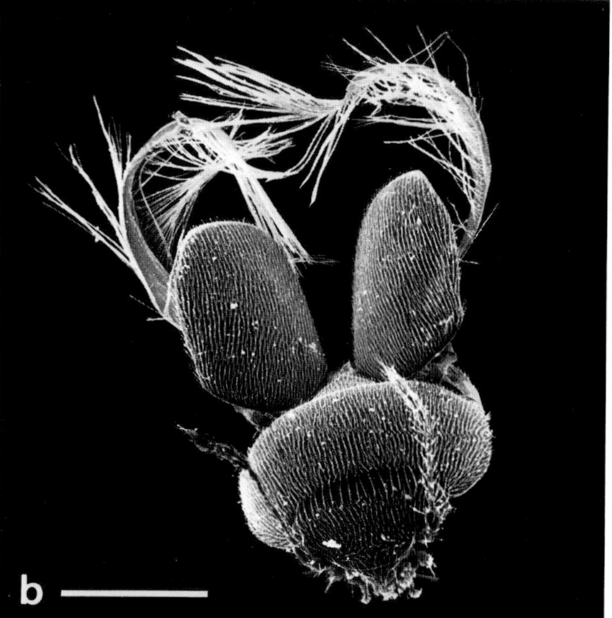
b
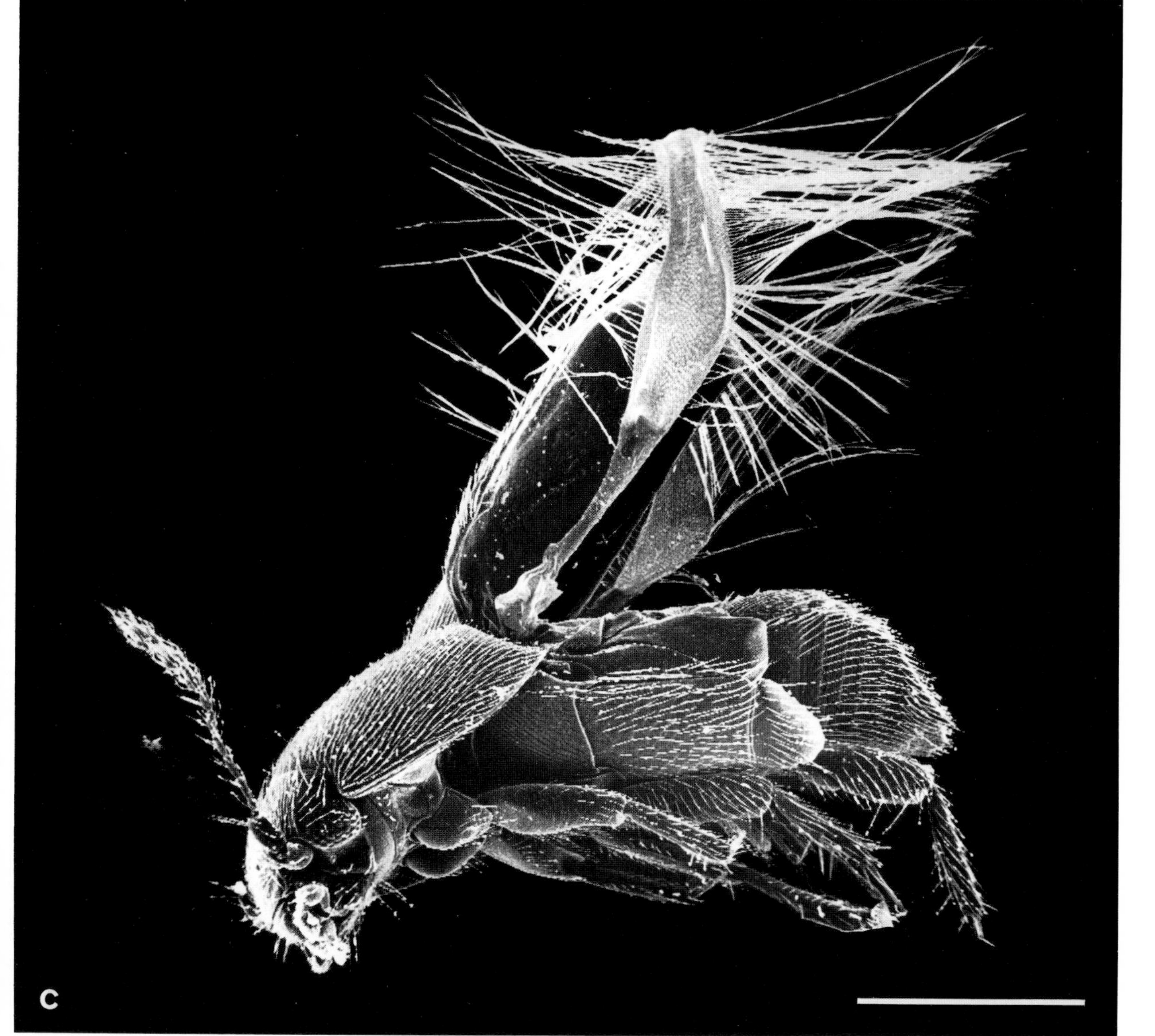
c

2.20.15 Structural Features of the Hind Wings of the Ptiliidae

The feather-winged beetles are named after their characteristic hind wings, which resemble the wings of the thrips (Thysanoptera) in structure (Fig. 171). In principle, a hind wing consists of a short quill followed by a thin, curved shaft which bears densely crowded branches on each side. The branches themselves are long, thin ciliated hairs (setae) (Plate 147). The area of the hind wings is considerably increased by the hair border, especially as the wings, despite the tiny size of the beetles, are often more than twice the length of the body. In the resting position the hind wings are folded twice so that three overlapping layers lie hidden under the small elytra.

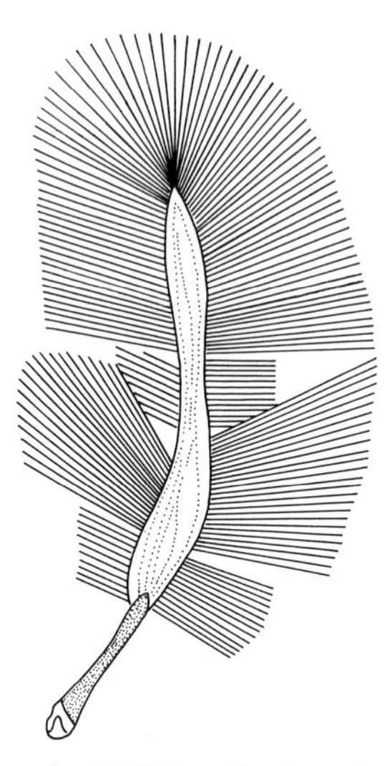

Fig. 171. Schematic illustration of the hind wing of a Ptiliidae (feather-winged beetle). (After Jacobs and Renner, 1974)

Plate 147. *Acrotrichis* sp. (Ptiliidae), feather-winged beetles.
a Base of a hind wing. 100 µm.
b Joint region of a hind wing. 25 µm.
c Unfolded hind wing, central portion. 100 µm.
d Folding joints of the wing setae. 10 µm.
e Hind wing, distal. 25 µm.
f Surface of the wing setae. 3 µm

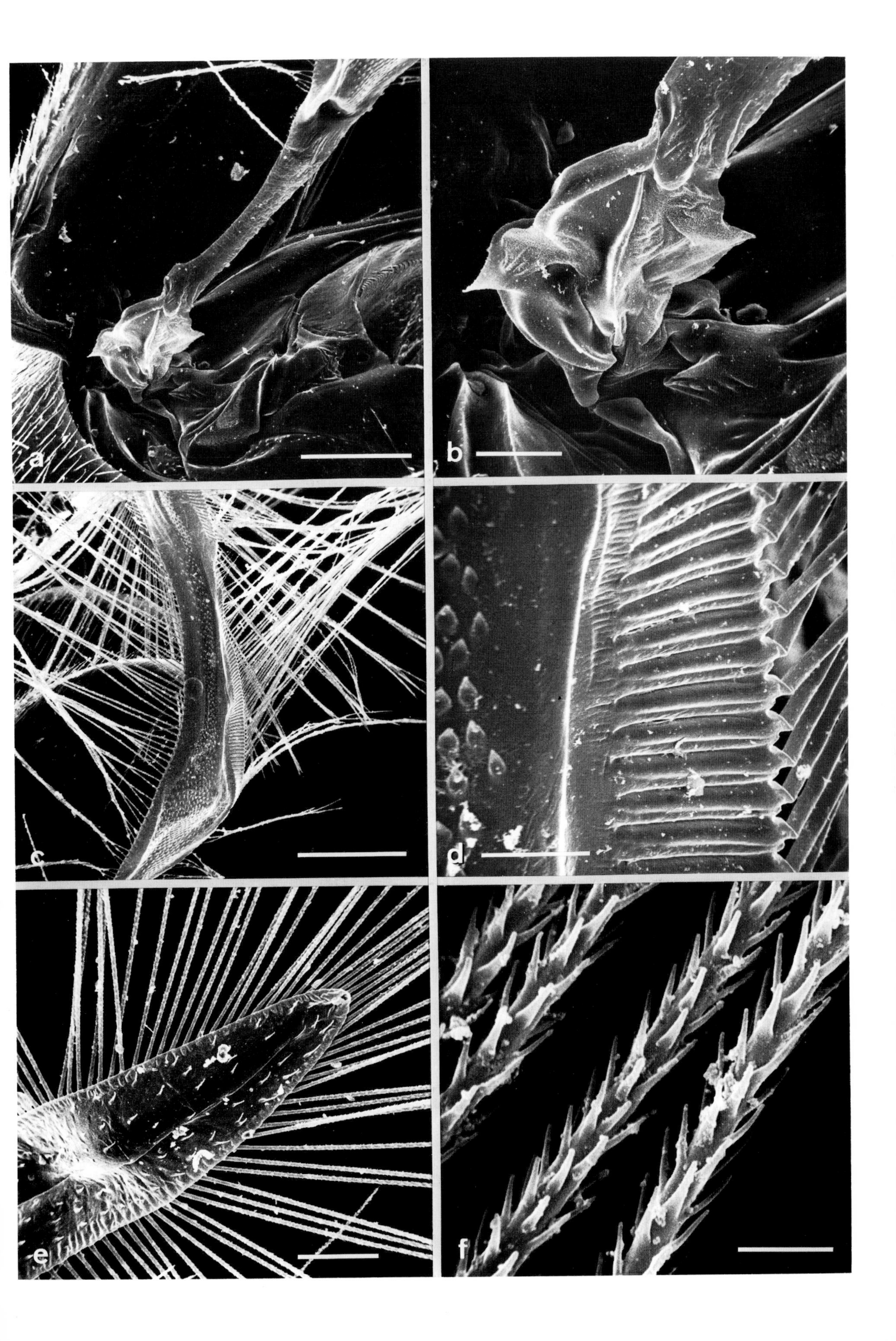
a
b
c
d
e
f

2.20.16 Life Forms of the Rove Beetles (Staphylinidae)

The external appearance of the soil-dwelling staphylinid imagines indicates their different adaptations to soil life (Fig. 172) (Topp, 1981).

Beetles of the Omaliinae subfamily, to which *Lathrimaeum* and other epedaphic soil-dwellers belong, are remarkable for their relatively long elytra. They live in the litter layer and in moss, but also occur in the herbaceous layer. In comparison the Tachyporinae have shortened elytra. Although they prefer the herbaceous layer they overwinter in the soil litter.

The Staphylininae and Xantholininae subfamilies comprise beetles of slender form with greatly shortened elytra. They not only occur on the soil surface but also live hemiedaphically in the system of spaces of the upper soil layers. The shortening of the elytra and the resulting increase in the active and passive mobility of the abdomen is favourable for their mode of life in the interstitial pore system which would otherwise be reserved only for the larvae (Blum, 1979). In some species the hind wings are completely reduced, hence they have lost the ability to fly. Members of the genus *Bledius* (Oxytelinae) are also hemiedaphic staphylinids which burrow deep into the soil.

The Leptotyphlinae represent the euedaphic life-form type. Their form is best adapted to the interstitial system of the soil due to the extreme reduction of the elytra and shortening of the extremities. They live permanently in the darkness of deeper soil layers and thus lack pigment and have degenerate eyes (Coiffait, 1958, 1959).

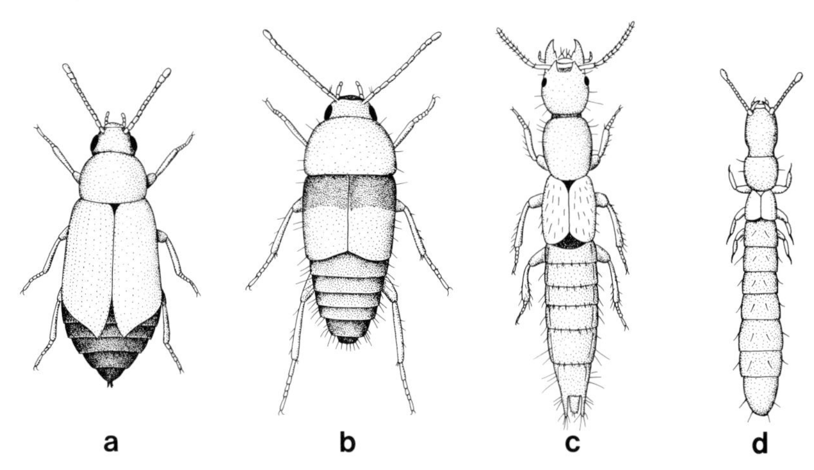

Fig. 172. Habitus forms of the rove beetles (Staphylinidae). [After Hansen, 1951, 1952 (**a, b**) and Coiffait, 1972 (**c, d**) from Topp, 1981]. **a** *Eusphalerum minutum* and **b** *Tachyporus obtusus* inhabit the herbaceous layer. **c** *Othius punctulatus*, an epedaphic species. **d** *Entomoculia occidentalis*, an euedaphic species

---

Plate 148. *Oxytelus rugosus* (Staphylinidae), rove beetles.
a Overview, dorsal. 1 mm.
b Overview, oblique frontal. 1 mm.
c Head and prothorax, oblique frontal. 250 µm.
d Labial palp, distal, with sensilla. 3 µm.
e Mouthparts, oblique frontal. 100 µm.
f Galea of the maxilla, distal, with cleaning brush. 25 µm

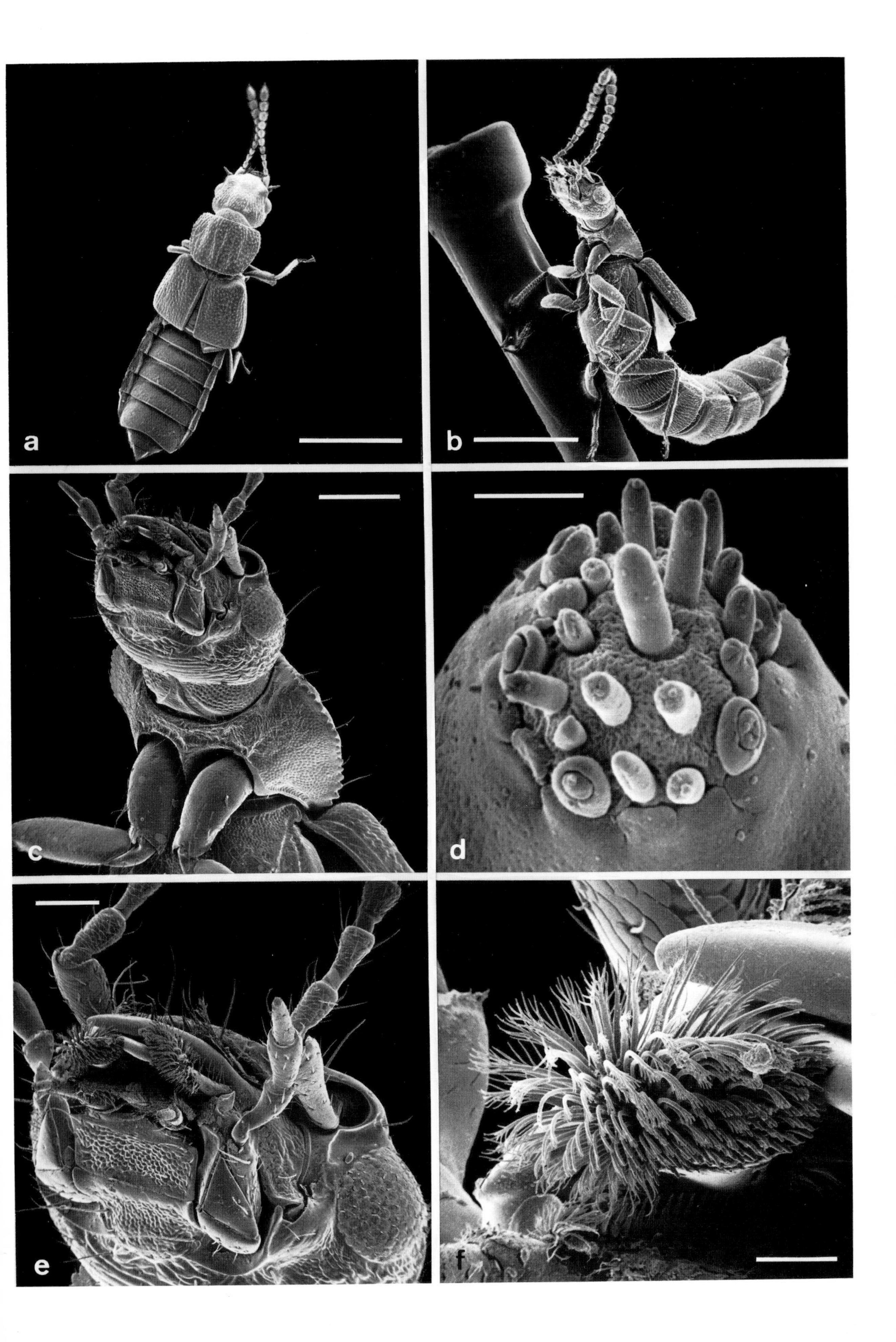
a
b
c
d
e
f

2.20.17 Staphylinidae in the Forest Beetle Community

Comparison of the annual emergence abundance and activity density of beetles in four types of forest (Fig. 173a) shows that the Staphylinidae play the dominant role. If appropriate "imagines production" data are compared, the proportion of Staphylinidae decreases considerably in favour of other beetles families (Fig. 173b). However, their proportion of biomass among the beetles remains high in each case. This emphasizes the important role played by the predominantly predatory rove beetles in the decomposer food web of woodland soil (Roth et al., 1983).

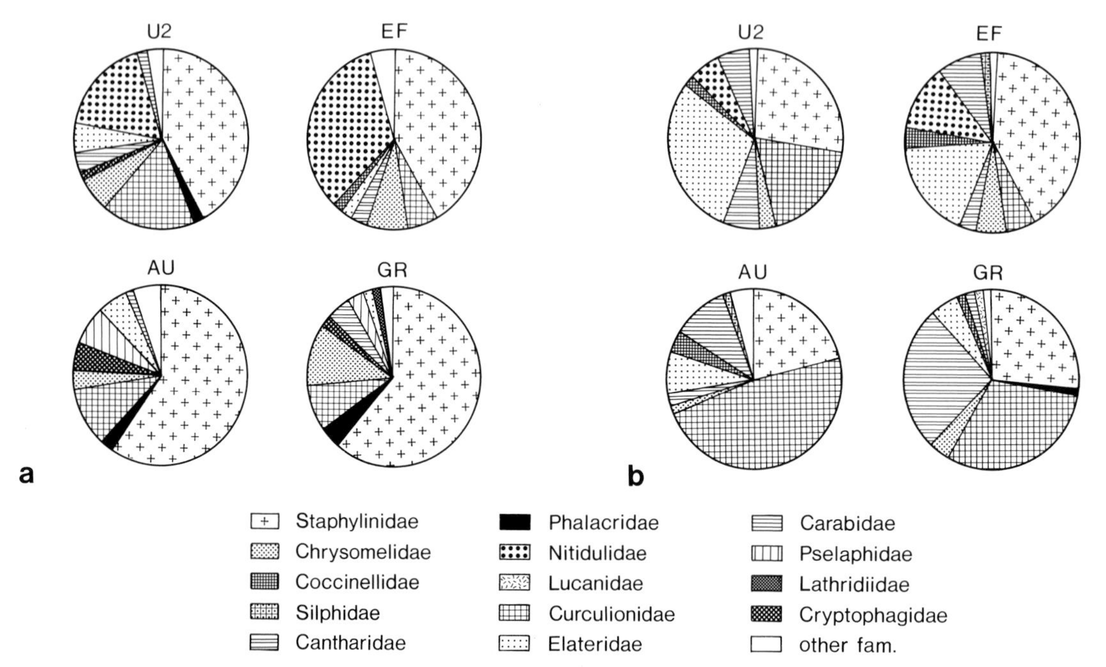

Fig. 173. Comparison of the annual beetle fauna in the Ulm-Günzburg region (W. Germany) for 1980 (Roth et al., 1983).
a Emergence abundance and activity density measured as ind/m^2 × year; percentage proportion within the Coleoptera. **b** "Imagines production" expressed as biomass in mg dry weight/m^2 × year; percentage proportion within the Coleoptera.
EF Limestone beech wood, Schwäbische Alb (Melico-Fagetum, 90-years-old); *U2* acid humus (raw humus) beech wood, Ulm (Luzolo-Fagetum, 100-years-old); *AU* water-meadow woodland on the Iller (Ulmo-Fraxinetum, 40-years-old); *GR* water-meadow woodland on the Danube (Querco-Carpinetum, 60-years-old)

Plate 149. *Oxytelus rugosus* (Staphylinidae), rove beetles.
a Antenna, proximal portion of the flagellum. 50 µm.
b Antennal segments from the middle portion of the flagellum. 25 µm.
c Abdominal segment with spiracle, oblique lateral. 100 µm.
d Abdomen, caudal. 100 µm.
e Spiracle. 25 µm.
f Tibio-tarsal joint. 20 µm

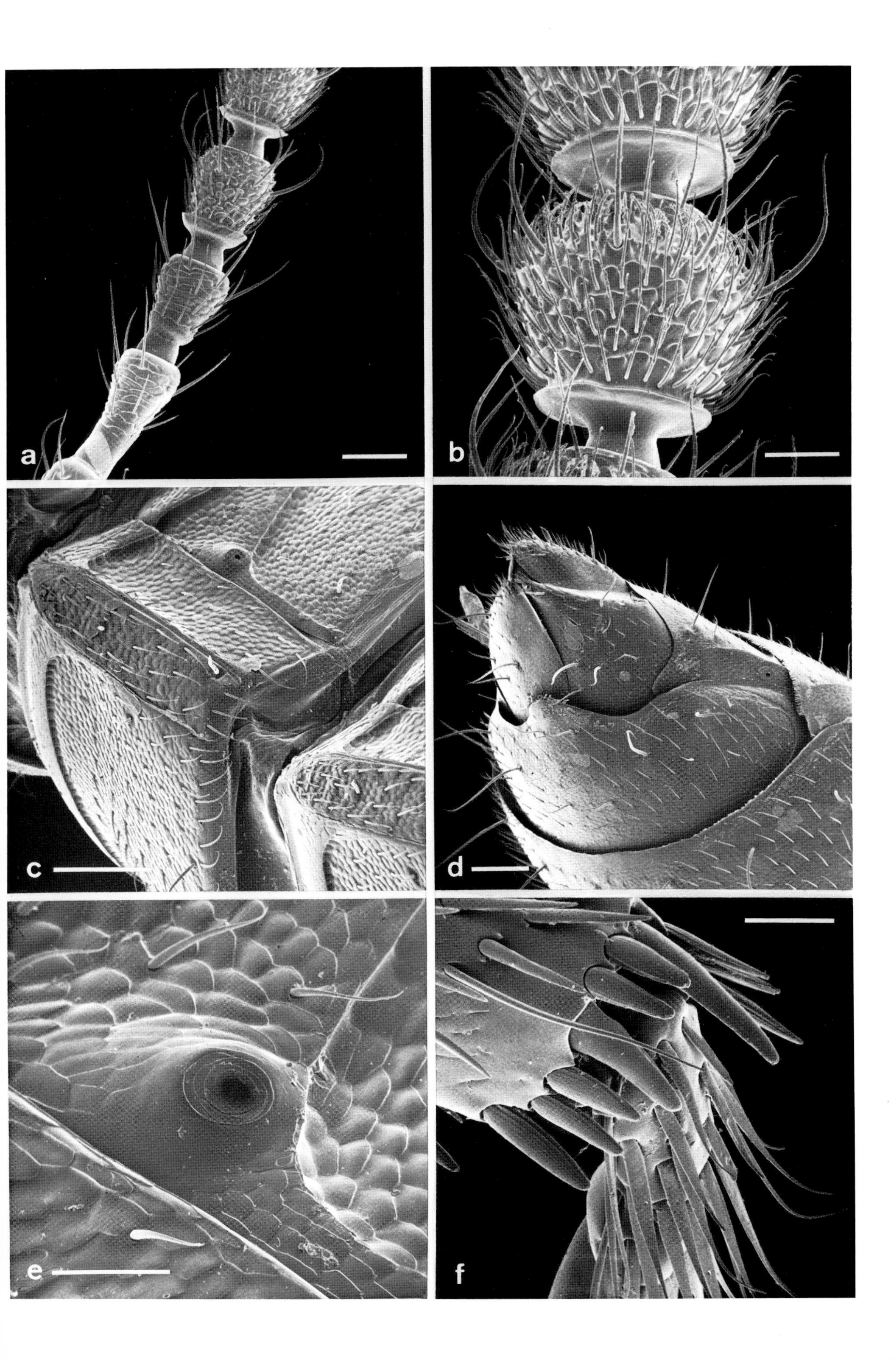

a
b
c
d
e
f

2.20.18 The Burrowing Rove Beetle: *Bledius arenarius* (Staphylinidae)

The 2–8-mm-long species of the staphylinid genus *Bledius* live in clay or sandy soil, primarily in the littoral zone (Topp, 1975), and on the banks of inland lakes and rivers. They excavate deep burrows to serve as dwellings and brood chambers using their strong mandibles and front extremities which are modified to fossorial legs (Plate 150). They feed on algae.

Bledius arenarius digs a vertical shaft with two sloping passageways leading to it. The dwelling shaft can reach a depth of 40 cm. In its immediate vicinity egg chambers are established, in which eggs are deposited on a socket of small sand grains (one egg per socket and chamber) (Fig. 174). In contrast to other *Bledius* species, there is no connecting passage between the egg chambers and the burrow. The emerging larvae must seek their own food (algae), as the female does not provide them with a food store (v. Lengerken, 1954).

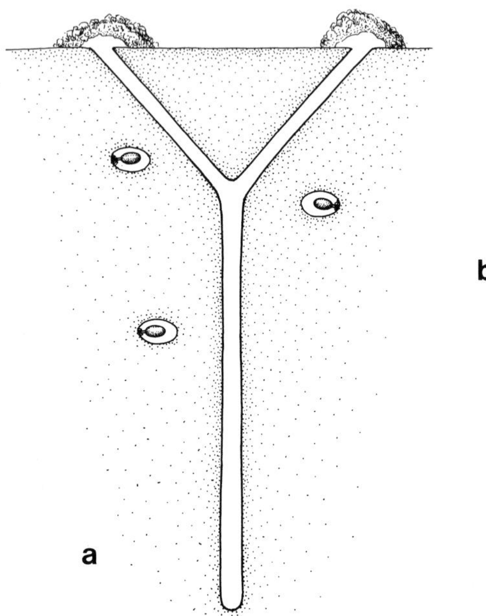

Fig. 174. Schematic section through the brood burrow of *Bledius arenarius* (Staphylinidae). (After v. Lengerken, 1954).
a Brood burrow with central dwelling shaft. The connecting passages to the egg chambers have been resealed. **b** Egg chamber. The stalked egg sits on a socked of sand grains

Plate 150. *Bledius arenarius* (Staphylinidae), rove beetles.
a Anterior of the body, dorsal. 0.5 mm.
b Anterior of the body, oblique ventral. 0.5 mm.
c Head with antennae and mouthparts, frontal. 200 µm.
d Antenna, distal. 25 µm.
e View of part of the mouthparts, dorsal, with mandibles and cleaning or filter brushes. 50 µm.
f Fossorial leg. The most effective part for excavation is the spiny tibia. 100 µm

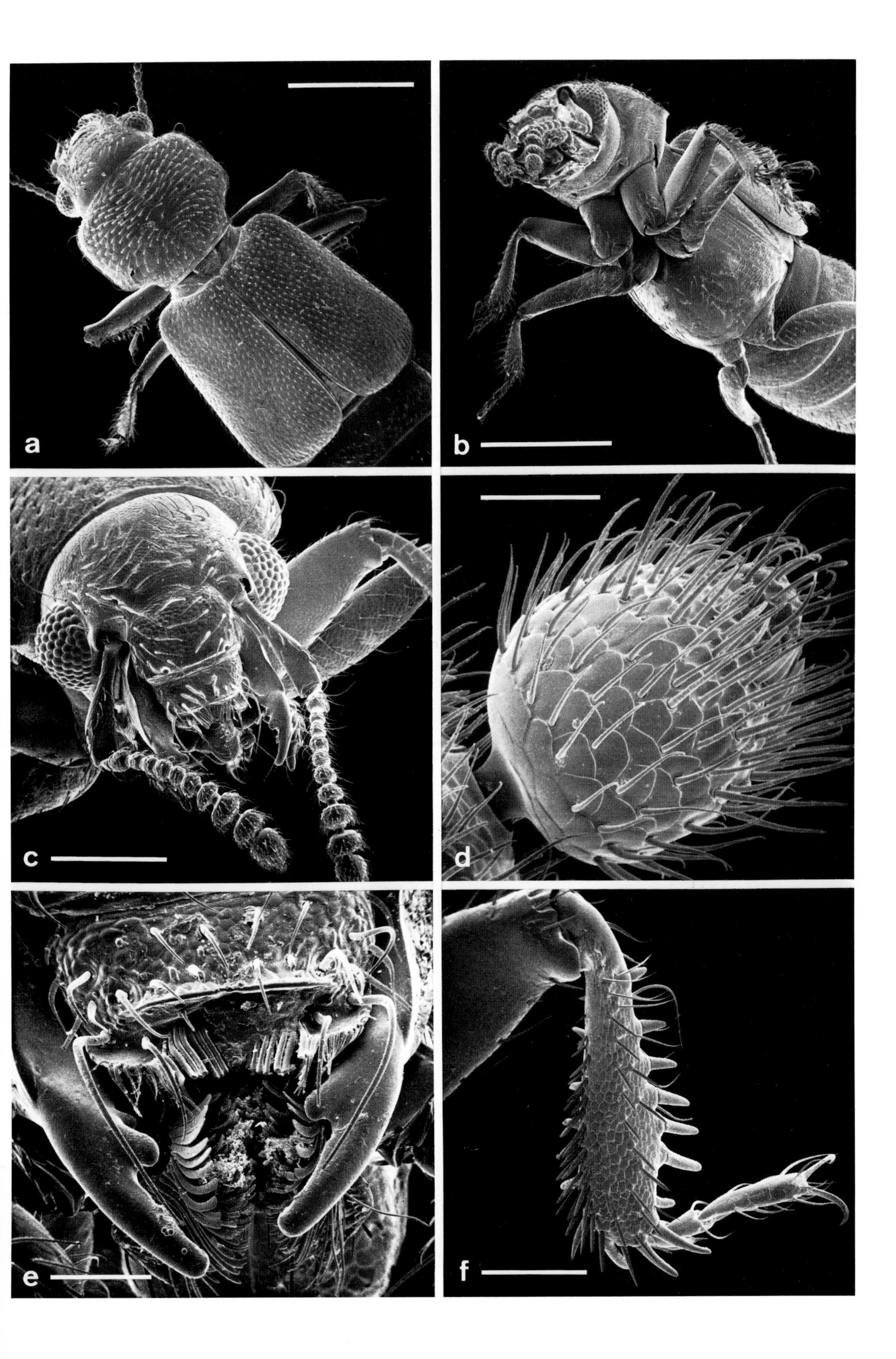
a
b
c
d
e
f

2.20.19 Phenology of the Epedaphic *Lathrimaeum atrocephalum* – Staphylinidae (Rove Beetles)

Lathrimaeum atrocephalum (Fig. 175; Plate 151) is a 3–3.5-mm-long beetle which is distributed throughout the Palaearctic and is a frequently encountered epedaphic representative of the Omaliinae subfamily. It prefers the moist litter of beech woods and alder marshes where large populations often occur (Hartmann, 1979; Späh, 1980; Rehage and Renner, 1981; Friebe, 1983). *Lathrimaeum atrocephalum* is active in winter. The first individuals are caught in pitfall traps in September. Their activity reaches its first peak at the end of November and, with interruptions, remains high till the middle of April when it slowly, but steadily, declines. They are absent from pitfall traps during the warm season (June to August) (Fig. 176). The phenology of this species is largely governed by their diapause (allopause) (Topp, 1979).

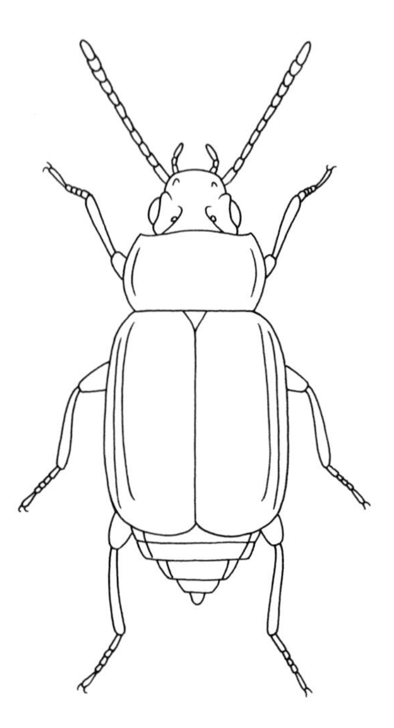

Fig. 175. Habitus of *Lathrimaeum atrocephalum* (Staphylinidae) from the dorsal side (Lohse, 1964)

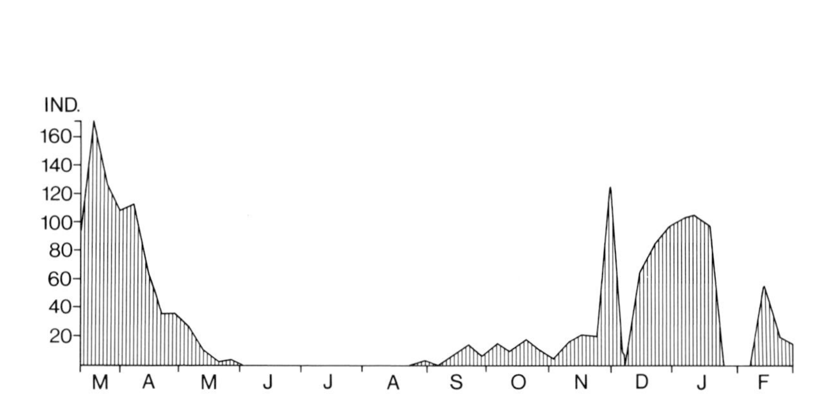

Fig. 176. The occurrence of *Lathrimaeum atrocephalum* (Staphylinidae) in an alder marsh in 1975. (Späh, 1980)

Plate 151. *Lathrimaeum atrocephalum* (Staphylinidae), rove beetles.

a Habitus, dorsal. 0.5 mm.
b Head and prothorax, lateral. 250 µm.
c Overview, frontal. Between the lateral compound eyes lie two ocelli. 250 µm.
d Ocellus. 10 µm.
e Abdomen of a female, caudal, with the telescope-like everted terminal segments and external genital appendages. 250 µm.
f Terminal segments with external female genital appendages. The *arrowhead* shows the vaginal palps. 100 µm

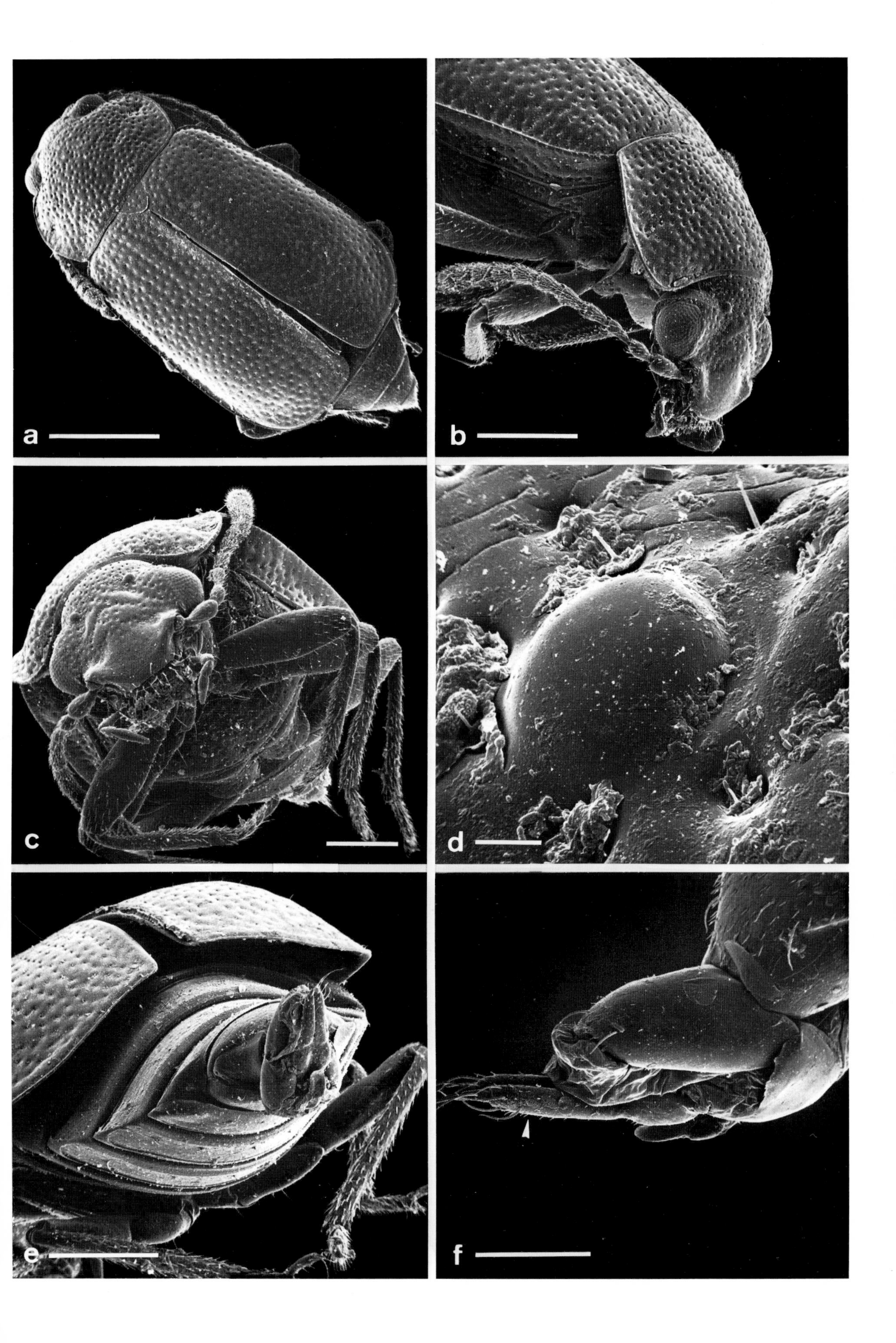
a
b
c
d
e
f

2.20.20. Phenology of the Epedaphic *Atheta fungi* – Staphylinidae (Rove Beetles)

Many members of the Omaliinae are typical soil beetles with an epedaphic mode of life. Several Oxytelinae and Aleocharinae also inhabit the litter layer of the soil. *Atheta fungi* (Fig. 177) is a member of this group and is often caught in pitfall traps. The seasonal activity maximum of *Atheta fungi* occurs during the summer months. In an alder marsh (Späh, 1980) the active phase begins with a gradual increase in April and reaches its peak between the second half of July and the beginning of August. Their abundance sharply decreases thereafter with only a few animals active between September and November (Fig. 178).

Topp (1975) observed an activity peak of *Atheta fungi* in an oak forest in September. The ovarian pause (diapause), stimulated by the decreasing daylight hours, begins at this time and the beetles seek their winter quarter in the soil litter. As a rule, the end of the diapause is also signalled by the photoperiod, together with an average temperature of around 16 °C.

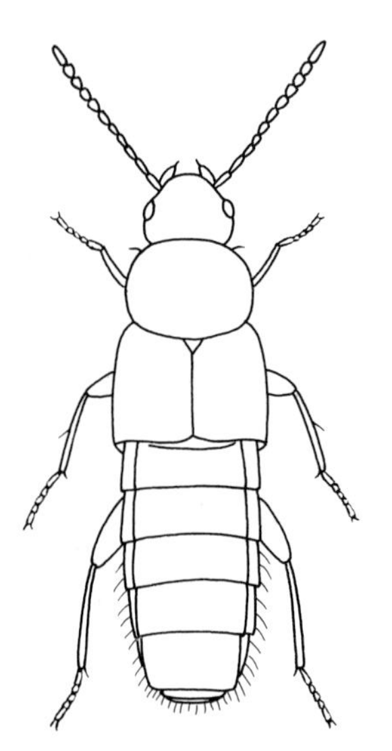

Fig. 177. Habitus of *Atheta fungi* (Staphylinidae), dorsal. (Lohse, 1964)

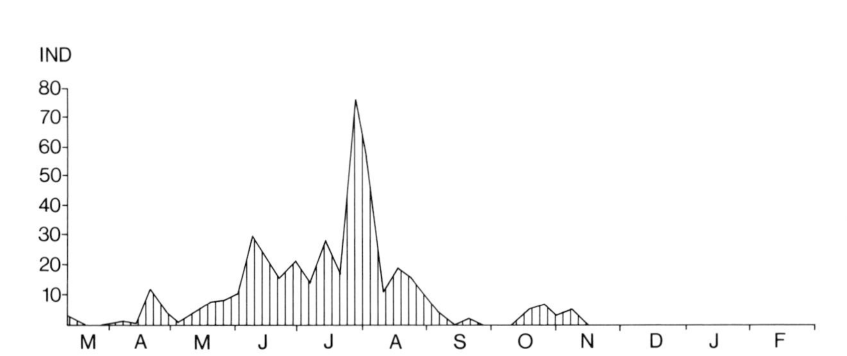

Fig. 178. The occurrence of *Atheta fungi* (Staphylinidae) in an alder marsh in 1975. (Späh, 1980)

Plate 152. *Atheta orbata* (Staphylinidae), rove beetles.
a Habitus, oblique frontal. 0.5 mm.
b Elytra with the folded hind wings beneath. Abdominal segments with spiracles, oblique dorsal. 200 µm.
c Overview, oblique caudal, with partly unfolded hind wings. 0.5 mm.
d Wings, partly unfolded. 0.5 mm

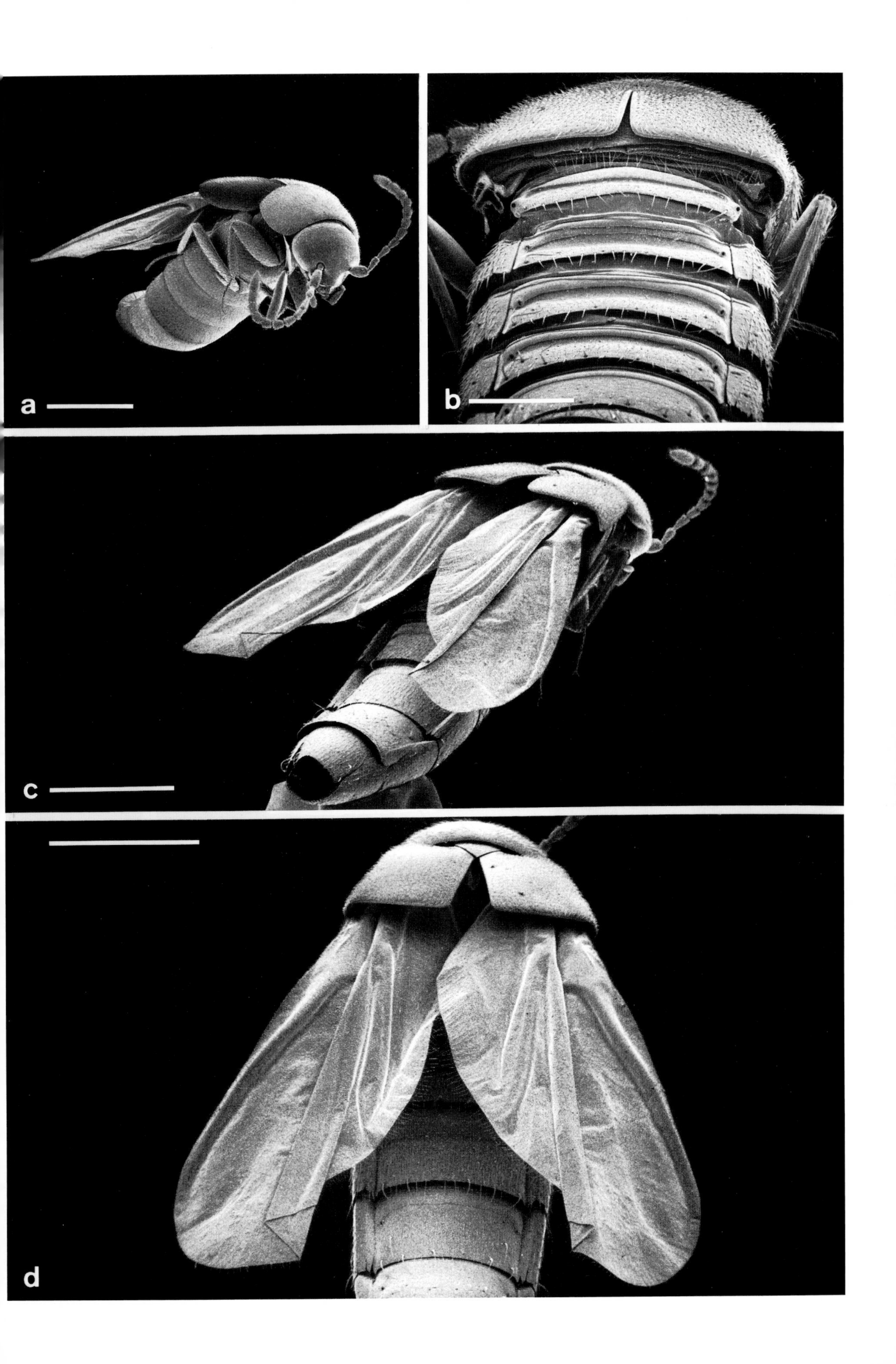
a
b
c
d

2.20.21 Larvae of the Rove Beetles (Staphylinidae)

Rove beetle larvae (Fig. 179; Plates 153 and 154) often occur together with the imagines. Among the soil-dwelling larvae are members of the Omaliinae, Aleocharinae as well as the Staphylininae and Xantholininae, etc. They are predatory like the most imagines.

The development from the egg to the imago of the mostly univoltine central European Staphylinidae takes place in spring and summer, as many beetle species are active in summer. As a rule, three larval stages are passed through, whereby some Staphylininae also overwinter as third instars. Winter-active beetles, to which many Omaliinae and Aleocharinae belong, also have larvae which are active in the cold months. Studies of staphylinid larvae are, however, only just beginning (cf. Kasule, 1966, 1968, 1970; Steel, 1966, 1970 and Topp, 1971, 1973, 1978).

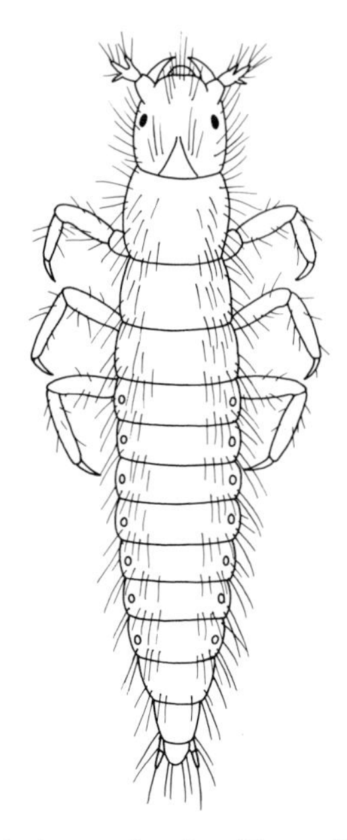

Fig. 179. Larva of *Atheta sordida* (Staphylinidae), rove beetles. (Topp, 1971 from Klausnitzer, 1978)

Plate 153. Larva of the Staphylinidae (rove beetles).
a Head and prothorax, lateral. 1 mm.
b Abdomen, dorsal, with uniform tergites. In the intersegmental and lateral region the cuticle is membranous. The *arrowhead* points in the caudal direction. 0.5 mm.
c Head and prothorax ventral. 1 mm.
d Abdominal segment, lateral, with dorsal tergite, lateral sclerites and ventral sternites. In the mainly membranous skin on the flank there is a spiracle. The arrowhead points in the caudal direction. 0.5 mm.
e Labial palp, distal, with digitiform sensilla. 10 μm.
f Abdominal segment, ventral, with one pre-, two latero- and the subdivided poststernite. The *arrowhead* points in the caudal direction. 0.5 mm

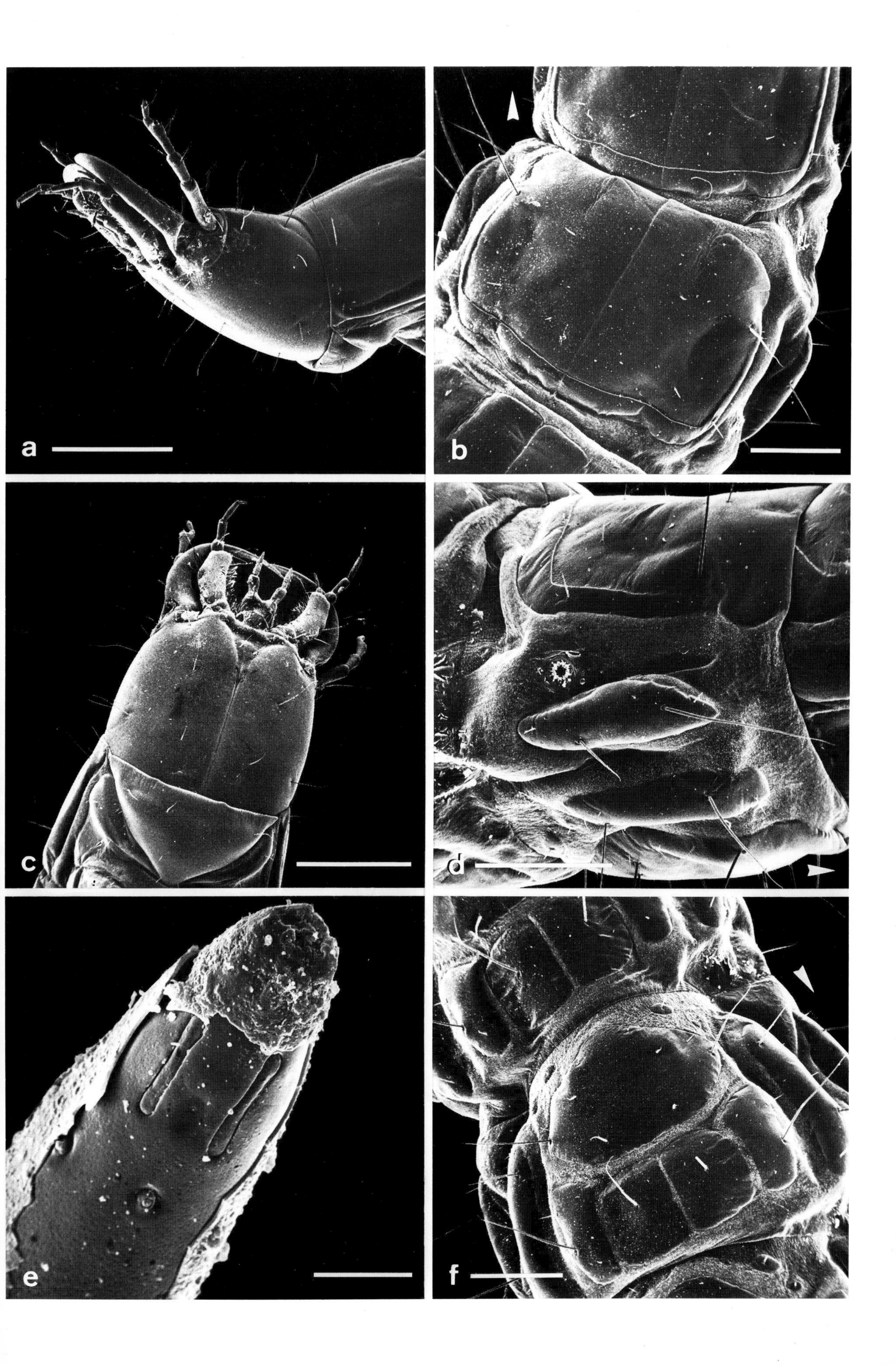
a
b
c
d
e
f

2.20.22 Digitiform Sensilla in Beetles and Their Larvae

Sensilla of insect larvae and imagines are often concentrated on the palps of the mouthparts and on the antennal segments. Digitiform sensilla of the maxillary and labial palps are widespread among the Coleoptera (Honomichl, 1980). Though their number and arrangement varies greatly between the individual species, their form is similar. The entire length of their shafts (20–30 µm) is inserted, or even deeply sunk, into a small cuticular pit (Fig. 180). The ultrastructure of the digitiform sensilla, which are also found in staphylinid larvae (Plate 153), were investigated in elaterid larvae (Zacharuk et al., 1977) and in the imagines of a variety of families (Guse and Honomichl, 1980; Honomichl and Guse, 1981; Mann and Crowson, 1984). Their structure is not consistent with that of either olfactory or chemoreceptive sensilla. For the time being, their sensitive function remains unknown.

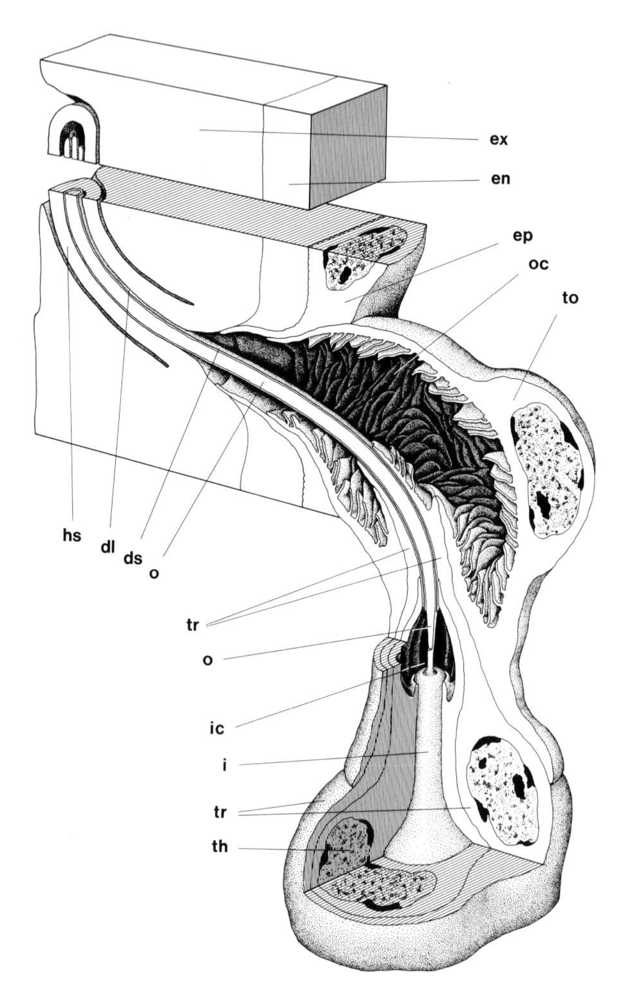

Fig. 180. Micromorphology of a digitiform sensillum, typically found on the mouthparts (palps) of beetles, exemplified by *Tenebrio molitor* (Tenebrionidae). The hair shaft is sunk into a cuticular pit. (After Honomichl and Guse, 1981). *dl* Hair lumen; *ds* dendritic sheath; *en, ex* endo-, exocuticle; *ep* epidermis; *hs* hair shaft; *i, o* inner, outer dendritic segment; *ic, oc* inner, outer receptor lymph cavity; *th, to, tr* thecogen, tormogen, trichogen envelope cells

---

Plate 154. Larva of the Staphylinidae (rove beetles).

- **a** Thorax, lateral. 1 mm.
- **b** Prothorax and mesothorax (*left*) with spiracle. 0.5 mm.
- **c** Abdomen, caudal, lateral, with anal tube and dorsal appendages (cerci, urogomphi). 100 µm.
- **d** Spiracle. 25 µm.
- **e** Surface of the metanotum with hair base. 10 µm.
- **f, g** Finely perforated cuticle of the tergum. 3 µm, 2 µm

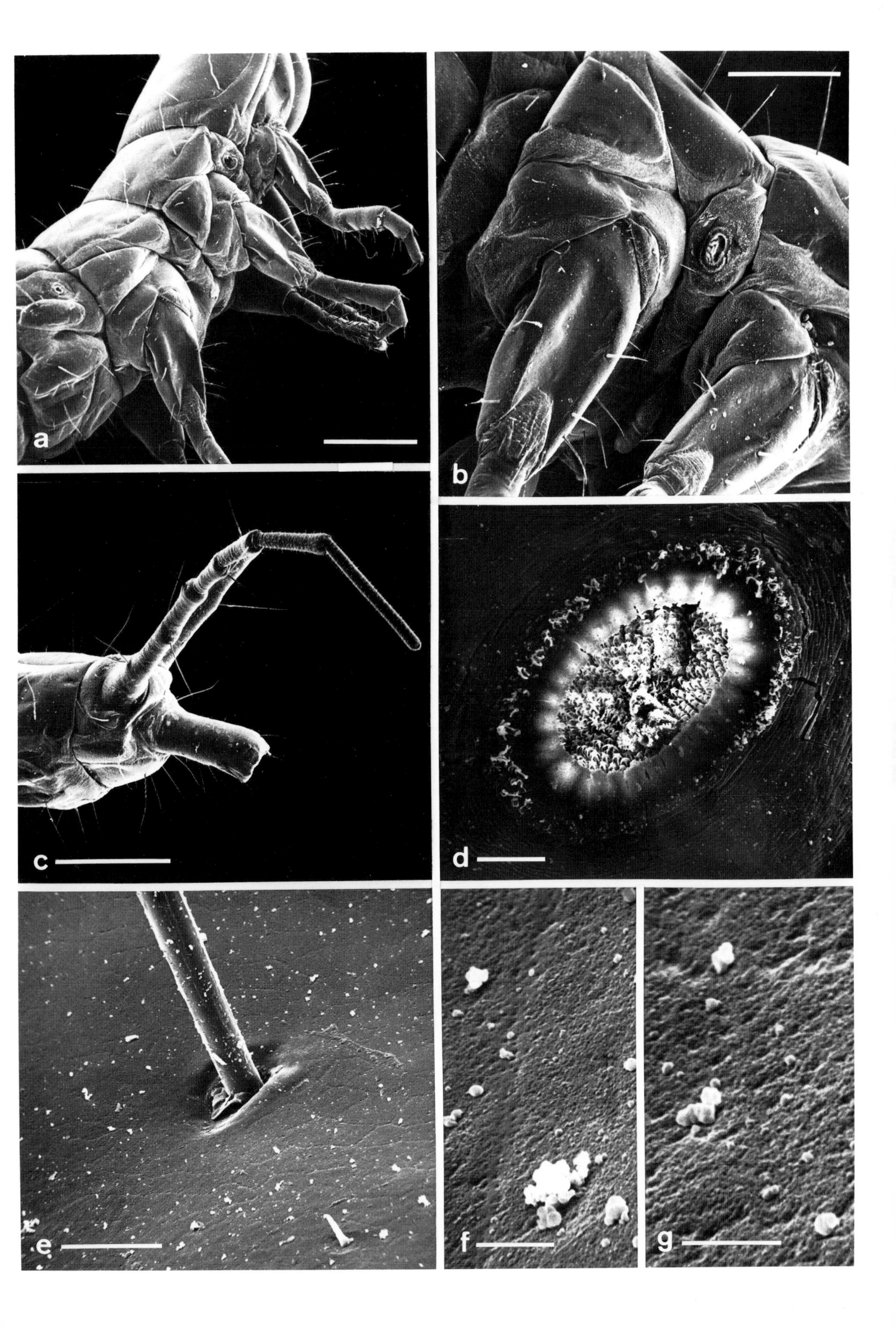
a
b
c
d
e
f
g

2.20.23 The Common Glowworm *Lampyris noctiluca* (Lampyridae)

Lampyris noctiluca (Fig. 181; Plates 155 and 156) is one of the most common species of glowworms belonging to the Lampyridae family. They are distinguished by two striking characteristics:

1. The adult beetles have a marked sexual dimorphism (Geisthardt, 1974, 1977). Whereas the males have normal, functional wings, the females are wingless, lacking both elytra and alae. Thus, the females are restricted to life on the ground. The larvae are similarly adapted to this habitat. Both are epedaphic, living in the upper soil layer.
2. Their sexual behaviour is determined by the female's emission of light, normally during their nocturnal period of activity before midnight. A luminous plate situated ventrally on each of the sixth and seventh abdominal segments and a pair of luminous dots on the eighth segment (Fig. 183) comprise the luminescent organs of the female of *L. noctiluca.* The light emitted by the female on the ground attracts males which are willing to copulate. In response, they fly towards the female. Contrary to the equally common males of *Lamprohiza splendidula* (large luminescent apparatus on the fifth and sixth segments), *Lampyris noctiluca* males only have two small luminescent organs on the seventh sternite (Geisthardt, 1979).

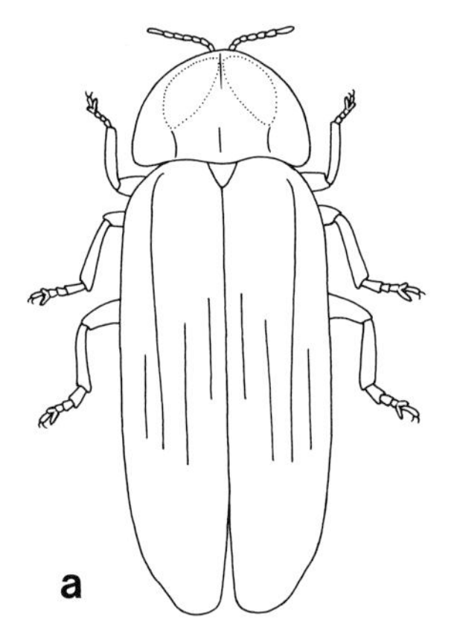

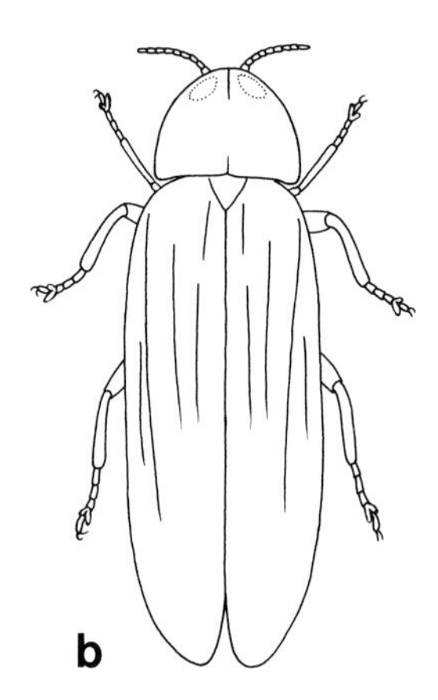

Fig. 181. Habitus of the Lampyridae (glowworms). (After Geisthardt, 1979).
a Male of *Lamprohiza splendida,* dorsal. Two large, transparent windows are situated parasagittally in the pronotum.
b Male of *Lampyris noctiluca*, dorsal. Two small, transparent windows, which are sometimes lacking, are shown in the pronotum

---

Plate 155. *Lampyris noctiluca* – Lampyridae (glowworms), common glowworm, male.
a Thorax and head, ventral. The head is borne in a ventral hollow of the pronotum. 1 mm.
b Abdomen, ventral. The luminous areas are not visible. 1 mm.
c Head with reduced mouthparts between the eyes. The terminal segments of the palps are axe-shaped; they bear numerous papillae-like sensilla 0.5 mm.
d Frontal view. At the anterior edge of the pronotum lie the parasagittally situated transparent windows through which light can reach the eyes dorsally. 50 μm.
e Mouthparts with pointed mandibles (*arrows*). 100 μm.
f Ommatidia surface. 10 μm

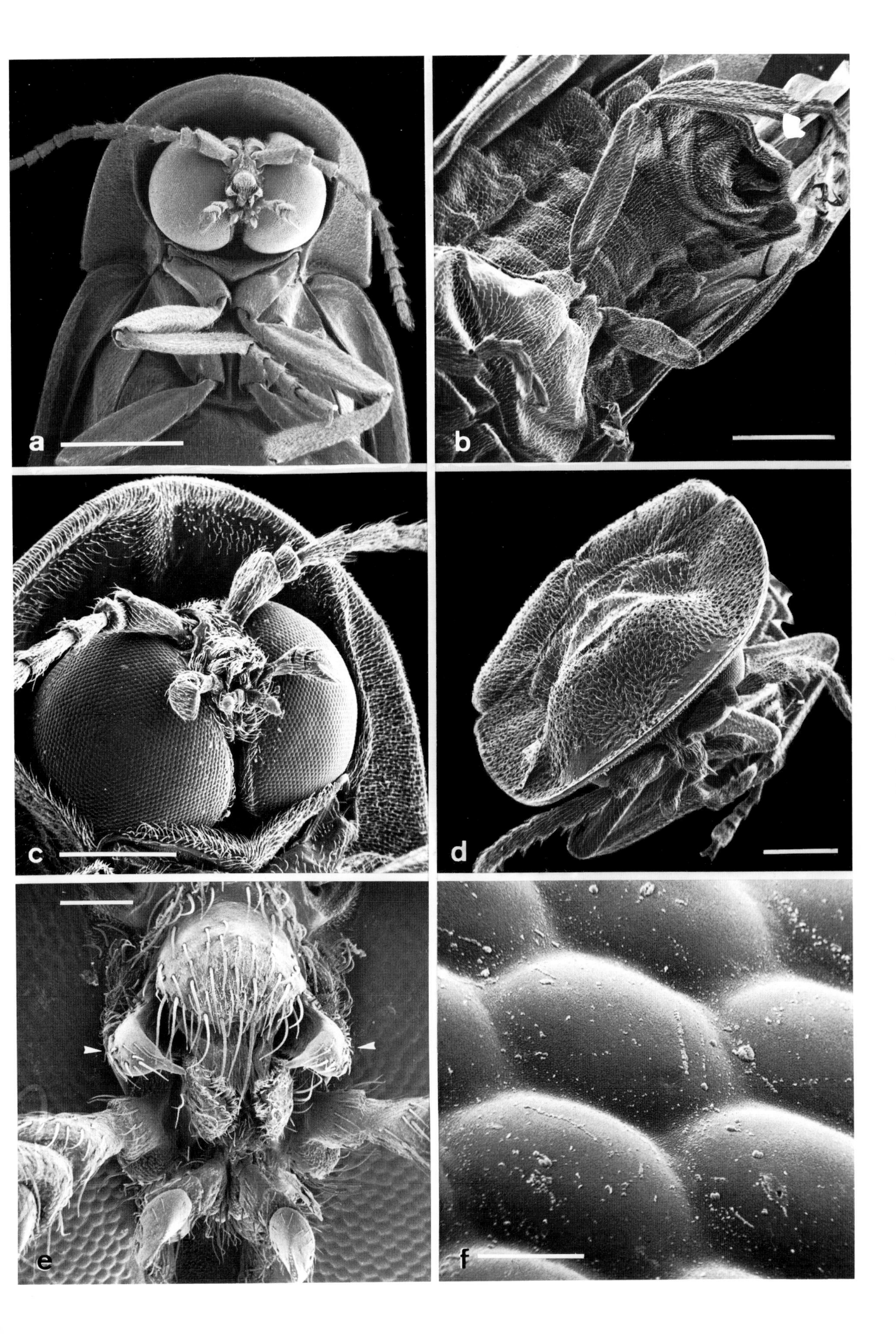
a
b
c
d
e
f

2.20.24 Larvae of the Glowworm (Lampyridae)

The life span of *Lampyris noctiluca*, lasting approximately 3 years, is divided into: 30 days of embryonic development, 2.6 years of larval development, 9 days at the pupal stage and 10–16 days as imagines. During the larval development the larvae are subject to annual and circadian activity rhythms. A (facultative) winter diapause alternates three times with the active periods during the remainder of the year. In the course of 24 h, a period of nocturnal activity succeeds the inactive phase during the day. This circadian rhythm conforms to the luminescent rhythm. At night the larvae emit light from their paired luminous areas on the abdomen.

The epedaphic larvae (Fig. 182; Plate 156) mainly feed on slugs and snails. They find their prey by following the slime trail left behind by the snails. Their olfactory sensory organs, which are capable of perceiving a trail even when it is 1–2-days-old, are assumed to lie on the feelers of the first maxilla. Their prey is killed by a bite from their dagger-shaped mandibles which simultaneously inject poison into the prey. In contrast to the predatory larvae, no food is consumed by the imagines (Schwalb, 1961).

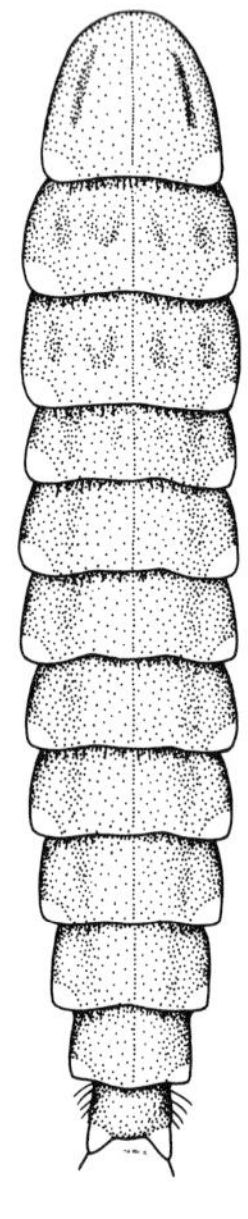

Fig. 182. Larva of *Lampyris noctiluca* – Lampyridae (glowworms), common glowworm, dorsal. (After Korschefsky, 1951 from Klausnitzer, 1978)

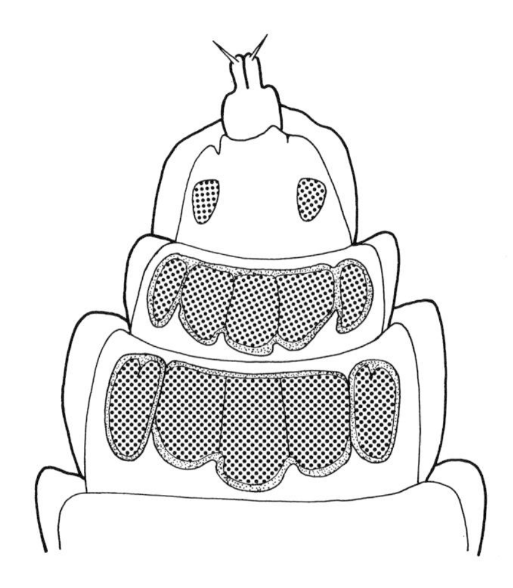

Fig. 183. Large luminescent organ of the female of *Lampyris noctiluca* – Lampyridae (glowworms), schematic. Complex luminous areas are situated on the sixth and seventh sternites, whereas only two small dots of light are emitted from the eighth sternite (Original)

Plate 156. *Lampyris noctiluca* – Lampyridae (glowworms), common glowworm, female (**a, b**) and larva (**c–f**).

a Anterior of the body, ventral. 0.5 mm.

b Abdomen, caudal, ventral. The luminous areas are not clearly visible (due to the gold plating during preparation of the specimen). 1 mm.

c Pro- and mesothorax, dorsal. 1 mm.

d Prothorax, lateral. The head is withdrawn. In life it can be extended like a telescope and retracted extremely swiftly when disturbed. 1 mm.

e Anterior of the body, ventral. The only visible parts of the head are the tips of the dagger-shaped mandibles (*arrow*). 1 mm

f Cuticle of the dorsal trunk. 100 µm

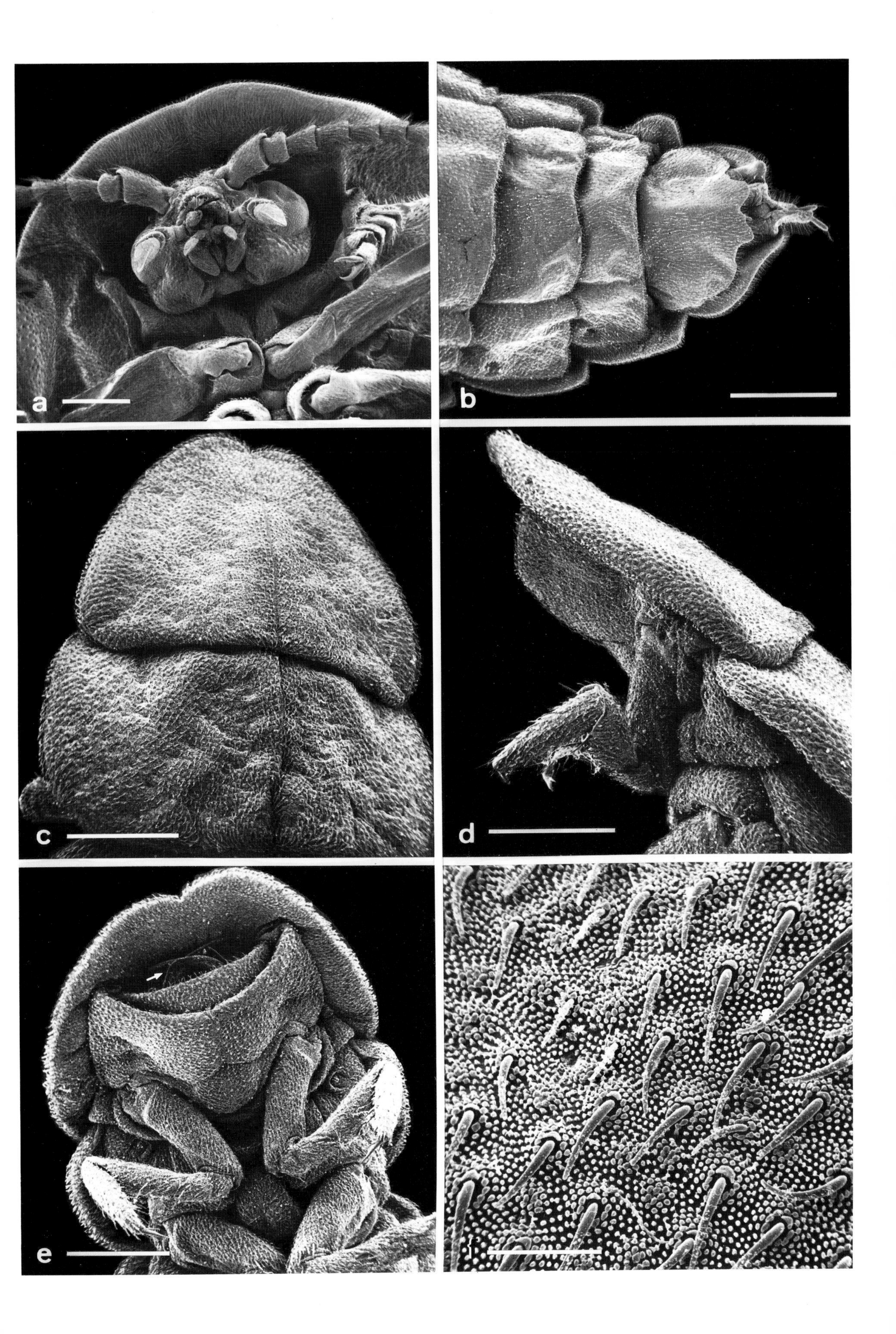
a
b
c
d
e
f

2.20.25 Larvae of the Soldier Beetles (Cantharidae)

Like many beetle larvae, surprisingly little is known about the biology and ecology of the larval stage of the soldier beetle compared to the well-known adult. Whereas the beetles are predominantly found on flowers and leaves in the herbaceous and shrub layers, the larvae (Fig. 184; Plates 157 and 158) live in the upper soil layer under stones, in litter and rotting material. They are obviously predators.

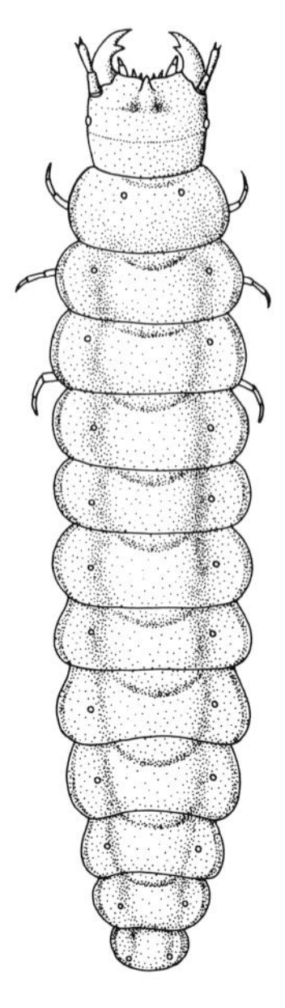

Fig. 184. Larva of *Cantharis* sp. – Cantharidae (soldier beetles), habitus dorsal. (After Larsson, 1941 from Klausnitzer, 1978)

Plate 157. Larva of the Cantharidae (soldier beetles).
a Anterior of the body, frontal. 0.5 mm.
b Abdomen, lateral. 1 mm.
c Mouthparts. 250 μm.
d Head, ventral. 0.5 mm.
e Antenna and eye (stemma; *arrowhead*), dorsal. On the *upper right* lies the base of a mandible. 100 μm.
f Antenna, distal, with different types of sensilla. 25 μm

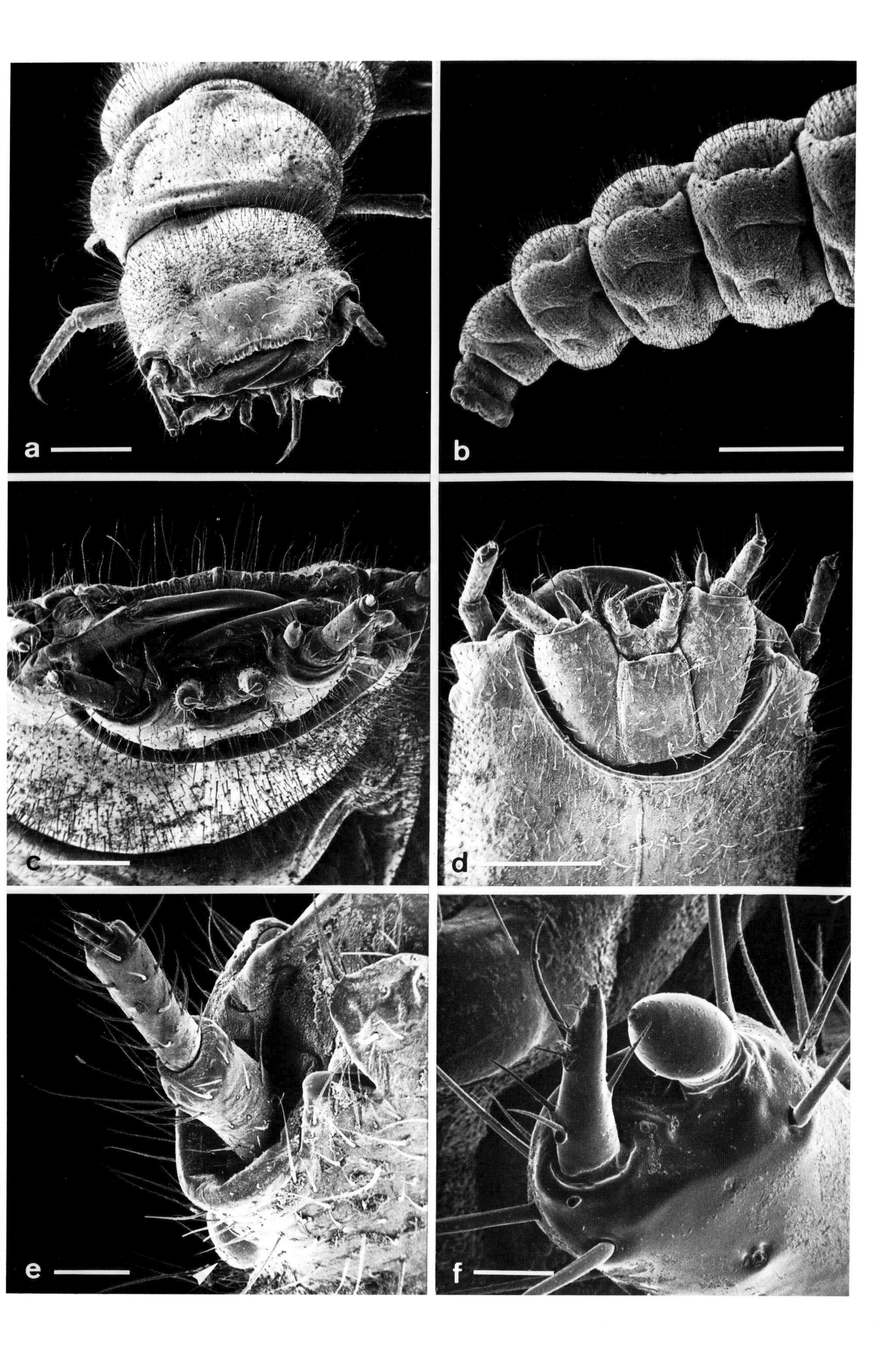

a
b
c
d
e
f

2.20.26 Structural Features of the Body of Soldier Beetle Larvae (Cantharidae)

The anatomy of the larva is appropriately adapted for life in the soil. The anterior of the head is broad and wedge-shaped (Plate 157). The powerful, projecting mouthparts with their sickle-shaped mandibles have a biting-chewing function. Thorax and abdomen are almost worm-shaped and slightly dorso-ventrally flattened, forming a functional-morphological unit. Under the microscope the soft velvety skin proved to consist of a thick coat of fine cuticular trichomes (false hairs) which are only absent where there are spiracles for tracheal respiration, defensive glands to repel enemies and sensory hairs for orientation (Plate 158). This hairy surface is assumed to increase the resistance of the integument to wetting.

The larvae are able to evert their terminal hindgut segment (Plate 158b) which indicates that water intake via the anus is possibly a common mode of water regulation in these soil-dwellers (cf. Chap. 2.20.13).

Plate 158. Larva of the Cantharidae (soldier beetles).
a Abdomen, ventral, with everted anal lobes. 0.5 mm.
b Anal lobes of the hindgut. 200 μm.
c Abdominal segment, lateral. 250 μm.
d Thoracic segment, ventral, with spiracles. 250 μm.
e Spiracle. 25 μm.
f Cuticle with hair base. 5 μm

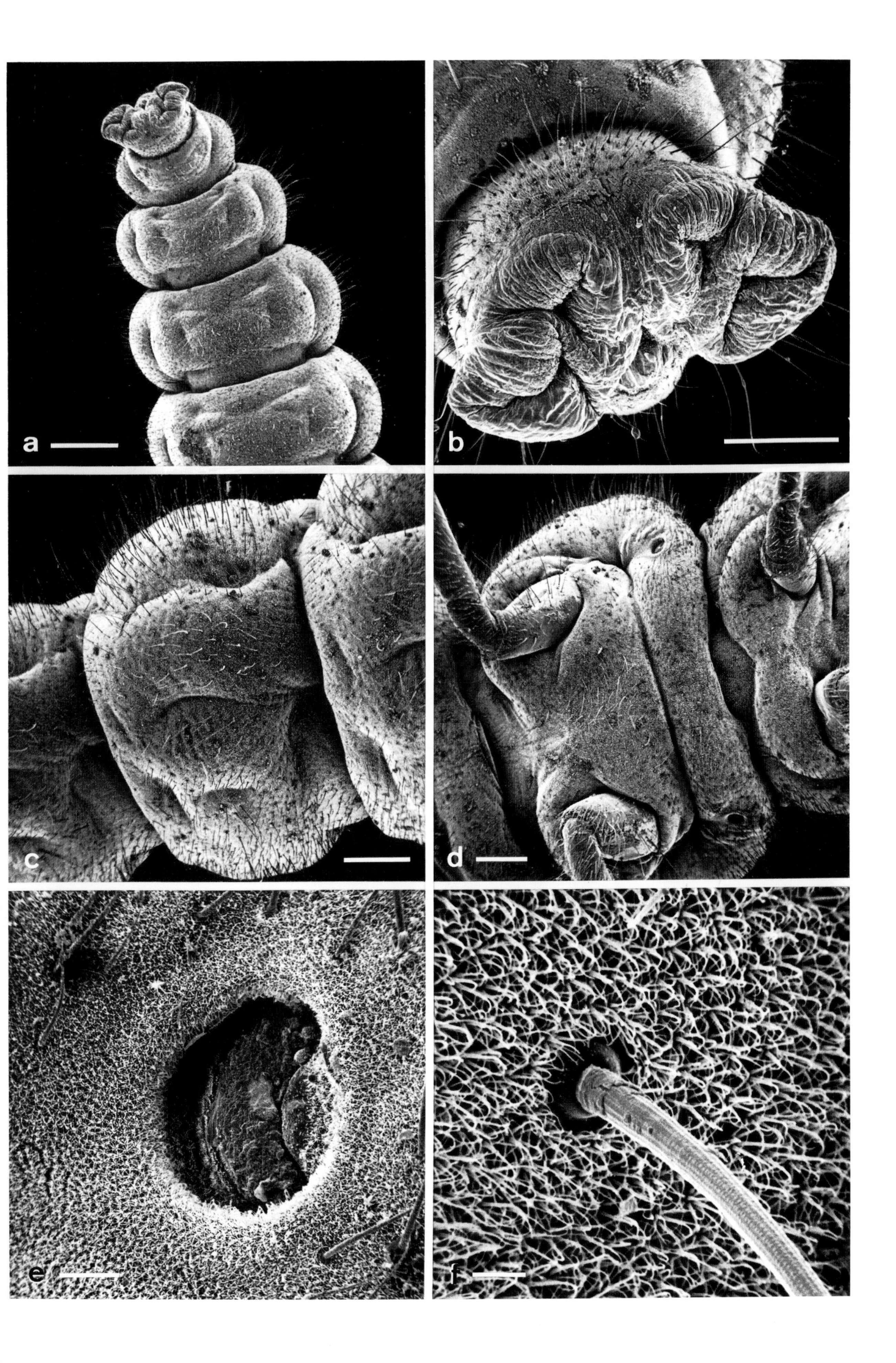
a
b
c
d
e
f

2.20.27 Larvae of the Click Beetles (Elateridae)

The shape of Elateridae larvae is well adapted to life in the soil. Their body is elongate, often cylindrical or dorso-ventrally flattened (Fig. 185; Plates 159 and 160). The body surface is rich in glands and, apart from the Cardiophorinae, strongly sclerotized, smooth and with sensory hairs projecting laterally. The legs are short and suitable for digging (Rudolph, 1970, 1974, 1978).

In studies of sense organs of the head based on electron microscopy, Zacharuk et al. (1977) and Zacharuk and Albert (1978) described digitiform sensilla on the labial palp and scolopophorous sensilla in the mandible of an elaterid larva, *Cternicera destructor*. On the labial palp there are typically six elongate pegs positioned in longitudinal grooves in the outer walls of the distal segment of the palp. They are termed digitiform sensilla on the basis of their form and function. Each has a subapical pore typical of a contact chemoreceptor and a terminally branched dendrite, but not the cuticular pores typical of a chemoreceptive porous hair. The dendritic endings are encased by the dendritic sheath, the subapical pore is inserted in the wall of the peg, and there is only one innervating neuron, all of which are typical of a tactile mechanoreceptor. These pegs respond electrophysiologically to contact and vibratory stimuli, but not to the amino acids, sugars, salts, and water tested, nor to changes of pressure in the cephalic hemocoel.

The mandibles of the larvae of *Cternicera destructor* have a large primary tooth with five, and a small medial tooth with one, scolopophorous sensilla of an amphinematic type. Each sensillum is innervated by two bipolar neurons, one with a large and the other with a small distal dendrite. These sensilla did not respond to chemical stimulation. Both nerve cells were stimulated by bending the mandibular tooth outward. One produced impulses with a higher amplitude and at a higher rate than the other. These sensilla are proprioreceptive mechanoreceptors.

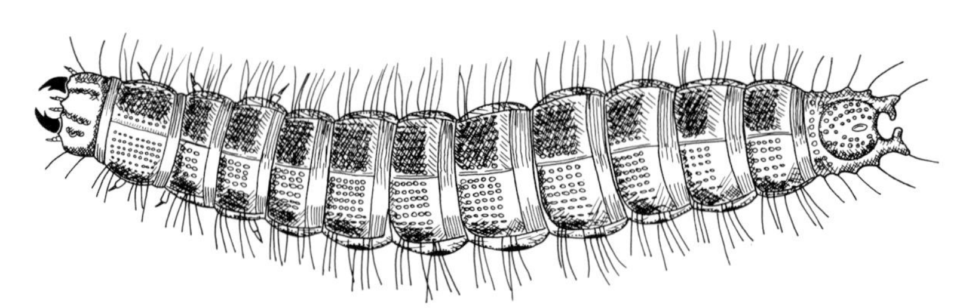

Fig. 185. Habitus of an Elateridae larva (*Athous* sp.), dorsal. (After Reiter, 1911)

Plate 159. Larvae of the Elateridae (click beetles).
a Larva, dorsal. The head is situated at the *right* edge of the picture. 1 mm.
b Anterior of the body, ventral. 1 mm.
c Head and prothorax, dorsal. 200 μm.
d Head and prothorax, oblique ventral. 200 μm.
e Labial palp, distal, covered with secretion. 10 μm.
f Antenna, dorsal. 25 μm

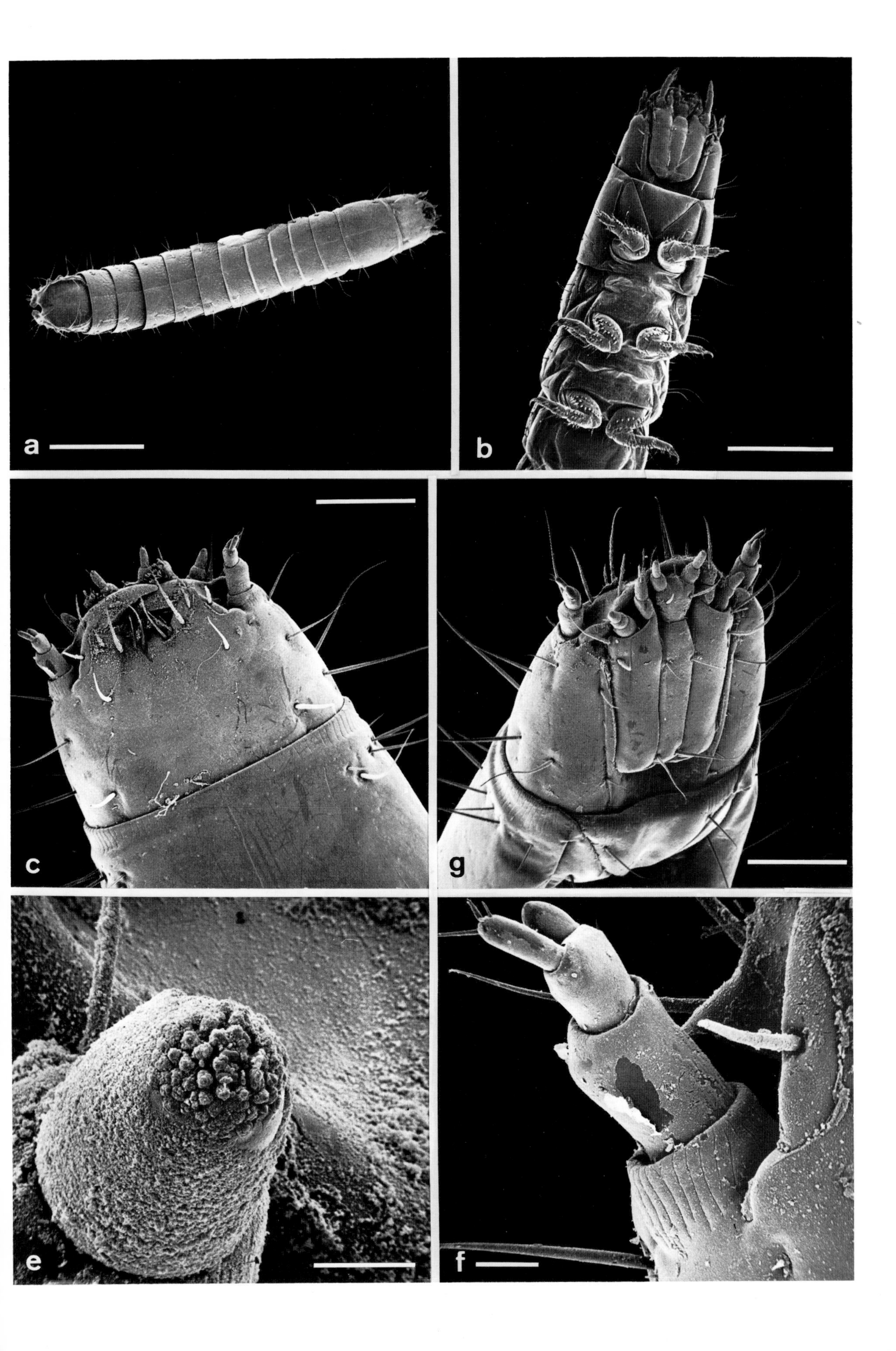
a
b
c
g
e
f

2.20.28 Importance of the Elateridae Larvae to Soil Biology

The development of the Elateridae from the egg to the imago often takes 5, sometimes up to 7, years. This period differs very much from species to species, and even intraspecifically. Besides species-specific endogenous factors, the duration is probably influenced by environmental conditions, such as climatic and nutritional factors. Therefore, the number of instars varies greatly, generally between 9 and 15 (Kosmaschewski, 1958; Strey, 1972; Rudolph, 1974).

Wireworms, as elaterid larvae are also called, live in rotting material, dead wood and epedaphically in the litter layer of woodland, but they are also regularly encountered in the cultivated soils of gardens and fields. They have diverse feeding habits: macrophytophagous and saprophytophagous, though some prey on insect larvae and pupae (Schaerffenberg, 1942; Subkew, 1934).

Strey (1972) drew attention to the bioproductive significance of polyphagous larvae, exemplified by *Athous subfuscus* larvae, which have a particularly broad spectrum of nutritional preferences, ranging from plants to small soil animals. Results of investigations in beech woods in Solling (W. Germany) show that these larvae are clearly of superior importance compared to the larvae of other soil beetles. The energy turnover of the population of these larvae is estimated to be 121 kcal $\times 10^3$/ha and is three to eight times higher than phytophagous beetles (Grimm, 1973; Schauermann, 1973) and about twelve to fifteen times higher than predatory carabids (Weidemann, 1972).

The larvae of the genus *Agriotes* often prefer cultivated soils which are enriched with humus and vegetable mould. Some species must be regarded as pests because they damage crops by feeding on plant roots, e.g. *A. lineatus* and *A. sputator*, in fields and gardens.

Plate 160. Larva of the Elateridae (click beetles).
a Pygidium (anal segment), dorsal. 250 μm.
b Abdomen, terminal, ventral, with pygidium and anal tube. 400 μm.
c Pygidium, dorsal, with glandular pores. 0.5 mm.
d Abdomen, terminal, lateral. 1 mm
e, f Glandular pores on the pygidium. 50 μm, 20 μm

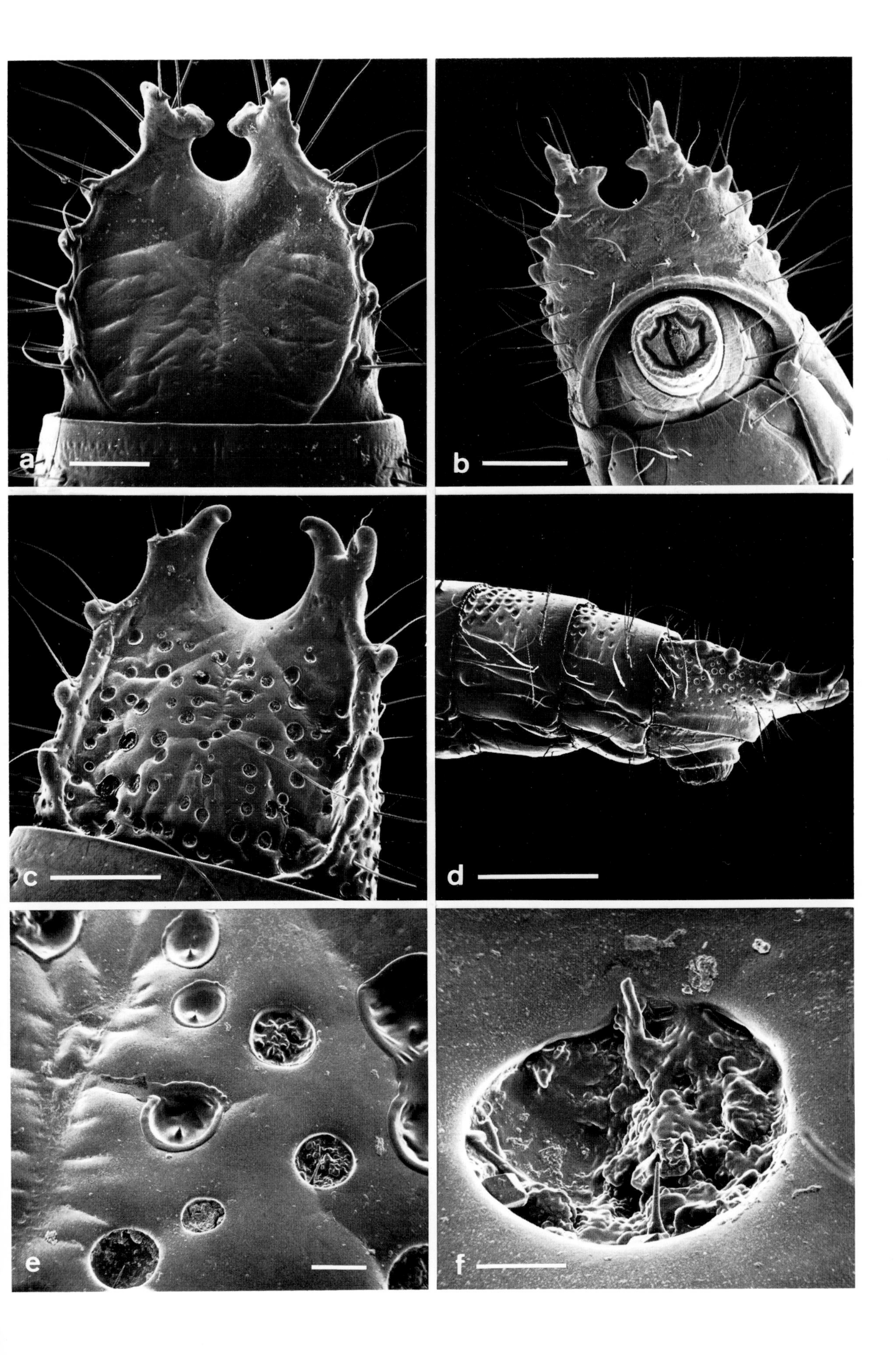

a
b
c
d
e
f

2.20.29 Burrowing Heteroceridae (Hide Beetles)

The few species, only 3–8-mm-long, which belong to the Heteroceridae family are burrowers which use their powerful mandibles and thorned fossorial legs as excavating equipment (Plate 161). They often live in communities with *Bledius* species (Staphylinidae) on the shore of the sea and of inland lakes and dig burrows similar to those of *Bledius*. Their common enemy are the predatory carabids of the genus *Dyschirius* (cf. Chaps. 2.20.10 and 2.20.18 and Plates 142 and 150).

Plate 161. *Heterocerus flexuosus* (Heteroceridae), hide beetles.
a Overview, lateral. 1 mm.
b Head, frontal. 250 μm.
c Head and thorax with fossorial legs, oblique, frontal. 0.5 mm.
d Structure of the cuticle. 25 μm.
e Fossorial legs with spiny tibiae. 200 μm

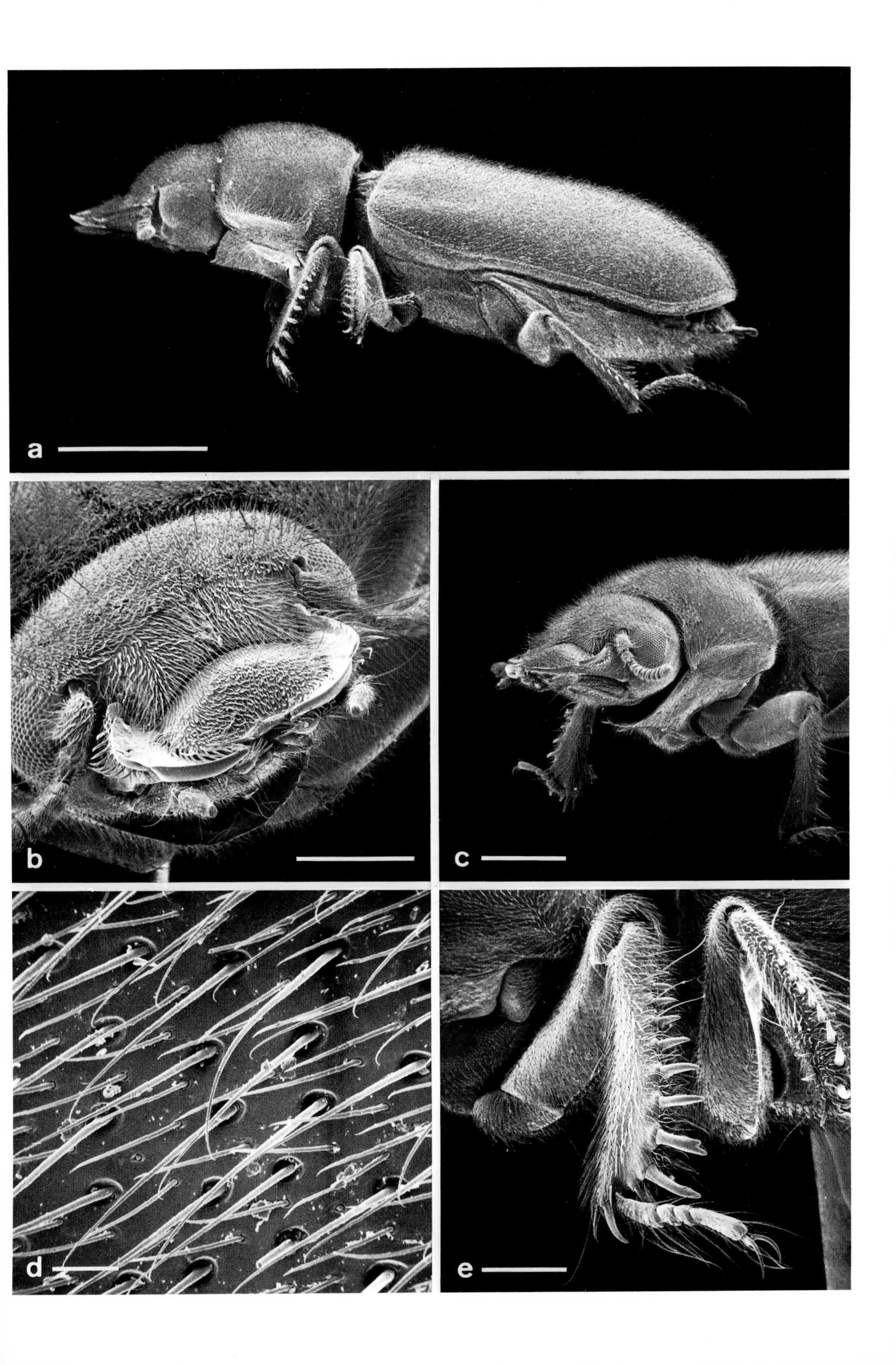
a
b
c
d
e

2.20.30 Coprophagous Dung Beetles (Scarabaeidae)

Diptera and beetles play an important role in the degradation of excrement; initially the Hydrophilidae, then later coprophagous dung beetles (*Aphodius, Copris, Geotrupes, Onthophagus,* etc.), feed on excrement with its flourishing fungal culture.

The shining metallic blue or green dung beetles of the genus *Geotrupes* (Plate 162) are attracted to the dung by its odour (anemotaxis). In the immediate vicinity they dig a brood shaft to a depth of 60 cm in the soil. From the main shaft the beetles excavate 15–20-cm-deep lateral shafts, into which the female stuffs up to 10 cm of dung; in each of these brood chambers she deposits one egg. Afterwards the entrances from the main shaft to the side shafts are sealed off with sand or earth. The emerging larvae feed on the rich supply of nutrition in these closed brood tunnels. They pupate in summer and subsequently emerge from the ground as young beetles via the main shaft (Fig. 186) (v. Lengerken, 1954).

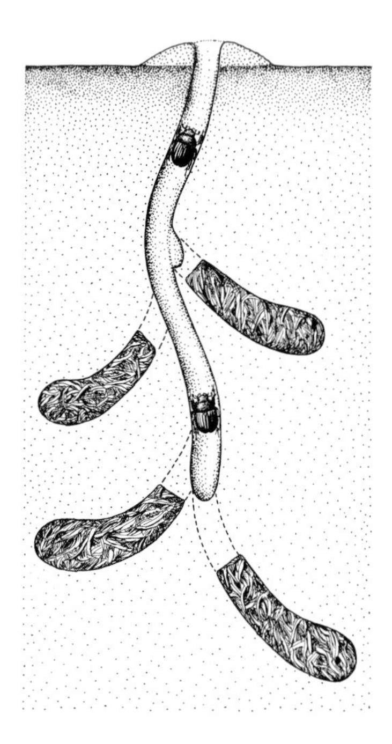

Fig. 186. Brood chambers of *Geotrupes stercorarius* (Scarabaeidae), dung or dor beetles, cockchafers. (After v. Lengerken, 1954). After completion of the brood chambers the lateral shafts are resealed

Plate 162. *Geotrupes stercorarius* (Scarabaeidae), dung or dor beetles, cockchafers.
a Anterior of the body, ventral. 3 mm.
b Head, dorsal. 1 mm.
c Antennal club, dorsal. 200 µm.
d Head and prothorax, lateral. 1 mm.
e Tibia and tarsus of the fossorial leg (front leg). 800 µm.
f Surface of the ommatidia. 10 µm.
g Sensory hairs on the antennal tip. 20 µm

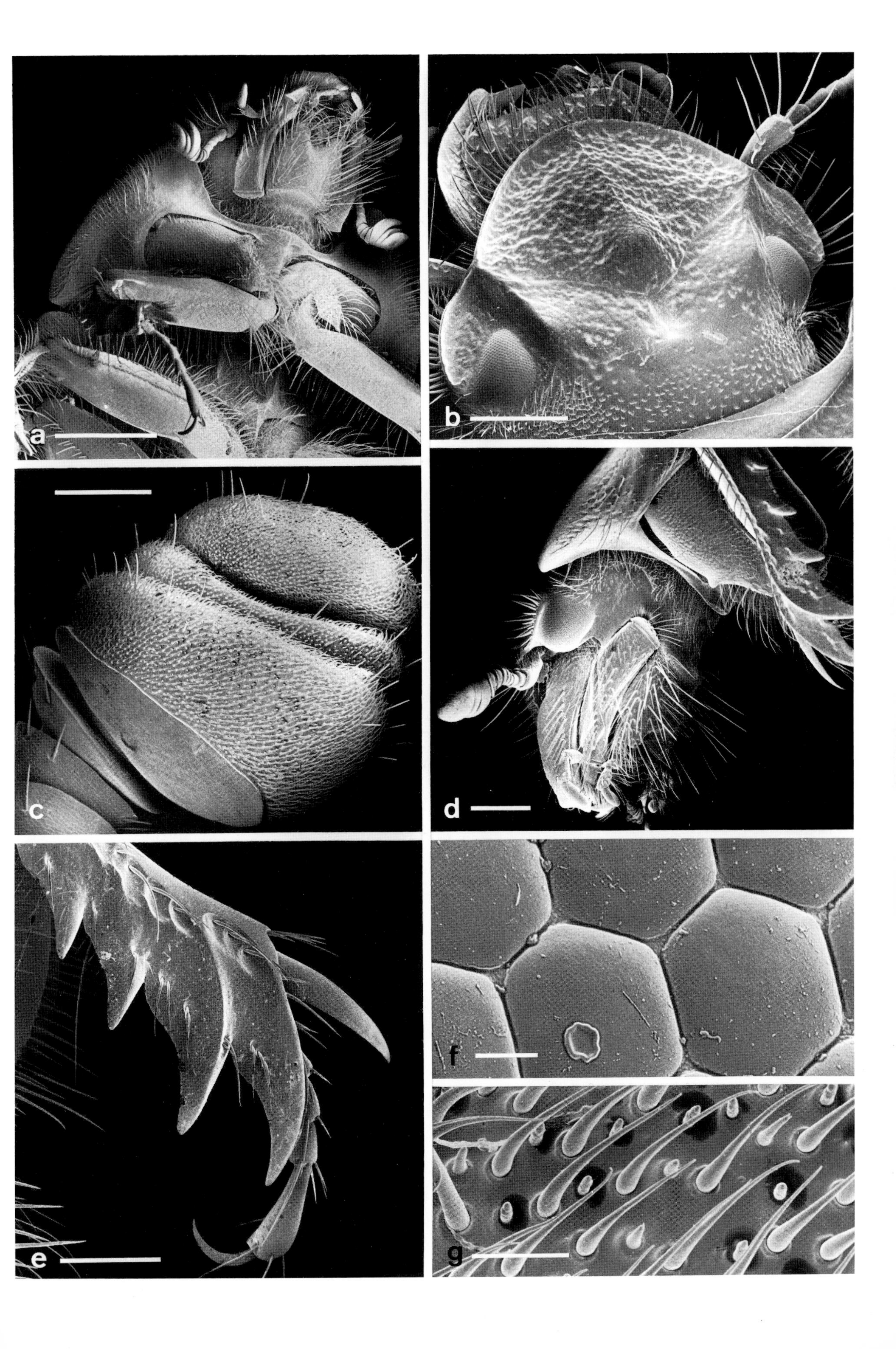
a
b
c
d
e
f
g

2.20.31 Rhizophagous Weevils (Curculionidae)

The larvae of the weevils, a family rich in species, feed on plants and are often restricted to certain host and food plants (Scherf, 1964). Weevil larvae often feed on roots and are endophytic, rarely ectophytic. The only indirectly belong to the community of the soil arthropods. As rhizophagous consumers they do not generally contribute to the decomposition of plant debris, but are merely associated with the soil habitat through their consumption of roots.

However, considering the total energy turnover and especially the total biomass dynamics of the soil, particularly with regard to the soil-emerging pterygote insects, the rhizophagous curculionids show a remarkable emergence, or "production of imagines" (Funke, 1971). With an abundance of 128–463 ind/m^2 and a biomass of 125–316 mg dry weight/m^2, they comprise 12.4% of the soil-emerging pterygote insects of a beech forest in Solling (W. Germany) (Fig. 187) (Grimm, 1973; Schauermann, 1973, 1977).

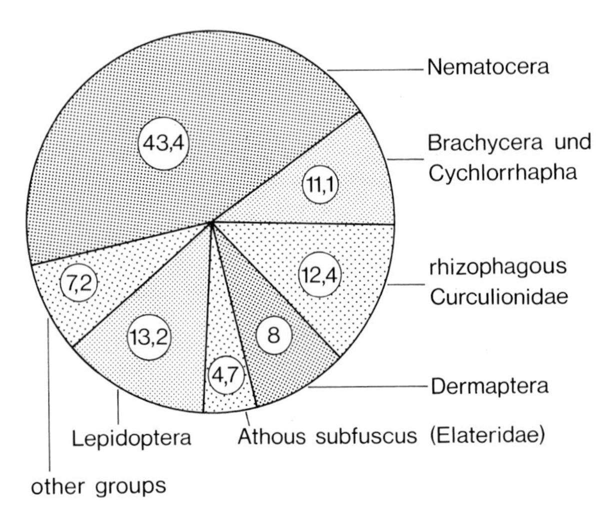

Fig. 187. The percentage proportion of pterygote insects which emerge from the soil of an acid humus beech woodland (old beech trees) expressed as "production of imagines" and calculated from the biomass caught in photo-eclectors/m^2/year. (After Schauermann, 1977)

Plate 163. *Apion* sp. (Curculionidae), weevils.
a Overview, lateral. 250 µm.
b Overview, frontal. 250 µm.
c Proboscis, distal, with mouthparts. 50 µm.
d Antenna, distal. 25 µm.
e Compound eye. 50 µm

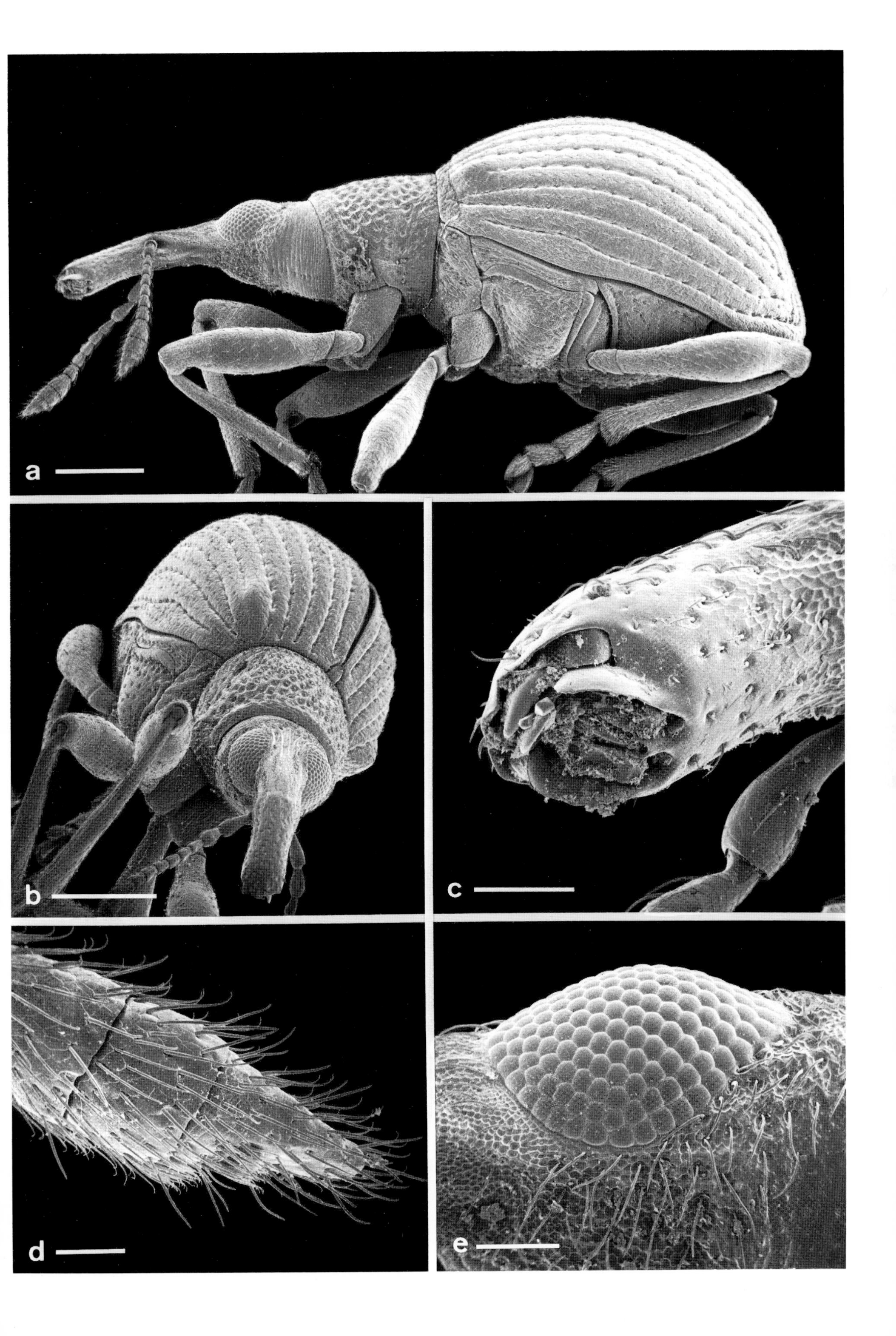
a
b
c
d
e

2.21 Order: Hymenoptera (Insecta)

General Literature: Forel, 1874; Escherich, 1917; Dornisthorpe, 1927; Bischoff, 1927; Stitz, 1939; Gösswald, 1951; Otto, 1962; Bernard, 1968; Brian, 1977; Dumpert, 1978; Gösswald, 1985

2.21.1 Soil-Dwelling Hymenoptera

Among the Hymenoptera some groups can be regarded as important soil-dwellers, especially those groups which generally construct their nests in the earth, e.g. vespoid wasps (Vespoidea), spider wasps (Pompiloidea), sphecoid wasps (Sphecoidea), bees (Apoidea) and ants (Formicoidea).

Of undoubted importance to soil biology are ants, which form colonies. Their nests have a high individual population, so that there is an enormous turnover of matter within the radius of activity around the nest.

The usefulness of the wood ants (*Formica rufa* and *F. polyctena*) (Fig. 188; Plates 164–166) has been repeatedly emphasized (Gösswald, 1951, 1985; Brauns, 1968; Dunger, 1983). They play a significant role in the decimation of growing populations of insect pests like the pine lappet (*Dendrolimus pini*), the bordered white beauty moth (*Bupalus piniarius*), the pine beauty (*Panolis flammea*) and the black arches (*Lymantria monacha*). In addition, common sawflies, cockchafer, weevils, scolytid beetles, leaf rollers, etc. are all brought regularly into the nest as prey. At least 100000 insects are killed daily in a large nest of wood ants (Escherich, 1917).

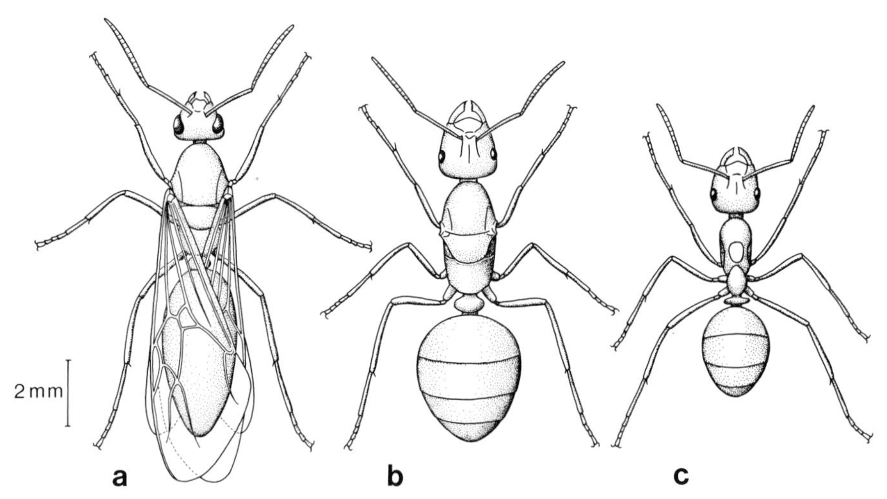

Fig. 188. Castes of the wood ant *Formica polyctena* (Formicidae) (Dumpert 1978). **a** Male, winged. **b** Queen, after shedding her wings. **c** Worker

Plate 164. *Formica polyctena* (Formicidae), wood ant, worker.

a Anterior of the body with caput (head), thorax, petiolus with lobe and anterior part of the gaster (abdomen), lateral. 1 mm.

b Anterior of the body, dorsal. 1 mm.

c Epinotum (*arrow*), lobed petiolus and anterior part of the gaster, dorsal. 250 μm.

d Epinotum (*arrowhead*), petiolus with lobe and the anterior part of the gaster, lateral. 250 μm.

e Epinotum and mesonotum (*arrowhead*) with spiracles, dorsal. 250 μm.

f Spiracle on the posterior margin of the mesonotum. 25 μm.

g Spiracle on the epinotum with spiracular hairs forming a filter in the opening leading to the atrium. 25 μm

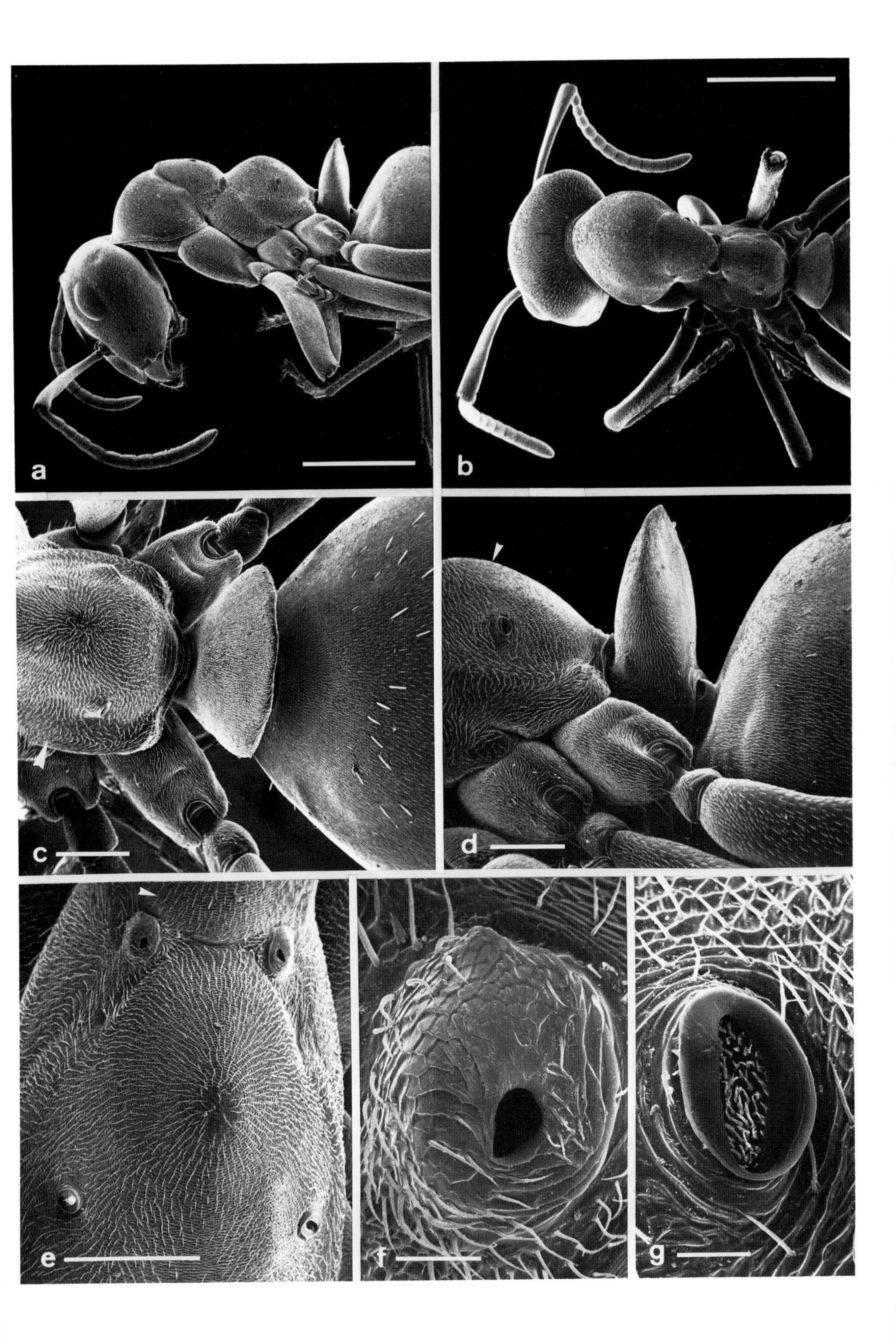
a
b
c
d
e
f
g

2.21.2 The Wood Ant in the Woodland Ecosystem

The wood ant often plays a key role with regard to the turnover of matter in the forest ecosystem. The dense population in a nest – a middle-sized colony comprises 500000 to 800000 individuals – concentrates a considerable amount of organic substances into a small area. As gatherers, harvesters or predators they bring honeydew from lachnids, seeds, leaves, etc. or animal prey from a large area into the nest. Plant debris is produced as a by-product of the construction and constant alteration of their nests using plant material. Together with carrion and excrement this plant debris, which is mechanically fragmented by the ants, forms the concentrate which is rapidly and completely mineralized by soil bacteria and fungi at the favourably warm temperatures inside the wood ant nests (Fig. 189) (Kloft, 1959, 1978; Otto, 1962; Kneitz, 1974).

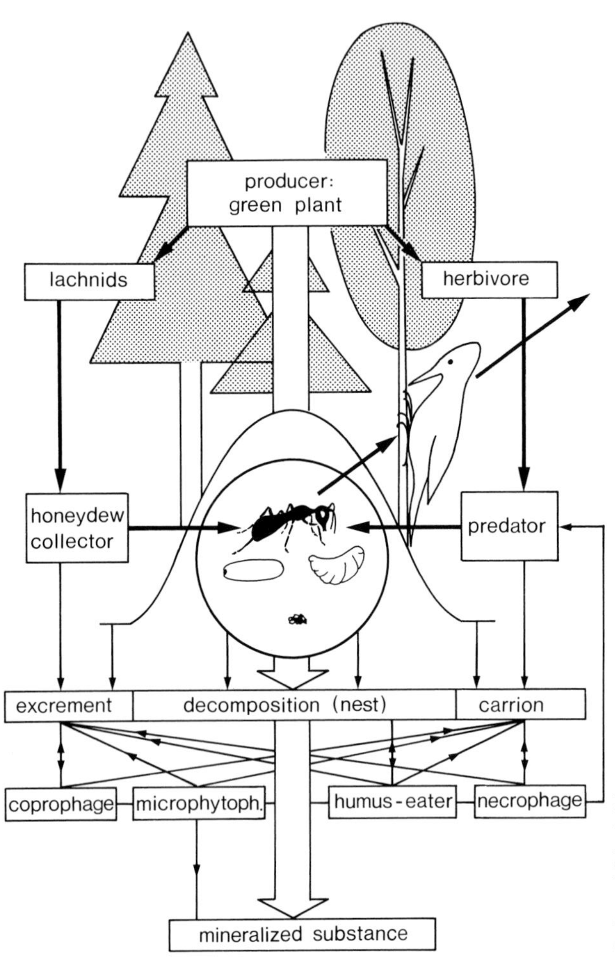

Fig. 189. Schematic illustration showing the influence of a wood ant colony on the turnover of matter of a central European forest ecosystem. The organic substance produced by plants is initially processed by the primary consumers. Ants act as secondary consumers, e.g. as food gatherers (honeydew) or as predators. (After Kloft, 1978)

Plate 165. *Formica polyctena* (Formicidae), wood ant, worker.

- **a** Antenna with long scapus and poly-segmented flagellum (funiculus). 0.5 mm.
- **b** Portion from the middle part of the funiculus. 50 μm.
- **c** Joint between the scapus and the head capsule with cushion of proprioreceptive hairs to perceive the position of the feeler. 50 μm.
- **d** Funiculus, distal. 100 μm.
- **e** Scapus-funiculus joint. 100 μm.
- **f** Pattern of sensilla with squamiform and hair sensilla on the funiculus, distal. 10 μm

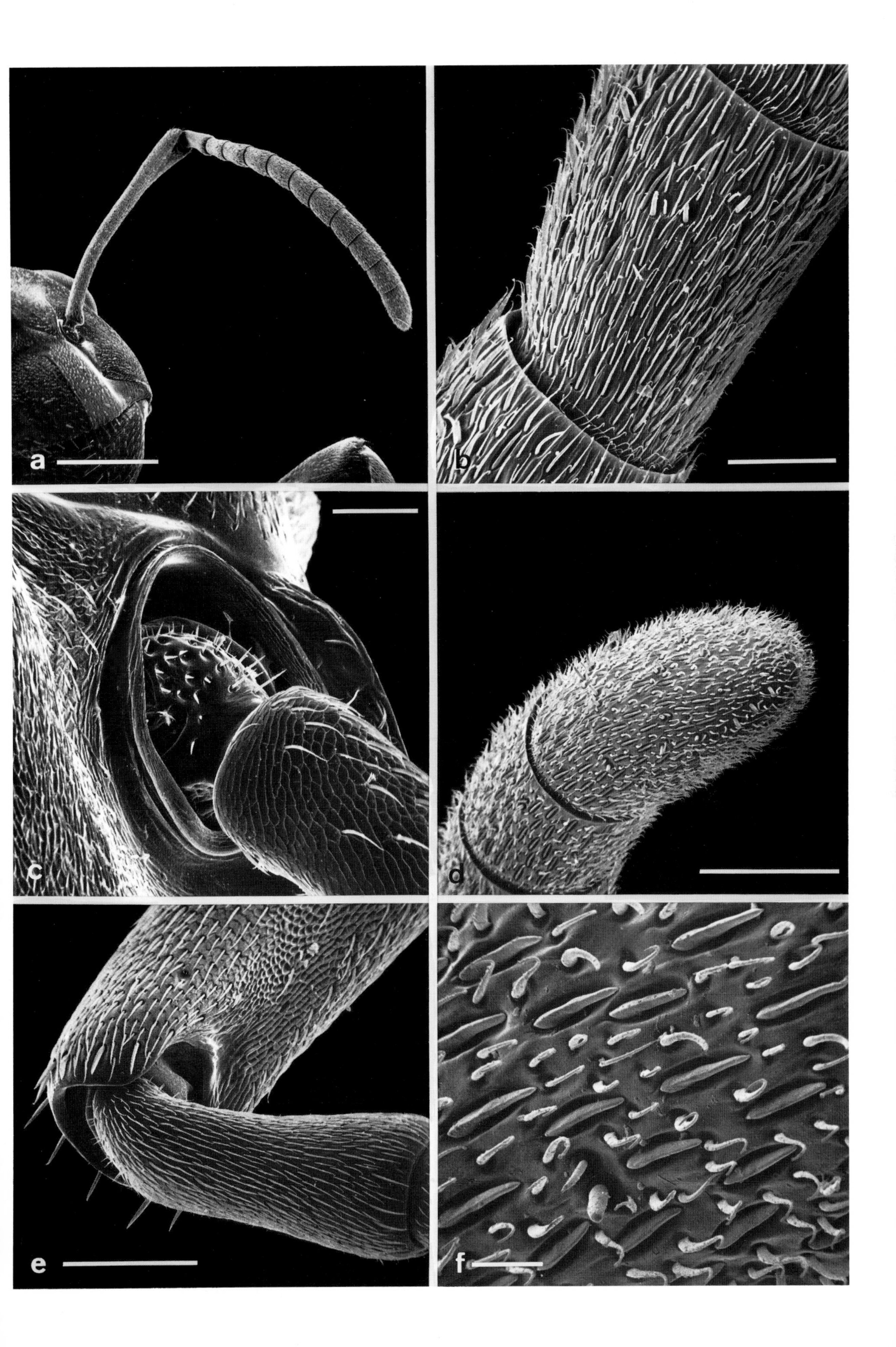
a
b
c
d
e
f

2.21.3 Subterranean Ant Nests

A continuous development of nest shapes from simply constructed subterranean nests, through domed anthills and wood nests, to carton nests can be observed. All these types are associated with the soil. However, certain carton nests, which hang freely from twigs and sit on leaves (*Azteca* and *Crematogaster*) can be regarded as the most highly developed forms.

Subterranean nests are often constructed without any recognizable plan and consist of a maze of more or less irregular tunnels which are connected to the surface by several exits. The passageways are occasionally enlarged to form chambers which the ants use as nurseries and storehouses. Thus, the ant *Ponera coarctata* lives hidden in narrow tunnels deep underground and feeds on small insects and mites.

In domed nests the subterranean nest forms a mound above ground level. Construction material from the soil and plant fragments, such as blades of grass, twigs, leaves and needles, are piled up to form a mound (*Lasius niger*) or the large hill of *Formica rufa* which sometimes reaches a height of 1.5 m (Figs. 191 and 192). The size and form of the hill of the red wood ant is not only dependent on the population, but also on the microclimate, especially humidity and warmth. Pointed, conical anthills absorb warmth more readily in the shade of a forest than flatter constructions (Gösswald, 1951; Kloft, 1959; Otto, 1962).

Camponotus herculeanus, the giant wood ant, makes tunnels and chambers in the trunks of living spruce and pine conifers by gnawing spring wood and letting late wood form the walls of its chambers. The nests in the root system are connected over a large area at the contact points between the main roots of the trees (Kloft et al., 1965).

Other ants (e.g. *Lasius fuliginosus*, jet ant) extend their subterranean nest under hollow tree stumps. It is converted to a winter nest by cementing together wood fragments to construct a bizarre carton nest in the hollow stump (Maschwitz and Hölldobler, 1970).

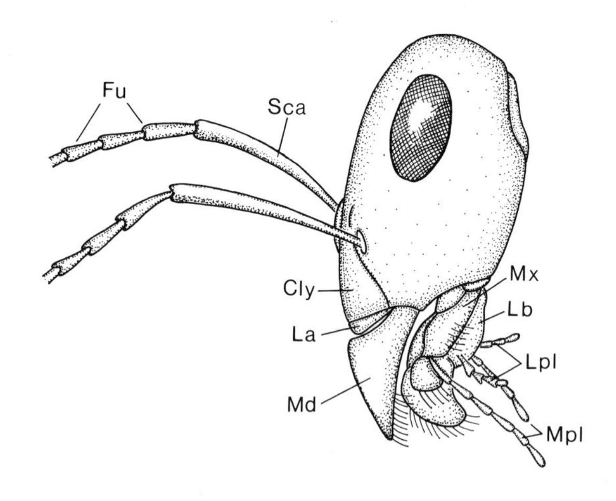

Fig. 190. Structure of the head in *Formica lugubris,* Formicidae. (After Sudd, 1967 from Dumpert, 1978). Head of a worker, schematic. *Cly* Clypeus; *Fu* funiculus; *La* labrum; *Lb* labium; *Lpl* labial palps; *Md* mandible; *Mpl* maxillary palps; *Mx* maxilla; *Sca* scapus

Plate 166. *Formica polyctena* (Formicidae), wood ant, worker.

a Head, lateral. 250 µm.

b Frontal view of head with three ocelli on the frons. 250 µm.

c Mandibles and palps, lateral. 250 µm.

d Mouthparts, ventral. 250 µm.

e Compound eye with single hairs between the ommatidia. 100 µm.

f Corneal surface of the ommatidia. 5 µm

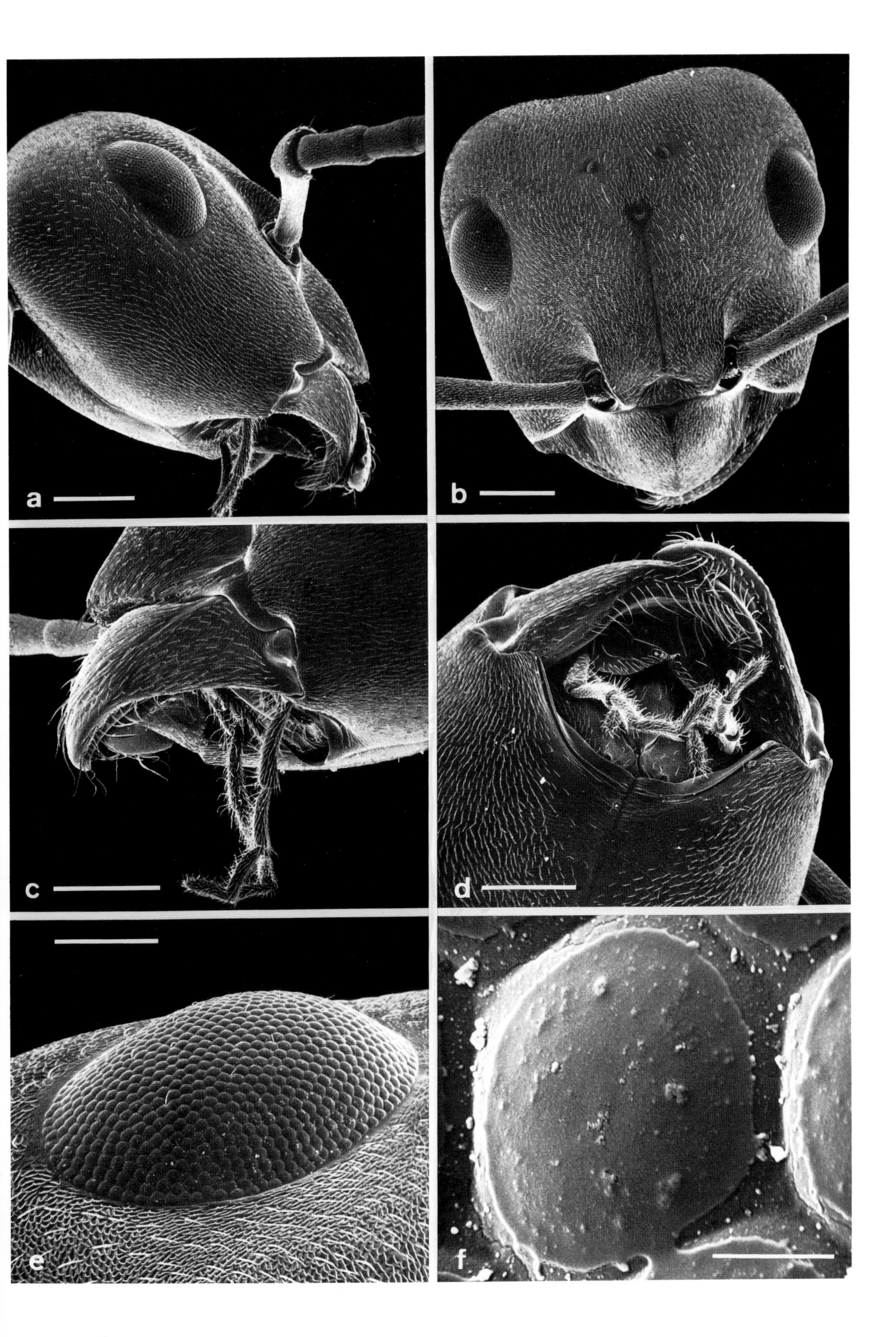
a
b
c
d
e
f

2.21.4 The Erratic Ant *Tetramorium caespitum*

In contrast to the wood ant, the erratic ant prefers dry, sunny biotopes, but does not always avoid moist soil and also occurs in meadows, fields and gardens (Stitz, 1939).

In principle, wood ants and erratic ants have the same shape of nest, but the mounds differ in their prominence. Wood ants pile up an impressively high anthill (Fig. 192), while the major part of the erratic ant nest lies underground so that only an insignificant mound is formed, often only a few centimetres higher than ground level. Furthermore, the nests are adapted to local surroudings, frequently lying under stones which are warmed by the sun. The nests of the erratic ant often resemble the small ones constructed by the small black ant, *Lasius niger* (Fig. 192).

Fig. 191. Section of the nest constructed by wood ants. It has a mound above ground and a subterranean system of tunnels. (After Dircksen and Dircksen, 1968).
1 Two workers expose the brood to the sun. **2** Two workers stroke each other's antennae as a means of communication. **3** Guarded egg chamber. **4** Prey (a grasshopper) is transported into the nest. **5, 6** Subterranean nurseries with maggot-like larvae (**5**) and pupae (**6**).

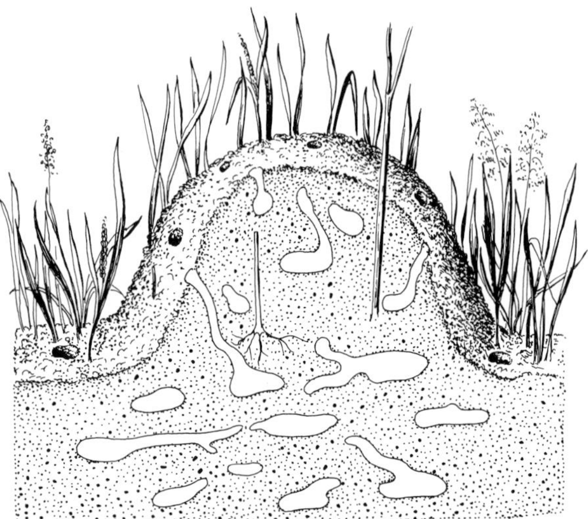

Fig. 192. Domed nest of the small black ant, *Lasius niger* (Formicidae), in a meadow. (After Goetsch, 1953)

Plate 167. *Tetramorium caespitum* – Mirmicidae (knot ants), erractic ant, worker.

a Habitus, dorsal. Thorax and gaster are connected by a double segmented, knotty petiolus. 0.5 mm.
b Overview, lateral. 0.5 mm.
c Head, oblique, frontal. 250 µm.
d Head, dorsal. 250 µm.
e Head with mandibles and antennal notches. 100 µm.
f Petiolus with postpetiolus (*arrow*) and gaster. 100 µm.

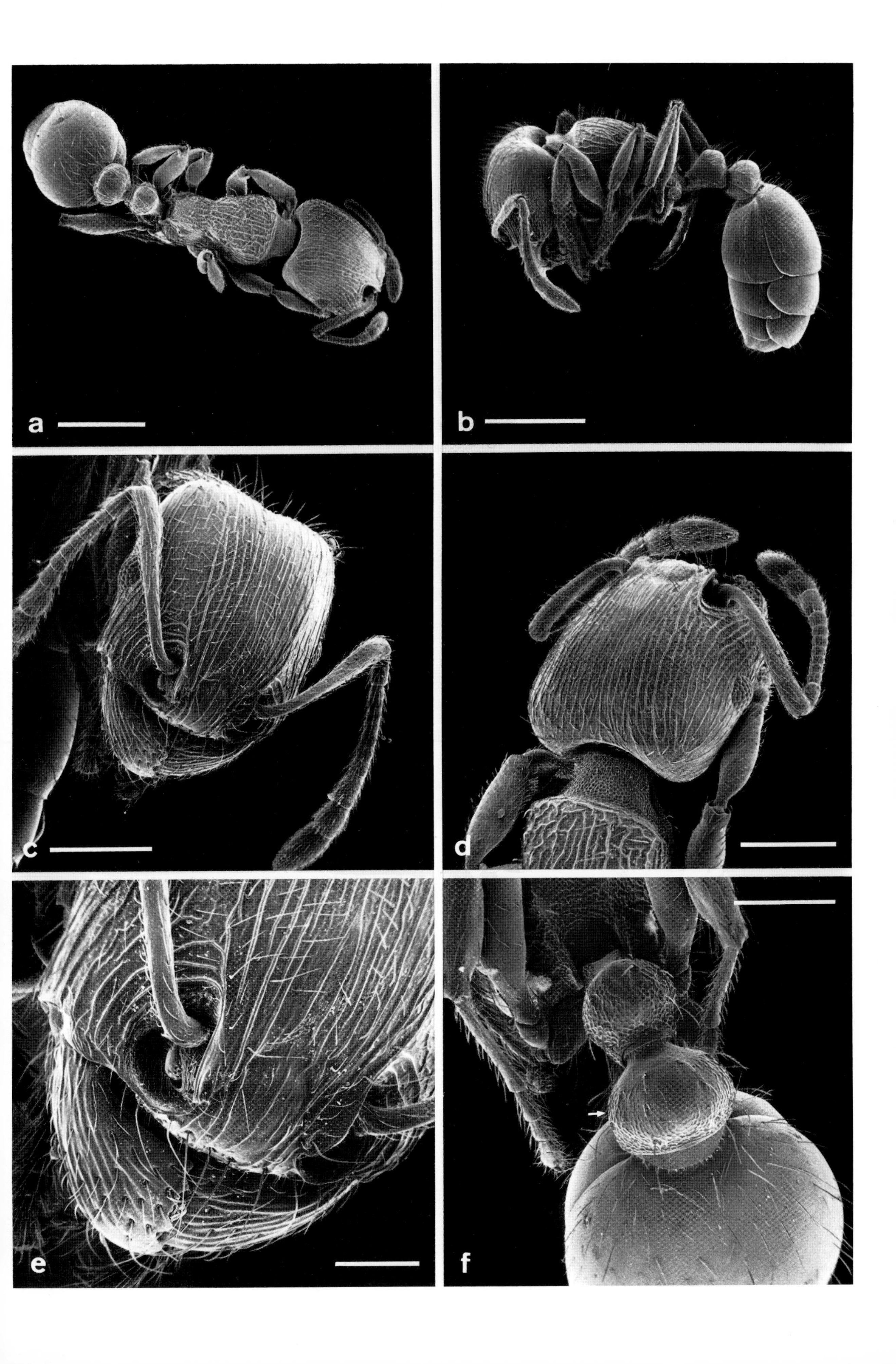
a
b
c
d
e
f

2.22 Order: Trichoptera – Caddisflies (Insecta)

GENERAL LITERATURE: Hickin, 1967; Malicky, 1973, 1983; Wichard, 1978; Wiggins, 1977

2.22.1 The Terrestrial Caddisfly: *Enoicyla pusilla* (Limnephilidae)

Only a few caddisflies are terrestrial, most are aquatic. The caddisfly *Enoicyla pusilla* (Plates 168–170) is a soil-dwelling Limnephilidae distributed in western European regions influenced by an Atlantic maritime climate. As a primary decomposer it participates in the degradation of plant debris in the soil.

Besides its terrestrial life, *Enoicyla pusilla* is distinguished by its marked sexual dimorphism (Fig. 193). The female, having only inconspicuous rudimentary wings, is relatively immobile and is restricted to crawling on the ground. Although the males have normal wings, they show a limited ability for dispersal, merely flying short distances immediately above the soil surface from one blade of grass to the next (Rathjen, 1939; Kelner-Pillault, 1960; Mey, 1983). In the Nearctic region a few limnephilids have a short terrestrial phase during their larval development. This terrestrial mode of life has been observed in *Ironoquia parvula* (Flint, 1958), in *Philocasca demita* (Anderson, 1967), and in *Desmona bethula* (Erman, 1981). The caddisfly larva, *Desmona bethula*, lives in slowly flowing, unshaded stretches of small spring streams in the Sierra Nevada, California. On nights in early summer the fifth instar leaves the water and feeds on several species of semiaquatic plants. It returns to the water when air temperatures approach freezing and on warm, still nights it stays out until dawn. These larvae may represent an evolutionary link between the few secondarily terrestrial Limnephilidae larvae and the majority of aquatic caddisfly larvae (Erman, 1981).

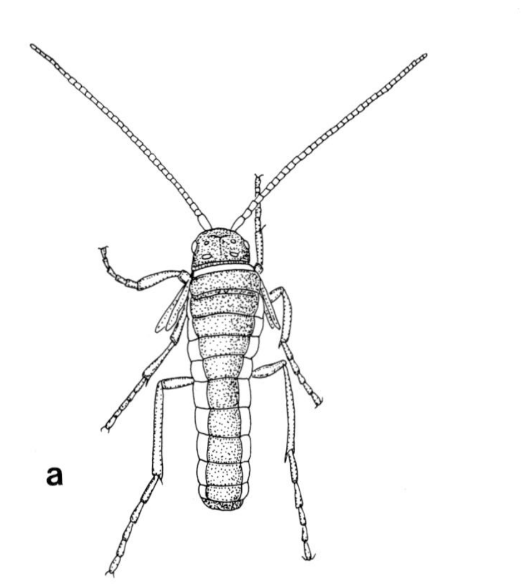

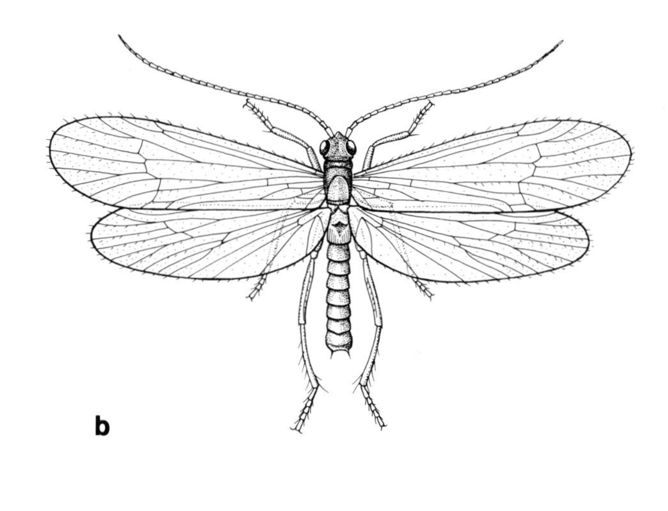

Fig. 193. *Enoicyla pusilla* (Limnephilidae). (After Rathjen, 1939). **a** Female, after oviposition; **b** male

Plate 168. *Enoicyla pusilla* (Limnephilidae), larva.
a Case. 1 mm.
b Larva in case, frontal view. 0.5 mm.
c Head, oblique, ventral, with larval eye, upper and lower lip. 200 µm.
d Upper lip (labrum, *arrowhead*) and maxillo-labium with palps. 50 µm.
e Larval eye with stemmata. 25 µm.
f Bulbiform sensilla on the maxillary palp. 1 µm

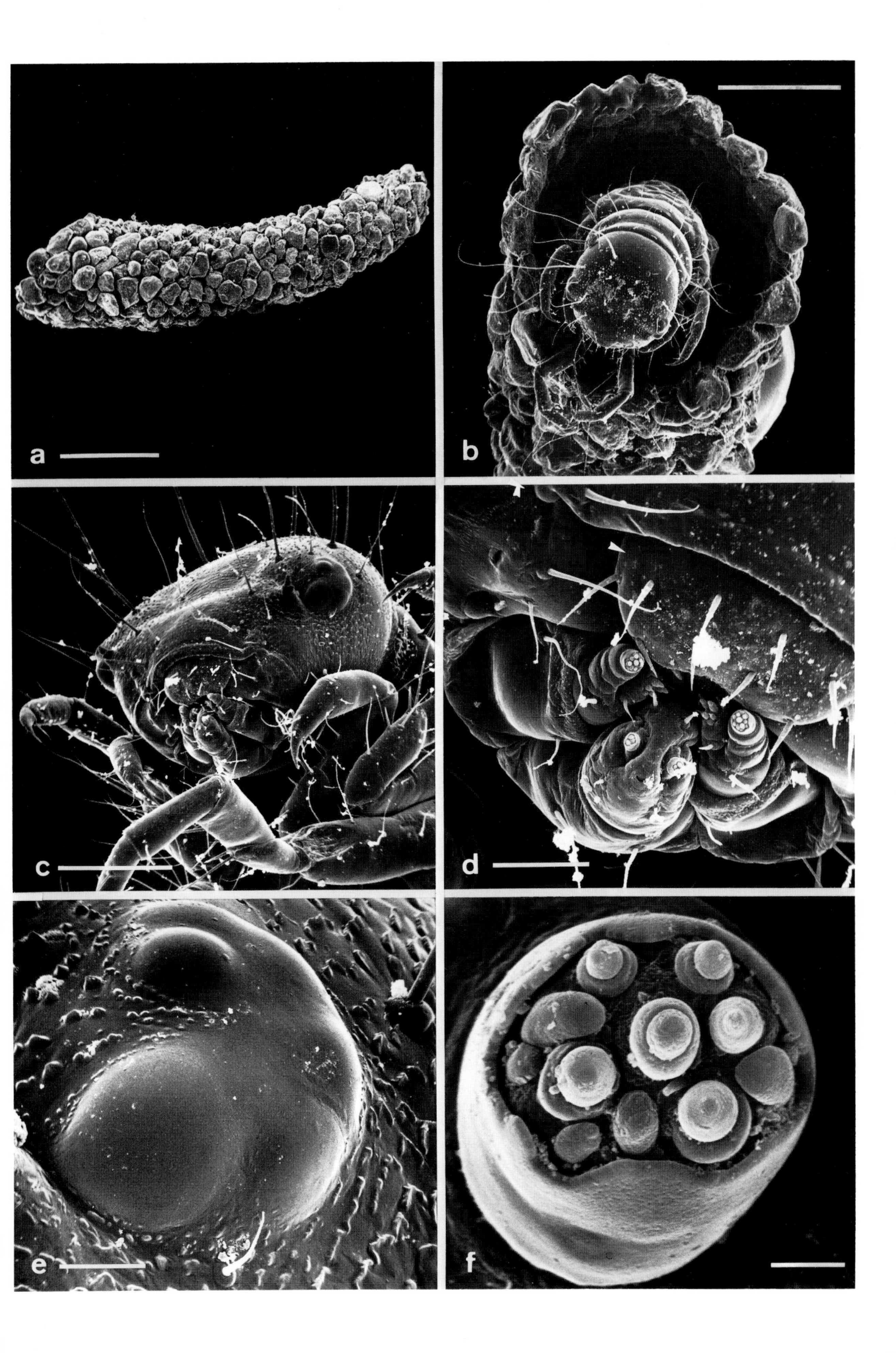
a
b
c
d
e
f

2.22.2 Importance of *Enoicyla pusilla* to Soil Biology

In early autumn the females deposit their "spawn" in moist places in the leaf litter or in moss. There, the gelatinous mass swells to form a clutch approximately 5 mm in size which generally contains 30–50, and occasionally up to 100 eggs. The larvae emerge about 3 weeks later and begin constructing conical cases out of small stones cemented together with threads from the labial and spinning glands. The interior of the case is also lined with silk (Fig. 196; Plates 168 and 169).

The larvae pass through five stages in the course of their development which lasts from October to July (Fig. 194). During this period their activity (measured with pitfall traps, Späh, 1978) and their food requirements constantly increase with their growth and with successive instars. They feed mainly on fallen leaves, leaving traces like many other primary decomposers, which consume the soft parts of the leaves and leave the vascular strands intact (Fig. 195). Moreover, the abundance of the larvae in soil litter is remarkable. According to Späh (1978) the average abundance of living larvae in an alder marsh in Westphalia (W. Germany) is 165 ind/m^2 with 720 empty cases/m^2; Drift and Witkamp (1960) reported abundances of 200–1200 larvae/m^2 in oak woodlands in the Netherlands.

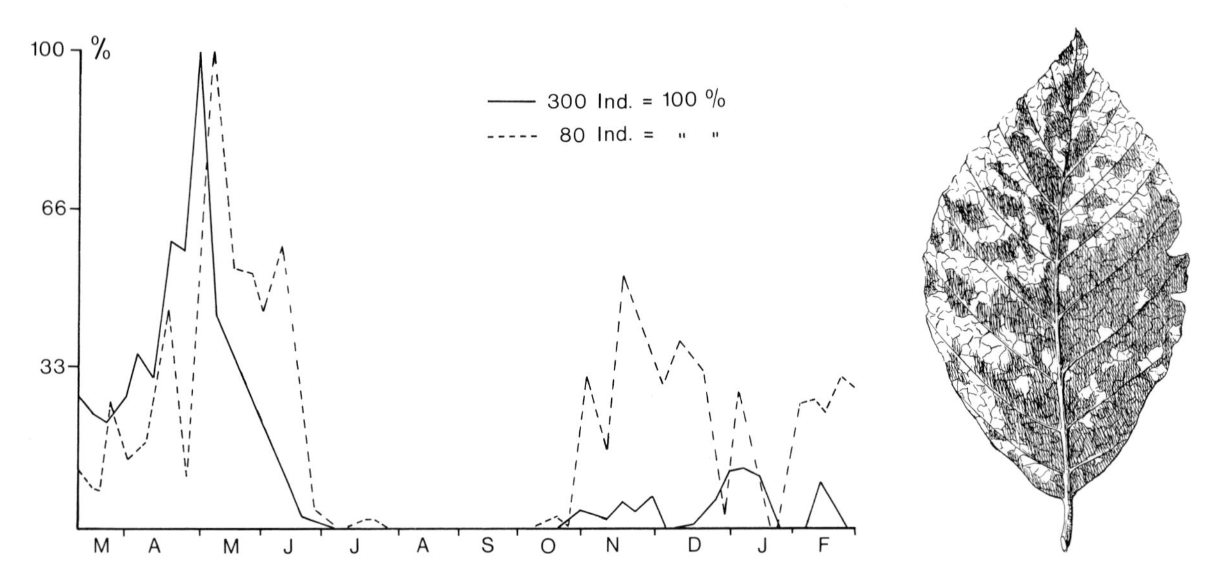

Fig. 194. *Enoicyla pusilla* (Limnephilidae). Phenology of a larval population (1975–1977) in the litter of an alder marsh. The activity of the larvae is limited to the period from October to July. Trapping method: pitfall traps. (After Späh, 1978)

Fig. 195. Traces of feeding left by *E. pusilla* larvae on a beech leaf from the litter layer. (After Brauns, 1964)

Plate 169. *Enoicyla pusilla* (Limnephilidae), larva.
a Outer surface of the case. 200 μm.
b Inner surface of the case. 200 μm.
c Outer surface of the case with cement material. 100 μm.
d Interior of the case with silk lining. 20 μm.
e Outer surface of the case with cementing threads. 20 μm.
f Interior of the case with silk lining. 10 μm

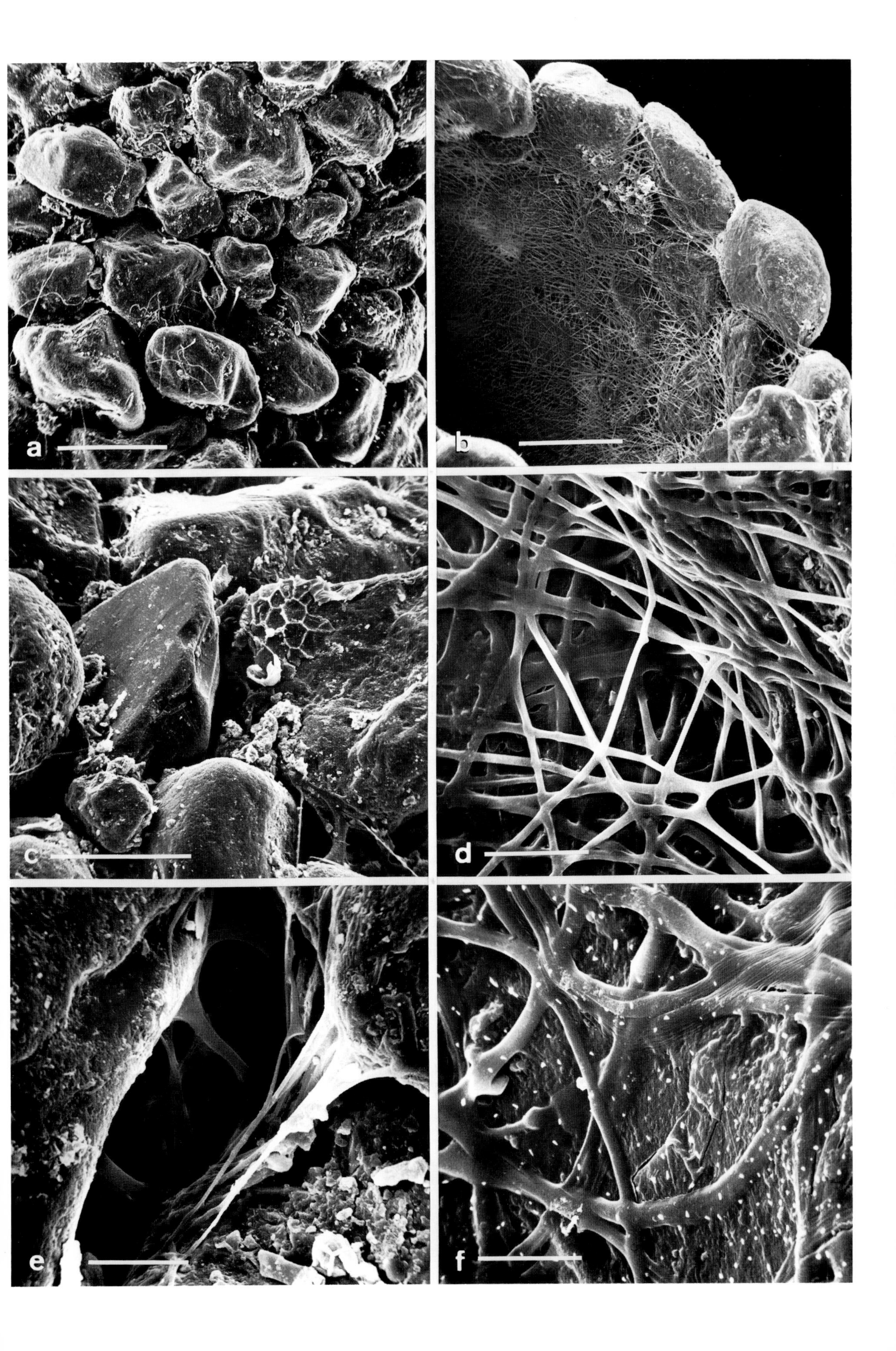

a
b
c
d
e
f

2.22.3 Adaptation of *Enoicyla pusilla* to the Terrestrial Biotope

Osmoregulation and respiration are vital physiological mechanisms for the ecological adaptation of aquatic insects. The terrestrial *Enoicyla pusilla* has lost these special aquatic adaptation mechanisms. The caddisflies which live in water have a closed tracheal system. The limnephilids, to which *Enoicyla pusilla* belongs, are especially suited to aquatic life due to the number and distribution of their abdominal tracheal gills, as well as the functional morphology of their respiratory epithelium (Wichard, 1973, 1974). *Enoicyla* larvae lack tracheal gills (Plate 170); but the tracheal system is closed as in the aquatic limnephilids. Thus, they exhibit cuticular tracheal respiration.

Furthermore, aquatic limnephilid larvae possess ion-absorbing chloride epithelia which participate in ionic and osmotic regulation. These are recognizable as round or oval areas on the sternites of the abdominal segments, varying in number and distribution according to the group. When the aquatic biotope is forsaken these structures for osmoregulation become superfluous. Consequently *Enoicyla pusilla* does not possess chloride epithelia (Wichard and Komnick, 1973; Wichard and Schmitz, 1980).

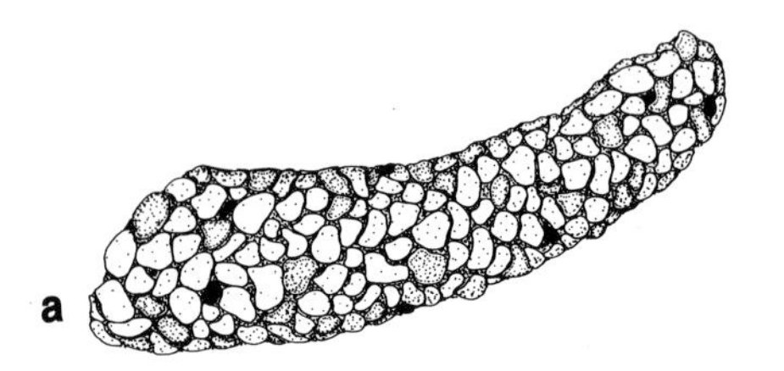

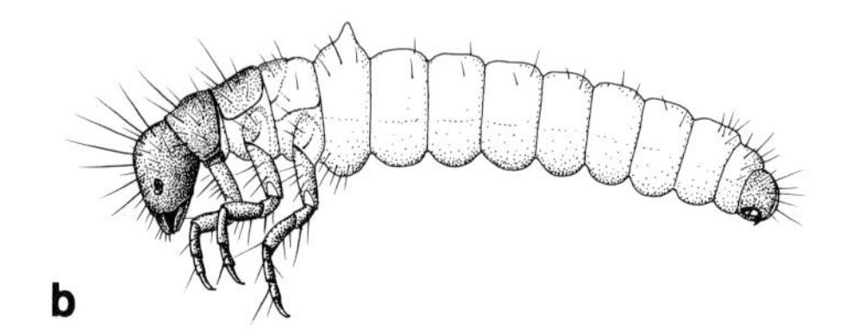

Fig. 196. *Enoicyla pusilla* (Limnephilidae). **a** Case; **b** Larva, lateral

Plate 170. *Enoicyla pusilla* (Limnephilidae), larva.
a Larva extracted from its case, lateral. 1 mm.
b Head and thorax, lateral, with larval eye. 200 µm.
c Overview from the caudal side with dorsal hump and anal valves. 250 µm.
d Dorsal hump on the first abdominal segment, lateral. 100 µm.
e Abdomen, caudal, lateral, with anal valves and abdominal leg. 200 µm.
f Abdominal leg (pygopodium with terminal claw). 50 µm

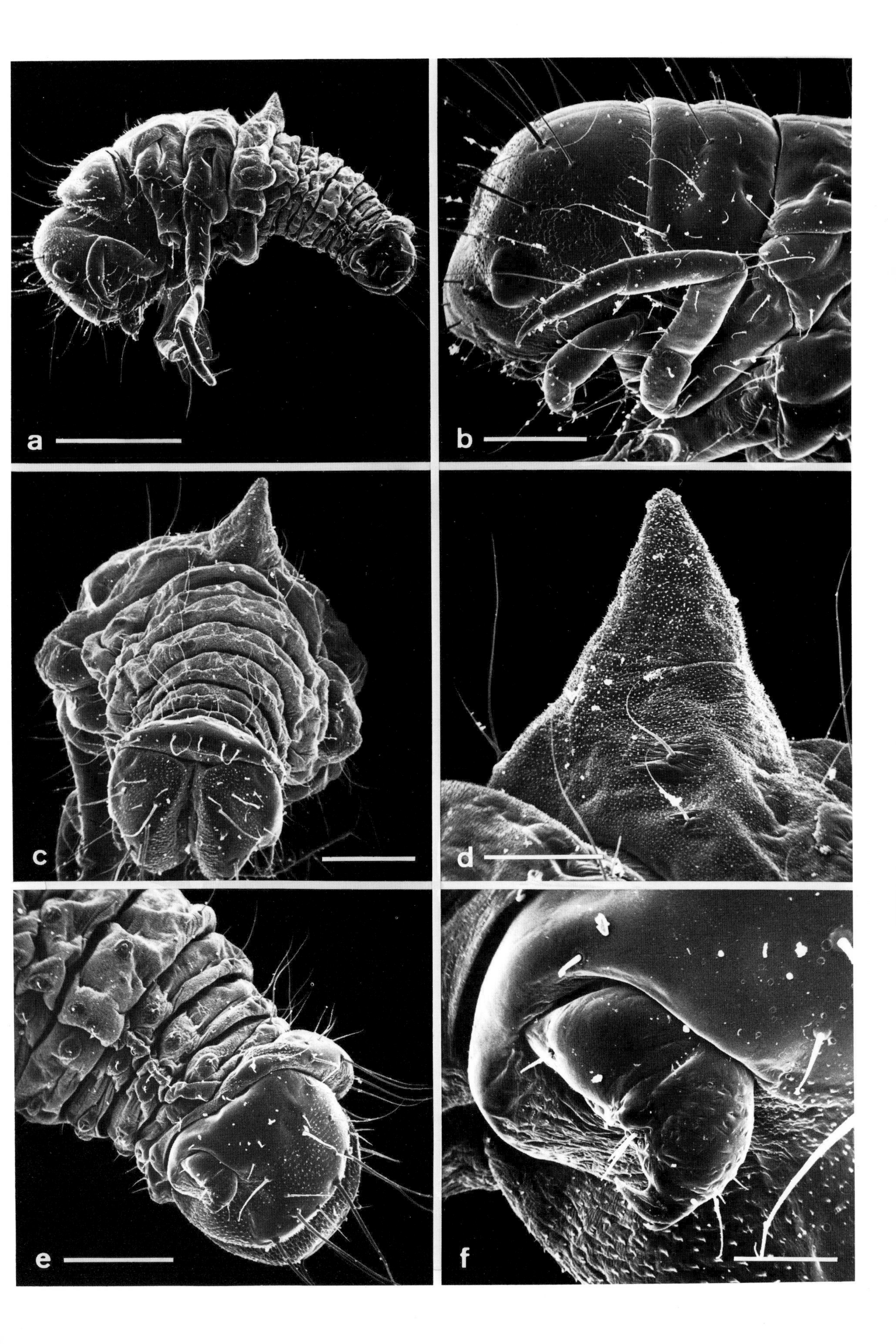

a
b
c
d
e
f

2.23 Order: Lepidoptera – Butterflies and Moths (Insecta)

GENERAL LITERATURE: Hering, 1926; Forster and Wohlfahrt, 1954–1982

2.23.1 The Root-Boring Swift Moths (Hepialidae) and Earth-Dwelling Noctuids (Noctuidae)

Only some of the well-known Lepidoptera belong to the soil arthropods, as the majority of caterpillars are restricted to their host and food plants. In general, caterpillars only reach the ground by chance, e.g. with falling leaves.

But the root-boring Hepialidae and the soil-dwelling Noctuidae in the subfamily Noctuinae (= Agrotinae) belong to the Lepidoptera which can occur as larvae and pupae in the soil during their postembryonic development. The caterpillars of the root-borer mainly live within the roots, but pupate outside them in tunnels which the larvae construct and line with spun threads prior to pupation. Noctuid caterpillars prefer to lie hidden in the soil during the day and feed above ground only when it is dark, or to drag plant fragments into their tunnels. They also pupate in the soil.

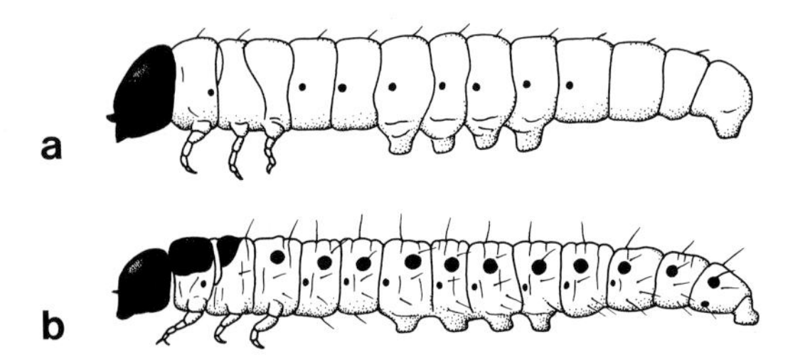

Fig. 197. Habitus of lepidopterous larvae. (Dunger, 1983). **a** Caterpillar of the soil-dwelling Noctuidae (Noctuinae); **b** Root-boring caterpillar of the Hepialidae

Plate 171. *Epichnopteryx ardua* (Psychidae), bagworm moths.

a Overview, lateral. The so-called bag is made of pine needle fragments and reinforced longitudinally with needles. 2 mm.

b Head and thorax, oblique frontal. 250 μm.

c Head and thorax, ventral. 100 μm.

d, e Bag surface. Materials are held together by spun threads. 200 μm, 50 μm

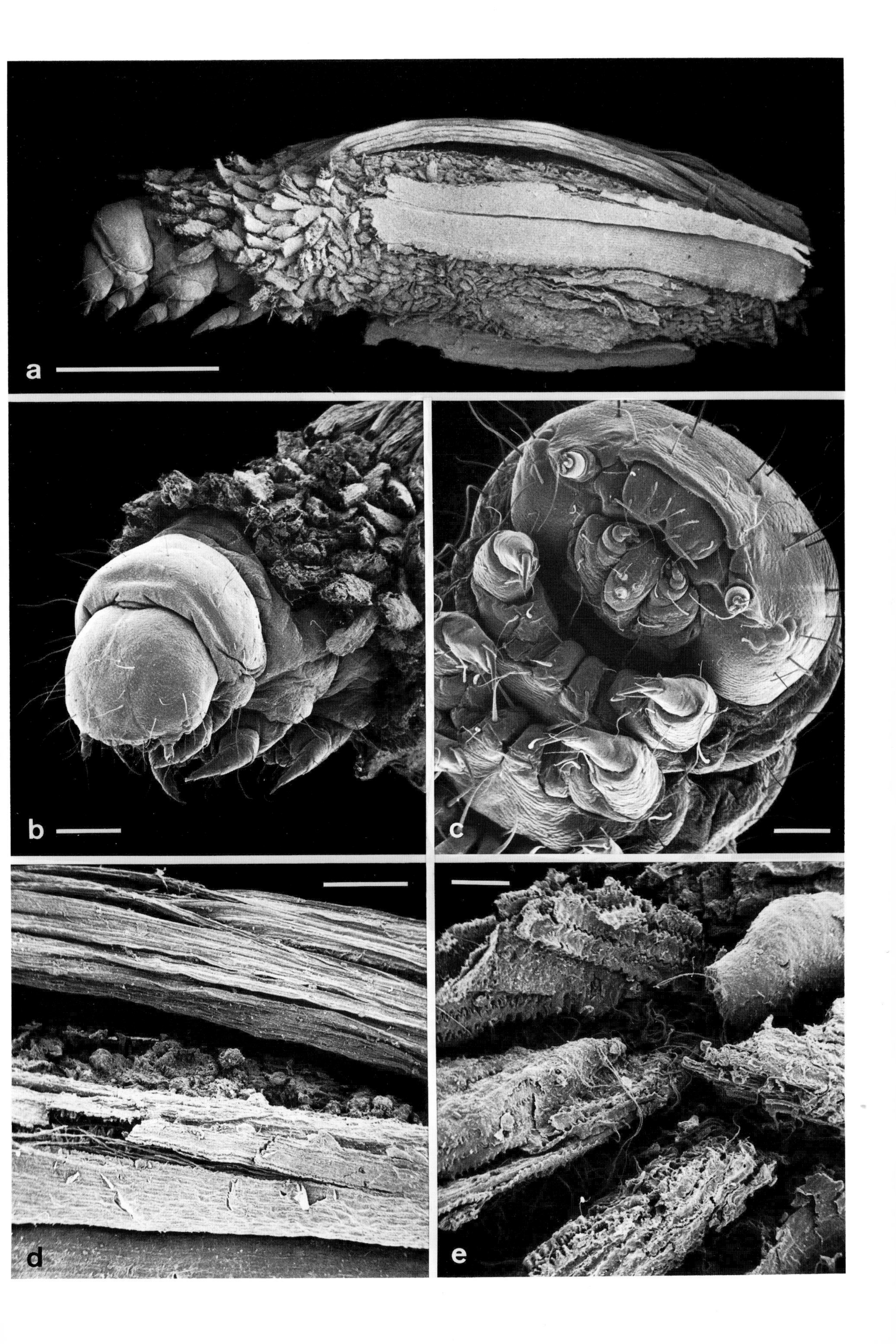
a
b
c
d
e

2.23.2 Bagworm Moths (Psychidae)

Many species of the remarkable bagworm moth or Psychidae family initially belong to the soil arthropods as larvae. These mainly polyphagous caterpillars feed on algae, lichen, moss and small pieces of plants. Their soft-skinned trunk is protected by diverse, often bizarre bags of small stones or pieces of plants, with each species having its characteristic construction (Fig. 198; Plates 171 and 172). They are generally of little importance to soil biology, but the population density of psychids can sometimes be considerable. Mature larvae of *Epichnopteryx ardua*, occurring at high altitudes in the Alps, were found to have a density of 365 ind/m^2 with a biomass of 341 mg dry weight/m^2 in a sedge (*Carex curvula*) meadow in the Hohe Tauern, Austria (Meyer, 1983).

The inconspicuous and small psychid moths exhibit sexual dimorphism. The males are winged, but their adult life lasts only a few hours or days. During this time they go in search of their sexual partners, attracted by the female's odourous substances (pheromones). The females are wingless, maggot-shaped, often with reduced extremities and mouthparts, and without eyes (Fig. 198). They await the males in or on the original larval bag, which is often attached to tree trunks, rocks and posts above ground. Mating and oviposition frequently take place in the bag (Forster and Wohlfahrt, 1960; Davis, 1964; Kozhanchikov, 1956).

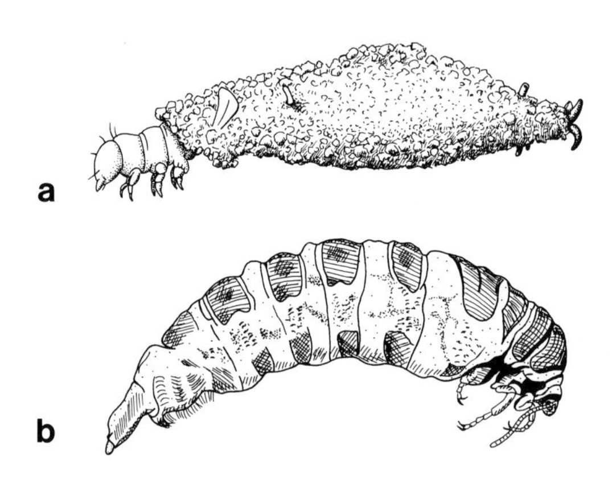

Fig. 198. Larva (**a**) with bag and female (**b**) of *Solenobia triquetrella* (Psychidae), bagworm moths. (Forster and Wohlfahrt, 1960)

Plate 172. *Epichnopteryx ardua* – Psychidae (bagworm moths), larva.

a, b Larva removed from the "bag", lateral and ventral. 1 mm, 0.5 mm.

c Head and thorax, ventral. 250 µm.

d Abdominal leg (plantula) viewed from above, with ring of hooks. 25 µm.

e Head, lateral. 100 µm.

f Anal region, ventral. 100 µm

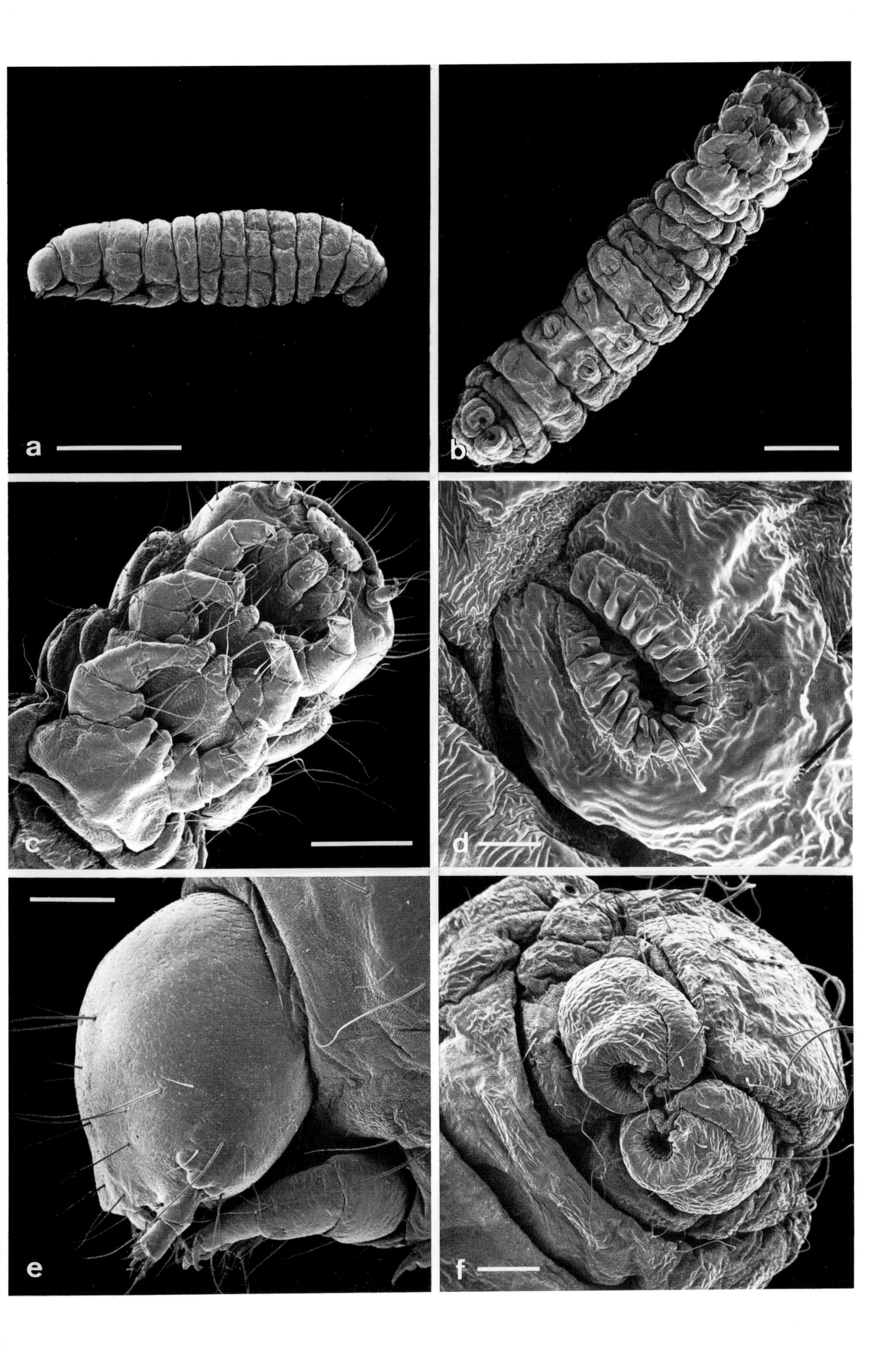
a
b
c
d
e
f

2.24 Order: Mecoptera – Scorpionflies (Insecta)

General Literature: Stitz, 1926; Strübing, 1958; Kaltenbach, 1978

2.24.1 Snow Scorpionflies (Boreidae)

Some members of the Boreidae family in the Mecoptera order are small, only 2–7-mm-long insects with wings reduced to scales in the female and to hook-shaped bristles in the male. Besides their lack of wings, the family is remarkable for the phenology and geographic distribution of its species. Boreids, with the Holarctic genus *Boreus* and the Nearctic genus *Hesperoboreus*, are distributed in the far north and especially in boreo-alpine regions. They are known as snow scorpionflies because the imagines appear at the beginning of the winter, and can be observed from October to March. After mating the eggs are deposited in moss or the upper soil layer. Caterpillar-like larvae with broad heads emerge and live in self-made tunnels all summer until pupation in September. The species which are distributed in central Europe are *Boreus westwoodi* and *Boreus hyemalis* (Fig. 199; Plates 173–175) (Aubrook, 1939; Strübing, 1950; Sauer, 1966).

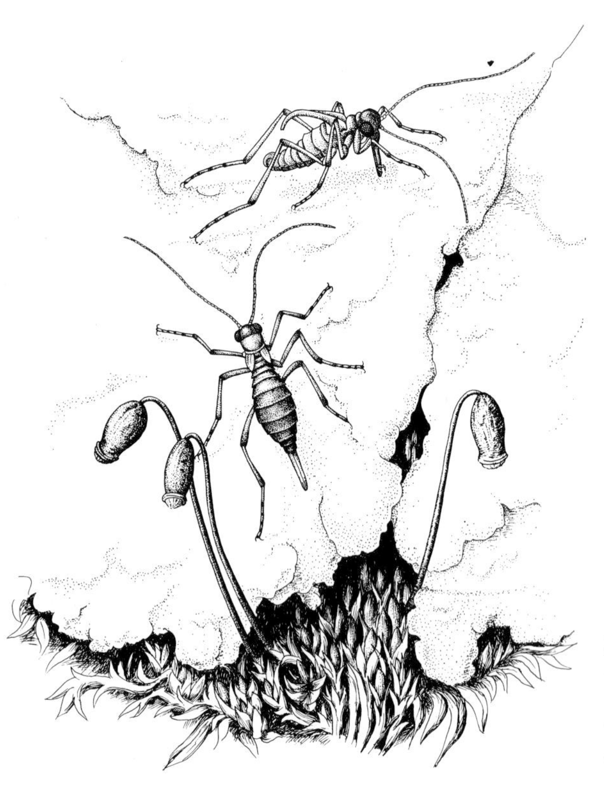

Fig. 199. A pair of snow scorpionflies, *Boreus hyemalis* (Boreidae), on snow above a cushion of moss. Their main food consists of moss and small, dead animals. Oviposition takes place in the upper soil layer where the eggs overwinter. (After Engel, 1961)

Plate 173. *Boreus westwoodi* – Boreidae (snow scorpionflies), female.

a Overview, lateral. 1 mm.
b Overview, dorsal. 1 mm.
c Overview, frontal. The mouthparts are at the distal end of the beak-like head. 0.5 mm.
d Thorax and head, dorsal, with rudimentary wings. 250 μm.
e Abdomen, lateral, terminal. The strongly tapering terminal segments and the ventral valves form the ovipositor. 250 μm.
f Tarsus, distal, with claws. 25 μm

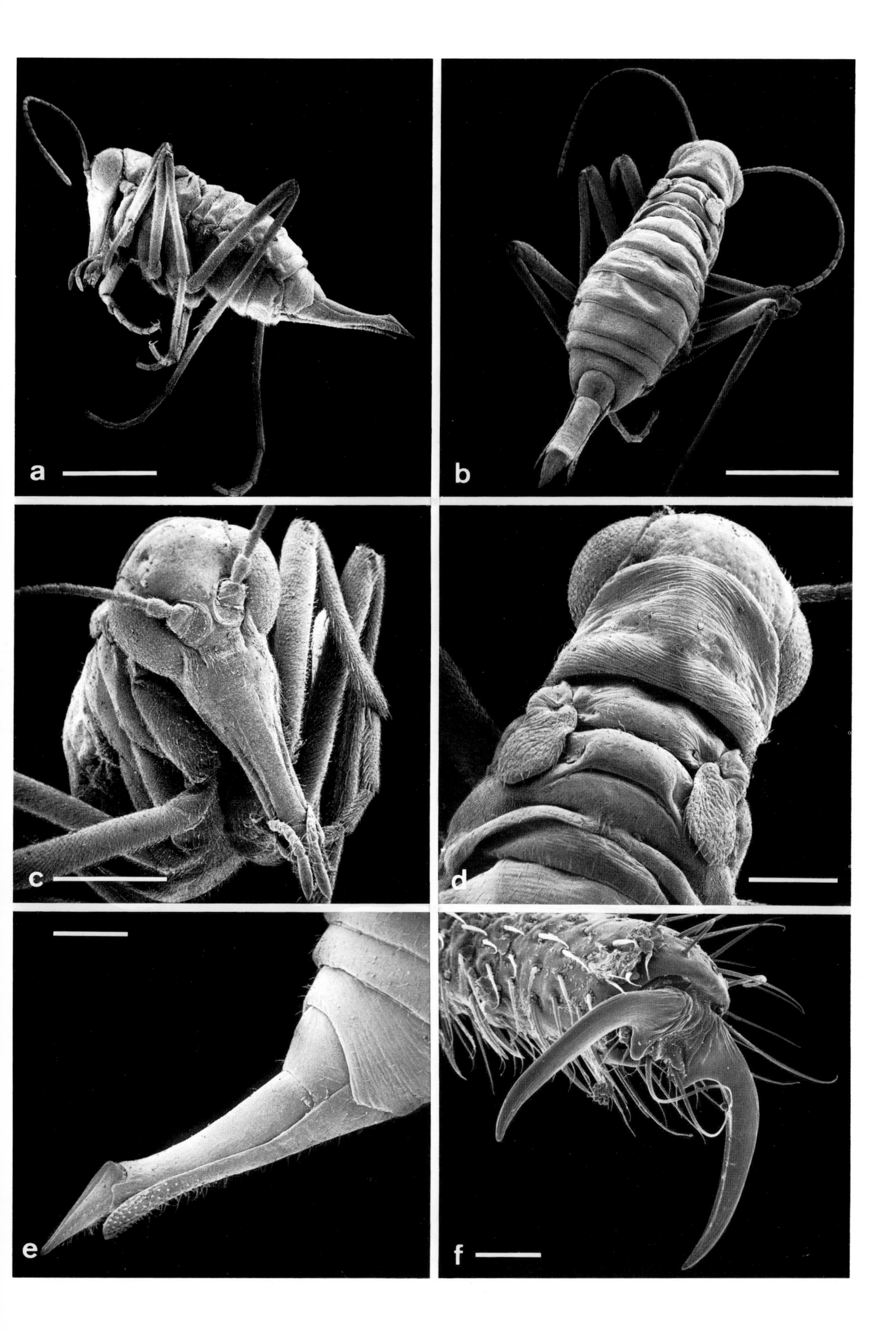
a
b
c
d
e
f

2.24.2 Mating Behaviour of Snow Scorpionflies (Boreidae)

Mating, which takes place in extreme, wintery conditions, has been described in *Boreus hyemalis* (Steiner, 1937) and in *Boreus westwoodi* (Sauer, 1966; Mickoleit and Mickoleit, 1976). Both species exhibit similar mating behaviour.

When ready to mate, a *Boreus westwoodi* male grasps the extremities of a passing female with his gonostylets and attempts, despite the often violent struggles of the female, to wedge her between his hook-like wing rudiment and his dorsal abdomen (Fig. 200a). If he succeds the female is incapable of freeing herself from his grip because the tergal apophyses of the second and third abdominal segments press into the female from beneath and the hook-like, wing bristles of the male are curved to fit the shape of the female's abdomen from above. Then the gonostylets are extended from the extremities and clasp the genital plates of the female so that the two abdomina are firmly attached. Afterwards the male releases his partner from the grip of his wing and manoeuvres her into the correct copulation position above him by stretching and bending his abdomen (Fig. 200b; Plate 174).

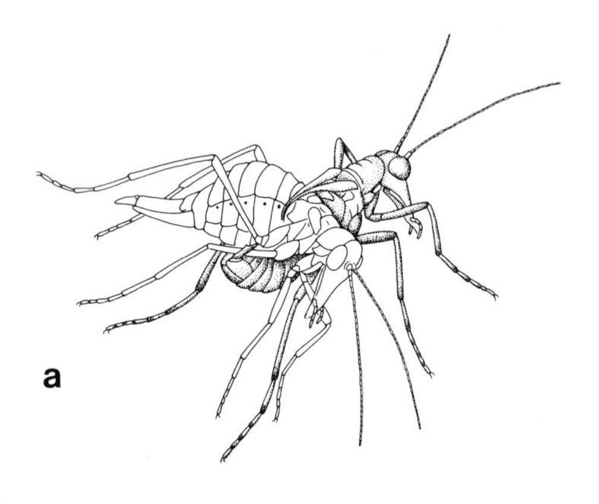

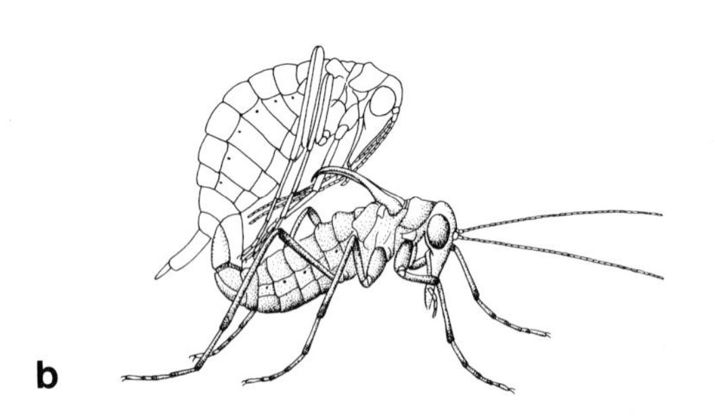

Fig. 200. Mating of the snow scorpionfly, *Boreus westwoodi* (Boreidae). (Mickoleit and Mickoleit, 1976).
a Pair in clasping position. The male (*below*) grips the female between his abdomen and his wing stumps. He then tries to push her onto his back.
b Circular copulatory position. The male (*below*) has inserted the membranous penis into the ovipositor of the female (*above*). Sperm transfer is achieved by means of a spermatophore. Copulation can last from several hours to 2 days

Plate 174. *Boreus hyemalis* (Boreidae), snow scorpionflies.
a Copulatory position, lateral. The female is *above*, the male *below*. 1 mm.
b Copulatory position, caudal view. The penis (*arrow*) is inserted into the female genital aperture. The terminal segments of the female are opened upwards. 0.5 mm.
c Copulatory position, frontal. 1 mm.
d Open ovipositor of the female. 250 µm.
e Ovipositor, lateral. 250 µm.
f Terminal segment of the female with ventro-caudal position of the anal plate. 50 µm

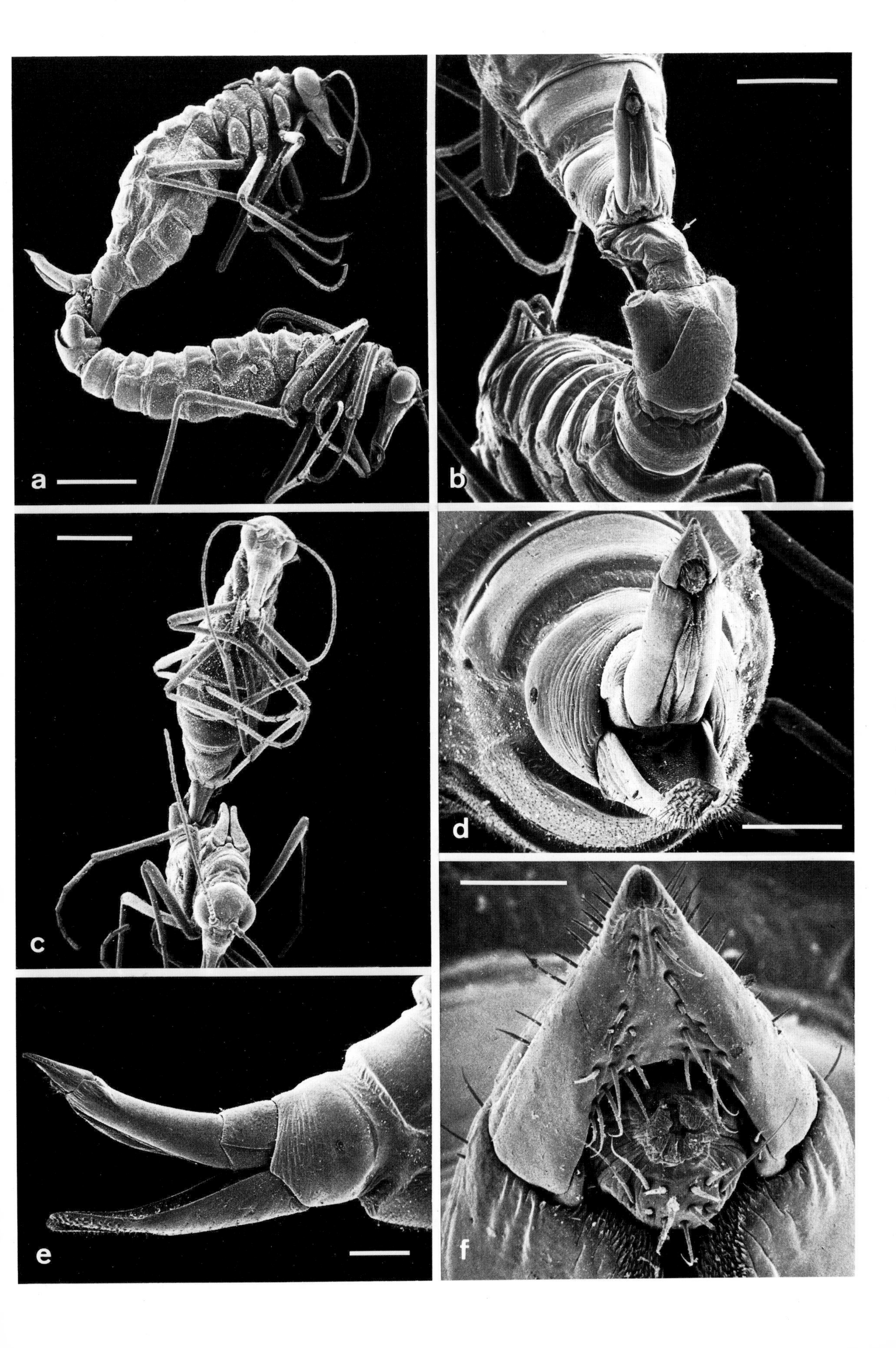
a
b
c
d
e
f

2.24.3 Wingless Soil-Dwelling Insects

Besides the primitive wingless insects, which were formerly united systematically and known as the Apterygota, there are also wingless insects, belonging to the soil arthropods, which underwent a secondary degeneration of the wings, but are phylogenetically derived from pterygote insects. These insects remain wingless even after completing their developmental stages and belong to diverse systematic groups.

The inability to fly restricts the dispersal of apterous insects, confining them mainly to the soil. But only a few leptotyphlins, a subfamily of the Staphylinidae (Coleoptera), are completely adapted to euedaphic life in the soil as a result of the reduction of their wings and progressive modification of their body form (Fig. 172d).

In addition, certain epedaphic insects exhibiting sexual dimorphism are wingless. In this case, the female is apterous while the male is winged and capable of flight. Beetles belonging to the Lampyridae (Plates 155 and 156), the terrestrial caddisflies of the genus *Enoicyla* (Limnephilidae) (Fig. 193) and the bagworm moths (Psychidae) are all included in this group. In the widest sense, the ants must also be mentioned in the category of apterous insects with winged partners, as they have both wingless (workers) and winged (queens and males) castes in their social system.

In contrast, both sexes of the dipteran snowflies of the genus *Chionea* (Limoniidae) (Plate 179) and the mecopteran snow scorpionflies (Boreidae) (Plate 173) are wingless and epedaphic inhabitants of the soil. In addition to this common character, they live gregariously together and are both cold, stenothermic soil arthropods.

Plate 175. *Boreus hyemalis* – Boreidae (snow scorpionflies), male.
a Overview, frontal. 0.5 mm.
b Rudimentary hook-like wings. 150 µm.
c Wings, terminal, lateral. 100 µm.
d Hind and front wing, lateral. The lower hind wing (*arrow*) is hardly more than a comb border. 25 µm.
e Mouthparts, lateral. The *arrows* mark the maxillary palps (*above*), the right galea and the mandible. 100 µm.
f Mouthparts, frontal, with maxillary palps, galeae and mandibles. 100 µm

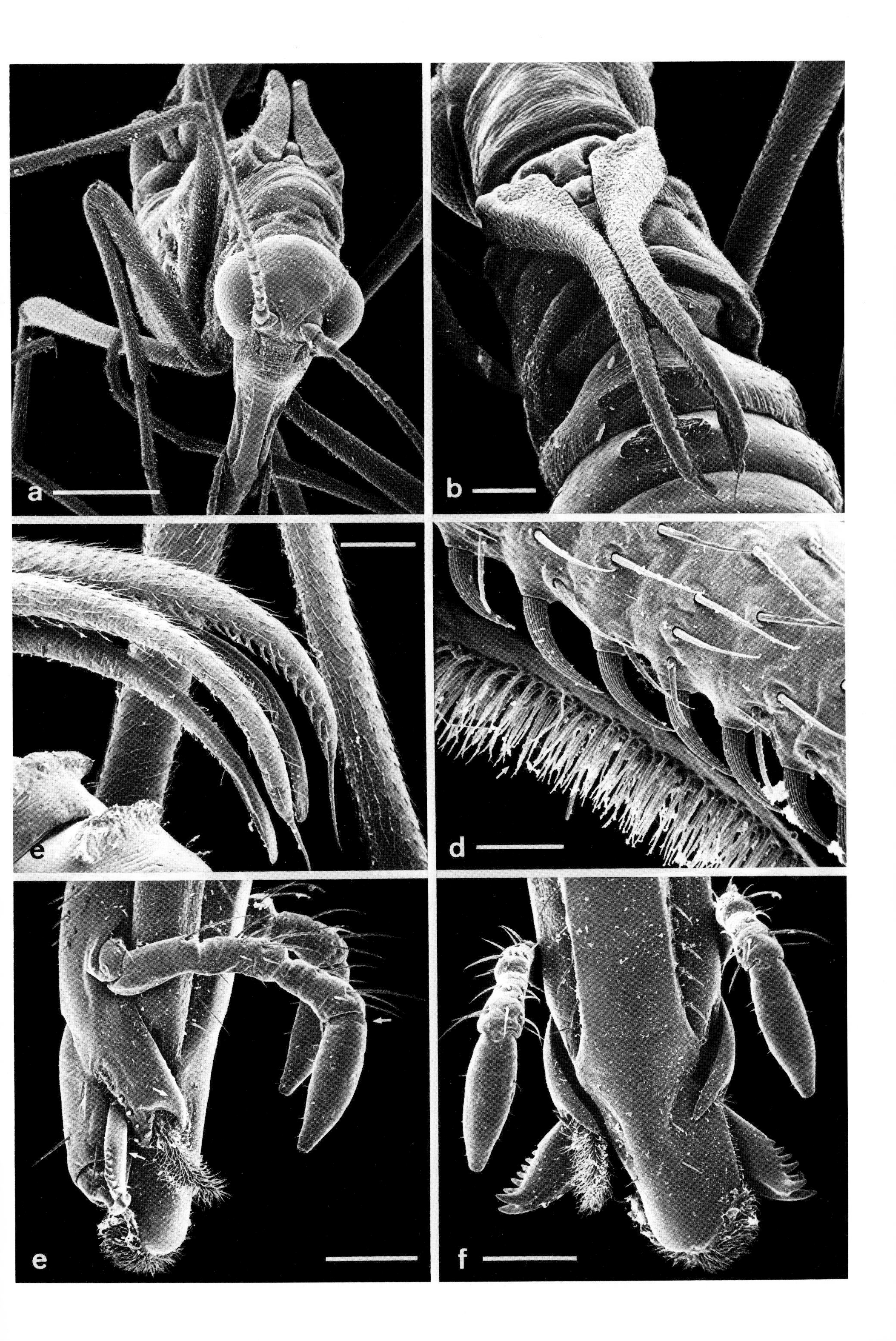
a
b
e
d
e
f

2.25 Order: Diptera – Flies (Insecta)

General Literature: Hennig, 1948–1952, 1973; Brauns, 1954; Séguy, 1950; McAlpine et al., 1981

2.25.1 Soil-Dwelling Diptera

The larvae of Diptera or flies make a considerable contribution to the soil community (Fig. 201). Although in many cases we have insufficient knowledge of the ecology and physiology, biology and classification of the soil-dwelling dipteran larvae, their significance to soil biology is undisputed. Besides their diversity of form, the size of their larval populations and their feeding activities as primary and secondary consumers engaged in the decomposition of plant material make them decisively important soil-dwellers.

Fundamental research on terrestrial dipteran larvae was published by Hennig (1948–1952, 1973) and especially by Brauns (1954a, b, c, 1955). This research is being pursued to an increasing degree in more recent soil biological investigations, e.g. by Altmüller (1979), Volz (1983) and Hövemeyer (1984).

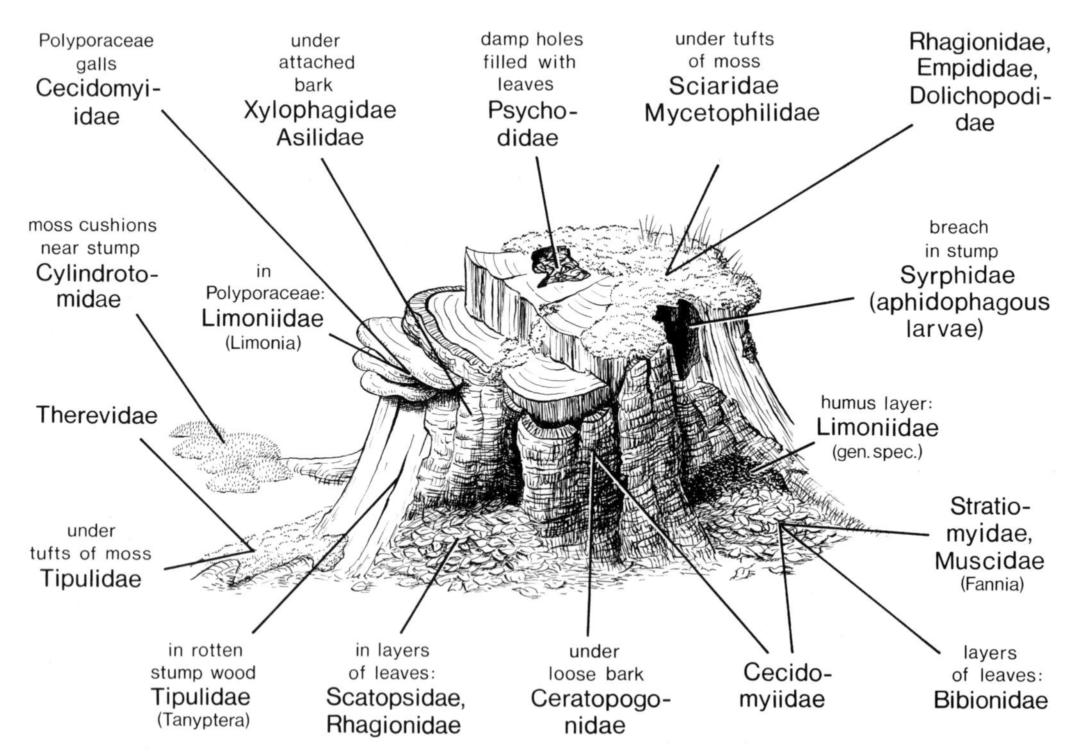

Fig. 201. Habitats of the most common terricolous dipteran larvae in a 6–8-year-old beech stump. (After Brauns, 1954)

Plate 176. Larvae of the Tipulidae (crane flies).

- **a** Overview, lateral. The anterior of the body is on the *upper left* of the picture. 0.5 mm.
- **b** "Head region", oblique, frontal. 200 μm.
- **c, d** Mouthparts and antennae (*arrows*). The apical antennal bulb is everted in **d.** 250 μm, 50 μm.
- **e** Maxillary palp, distal, with bulbiform sensilla. 5 μm.
- **f** Antenna, distal, with retracted apical bulb and sensilla. 20 μm

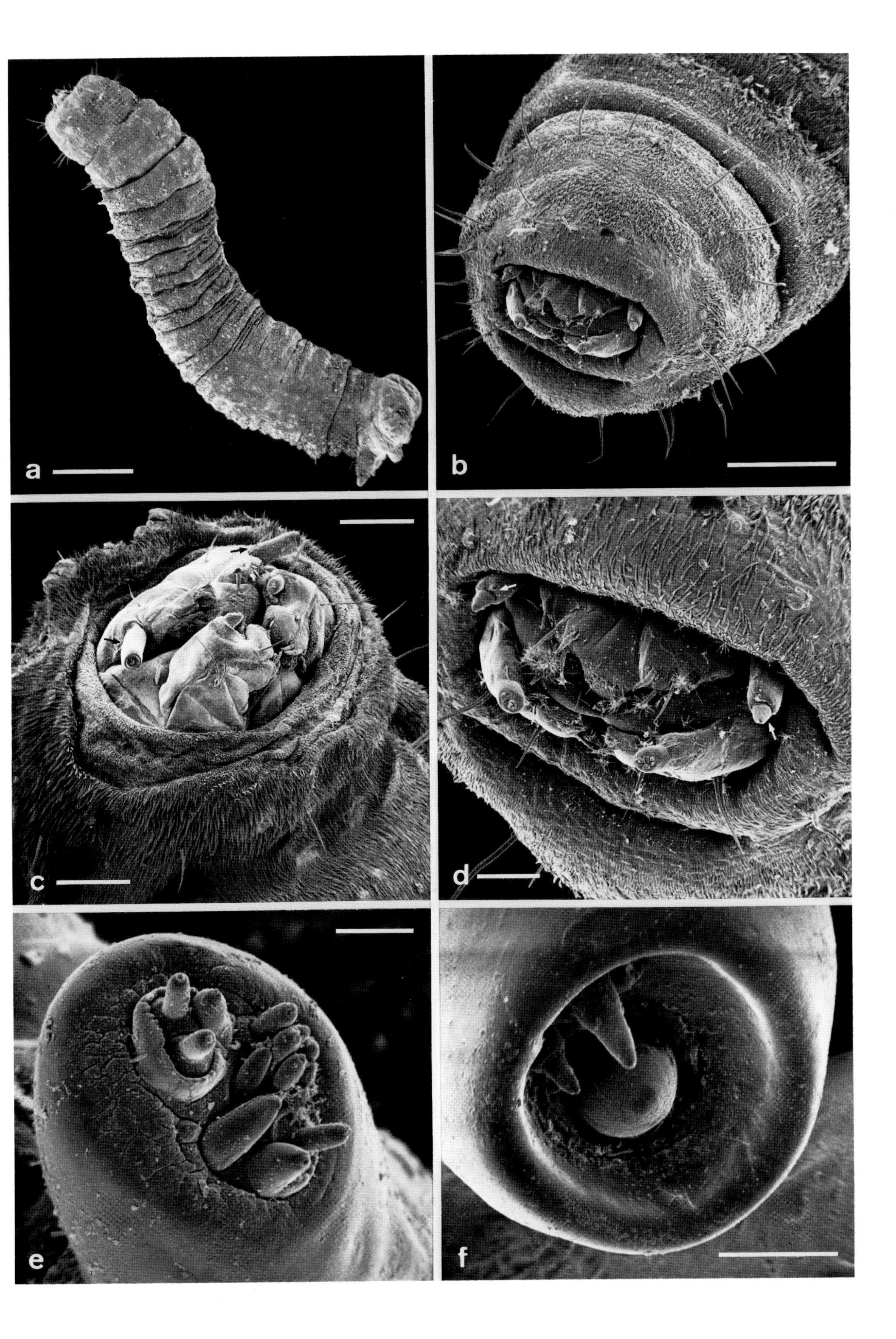

a
b
c
d
e
f

2.25.2 Tipulidae Larvae (Crane Flies)

The fat, cylindrical or caterpillar-shaped larvae are soft-bodied with dense, fine hair-like surface structures (Plates 176 and 178).

Their average length is 2 cm, but they can sometimes reach a size of 4 cm. The retractile hemicephalic "head" is often sunk into the prothorax so that only the powerful, toothed mandibles of the chewing mouthparts and the antennae are visible. The larvae are metapneustic (Fig. 212); the pair of functional spiracles is situated on the eighth abdominal segment in a spiracular area from which six bordering lobes radiate outwards (Fig. 202; Plate 177). Beside them on the anus lie the anal papillae which collapse when retracted and function as osmotic and ionic regulatory organs in the everted state. In terrestrial larvae, which live in moist soils, the anal papillae are often no more than short protuberances on the right and left of the anal aperture (Brauns, 1954a, Theowald, 1967).

The larvae are mainly semi-aquatic, but also purely terrestrial or aquatic. They are macrophytophagous and saprophytophagous. Terrestrial larvae prefer deciduous woodland and make a notable contribution to the decomposition of leaf litter (Perel et al., 1971; Priesner, 1961).

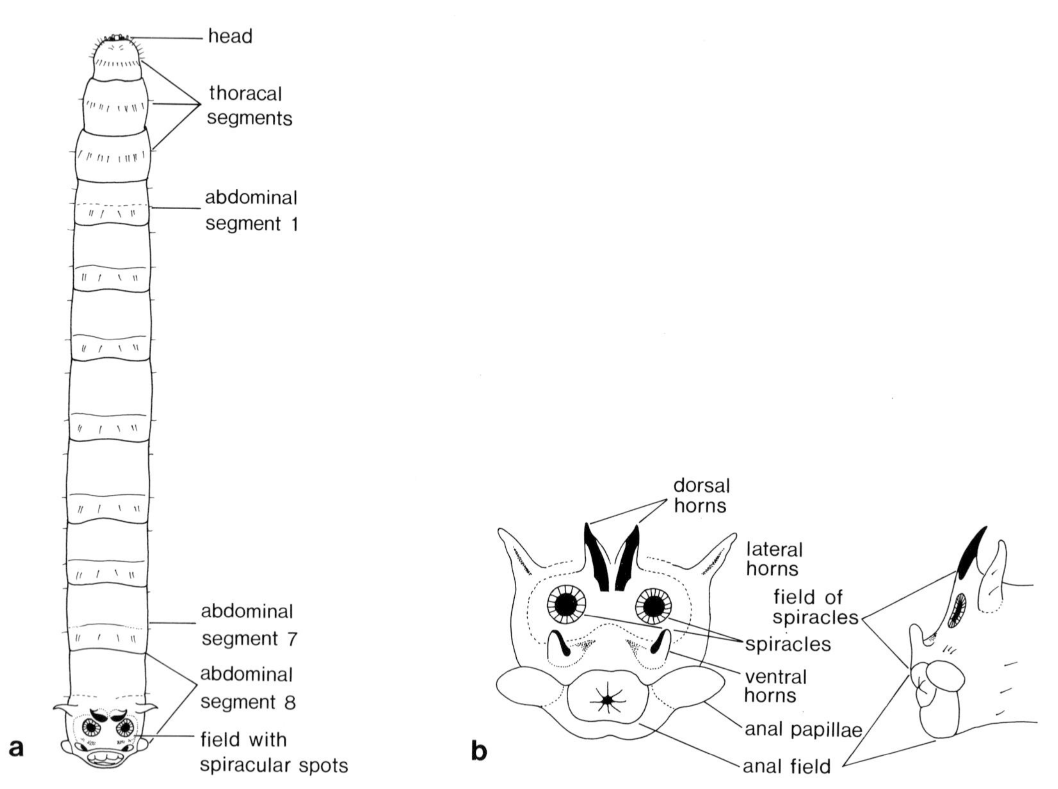

Fig. 202. Larva of *Tipula scripta* (Tipulidae), crane flies. (Theowald, 1967).
a Habitus, dorsal. **b** Arrangement of the anal region ("devil's mask"), caudal and lateral

Plate 177. Larvae of the Tipulidae (crane flies).
a, b Anal region, oblique caudal. This is the so-called devil's mask which gets its characteristic appearance from the anal horns, anal papillae and the jet-black spiracular rosettes. 200 µm, 0.5 mm.
c Anal region, lateral. 250 µm.
d Rosette-like spiracular valve. 100 µm.
e Anus. 100 µm.
f Anal horn. 100 µm

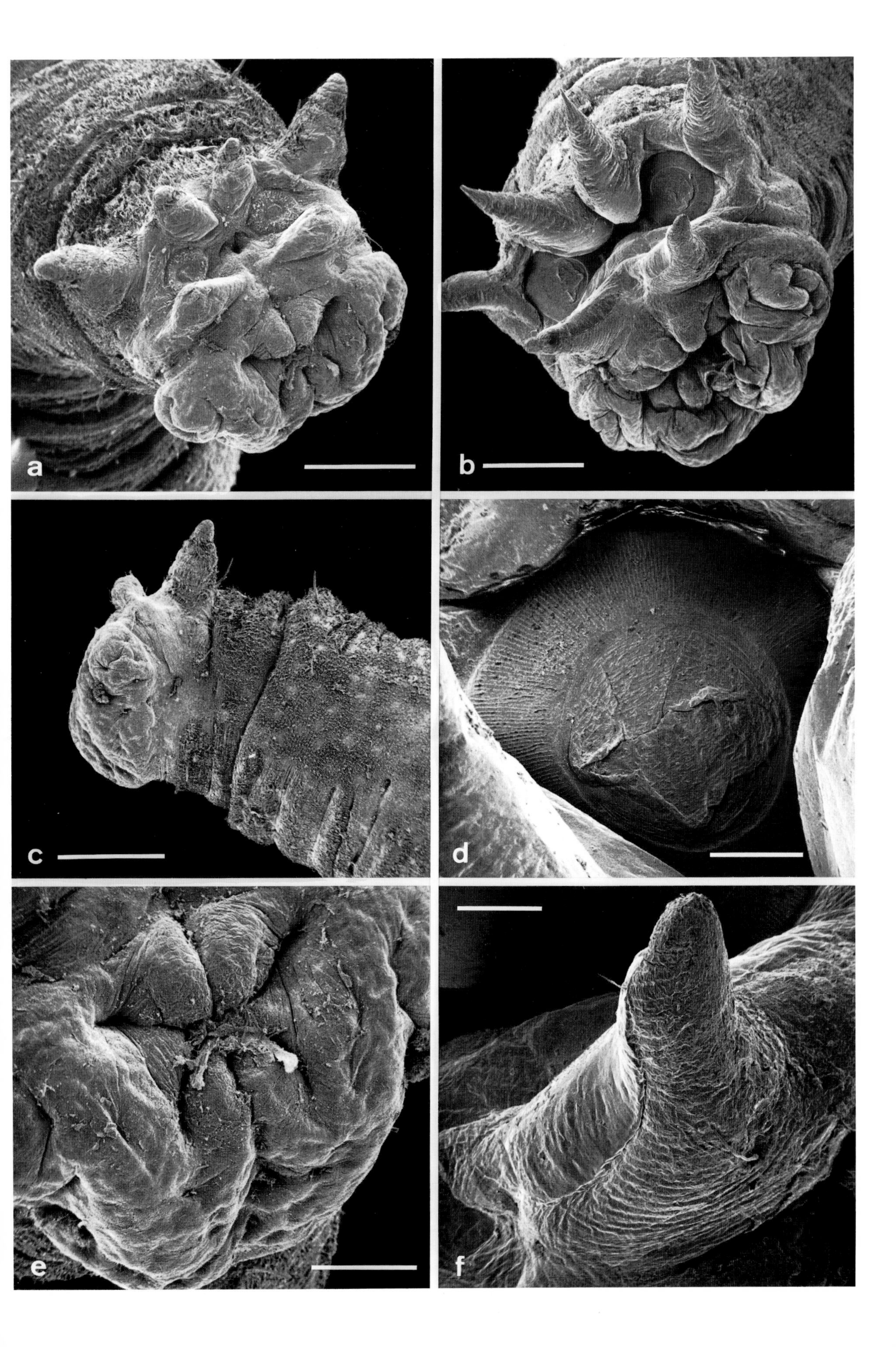

a
b
c
d
e
f

2.25.3 Bioproductive Aspects of Tipulidae Larvae

The larvae of *Tipula maxima* colonize wet and marshy woodland soils, but are also encountered in the riparian zones of small ponds and rivers. Caspers (1980) investigated the bioproduction of these larvae and measured the food consumption, defaecation and respiration of autumnal larvae (fourth instar) at various temperatures (Fig. 203). Even at temperatures of around 10 °C, which are realistic for autumn, the larvae almost reach their maximal metabolic activity; further temperature increases only slightly enhance the rates (cf. also Hofsvang, 1973; Zinkler, 1980, 1983).

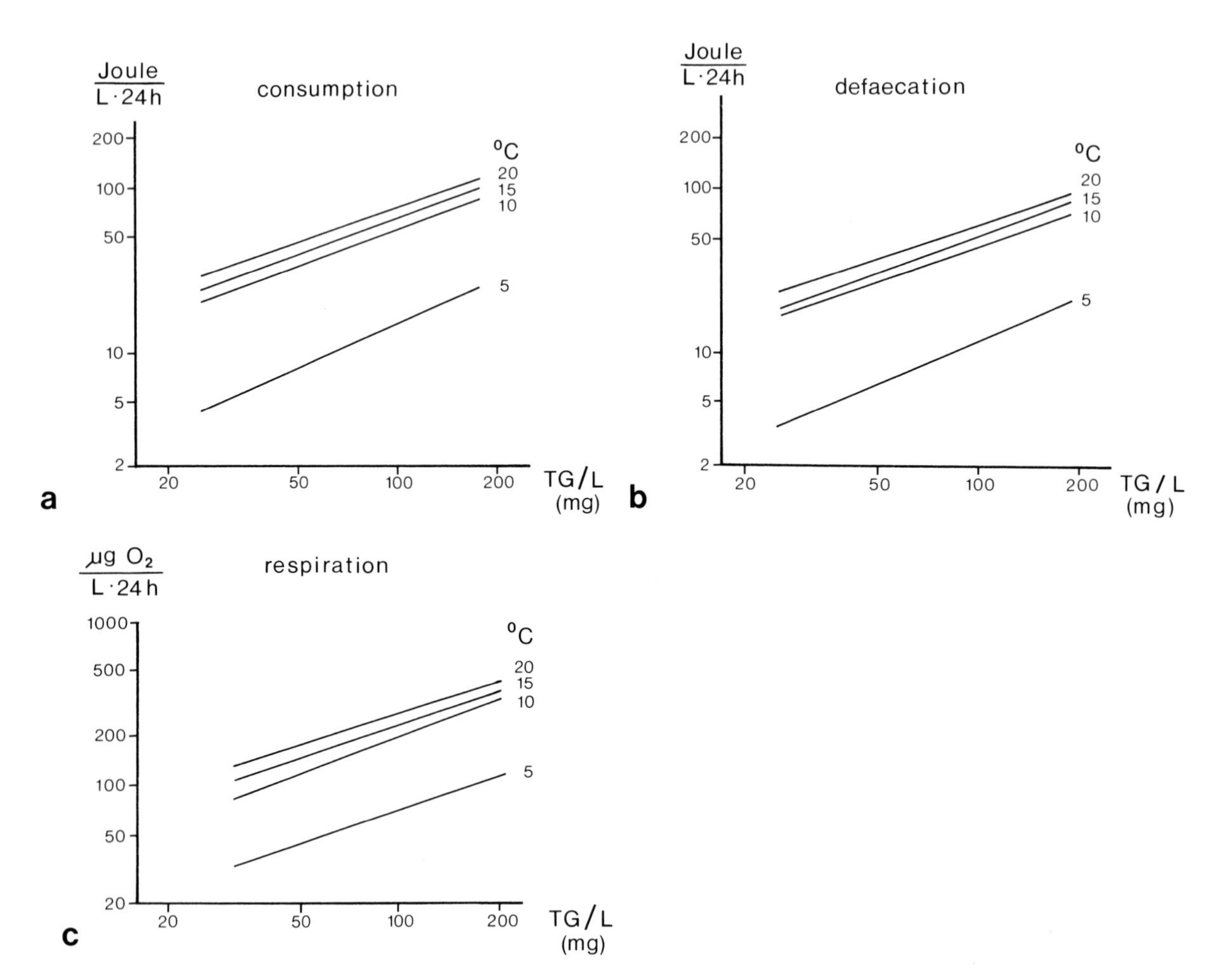

Fig. 203. Dependence of the food consumption **(a)**, defaecation **(b)** and respiration **(c)** of fourth instar larvae of *Tipula maxima* on temperature. (After Caspers, 1980). *TG/L (mg)* Dry body mass per larva in milligram

Plate 178. Larvae of the Tipulidae (crane flies).
a Abdomen, oblique dorsal. 0.5 mm.
b Anterior trunk segments, lateral. 0.5 mm.
c Abdominal segment, cuticle. 100 µm.
d, e Cuticle in the neck-head region. 50 µm, 25 µm.
f Cuticle in the dorsal region. 50 µm

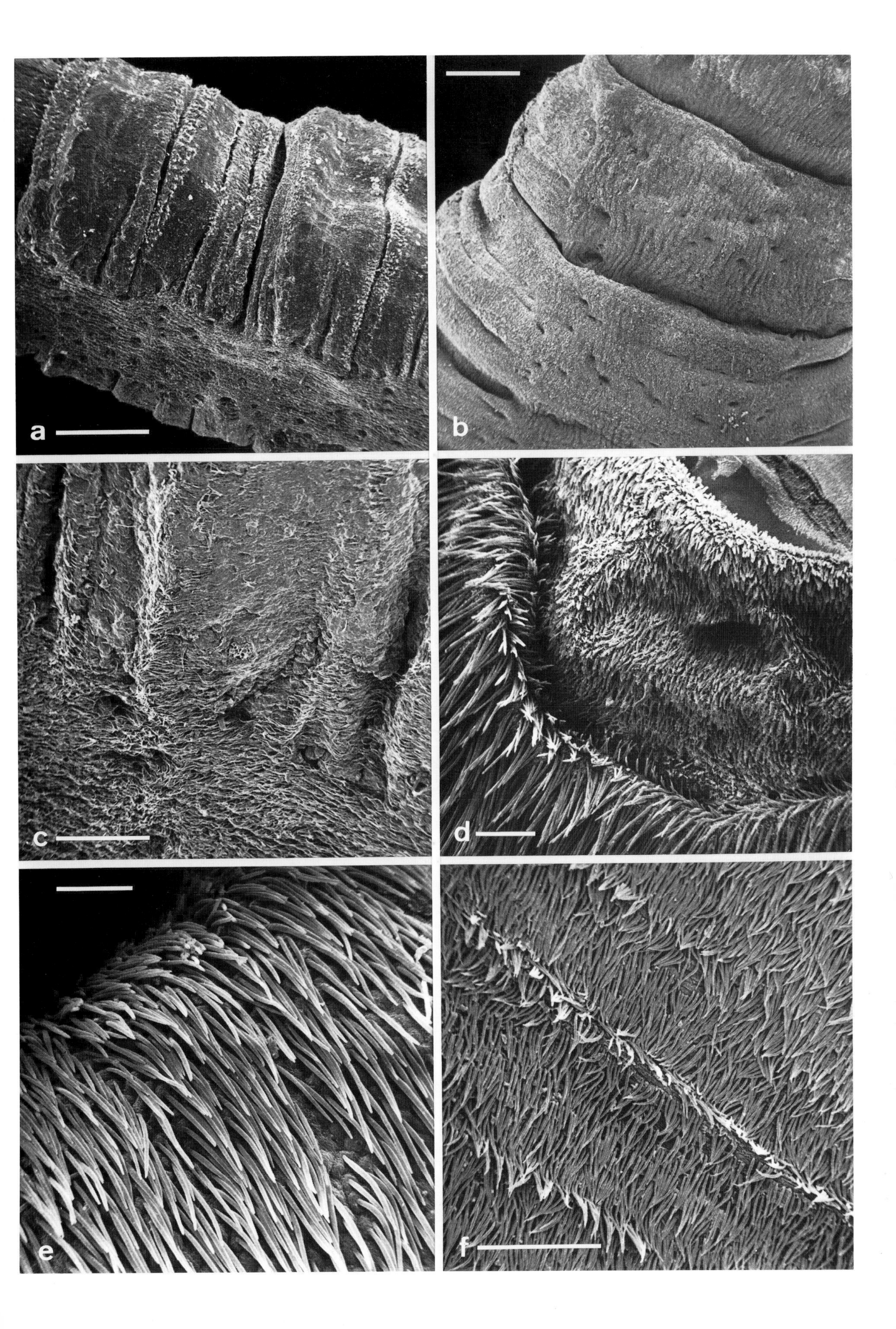
a
b
c
d
e
f

2.25.4 The Snowfly *Chionea lutescens* (Limoniidae)

The wingless snowflies of the genus *Chionea*, which belong to the Limoniidae family (Eriopterinae), are peculiar among the soil-dwelling Diptera. The most widely distributed species in central Europe is *Chionea lutescens* (Bezzi, 1917) (Plates 179 and 180), whereas the genus is represented in the mountains of northern and north-eastern Europe by *Chionea araneoides* (Fig. 204) and *Ch. crassipes* and in the Alps by *Chionea alpina* (Hågvar, 1971; Strübing, 1958). The sparse evidence of these species is not only attributable to their rarity, but also to the seasonal appearance of the flies, which are found exposed on snow in the winter months at temperatures of 0°–10 °C, while they seem to hide in the summer (Séguy, 1950). The annual activity peak occurs in January, with 58% of all individuals annually caught in soil traps, whereas the animals have already disappeared by March (Feldmann and Rehage, 1973); according to Erber (1972) the winter activity period lasts from mid-September to the end of February.

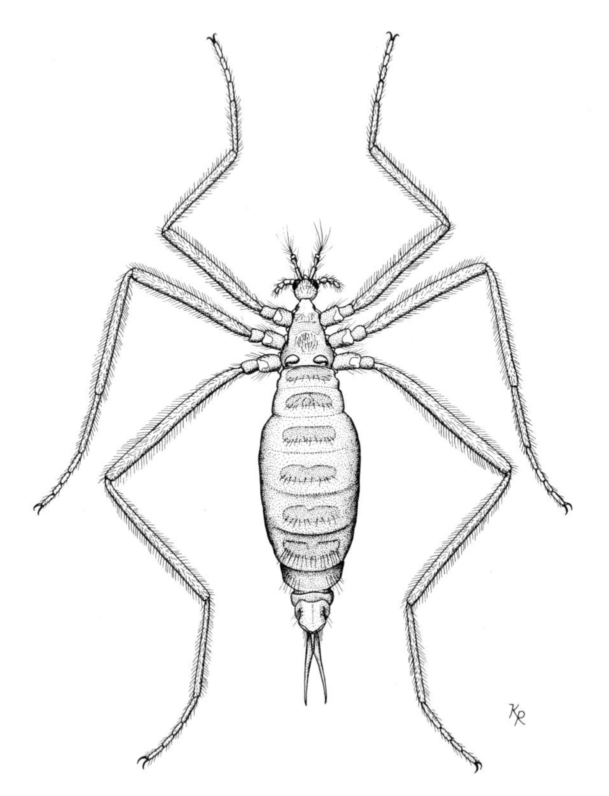

Fig. 204. *Chionea lutescens* – Limoniidae (crane flies), snowfly, habitus, dorsal view

Plate 179. *Chionea lutescens* – Limoniidae (crane flies) snowfly.
a Overview, dorsal. 1 mm.
b Overview, oblique frontal. The halteres and the fracture points of the front wings (*arrows*) are situated on the thorax. 400 µm.
c Head, ventral, with sucking proboscis. 100 µm.
d Proboscis, frontal. 50 µm.
e Filtration bristles at the tip of the proboscis. 5 µm.
f Tergum of the mesothorax with fracture point of a front wing. 25 µm

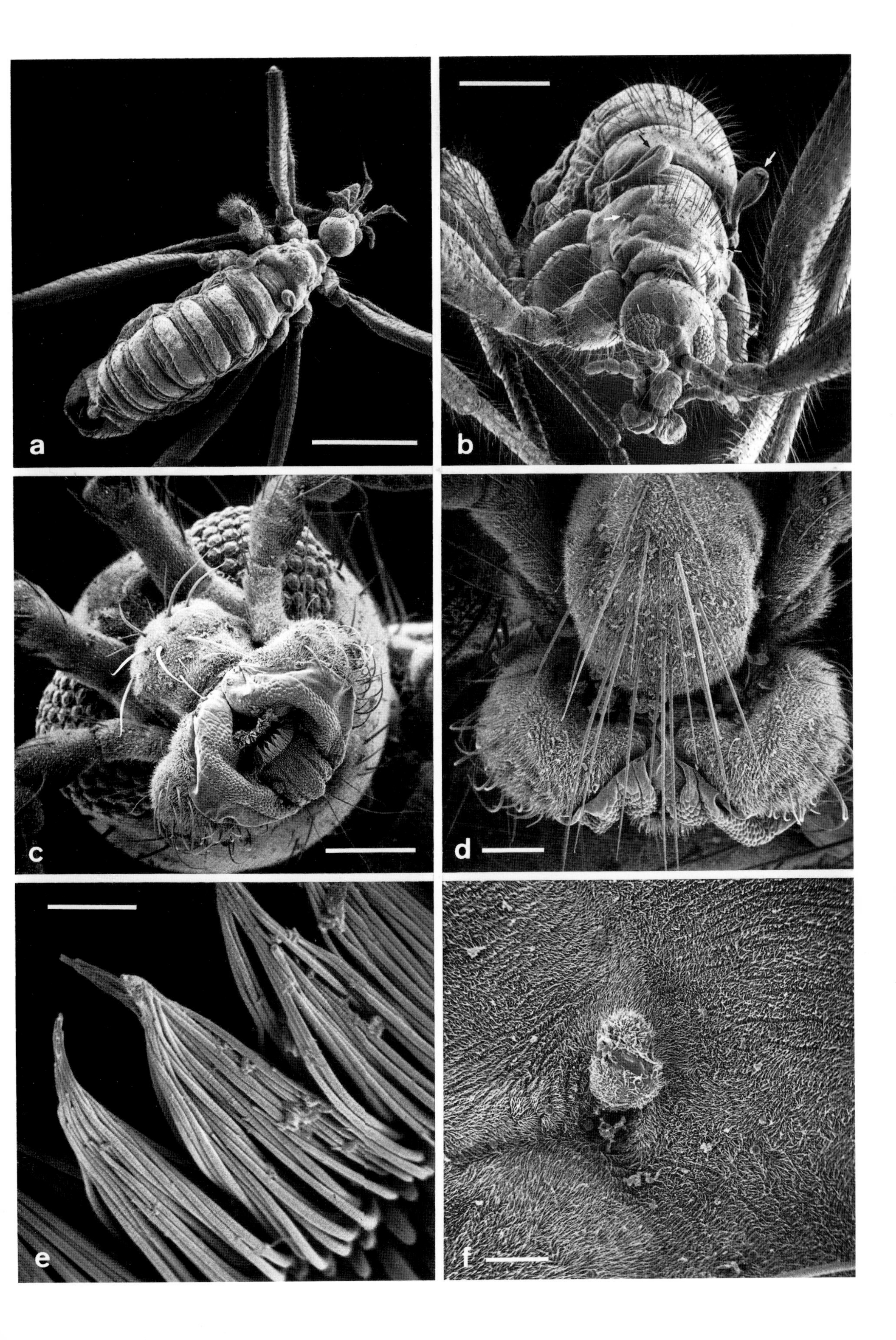
a
b
c
d
e
f

2.25.5 The Structural Features of the Cold Stenothermic Snowflies

The body and extremities of the approximately 4-mm-long snowflies are covered with hair, which, together with their long bent legs and lack of wings, gives them a spider-like appearance (cf. Bürgis 1982). The first pair of wings is completely reduced, but the halteres still remain, possibly with a modified function. Reduction of the wings is also found in the snow scorpionflies (Mecoptera) which frequently live in communities with *Chionea*. Strübing (1958) attributes their common wing reduction to the low temperatures and other winter-dependent factors of their habitat. Adaptation to cold is thought to have progressed so far that the animals are only capable of living at low temperatures, preferably in the boreo-alpine regions, and in the course of evolution have continuously specialized their shape by reduction of the wings, as well as their mode of life.

Plate 180. *Chionea lutescens* – Limoniidae (crane flies), snowfly.

- **a** Haltere, distal (the spherical granules on the cuticle are presumably secretion). 25 μm.
- **b** Caudal view of an antenna. 50 μm.
- **c, d** Surface of the second antennal segment. 25 μm, 5 μm.
- **e** Abdomen of a male, caudal. 250 μm.
- **f** Abdomen of a female, terminal, lateral, with ovipositor. 0.5 mm

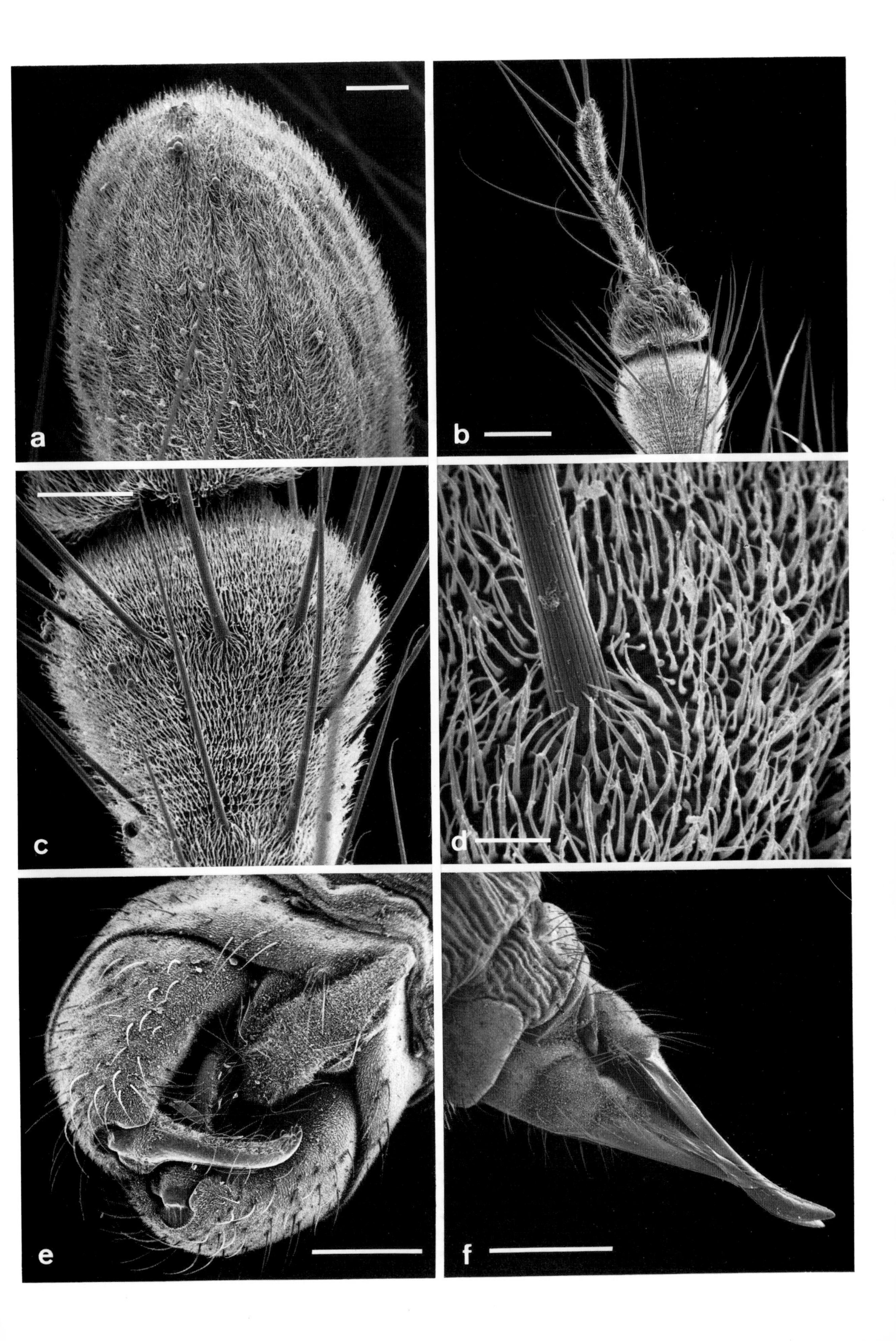
a
b
c
d
e
f

2.25.6 The Larvae of the Psychodidae (Moth Flies)

The small, only 4-mm-long psychodid larvae are worm-shaped or, rarely, isopod-like (*Sycorax*) and are generally distinguished by a conspicuous pseudosegmentation on the thorax and abdomen. In this manner their habitus is adapted to the interstitial system of the soil. Besides the surface texture formed by secondary segmentation, many larvae have dorsal rows of bristles which are interpreted as sensory hairs, but also serve to anchor small particles of earth emphasizing the close association of many larvae (*Pericoma*) to the soil.

The sessile, peritrichous ciliates (Plate 181) found on the head region of the animals are an indication of their preferred habitat. Psychodid larvae are primarily aquatic and only occur in very moist (semi-aquatic) woodland soil, between decaying leaves or in damp parts of rotten tree stumps.

They are zoophagous and saprophytophagous, but are also microphytophagous, consuming algae and the lower forms of fungi. In general, however, these larvae are of comparatively little significance to soil biology (Brauns, 1954).

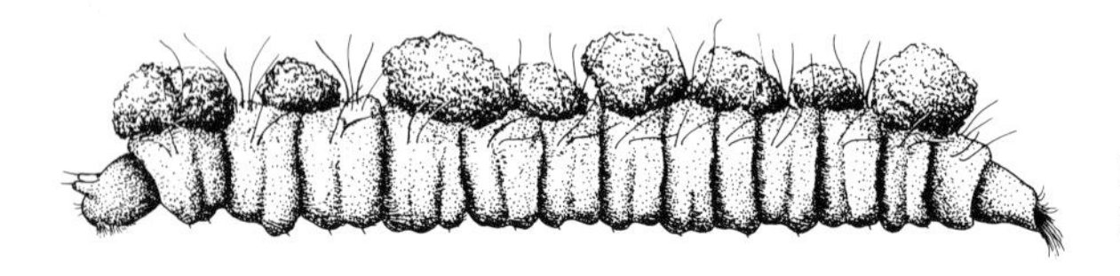

Fig. 205. Larva of the Psychodidae (moth flies), habitus lateral

Plate 181. Larvae of Psychodidae (moth flies).

a, b Overview, lateral. Earth particles are stuck on to the back of the larvae, presumably as camouflage. 0.5 mm, 0.5 mm

c Head and trunk, oblique, frontal. The *arrow* marks the position of sessile peritrichous ciliates, an indicator of the aquatic or semi-aquatic way of life of the larvae. 100 μm.

d Overview, frontal. 250 μm.

e Head, oblique, frontal. 50 μm.

f Overview, oblique, caudal. 0.5 mm.

g Antenna, distal, with sensilla. 10 μm

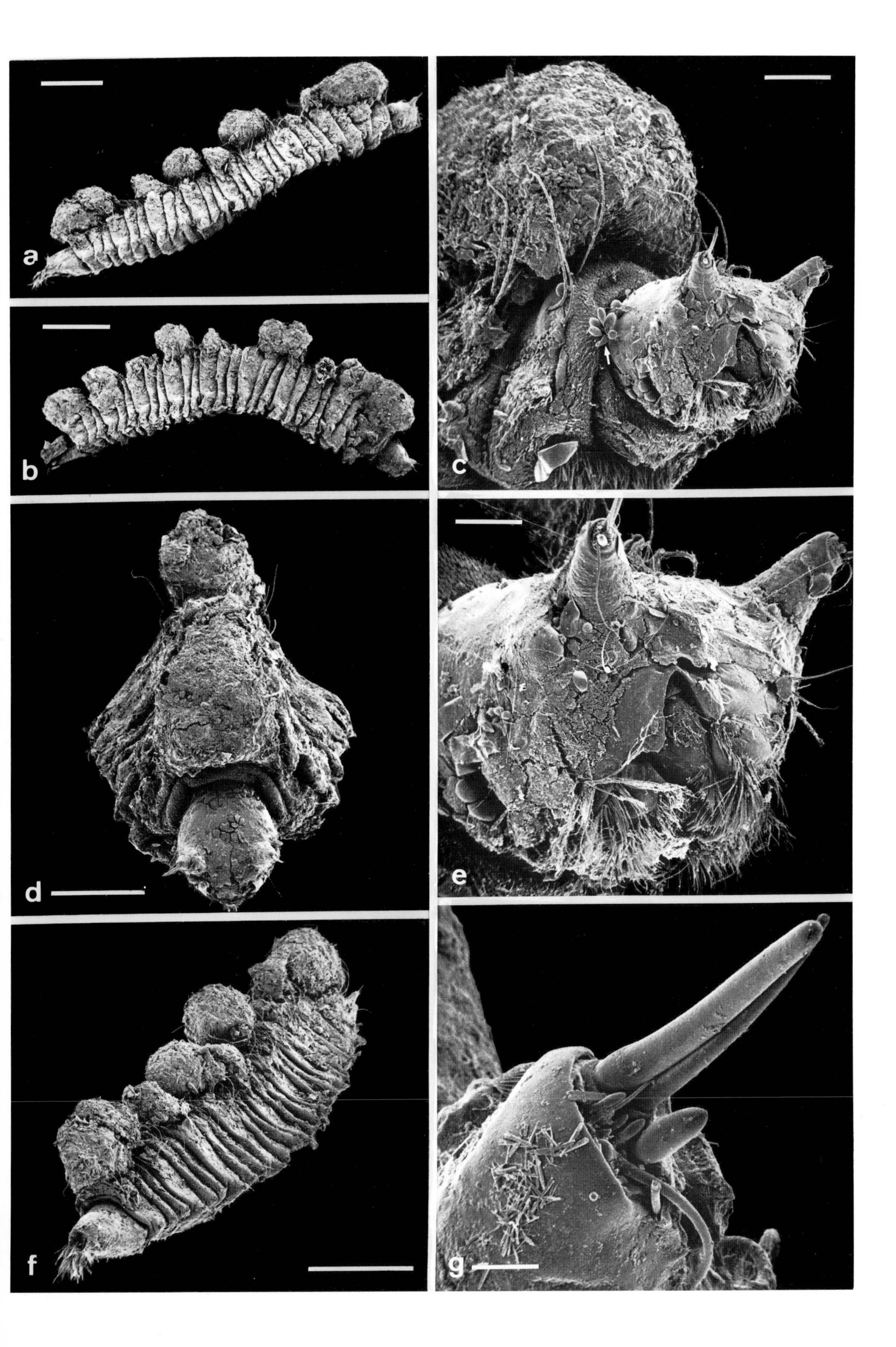

a
b
c
d
e
f
g

2.25.7 The Larvae of the Ceratopogonidae (Biting Midges)

The larvae of the ceratopogonids are predominately aquatic and semi-aquatic; terrestrial forms belong to the Forcipomyiinae subfamily. The elongated and distinctly segmented larvae have anterior and posterior leg stumps bearing hooks (Fig. 206; Plate 182). They have a closed tracheal system (apneustic type). The anus is bordered by anal papillae for osmotic and ionic regulation. The larvae (*Forcipomyia*) are found in the litter layer and under rotting tree bark in deciduous woodland.

Other ceratopogonid larvae with a distinctly different habitus from *Forcipomyia* also live in the litter layer and belong to the genus *Bezzia*. This genus is placed in the Culicoidinae subfamily, the larvae of which lack the anterior and posterior leg stumps. The extremely thin *Bezzia* larvae tend to be aquatic. Unlike the terrestrial *Forcipomyia* larvae, they prefer moist soil litter or moist moss cushions (Strenzke, 1950; Brauns, 1954).

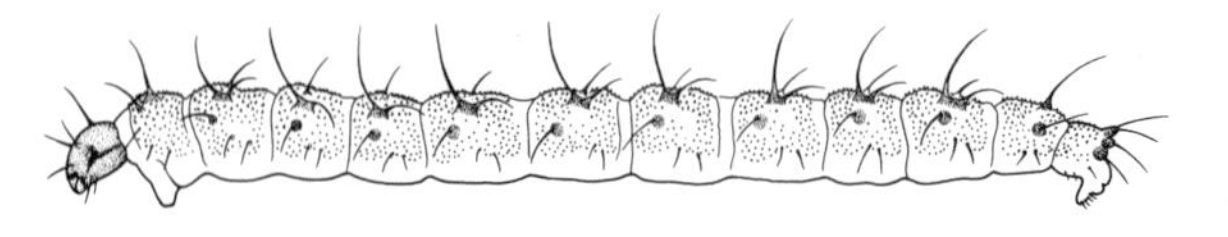

Fig. 206. Larva of the Ceratopogonidae (biting midges), *Forcipomyia* sp., habitus lateral. (After Brauns, 1954)

Plate 182. Larva of the Ceratopogonidae (biting midges).
a Overview, lateral. The head is situated on the *right* of the picture. 1 mm.
b Head and thorax with leg stumps, lateral. 200 μm.
c Abdominal segments, lateral. 100 μm.
d Head with leg stumps, ventral. 100 μm.
e Anal region with papillae. 50 μm.
f Mouth area. 20 μm

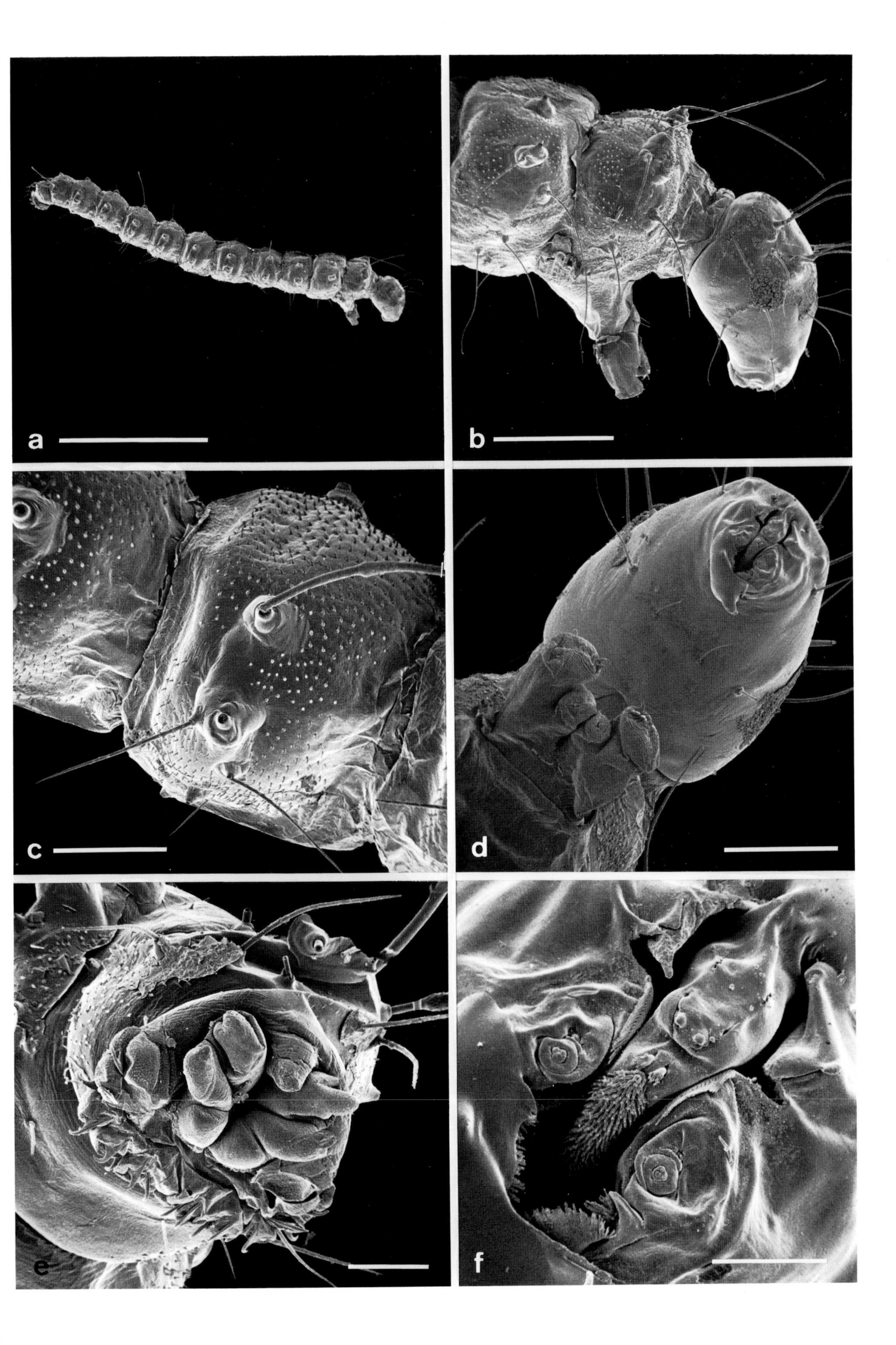
a
b
c
d
e
f

2.25.8 The Larvae of the Chironomidae (Non-Biting Midges)

The majority of the Chironomidae family, which is extremely rich in species, have a merolimnetic mode of life. The larvae can be up to 10-mm-long and are aquatic, mostly living in freshwater, but some are marine (Thienemann, 1954; Oliver, 1971). In addition to aquatic species, the Orthocladiinae subfamily includes terrestrial larvae which have obviously made the transition from water to the land. They do not form a monophyletic group, as they possess convergent morphological structures which evolved independently as adaptations to life in the soil (Strenzke, 1950; Brundin, 1956; Hennig, 1973).

The body surface of the larvae (Fig. 207; Plates 183 and 184) is smooth and uniformly segmented. Bristles on their head and body are strongly or completely reduced and their antennae shortened in favour of enhanced mobility. With increasing adaptation the elongated, cylindrical form of the larvae necessary for life in the interstitial system of the soil is emphasized by fusion and reduction of the front, and modification of the posterior leg stumps (Strenzke, 1950, 1959; Brauns, 1954).

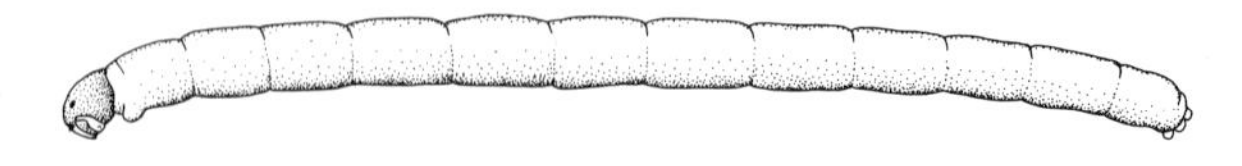

Fig. 207. Larva of the Chironomidae (non-biting midges), *Pseudomittia simplex*, habitus lateral. (Brauns, 1954)

Plate 183. Larva of the Chironomidae (non-biting midges).
a Overview, ventral. 1 mm.
b Anterior of the body, lateral, with anterior leg stumps. 100 μm.
c Head, frontal. 40 μm.
d Anterior leg stumps with tuft of hooked bristles. 20 μm.
e Antenna, medio-lateral. 10 μm.
f Antenna, distal. 3 μm

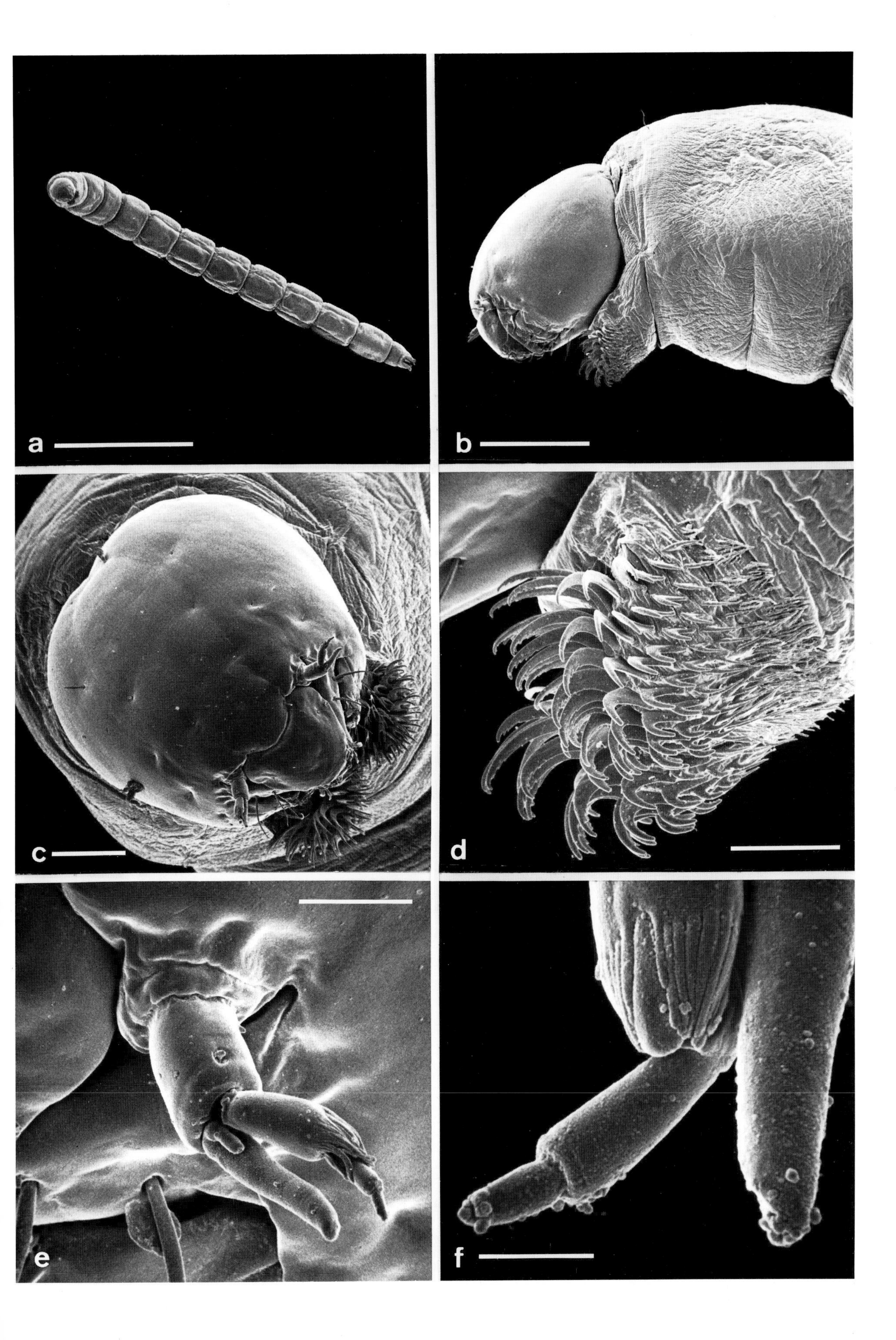
a
b
c
d
e
f

2.25.9 Vertical Distribution of Terrestrial Chironomidae

Many Chironomidae live gregariously in synusia, preferably in the uppermost, humus-rich soil layers where the larvae are generally microphytophagous. Depending on the season and microclimatic conditions, vertical migrations of certain larvae can be observed (Strenzke, 1950). Thus, the larvae of the *Smittia aquatilis* group are sensitive to temperature changes. When temperatures sink and during the cold winter period they burrow deeper into the soil (Fig. 208).

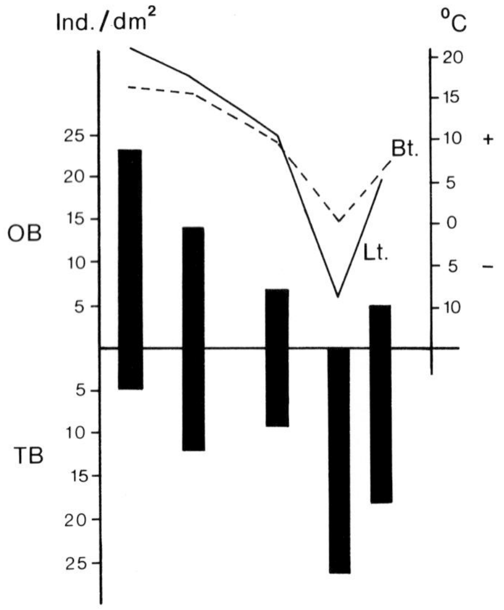

Fig. 208. Temperature-induced vertical migration of the larvae of *Smittia* sp. (aquatilis group), Chironomidae. (After Strenzke, 1950).
OB Uppermost soil layers; *TB* deeper soil layers; *Bt* soil temperature; *Lt* air temperature

Plate 184. Larva of the Chironomidae (non-biting midges).
a Abdominal segments, ventral. 100 μm.
b Anal region with papillae and caudal leg stumps. 50 μm.
c Caudal leg stumps with hook apparatus. 25 μm.
d Everted anal papillae between the caudal leg stumps. 20 μm.
e Caudal leg stump with a ring of hooks. 10 μm.
f Anal papilla, distal. 3 μm

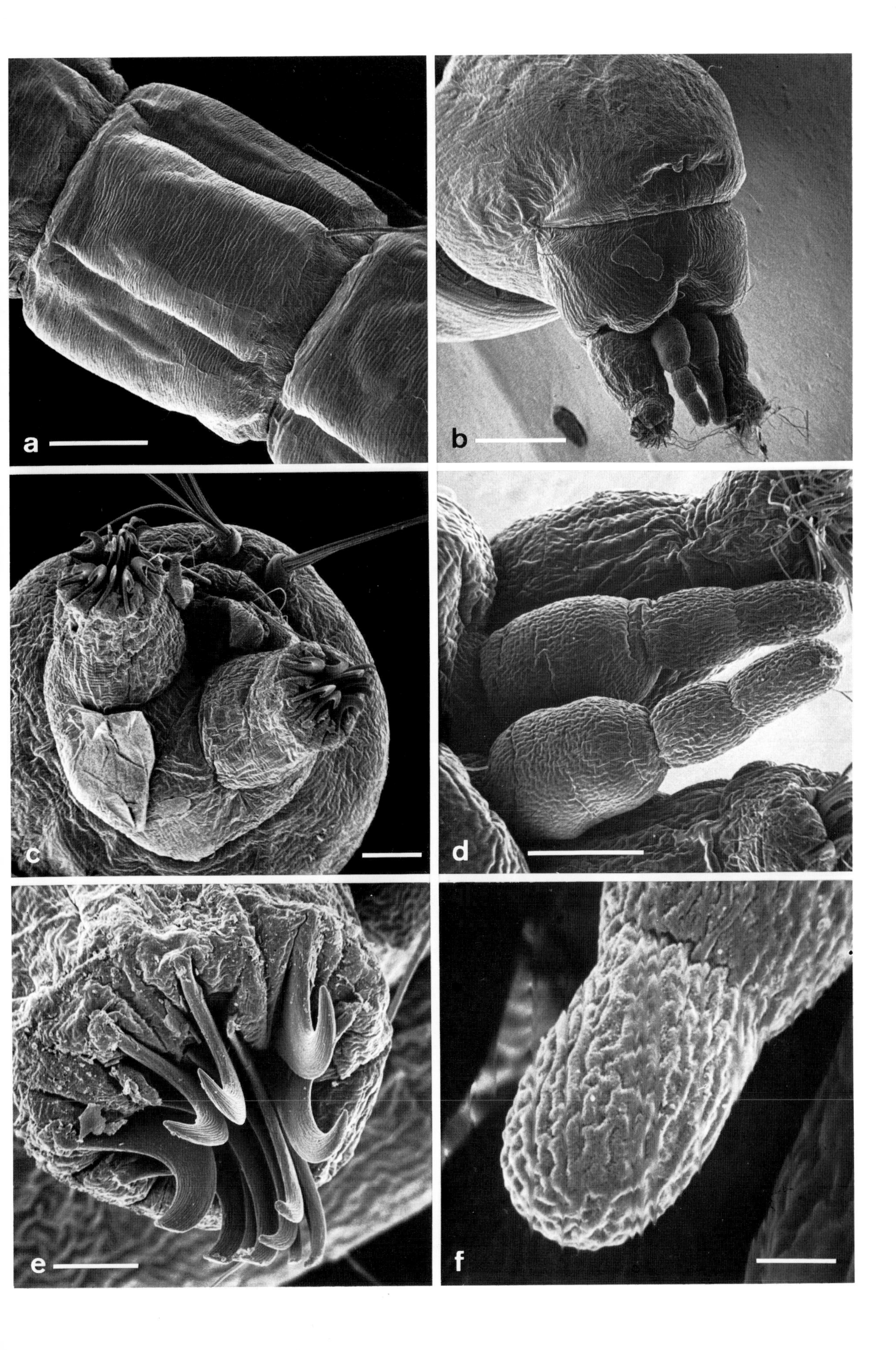
a
b
c
d
e
f

2.25.10 Larvae of the Bibionidae (March Flies)

The blackish-brown larvae of the march fly, *Penthetria holosericea*, live preferably in the fallen leaves of *Corylus avellana, Acer pseudoplatanus* and *Prunus padus*, as well as under alder trees (Brauns, 1954).

The form of the larvae is remarkable for the longitudinal rows of fleshy, thorn-like processes, which are covered with fine, false setae (Fig. 209; Plate 185). The number of body segments is increased by pseudosegmentation. The larvae are holopneustic; ten pairs of spiracles are situated on short, hairless protuberances (respiration tubes), which are easily distinguishable between the segmental ridges and the hairy processes (Plate 185c).

Penthetria belongs to the Pleciinae subfamily; other soil-dwelling species, of a different habitus, are in the genus *Bibio* and *Dilophus* (= *Philia*) of the Bibioninae subfamily (Fig. 210).

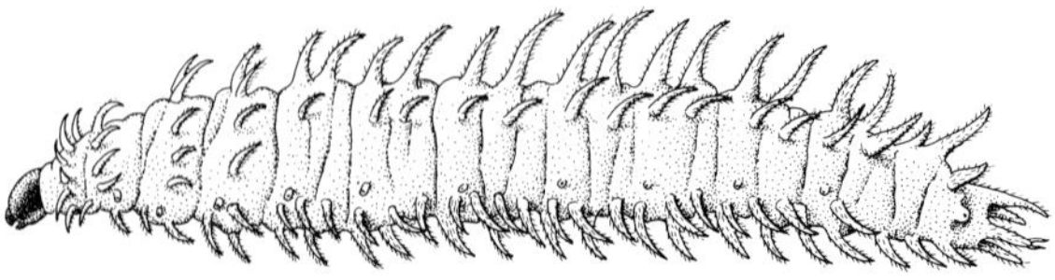

Fig. 209. Larva of the Bibionidae (march flies), *Penthetria holosericea.* (Brauns, 1954)

Plate 185. Larva of the Bibionidae (march flies).
a Overview, lateral. The head is situated on the *right* side. 1 mm.
b Head, frontal. 50 µm.
c Trunk segments (abdomen), oblique dorsal, with spiracular tubercles (*arrows*). 200 µm.
d Head, ventral. 100 µm.
e Surface of a lateral segmental process. 10 µm.
f Spiracle. 25 µm

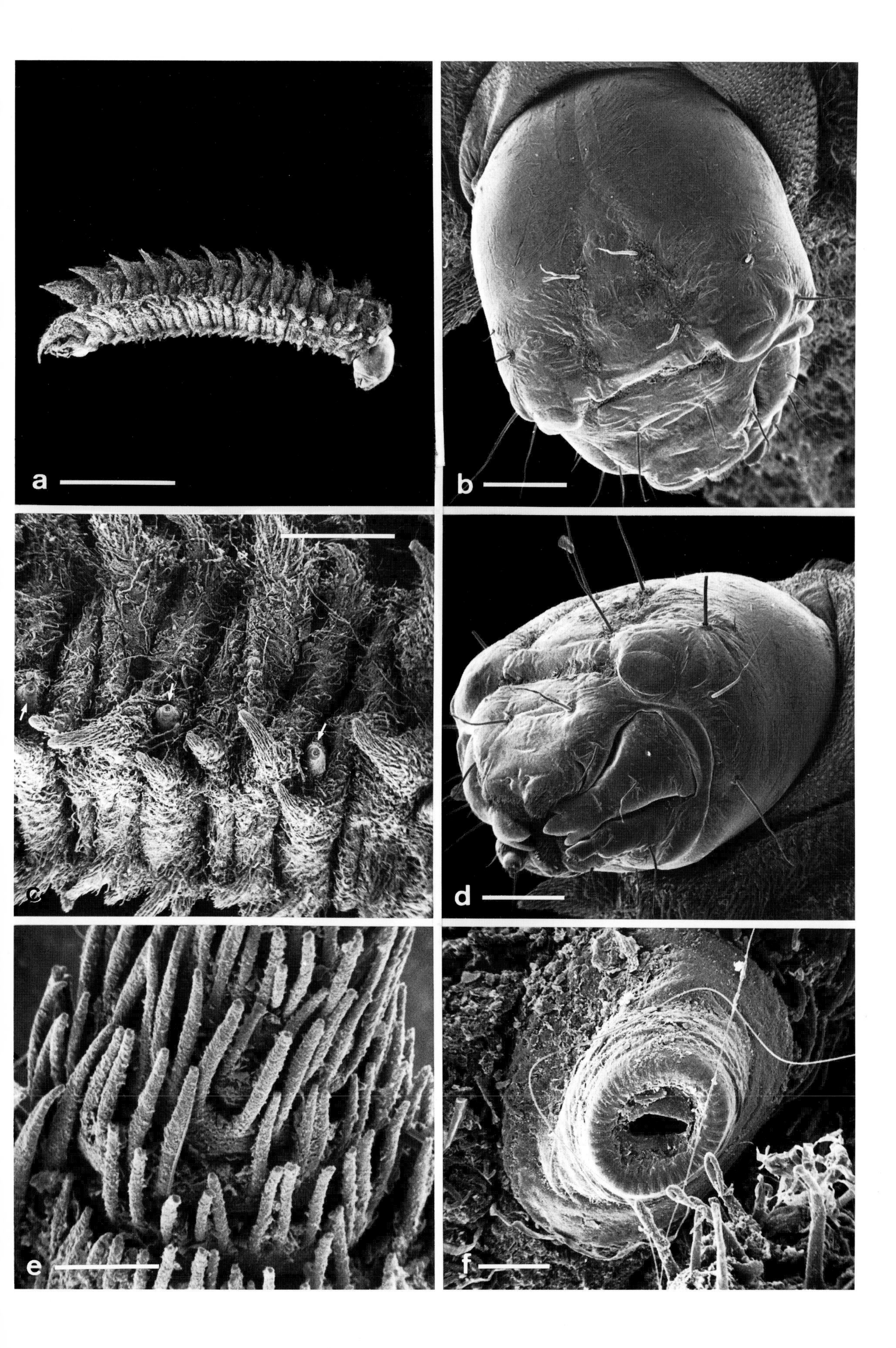
a
b
c
d
e
f

2.25.11 Importance of Bibionidae Larvae to Soil Biology

Besides the Tipulidae larvae, the larvae of the Bibionidae are of considerable importance to soil biology as they often occur in large populations and participate in humification by intensive consumption of fallen leaves with their chewing mouthparts. Dunger (1983) reported that the larvae of the garden march fly (*Bibio hortulanus*) were found in populations of 3000 to 12000 individuals per m^2. When such high densities are reached, great damage to forestry and agricultural crops can be caused by their feeding on roots. Under normal conditions of several hundred larvae per m^2, they make a considerable contribution to the decomposition of plant debris by feeding preferably on decaying leaves (saprophytophages). They colonize a range of different habitats from deciduous and mixed woodland to conifer forest of pine trees and spruce monocultures (Brauns, 1954).

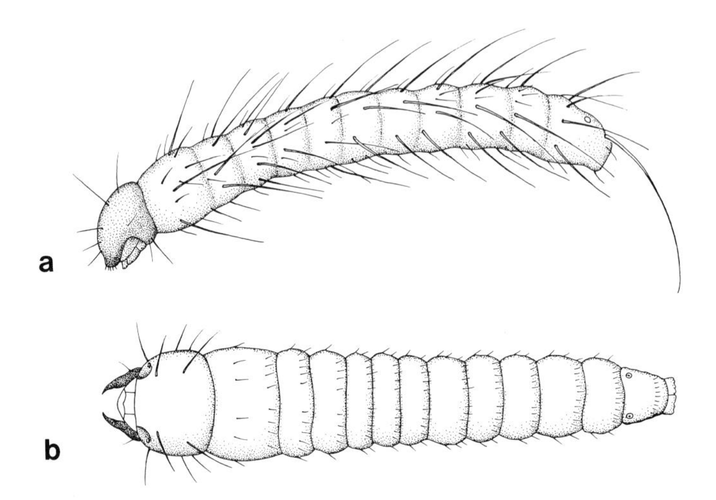

Fig. 210. Larvae of the Bibionidae (march flies), first instar. (Brauns, 1954).
a *Bibio marci,* **b** *Dilophus febrilis*

Plate 186. Larva of the Bibionidae (march flies).
a Anterior of the body, ventral. 100 μm.
b Mouthparts. 25 μm.
c Maxillary palp, distal, with sensilla. 5 μm.
d Part of the cleaning apparatus on the upper lip (labrum). 5 μm.
e Hair base on the head capsule. 5 μm.
f Cuticle on the vertex of the head capsule along the sutura medialis. 25 μm.
g Pit sensillum of the upper lip. 5 μm

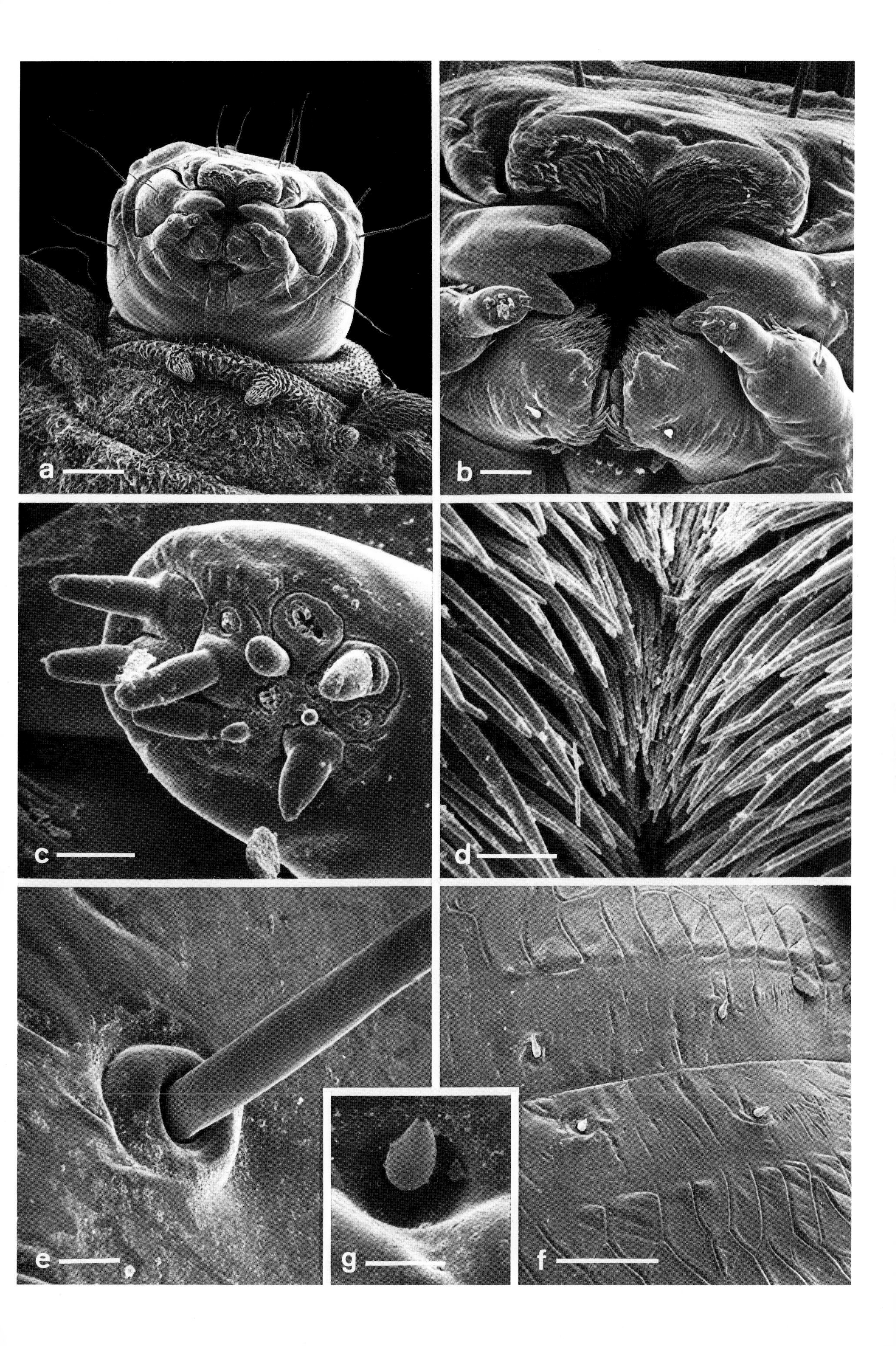

a
b
c
d
e
g
f

2.25.12 Larvae of the Scatopsidae (Black Scavenger Flies)

The terrestrial Scatopsidae larvae (Fig. 211), which can be up to 7-mm-long, occasionally occur in large, isolated populations in leaf litter of deciduous woods. Apart from that, they are primarily encountered in excrement.

In the first instar their respiratory spiracles have a metapneustic, in the second an amphipneustic and in the third and fourth a peripneustic distribution. The spiracles protrude from the body surface, the pair on the thorax being especially conspicuous, but even more distinctive are the terminal spiracles at the tip of respiration tubes (Fig. 212; Plate 187), the length of which is dependent on the moisture of the preferred habitat (Brauns, 1954).

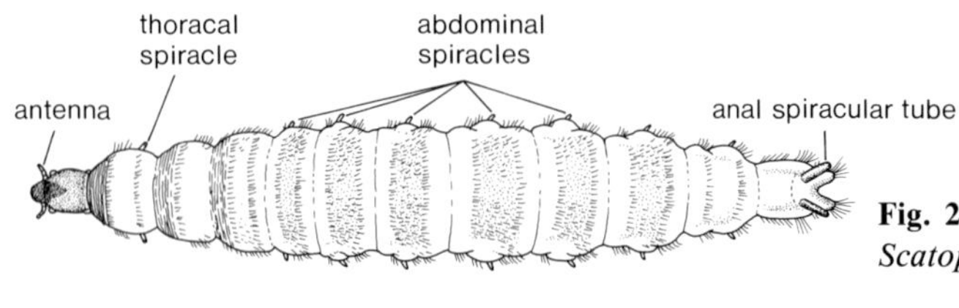

Fig. 211. Larva of the Scatopsidae (black scavenger flies), *Scatopse* sp., habitus dorsal. (Brauns, 1954)

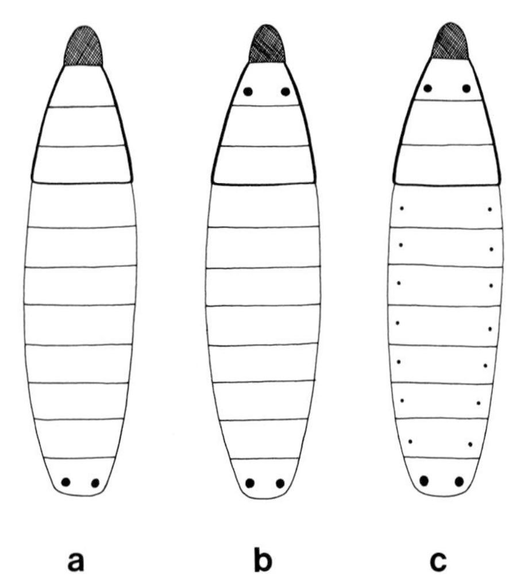

Fig. 212. Three types of spiracular distribution in Diptera. (After Keilin, 1944 from Hennig, 1973).
a Metapneustic; **b** amphipneustic; **c** peripneustic

Plate 187. Larva of the Scatopsidae (black scavenger flies).
a Overview, oblique dorsal, the head is situated on the *right*. 0.5 mm.
b Anterior of the body, oblique, dorsal. The *arrow* shows a thoracic, spiracular process. 100 μm.
c Anal region with dorsal spiracular tubes. 50 μm.
d Antenna with sensilla. 10 μm.
e Anal region, dorsal view, with spiracular tubes. 100 μm.
f Spiracular tube, distal. 10 μm

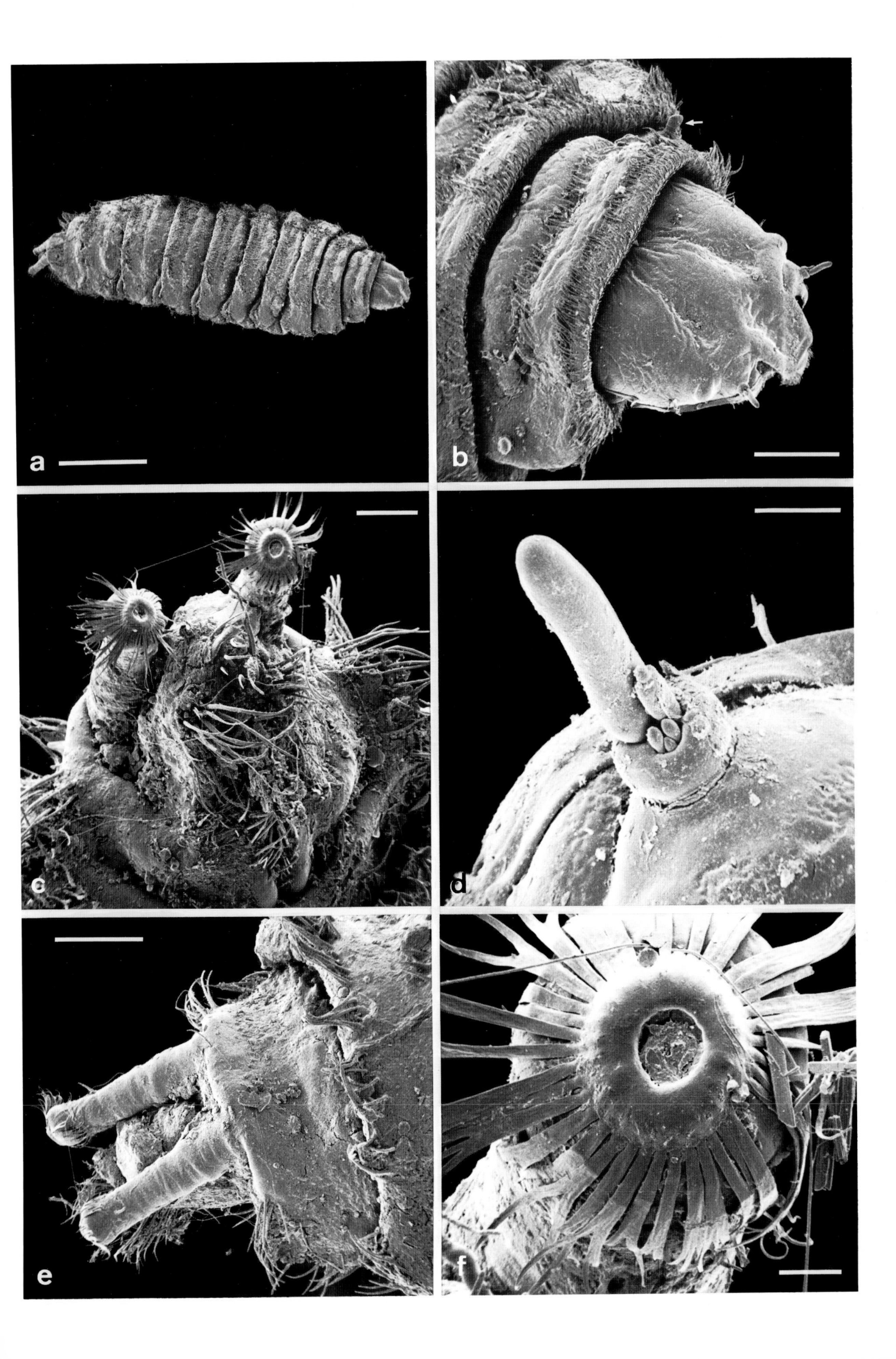
a
b
c
d
e
f

2.25.13 Larvae of the Stratiomyidae (Soldier Flies)

Soldier flies are frequently found on composites and umbellifers on sunny days. Their larvae are terrestrial, hygropetric and aquatic (Fig. 213). Aquatic forms are distinguished from terrestrial forms firstly by the large ring of hairs on the terminal pair of spiracles on the last elongated abdominal segment and secondly by the well-functioning ciliary organs on their mouthparts.

The larvae of the stratiomyids are 20–25-mm-long, spindle-shaped, often slightly dorso-ventrally flattened and distinctly segmented. Their body surface is opaque and coloured deep black. It is also calcareous and has a rought leathery appearance. Electron microscopy showed this shagreen-like surface to be a highly ordered surface structure (Plates 188 and 189).

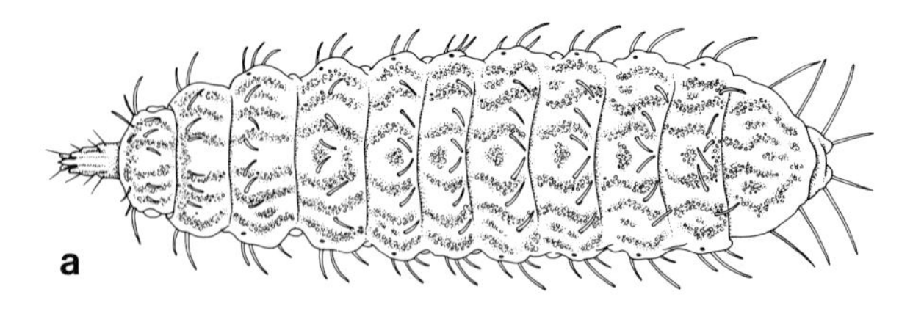

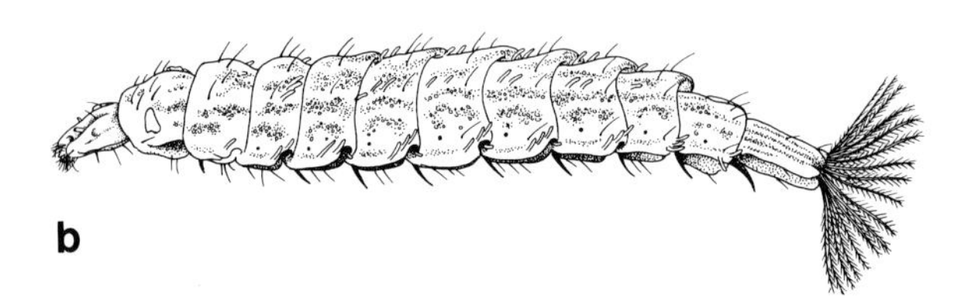

Fig. 213. Larvae of the Stratiomyidae (soldier flies). (Brauns, 1954).
a Terrestrial life form; **b** semi-aquatic life form

Plate 188. Semi-aquatic larva of the Stratiomyidae (soldier flies).
a Overview, lateral. The head region is situated on the *left.* 0.5 mm.
b Anal region, lateral, with spiracular tube (dorsal) and disarrayed circular hair organ. 100 µm.
c Anterior of the body, ventral. The *arrows* mark the spiracular tubercles. 200 µm.
d Anal region, oblique ventral, with spiracular tube (on the *left*) and disarrayed circular hair organ. 100 µm.
e Head, frontal. 50 µm

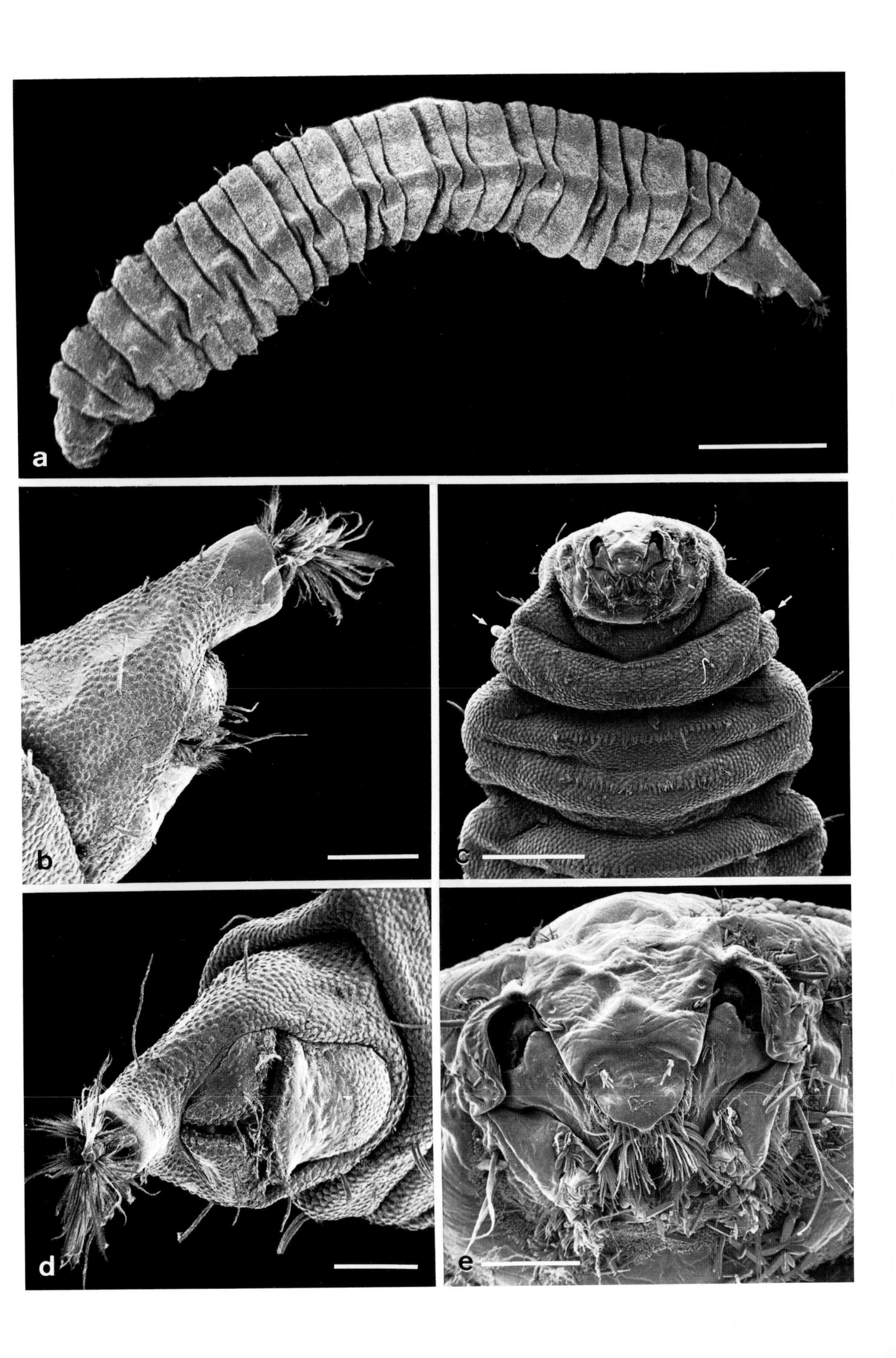
a
b
c
d
e

2.25.14 Life of the Stratiomyidae Larvae

The larvae colonize different biotopes. An important role in their obvious tolerance with respect to many ecological factors is played by their tough, but finely structured body surface (Plates 188 and 189). It protects them from mechanical influences and makes the larvae largely independent of moisture fluctuations. Thus, terrestrial larvae can live under semi-aquatic conditions or survive a drought. Their preferred biotope is in the soil of deciduous woodland. The larvae are microphytophagous feeders on algae or saprophages, so that stratiomyid larvae are also, to a modest extent – depending on the population density – of soil biological importance.

Plate 189. Semi-aquatic larva of the Stratiomyidae (soldier flies).
a Anterior of the body, lateral. 100 μm.
b Trunk segments, lateral. 100 μm.
c Thoracic, spiracular tubercle. 10 μm.
d Surface of the trunk with calcareous grains. 25 μm.
e Anal spiracular tube, distal, with circular hair organ. 25 μm.
f Cuticle with calcareous grains. 5 μm

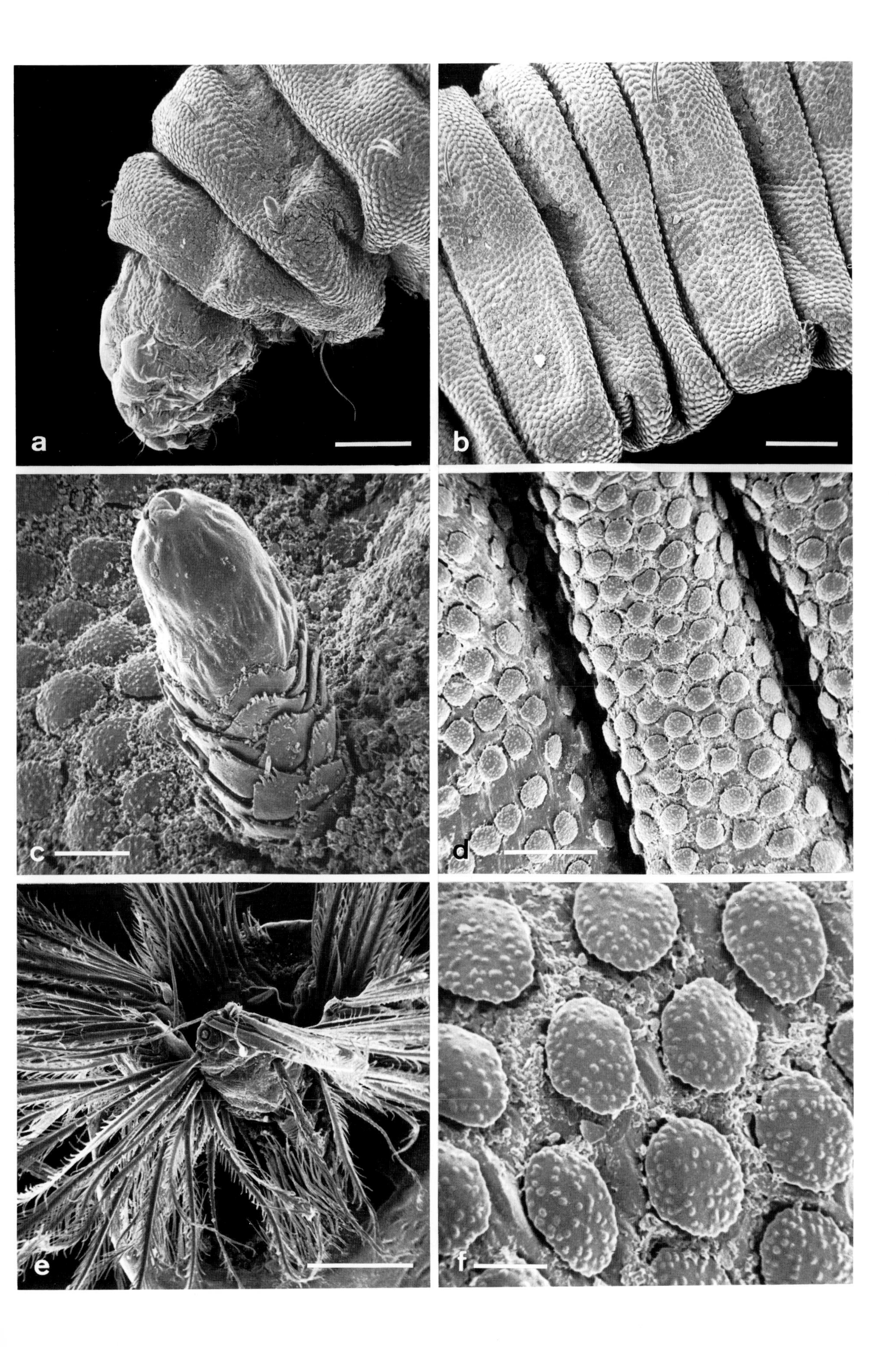
a
b
c
d
e
f

2.25.15 Larvae of the Asilidae (Robber Flies)

The whitish, often transparent larvae of the asilids have a thin, cylindrical form with distinct segmentation (Fig. 214; Plate 190). They crawl under bark and in soil litter using their transverse callosities situated ventrally on the second to the seventh segments. The asilid larvae possess an amphipneustic tracheal system (Fig. 212); they breathe through two pairs of spiracles found laterally on the prothorax and on the penultimate abdominal segment. The sclerotized, pointed head bears unisegmented, short feelers, usually small eyes and prominent mouthparts, consisting of a "hook-like labrum, knife-like mandibles and large, broad maxillae with cylindrical maxillary palps" (Brauns, 1954) (Plate 190). They are primarily zoophagous feeders (Fig. 215) (Musso, 1983).

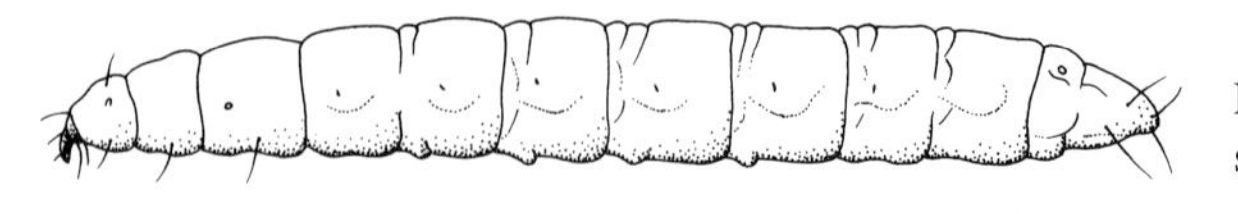

Fig. 214. Larva of the Asilidae (robber flies), *Dysmachus* sp., habitus lateral. (Brauns, 1954)

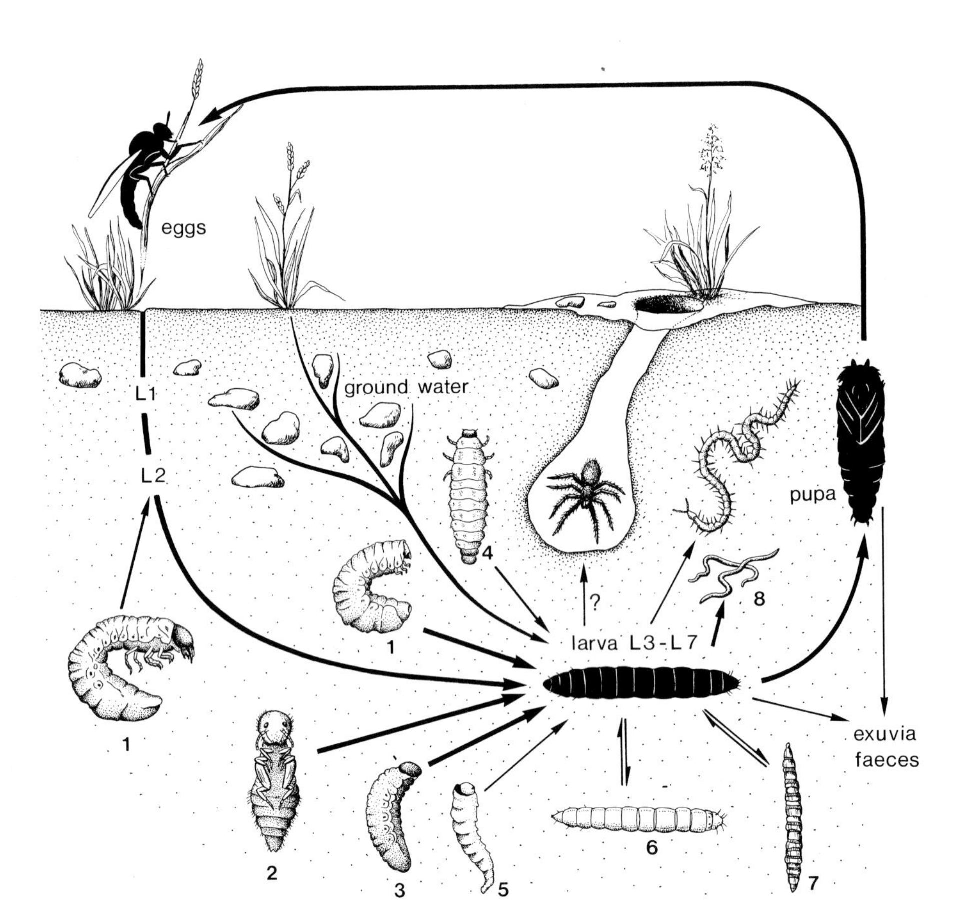

Fig. 215. Life cycle and trophic relationships of the larvae of *Machimus rusticus* (Asilidae). (After Musso, 1983). The larva *L1* lives on egg reserves, *L2* and especially *L3–L7* are predatory, feeding especially on insect larvae and additionally absorbing organic and inorganic substances from groundwater. *1–4* Larvae and pupae of the Coleoptera (Scarabaeidae, Curculionidae, Chrysomelidae); *5–7* larvae of the Diptera (Muscidae, Asilidae, Tabanidae). Their enemies are probably cave spiders, geophilids and worms (*8*)

Plate 190. Larva of the Asilidae (robber flies).
a Anterior of the body, ventral. 0.5 mm.
b Abdomen, terminal, dorsal. 1 mm.
c Anterior of the body, lateral. 100 μm.
d Trunk segments, lateral (the dorsal side is on the *right*). 250 μm.
e Head, oblique frontal, with pointed mouthparts. 100 μm.
f Anal region, ventral. 250 μm

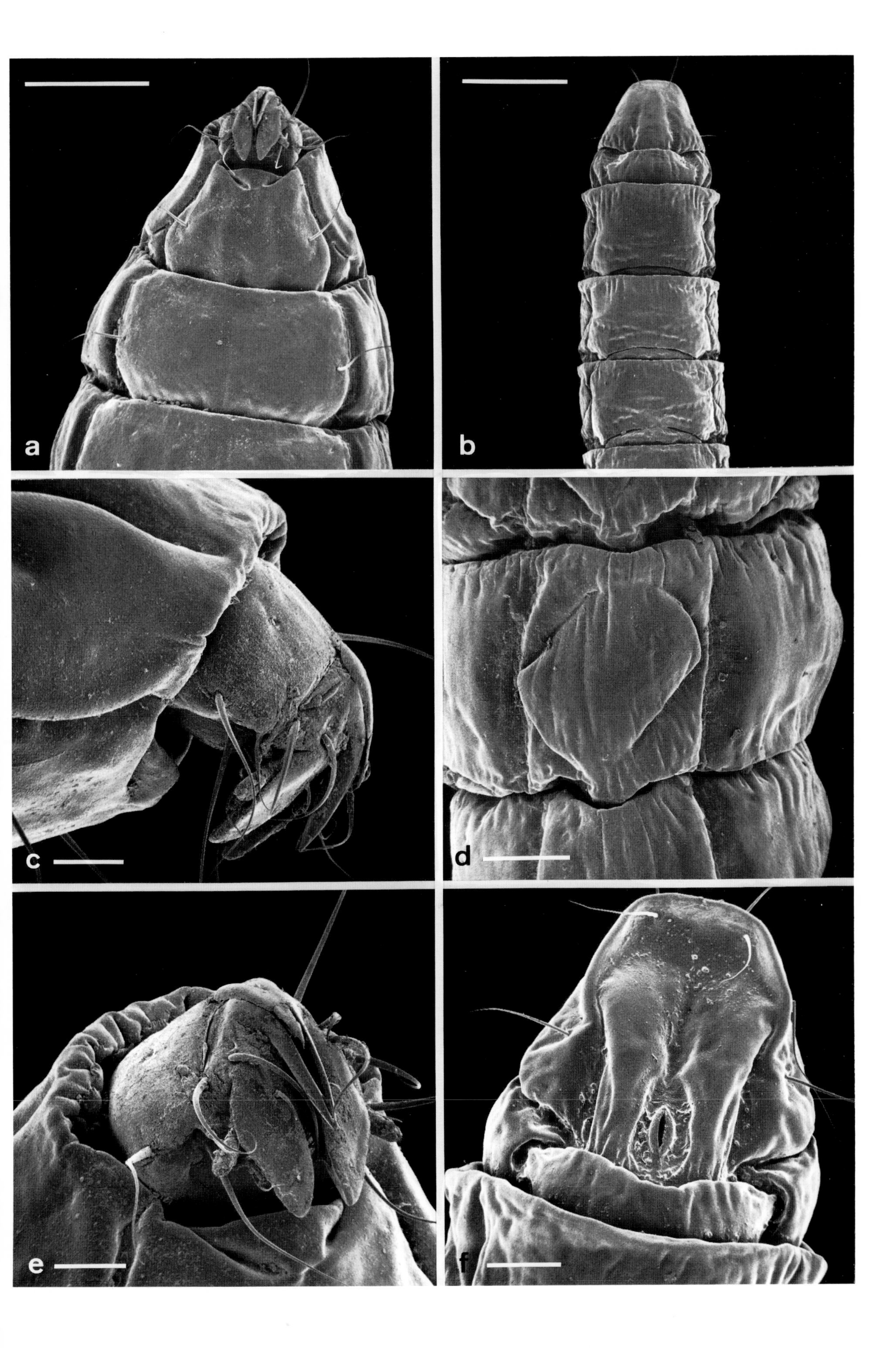
a
b
c
d
e
f

2.25.16 Larvae of the Muscidae (Flies)

The species of the genus *Fannia* (Fig. 216) belong to the terrestrial fly larvae which live gregariously in the soil and are thus of importance to soil biology. The maximum length of the larvae is 9 mm. Striking feathery marginal processes (Plates 191 and 192) emphasize the dorso-ventral flattening of the larval body. These bizarre processess are arranged along the lateral edges of both sides of the body from the mesothorax to the seventh abdominal segment. The terminal abdominal segment bears three pairs of processes in a semi-circle which, together with the lateral processes, form a complete border (Plate 192a). Only the prothorax lacks the feathery processes. Instead it bears two long spiny processes (presumably the remaining shafts of feathery processes) which point forwards and extend beyond the head when it is retracted into the thorax.

Brauns (1954) described the *Fannia* larvae as a characteristic form of the F-layer in deciduous and mixed woodland. Within this woodland biocenosis they feed saprophytophagously on rotting leaves in the leaf litter.

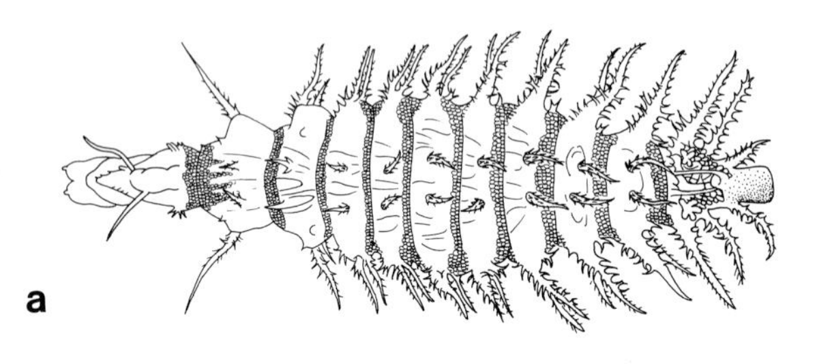

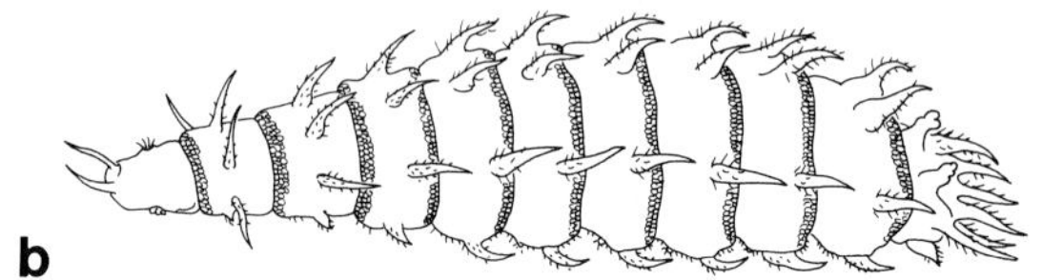

Fig. 216. Larvae of the Muscidae (flies). (Brauns, 1954).
a *Fannia fuscula* (Fanniinae), habitus dorsal. **b** *Fannia canicularis* (Fanniinae), habitus lateral

Plate 191. Larva of the Muscidae (*Fannia*), flies.
a Overview, oblique ventral. The head (*lower right* edge of the picture) is retracted. 0.5 mm.
b Overview, lateral. The dorsal side is below. 0.5 mm.
c Anterior of the body, ventral. 250 μm.
d Anterior of the body, lateral. 200 μm.
e Prothorax, ventral. Except for the antennae the head is retracted. 100 μm.
f Head pit. All that is visible of the head are the antennae and a sensory plate (*arrow*). 25 μm

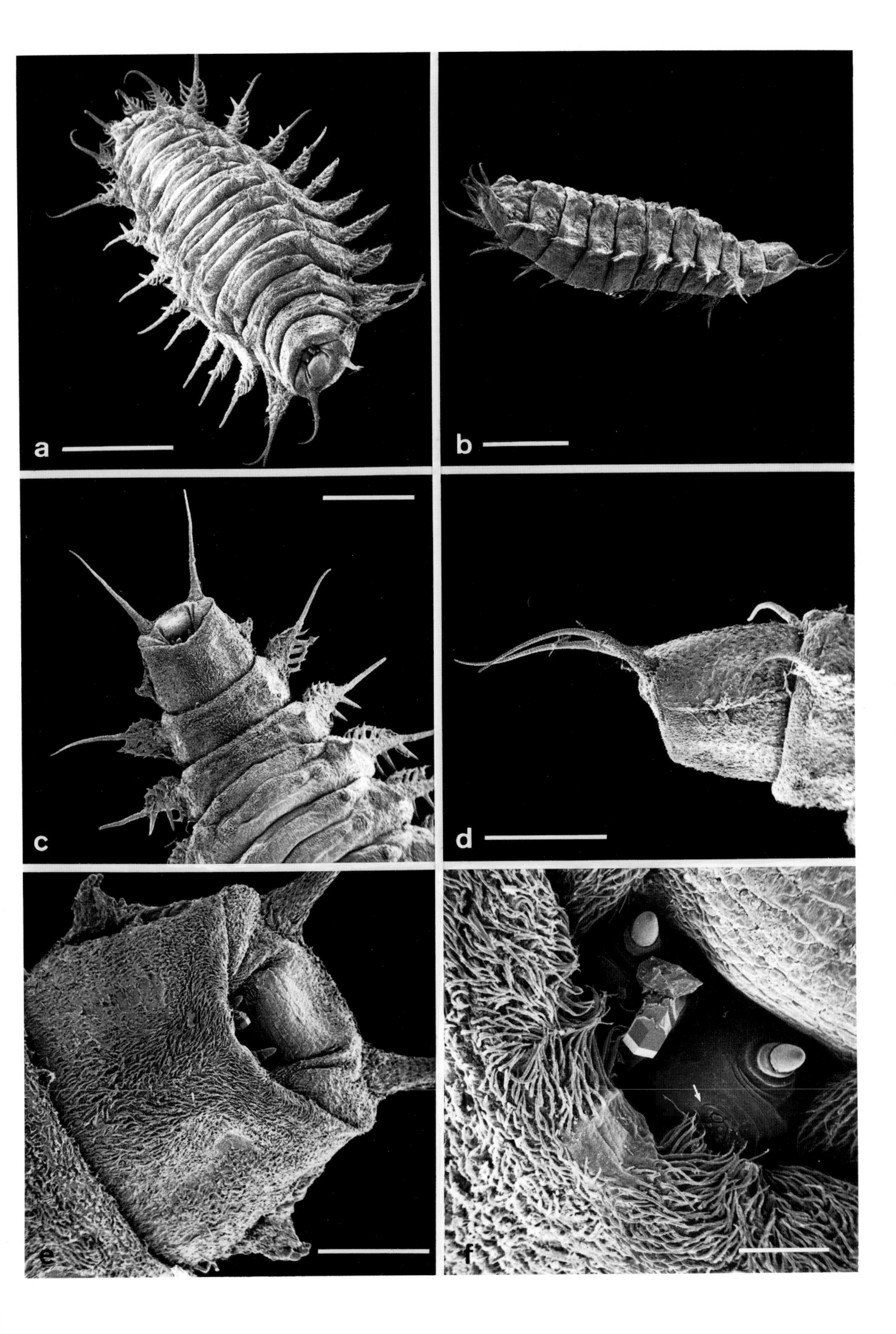
a
b
c
d
e
f

2.25.17 Energy Turnover of *Fannia polychaeta* (Muscidae)

The energy flow of soil-dwelling dipteran larvae was investigated in a beech forest (Luzulo-Fagetum) in Solling (W. Germany) (Altmüller, 1979). Nineteen nematoceran and brachyceran families were determined. Larval abundance ranged from 578 ind/m^2 in March to 14740 ind/m^2 in October.

Fannia larvae (*F. polychaeta*) showed their highest increase in biomass between the second and third instars. During this phase the energy content of a larva multiplied from an average of 17% to 73% of its maximal value (Fig. 217). Like most other soil-dwelling dipteran larvae, the decisive growth period takes place in autumn. At this time the larvae assimilate around 76% of their total annual turnover of energy. Towards the end of winter and in spring their respiration rate is low, and food is hardly consumed.

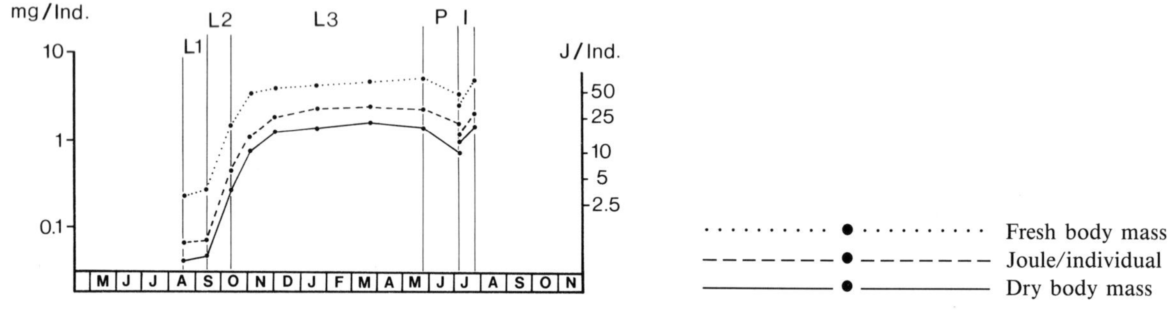

Fig. 217. Growth diagram for *Fannia polychaeta* (Muscidae). (After Altmüller, 1979). *L1–L3* Larval instars; *P* pupa; *I* imago

Plate 192. Larva of the Muscidae (*Fannia*), flies.

a Abdomen, terminal, dorsal, with marginal feathery processes. 0.5 mm.

b, c Cuticle on the ventral side. The hair in **b** is covered by secretion. 50 μm, 200 μm.

d Marginal, feathery process, dorsal. 200 μm.

e Cuticle on the ventral side. 5 μm.

f Anal region, ventral, with everted hindgut. 100 μm.

g Tuft of false bristles on the venter. 10 μm

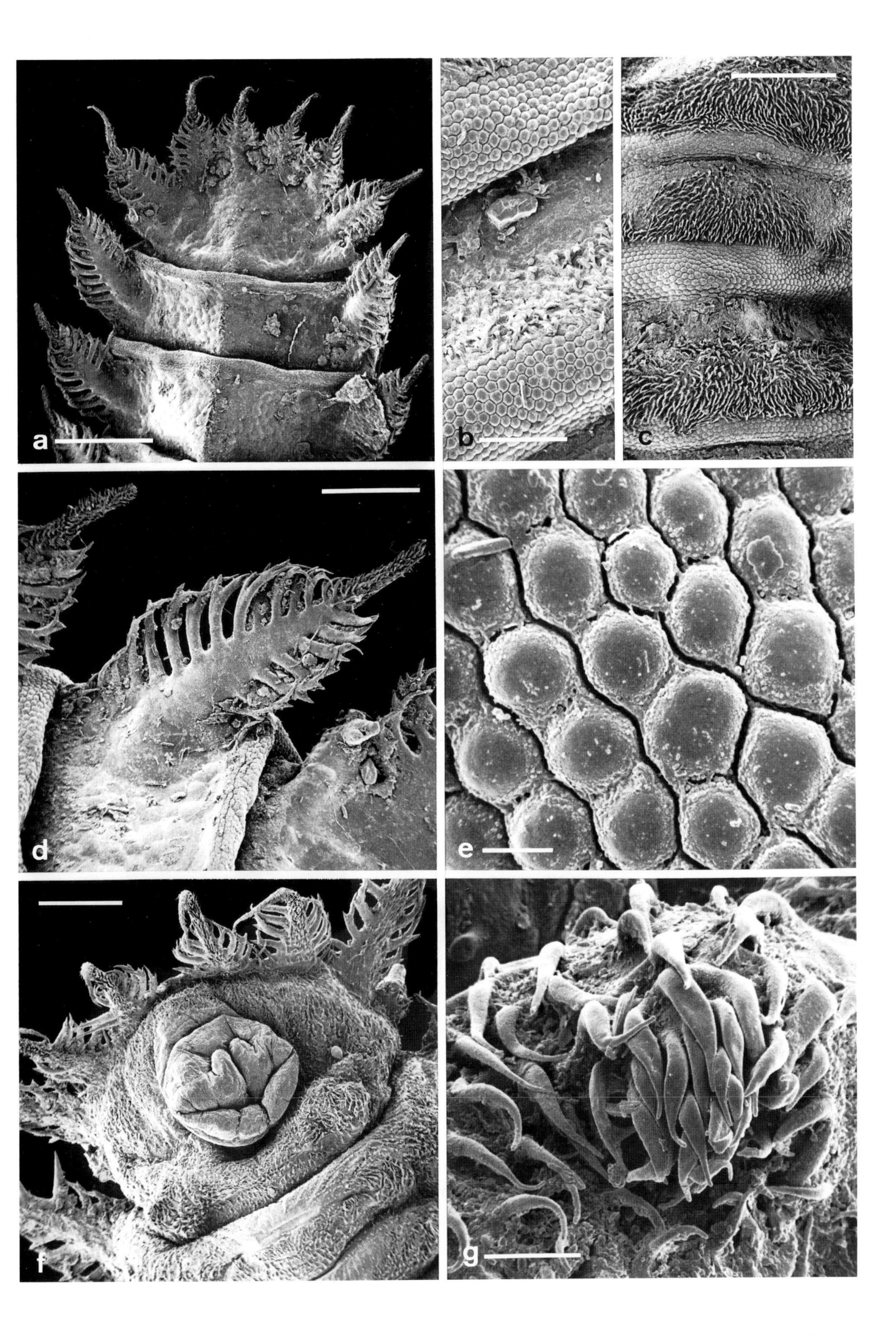

a
b
c
d
e
f
g

3 Preparation Techniques for Scanning Electron Microscopy

3.1 Selection of Specimens

Animals should be carefully removed from their microhabitats, if possible by hand. This firstly ensures that specimens are undamaged, and secondly reduces the amount of accompanying contamination. Soil particles generally adhere to those samples obtained from Berlese funnels.

3.2 Fixation and Dehydration of the Specimens

Many soft-bodied soil animals show a high degree of shrinkage as a result of dehydration. In this case, chemical fixation is essential.

The least complicated method of fixation is in 70% – 80% alcohol. The animals become slightly hardened, but still remain elastic. Delicate, soft-bodied forms should be treated with an osmium tetroxide fixation which stabilizes and hardens the tissues (e.g. 2% OsO_4 in Michaelis buffer at pH 7.2; in principle, all the fixatives used for transmission electron microscopy are also useful for scanning electron microscope fixation). In order to prevent detachment of any surface layers, fresh, green fir needles and beech leaves were neither fixated nor dehydrated in alcohol. In this case rapid drying in a water jet vacuum proved suitable. Freeze drying can also be used in such cases.

Fixated specimens were dehydrated in an ascending series of alcohol concentrations (50, 70, 96, 100%; non-methylated alcohol, pure ethanol) and pure acetone. The objects were then transferred to the critical point apparatus and dried after exchanging the acetone for liquid CO_2 at 37 °C under pressure. This method prevents an increase in the surface tension which causes strong shrinkage of the specimens in the final phase of drying. The average duration for the treatment of the specimens at room temperature is:

1. Osmium tetroxide fixation	2 h
Washing with distilled water	2×15 min
50, 70, 96% alcohol	2×15 min
100% alcohol	2×20 min
acetone	1 – 3×30 min
critical point drying	1 – 2 h
2. Alcohol fixation (70 – 80%)	>12 h, mostly overnight, then dehydration and drying as in 1

For delicate, soft-bodied animals the times can be halved, larger specimens with an impermeable cuticle require longer treatment. Warming the alcohol to 30°– 40 °C ensures faster penetration. Specimens which are stored in 70% alcohol for longer periods should, if possible, be rapidly dehydrated to prevent excessive shrinkage.

3.3 Mounting and Sputtering of the Specimens

Contrary to the usual method, specimens were mounted above the surface of the aluminium sample plates (specimen holder with ∅ 1 cm). Pins were bent into shape and attached to the holder, usually at its edge, with nail varnish or colloidal silver adhesive (Fig. 218). Afterwards the pinheads were covered with the same adhesive and the dried specimen was transferred and attached to the pin using a slightly moistened, fine brush. It is better to use stable extremities or less important parts of the body to attach specimens. Adhesion itself must be carried out swiftly and it is often necessary to press the specimen somewhat deeper into the rapidly forming skin on the adhesive with the hairs of a soft brush. On the other hand, the adhesive must on no account be too liquid, to ensure that important parts of the body are not immersed in it. Mounting should always be carried out under an illuminated stereo-microscope. After a sufficient drying period has elapsed (at least overnight) the specimens are examined and if necessary "cleaned" with a slightly moistened brush.

Figure 218 shows a mounted specimen. A Collembola is attached to the pinhead by one of its right legs, leaving the rest of the body exposed. The specimens are subsequently sputtered with a thin layer of gold using a sputter coater in which the specimens are plated with gold by means of the cathode sputtering method in an argon-regulated fine vacuum. Thus, the surface becomes conductive and releases the secondary electrons necessary for the formation of a picture.

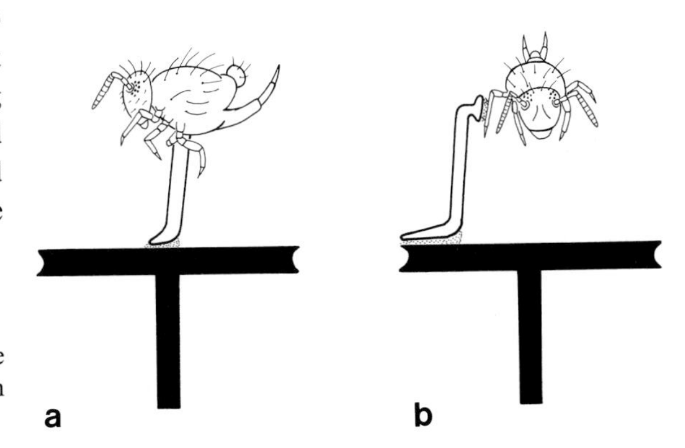

Fig. 218. Preparation of a Collembola on a "pin" in the lateral (**a**) and frontal (**b**) position. By rotating the specimen holder the animal is visible from almost every angle

3.4 Microscopy

The specimen holder was not exposed to the electron beam in the manner normally used for scanning electron microscopy. The holder surface was positioned parallel to the beam axis (Fig. 219). In the illustrated example the focussed electron beam strikes the right side of the specimen's body. The secondary electrons (dashed lines), produced by the primary beam, are diverted onto the scintillator tip by a collector grid, which converts them into photons and directs them through a light conductor to a photomultiplier and videoamplifier. The signal produced by the beam forms a dot on a screen. The scan coils, which are integrated into the beam pathway, direct the beam line by line over the object. A picture finally consists of 1000 lines.

The mounting of specimens as described in Chapter 3.3 as well as the vertical position of the holder has several advantages:

1. The attachment area between the specimen and the holder is small.
2. The specimen is accessible from almost every angle, as the plate can be rotated 360°. This is especially favourable for overview microscopy.
3. There ist no background interference from the mounting plate. When the specimen holder is vertically exposed the surroundings appear black (so-called cosmic effect). Although the mounting pin is also photographed it can easily be removed from the positive by retouching.

4. Contrary to widespread opinion, there is little tendency for the specimen to become statically charged. When the specimen is mounted directly on the surface of the plate, considerable interference occurs especially when the plate itself produces a strong electron signal. Static interference usually leads to worthless photographs. Only large objects, which must be in many cases halved should be vertically mounted onto the plate surface. Even in this case vertical exposure in the microscope is possible, permitting an all-round view against a dark background. The smaller the object, the more favourable it is to mount it on a pin above the specimen plate.

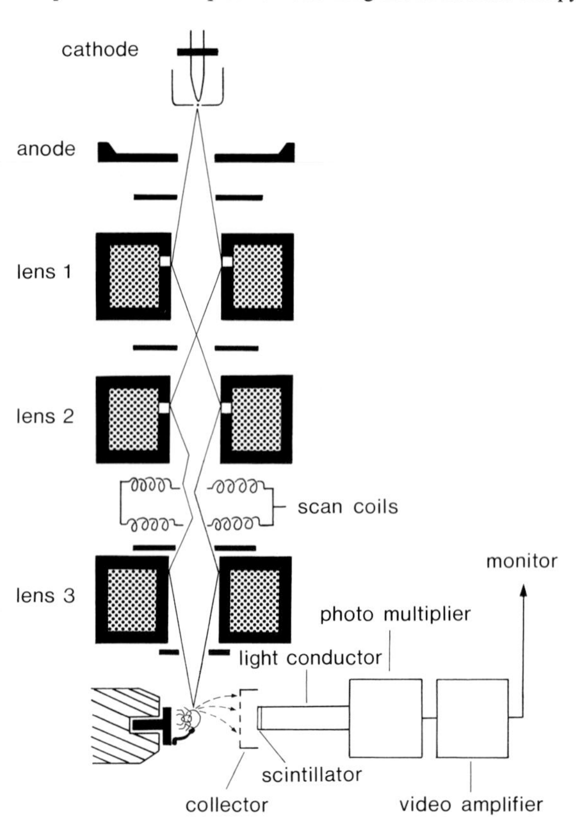

Fig. 219. Simplified schema of the electron pathway in a scanning electron microscope. The specimen holder plate is situated in the vertical position. The animal appears in the lateral position on the monitor screen

References

General Introduction

Beck L (1983) Zur Bodenbiologie des Laubwaldes. – Verh Dtsch Zool Ges Bonn 1983, 37–54

Beck L, Mittmann H-W (1982) Zur Biologie eines Buchenwaldbodens 2. Klima, Streuproduktion und Bodenstreu. Carolinea 40:65–90

Block W (1982) Cold hardiness in invertebrate poikilotherms. Comp Biochem Physiol 73A:581–593

Brauns A (1968) Praktische Bodenbiologie. Gustav Fischer Verlag, Stuttgart

Brown AL (1978) Ecology of soil organisms. Heineman, London

Burges A, Raw F (1967) Soil Biology. Academic Press, London New York

Dickinson CH, Pugh GJF (1974) Biology of plant litter decomposition. Academic Press, London

Drift J. van der (1951) Analysis of the animal community in a beechforest floor. Tijdschrift voor Entomologie 94:1–118

Dunger W (1958) Über die Zersetzung der Laubstreu durch die Boden-Makrofauna im Auenwald. Zool Jb Syst 86:139–180

Dunger W (1974, 1983) Tiere im Boden. Die Neue Brehm-Bücherei 327 (2., 3. Aufl.), Ziemsen Verlag, Wittenberg-Lutherstadt

Edney EB (1977) Water balance in land arthropods. Springer, Berlin Heidelberg New York

Eisenbeis G (1983a) Kinetics of water exchange in soil arthropods. In: Lebrun P et al.: New trends in soil biology, pp 417–425, Dieu-Brichart, Louvain-La-Neuve (Belgium)

Eisenbeis G (1983b) Kinetics of transpiration in soil arthropods. In: Lebrun P et al: New trends in soil biology, pp 626–627, Dieu-Brichart, Louvain-La-Neuve (Belgium)

Fitter AH, Atkinson D, Read DJ, Usher MB (eds) (1985) Ecological interactions in soil: plants, microbes and animals. Blackwell, Palo Alto

Funke W (1971) Food and energy turnover of leaf-eating insects and their influence on primary production. Ecol Studies 2:81–93

Ghilarov MS (1964) Die Bestimmung im Boden lebender Larven der Insekten (russisch) Wiss. Akademie UdSSR, Moskau

Gisin H (1943) Ökologie und Lebensgemeinschaften der Collembolen im Schweizerischen Exkursionsgebiet Basels. Rev Suisse Zool 50:131–224

Herlitzius H (1982) Zur Phänologie des Streuabbaus in Waldökosystemen. Verh Ges Ökol Mainz 1981, 10:27–34

Herlitzius R, Herlitzius H (1977) Streuabbau in Laubwäldern. Oecologia 30:147–171

Joosse ENG (1983) New developments in the ecology of Apterygota. Pedobiologia 25:217–234

Kevan DK McE (1962) Soil animals. Witherby Ltd, London

Kubiena W (1953) Bestimmungsbuch und Systematik der Böden Europas. Enke, Stuttgart

Kühnelt W (1950) Bodenbiologie. Herold, Wien

Kühnelt W (1961) Soil biology. Faber and Faber, London

Müller G (1965) Bodenbiologie. VEB Gustav Fischer Verlag, Jena

Palissa A (1964) Bodenzoologie. Wiss. Taschenbücher 17, pp 1–180, Akademie-Verlag, Berlin

Petersen H, Luxton M (1982) A comparative analysis of soil fauna populations and their role in decomposition processes. Oikos 39:288–388

Pimm SL (1982) Food webs, Chapman and Hall Ltd, London New York

Potts WTW, Parry G (1964) Osmotic and ionic regulation in animals. Pergamon Press, Oxford

Rahn H, Paganelli CV (1968) Gas exchange in gas gills of diving insects. Respir Physiol 5:145–164

Schaefer M (1982) Zur Funktion der saprophagen Bodentiere eines Kalkbuchenwaldes: ein langfristiges Untersuchungsprogramm im Göttinger Wald. Drosera 82:75–84

Schaller F (1962) Die Unterwelt des Tierreiches. Verständl. Wissenschaft 78, 1–126, Springer, Berlin Göttingen Heidelberg

Scheffer F, Schachtschabel P (1979) Lehrbuch der Bodenkunde. Enke, Stuttgart

Sømme L (1982) Supercooling and winter survival in terrestrial arthropods. Comp Biochem Physiol 73A:519–543

Topp W (1981) Biologie der Bodenorganismen. – UTB, Quelle und Meyer, Heidelberg

Trolldenier G (1971) Bodenbiologie. Franckh'sche Verlagshandlung, Stuttgart

Vannier G (1983) The importance of ecophysiology for both biotic and abiotic studies of the soil. In: Lebrun P et al: New trends in soil biology, pp 289–314, Dieu-Brichart, Louvain-La-Neuve (Belgium)

Verdier B (1975) Etude de l'atmosphère du sol. Rev Ecol Biol Sol 12:591–626

Wallwork JA (1970) Ecology of soil animals. McGraw-Hill, London New York

Wallwork JA (1976) The distribution and diversity of soil fauna. Academic Press, London New York

Zachariae G (1965) Spuren tierischer Tätigkeit im Boden des Buchenwaldes. Forstwiss Forsch 20:1–68

Zinkler D (1966) Vergleichende Untersuchungen zur Atmungsphysiologie von Collembolen (Apterygota) und anderen Bodenkleinarthropoden. Z Vergl Physiol 52:99–144

Zinkler D (1983) Ecophysiological adaptations of litter-dwelling Collembola and tipulid larvae. In: Lebrun P et al: New trends in soil biology, pp 335–343, Dieu-Brichart, Louvain-La-Neuve (Belgium)

Systematic Chapters

Araneae

Albert R (1977) Struktur und Dynamik der Spinnenpopulationen in Buchenwäldern des Solling. Verh Ges Ökologie, Göttingen 1976, pp 83–91

Baehr B (1983) Vergleichende Untersuchungen zur Struktur der Spinnengemeinschaften (Aranea) im Bereich stehender Kleingewässer und der angrenzenden Waldhabitate im Schönbuch bei Tübingen. Dissertation, Univ Tübingen, pp 1–199

Baehr B, Eisenbeis G (1985) Comparative investigations on the resistance to desiccation in Lycosidae. Hahniidae, Linyphiidae and Micryphantidae (Arachnida, Araneae). Zool Jb Syst 112:225–234

Braun F (1931) Beiträge zur Biologie und Atmungsphysiologie der *Argyroneta aquatica* Clerk. Zool Jb Syst 62:176–262

Bristowe WS (1958) The World of spiders. Collins, London

Collatz K-G, Mommsen T (1974) Lebensweise und jahreszyklische Veränderungen des Stoffbestandes der Spinne *Tegenaria atrica* C. L. Koch (Agelenidae). J Comp Physiol 91:91–109

Crome W (1956) Kokonbau und Eiablage einiger Kreuzspinnenarten des Genus *Araneus* (Araneae, Araneidae). Dt Ent Z 3:28–55

Dabelow S (1958) Zur Biologie der Schleimschleuderspinne *Scytodes thoracica* (Latreille). Zool Jb Syst 86:85–162

Duffey E (1966) Spider ecology and habitat structure (Arach., Araneae). Senck Biol 47:45–49

Dumpert K, Platen R (1985) Zur Biologie eines Buchenwaldbodens 4. Die Spinnenfauna. Carolinea 42:75–106

Dunger W (1983) Tiere im Boden. Die Neue Brehm-Bücherei 327 (3. Aufl.), A. Ziemsen Verlag, Wittenberg-Lutherstadt

Eakin RM, Brandenburger JM (1971) Fine structure of the eyes of jumping spiders. J Ultrastr Res 37:618–663

Edgar WE, Loenen M (1974) Aspects of the overwintering habitat of the wolf spider *Pardosa lugubris*. J Zool (Lond) 172:383–388

Engelhardt W (1964) Die mitteleuropäischen Arten der Gattung *Trochosa* C. L. Koch, 1848 (Araneae, Lycosidae). Morphologie, Chemotaxonomie, Biologie, Autoökologie. Z Morph Ökol Tiere 54:219–392

Foelix RF (1979) Biologie der Spinnen. G. Thieme Verlag, Stuttgart

Friedrich VL, Langer RM (1969) Fine structure of cribellate spider silk. Amer Zool 9:91–96

Gettmann WW (1976) Beutefang bei Wolfspinnen der Gattung *Pirata* (Arachnida: Araneae: Lycosidae). Ent Germ 3:93–99

Glatz L (1972) Der Spinnapparat haplogyner Spinnen (Arachnida, Araneae). Z Morph Tiere 72:1–25

Glatz L (1973) Der Spinnapparat der Orthognatha (Arachnida, Araneae). Z Morph Tiere 75:1–50

Harm M (1931) Beiträge zur Kenntnis des Baues, der Funktion und der Entwicklung des akzessorischen Kopulationsorgans von *Segestria bavarica* C. L. Koch. Morph Ökol Tiere 22:629–670

Heller G (1976) Zum Beutefangverhalten der ameisenfressenden Spinne *Callilepis nocturna* (Arachnida: Araneae: Drassodidae). Ent Germ 3:100–103

Higashi GA, Rovner JS (1975) Post-emergent behaviour of juvenile Lycosid spiders. Bull Brit Arach Soc 3: 113–125

Homann H (1971) Die Augen der Araneae. Anatomie, Ontogenie und Bedeutung für die Systematik (Chelicerata, Arachnidae). Z Morph Tiere 69:201–272

Jones D (1983) The Hamlyn guide to spiders of Britain and Europe. Hamlyn, Feltham

Kullmann E (1961) Der Eierkokonbau von *Cyrtophora citricola* Forskal. (Araneae, Araneidae). Zool Jb Syst 89:399–406

Kullmann E (1969) Spinnorgan mit 40000 „Drüsen“. Umschau Wiss Techn 3:82–83

Kullmann W, Kloft W (1969) Traceruntersuchungen zur Regurgitationsfütterung bei Spinnen (Araneae, Theridiidae). Zool Anz Suppl 32:487–497

Kullmann E, Stern H (1975) Leben am seidenen Faden. C. Bertelsmann Verlag, Gütersloh

Land MF (1969a) Structure of the retinae of the principal eyes of jumping spiders (Salticidae: Dendryphantinae) in relation to visual optics. J Exp Biol 51:443–470

Land MF (1969b) Movements of the retinae of jumping spiders (Salticidae: Dendryphantinae) in response to visual stimuli. J Exp Biol 51:471–493

Land MF (1972) Mechanism of orientation and pattern of recognition by jumping spiders. In: Wehner R (ed) Information processing in the visual systems of arthropods. Springer, Berlin Heidelberg New York

Lehmensick R, Kullmann E (1956) Über den Feinbau der Fäden einiger Spinnen. Zool Anz Suppl 19:123–129

Locket G, Millidge AF (1968) British Spiders. Vol. I–II. Johnson Reprint Corp, New York London

Nentwig W (1986) Ecophysiology of spiders. Springer, Berlin Heidelberg New York

Nentwig W, Heimer S (1984) Salticidae – Springende Spinnen: Vorstellung einer Spinnenfamilie. Biologie in unserer Zeit 14:1–5

Nørgaard E (1951) On the ecology of two Lycosid spiders (*Pirata piraticus* and *Lycosa pullata*) from a Danish Sphagnum bog. Oikos 3:1–21

Nørgaard E (1952) The Habitats of the Danish species of *Pirata*. Ent Medd 26:415–423

Peters HM (1955) Über den Spinnapparat von *Nephila madagascariensis*. Z Naturf 10b:395–404

Pötzsch J (1963) Von der Brutfürsorge heimischer Spinnen. Die Neue Brehm-Bücherei, A. Ziemsen Verlag, Wittenberg-Lutherstadt

Richter CJJ (1970) Morphology and function of the spinning apparatus of the wolf spider *Pardosa amentata* (Cl.) (Araneae, Lycosidae). Z Morph Tiere 68:37–68

Roberts MJ (1985–1986) The spiders of Great Britain and Ireland. Vol. 1–3, Brill, Leiden

Schaefer M (1972) Ökologische Isolation und die Bedeutung des Konkurrenzfaktors am Beispiel des Verteilungsmusters der Lycosiden einer Küstenlandschaft. Oecologia (Berlin) 9:171–202

Schaefer M (1974) Experimentelle Untersuchungen zur Bedeutung der interspezifischen Konkurrenz bei drei Wolfspinnen-Arten (Araneida: Lycosidae) einer Salzwiese. Zool Jb Syst 101:213–235

Schaefer M (1976) Experimentelle Untersuchungen zum Jahreszyklus und zur Überwinterung von Spinnen (Araneida). Zool Jb Syst 103:127–289

Tretzel E (1961a) Biologie, Ökologie und Brutpflege von *Coelotes terrestris* (Wider) (Araneae, Agelenidae). I Biologie und Ökologie. Z Morph Ökol Tiere 49:658–745

Tretzel E (1961b) Biologie, Ökologie und Brutpflege von *Coelotes terrestris* (Wider) (Araneae, Agelenidae). II Brutpflege. Z Morph Ökol Tiere 50:375–542

Vachon M (1957) Contribution à l'étude du dévelopement postembryonnaire des araignées. Première note. Généralités et nomenclature des stades. Bull Soc Zool France 82:337–354

Wiehle H (1949) Vom Fanggewebe einheimischer Spinnen. Die Neue Brehm-Bücherei, Akademische Verlagsgesellschaft Geest u. Portig, Leipzig

Wiehle H (1953) Spinnentiere oder Archnoidea (Araneae), IX: Orthognatha-Cribellatae – Haplogynae – Entelegynae. Die Tierwelt Deutschlands 42:1–150

Wiehle H (1960) Spinnentiere oder Arachnoidea (Araneae). XI: Micryphantidae – Zwergspinnen. Die Tierwelt Deutschlands 47:1–620

Wiehle H (1961) Der Embolus des männlichen Spinnentasters. Zool Anz Suppl 24:457–480

Pseudoscorpiones

Beck L (1983) Zur Biologie des Laubwaldes. Verh Dtsch Zool Ges Bonn 1983, pp 37–54

Beier M (1950) Zur Phänologie einiger *Neobisium*-Arten (Pseudoskorp.). Proc 8th Int Congr Ent, Stockholm 1948, pp 1002–1007

Beier M (1963) Ordnung Pseudoscorpionidea. Bestimmungsbücher zur Bodenfauna Europas 1, pp 1–313, Akademie-Verlag, Berlin

Braun M, Beck L (1986) Zur Biologie eines Buchenwaldbodens 9. Die Pseudoskorpione. Carolinea 44:139–148

Gabbutt PD, Vachon M (1965) The external morphology and life history of the Pseudoscorpion *Neobisium muscorum*. Proc Zool Soc London 145:335–358

Goddard SJ (1976) Population dynamics, distribution patterns and life cycles of *Neobisium muscorum* and *Chthonius orthodactylus* (Pseudoscorpiones: Arachnida). J Zool London 178:295–304

Goddard SJ (1979) The population metabolism and life history tactics of *Neobisium muscorum* (Leach) (Arachnida: Pseudoscorpiones). Oecologia (Berlin) 42:91–105

Janetschek H (1948) Zur Brutbiologie von *Neobisium jugorum* (L. Koch). Ann Nat Hist Mus Wien 56:309–316

Kaestner A (1927) Pseudoscorpiones, After- oder Moosscorpione. Biologie der Tiere Deutschlands 18:1–68

Karg W (1962) Räuberische Milben im Boden. Die Neue Brehm-Bücherei 296, A. Ziemsen Verlag, Wittenberg-Lutherstadt

Ressl F, Beier M (1958) Zur Ökologie, Biologie und Phänologie der heimischen Pseudoscorpione. Zool Jb Syst 86:1–26

Roewer CF (1940) Chelonethi oder Pseudoskorpione. In: Bronn HG: Klassen und Ordnungen des Tierreichs 5, Leipzig

Schaller F (1962) Die Unterwelt des Tierreiches. Verständl Wissenschaft 78:1–126, Springer, Berlin Göttingen Heidelberg

Vachon M (1949) Ordre des Pseudoscorpions. Traité de Zoologie 6:437–481

Wäger H (1982) Populationsdynamik und Entwicklungszyklus der Pseudoscorpiones im Stamser Eichenwald (Tirol). Examensarbeit, Univ Innsbruck

Weygoldt P (1965) Vergleichend-embryologische Untersuchungen an Pseudoscorpionen. III. Die Entwicklung von *Neobisium muscorum* Leach (Neobisiina, Neobisiidae). Z Morph Ökol Tiere 55:321–382

Weygoldt P (1966a) Moos- und Bücherskorpione. Die Neue Brehm-Bücherei 365, A. Ziemsen Verlag, Wittenberg-Lutherstadt

Weygoldt P (1966b) Vergleichende Untersuchungen zur Fortpflanzungsbiologie der Pseudoscorpione. Beobachtungen über das Verhalten, die Samenübertragungsweisen und die Spermatophoren einiger einheimischer Arten. Z Morph Ökol Tiere 56:39–92

Weygoldt P (1969) The biology of pseudoscorpions. Harvard Univ Press, Cambridge

Opiliones

Barth FG, Stagl J (1976) The slit sense organs of Arachnids. A comparative study of their topography on the walking legs (Chelicerata, Arachnida). Zoomorphologie 86:1–23

Blum MS, Edgar AL (1971) 4-Methyl-3-heptanone: Identification and role in Opilionid exocrine secretions. Insect Biochem 1:181–188

Engel H (1961) Mitteleuropäische Insekten. – Sammlung Naturkundlicher Tafeln. Kronen-Verlag Erich Cramer, Hamburg

Franke U (1985) Zur Biologie eines Buchenwaldbodens – 5. Die Weberknechte. Carolinea 42:107–114

Gnatzy W (1982) „Campaniforme" Spaltsinnesorgane auf den Beinen von Weberknechten (Opiliones, Arachnida). Verh Dtsch Zool Ges 75:248

Gruber J, Martens J (1968) Morphologie, Systematik und Ökologie der Gattung *Nemastoma* C. L. Koch (s. str.) (Opiliones, Nemastomatidae). Senckenbergiana biol 49:137–172

Hoheisel U (1983) Sensillen und Drüsen der Legeröhre von Weberknechten. Ein feinstruktureller Vergleich (Arachnida: Opiliones). Diss Univ Mainz

Immel V (1954) Zur Biologie und Physiologie von *Nemastoma quadripunctatum* (Opiliones, Dyspnoi). Zool Jb Syst 83:129–184

Juberthie C (1961) Structure des glandes odorantes et modalités d'utilisation de leur sécrétion chez deux opilions cyphophthalmes. Bull Soc Zool France 86:106–116

Juberthie C (1964) Recherches sur la biologie des Opiliones. Ann Spéléol 19:1–237

Juberthie C (1976) Chemical defence in soil opiliones. Rev Ecol Biol Sol 13:155–160

Kästner A (1928) Opiliones (Weberknechte, Kanker). Die Tierwelt Deutschlands 8:1–51

Kästner A (1935–1937) Opiliones Sundevall. Handb Zool 3 (2) 1:300–393

Martens J (1965) Verbreitung und Biologie des Schneckenkankers *Ischyropsalis hellwigi*. Natur und Museum 95:143–149

Martens J (1967) Bedeutung einer Chelicerendrüse bei Weberknechten (Opiliones). Naturwissenschaften 54:346

Martens J (1969a) Die Abgrenzung von Biospecies auf biologisch-ethologischer und morphologischer Grundlage am Beispiel der Gattung *Ischyropsalis* C. L. Koch 1839 (Opiliones, Ischyropsalididae). Zool Jb Syst 96:133–264

Martens J (1969b) Die Sekretdarbietung während des Paarungsverhaltens von *Ischyropsalis* C. L. Koch (Opiliones). Z Tierpsychol 26:513–523

Martens J (1973) Feinstruktur der Cheliceren-Drüse von *Nemastoma dentigerum* Canestrini (Opiliones, Nemastomatidae). Z Zellforsch 136:121–137

Martens J (1975a) *Ischyropsalis hellwigi* (Opiliones): Paarungsverhalten. Encyclopaedia Cinematographica E 2128, Beiheft – Göttingen

Martens J (1975b) *Ischyropsalis hellwigi* (Opiliones): Nahrungsaufnahme. Encyclopaedia Cinematographica E 2129, Beiheft – Göttingen

Martens J (1978) Weberknechte, Opiliones. Die Tierwelt Deutschlands 64:1–464

Martens J (1986) Die Großgliederung der Opiliones und die Evolution der Ordnung (Arachnida). Actas X Congr Int Aracnol, Jaca/España 1986 I:289–310

Martens J, Hoheisel U, Götze M (1981) Vergleichende Anatomie der Legeröhren der Opiliones als Beitrag zur Phylogenie der Ordnung (Arachnida). Zool Jb Anat 105:13–76

Martens J, Schawaller W (1977) Die Cheliceren-Drüsen der Weberknechte nach rasteroptischen und lichtoptischen Befunden (Arachnida: Opiliones). Zoormorphologie 86:223–250

Pabst W (1953) Zur Biologie der mitteleuropäischen Troguliden. Zool Jb Syst 82:1–46

Pfeifer H (1956) Zur Ökologie und Larvalsystematik der Weberknechte. Mitt Zool Mus Berlin 32:59–104

Rimsky-Korsakow AP (1924) Die Kugelhaare von *Nemastoma lugubre* Mull. Zool Anz 60:1–16

Rüffer H (1966) Beiträge zur Kenntnis der Entwicklungsbiologie der Weberknechte. Zool Anz 176:160–175

Steinböck O (1939) Der Gletscherfloh. Z Dtsch Ö Alpenverein 70:138–147

Wachmann E (1970) Der Feinbau der sog. Kugelhaare der Fadenkanker (Opiliones, Nemastomatidae). Z Zellforsch 103:518–525

Wasgestian-Schaller C (1958) Die Autotomie-Mechanismen an den Laufbeinen der Weberknechte (Arach., Opil.). Diss Univ Frankfurt

Acari

Berthet P (1964) L'activité des Oribatides (Acari, Oribatei) d'une chênaie. Mém Inst Roy Soc Nat Belg 152:1–152

Berthet P, Gérard G (1965) A statistical study of microdistribution of Oribatei (Acari). Part I. The distribution pattern. Oikos 16:214–227

Butcher JW, Snider R, Snider RJ (1971) Bioecology of edaphic Collembola and Acarina. Ann Rev Ent 16:249–288

Coineau Y (1974) Eléments pour une monographie morphologique, écologique et biologique des Caeculidae (Acariens). Mém Mus Hist Nat (NS) A81:1–299

Coineau Y, Haupt J, Delamare Deboutteville C, Théron P (1978) Un remarquable exemple de convergence écologique: l'adaption de *Gordialycus tuzetae* (Nematalycidae, Acariens) à la vie dans les interstices des sables fins. C R Acad Sc Paris 287D:883–886

Desender K, Vaneechoutte M (1984) Phoretic associations of carabid beetles (Coleoptera, Carabidae) and mites (Acari). Rev Ecol Biol Sol 21:363–371

Dindal D (1977) Biology of oribatids. Syracuse

Dunger W (1983) Tiere im Boden. Die Neue Brehm-Bücherei 327 (3. Aufl.), A. Ziemsen Verlag, Wittenberg-Lutherstadt

Forsslund KH (1938) Beiträge zur Kenntnis der Einwirkung der bodenbewohnenden Tiere auf die Zersetzung des Bodens. I: Über die Nahrung einiger Hornmilben (Oribatiden). Medd Statens Skogsförsöksanstalt 31:99–107

Forsslund KH (1939) Über die Ernährungsverhältnisse der Hornmilben (Oribatiden) und ihre Bedeutung für die Prozesse im Waldboden. 7. Int Kongr Entomol Berlin (1938), pp 1950–1957

Gérard G, Berthet P (1966) A statistical study of microdistribution of Oribatei (Acari). Part II: The transformation of the data. Oikos 17:142–149

Gjelstrup P (1979) Epiphytic cryptostigmatid mites on some beech- and birch-trees in Denmark. Pedobiologia 19:1–8

Hågvar S, Abrahamsen G (1980) Colonisation by Enchytryeidae, Collembola and Acari in sterile soil samples with adjusted pH levels. Oikos 34:245–258

Hågvar S, Amundsen T (1981) Effects of liming and artificial acid rain on the mite (Acari) fauna in coniferous forest. Oikos 37:7–20

Hågvar S, Kjøndal BR (1981) Effects of artificial acid rain on the microarthropod fauna in decomposing birch leaves. Pedobiologia 22:409–422

Hirschmann W (1966) Milben (Acari). Franckh'sche Verlagshandlung, Stuttgart

Karg W (1961) Ökologische Untersuchungen von edaphischen Gamsiden (Acarina, Parasitiformes) 1. u. 2. Teil. Pedobiologia 1:53–74, 77–98

Karg W (1962) Räuberische Milben im Boden. Die Neue Brehm-Bücherei 296, A. Ziemsen Verlag, Wittenberg-Lutherstadt

Karg W (1971) Die freilebenden Gamasina (Gamasides), Raubmilben. Die Tierwelt Deutschlands 59:1–475

Klima J (1956) Strukturklassen und Lebensformen der Oribatiden (Acari). Oikos 7:227–242

Knülle W (1957) Die Verteilung der Acari: Oribatei im Boden. Z Morph Ökol Tiere 46:397–432

Korn W (1982) Zur Eidonomie der *Poecilochirus*arten *P. carabi* G. und R. Canestrini (= *P. necrophori* Vitzthum), *P. austroasiaticus* Vitzthum und *P. subterraneus* Müller (Gamasida, Acari). Zool Jb Anat 108:145–224

Krantz GW (1978) A manual of Acarology. Oregon State University Book Stores, Corvallis

Lebrun P (1968) Ecologie et biologie de *Nothrus palustris* (C. L. Koch, 1839). Pedobiologia 8:223–238

Lebrun P (1969) Ecologie et biologie de *Nothrus palustris* (C. L. Koch, 1839). Densité et structure de la population. Oikos 20:34–40

Lebrun P (1970) Ecologie et biologie de *Nothrus palustris* (C. L. Koch, 1839). 3e note: Cycle de vie. Acarologia 12:193–207

Luxton M (1972) Studies on the Oribatid mites of a Danish beech wood soil. I. Nutritional biology. Pedobiologia 12:434–463

Luxton M (1975) Studies on the Oribatid mites of a Danish beech wood soil. II. Biomass, calorimetry and respirometry. Pedobiologia 15:161–200

Luxton M (1979) Food and energy processing by oribatid mites. Rev Ecol Biol Sol 16:103–111

Märkel K (1958) Über die Hornmilben (Oribatei) in der Rohhumusauflage älterer Fichtenbestände des Osterzgebirges. Archiv Forstwesen 7:459–501

Metz LJ (1971) Vertical movement of Acarina under moisture gradients. Pedobiologia 11:262–268

Mitchell MW, Parkinson D (1976) Fungal feeding in oribatid mites in an aspen woodland soil. Ecology 57:302–312

Mitchell MJ (1977) Population dynamics of Oribatid mites (Acari, Cryptostigmata) in an aspen woodland soil. Pedobiologia 17:305–319

Mitchell MJ (1979) Energetics of Oribatid mites (Acari: Cryptostigmata) in an aspen woodland soil. Pedobiologia 19:89–98

Mittmann HW (1980) Zum Abbau der Laubstreu und zur Rolle der Oribatiden (Acari) in einem Buchenwaldboden. Diss Univ Karlsruhe

Mittmann HW (1983) Einfluß von Oribatiden (Acari) auf den Abbau der Laubstreu in einem Buchenwaldboden. Verh Dtsch Zool Ges 1983, p 220, Gustav Fischer Verlag, Stuttgart

Pauly F (1956) Zur Biologie einiger Belbiden (Oribatei, Moosmilben) und zur Funktion ihrer pseudostigmatischen Organe. Zool Jb Syst 84:275–328

Schaller F (1962) Die Unterwelt des Tierreiches. Verständl Wissenschaft 78:1–126, Springer, Berlin Göttingen Heidelberg

Schatz H (1977) Ökologie der Oribatiden (Acari) im zentralalpinen Hochgebirge Tirols (Obergurgl, Innerötztal). Diss Univ Innsbruck

Schuster R (1956) Der Anteil der Oribatiden an den Zersetzungsvorgängen im Boden. Z Morph Ökol Tiere 45:1–33

Sellnick M (1960) Formenkreis: Hornmilben, Oribatei. Die Tierwelt Mitteleuropas 3 (4):45–134

Strenzke K (1952) Untersuchungen über die Tiergemeinschaft des Bodens: Die Oribatiden und ihre Synusien in den Böden Norddeutschlands. Zoologica 104:1–180

Thomas JUM (1979) An energy budget for woodland populations of orbiatid mites. Pedobiologia 19:346–378

Usher MB (1975) Some properties of the aggregations of soil arthropods: Cryptostigmata. Pediobiologia 15: 355–363

Wallwork JA (1959) The distribution and dynamics of some forest soil mites. Ecology 40:557–563

Wallwork JA (1969) Some basic principles underlying the classification and identification of cryptostigmatid mites. In: Sheals JG (1969) The Soil Ecosystem. The Systematic Association 8:155–168, London

Wallwork JA (1970) Ecology of soil animals. McGraw-Hill, London

Wallwork JA (1983) Oribatids in forest ecosystems. Ann Rev Entomol 28:109–130

Weigmann G (1967) Faunistisch-ökologische Bemerkungen über einige Oribatiden der Nordseeküste (Acari, Oribatei). Faun-Ökol Mitt 3:173–178

Weigmann G (1971) Collembolen und Oribatiden in Salzwiesen der Ostseeküste und des Binnenlandes von Norddeutschland (Insecta: Collembola – Acari: Oribatei). Faun-Ökol Mitt 4:11–20

Weigmann G (1973) Zur Ökologie der Collembolen und Oribatiden im Grenzbereich Land-Meer (Collembola, Insecta – Oribatei, Acari). Z Wiss Zool (Leipzig) 186: 295–391

Willmann C (1931) Moosmilben oder Oribatiden (Cryptostigmata). Die Tierwelt Deutschlands 22 (5):79–200

Wink U (1969) Die Collembolen- und Oribatidenpopulationen einiger saurer Auböden Bayerns in Abhängigkeit von der Bodenfeuchtigkeit. Z Angew Ent 64:121–136

Woas S (1986) Beitrag zur Revision der Oppioidea sensu Balogh, 1972 (Acari, Oribatei). Andrias 5:21–224

Isopoda

Babula A, Bielawski J (1976) Ultrastructure of respiratory epithelium in the terrestrial isopod *Porcellio scaber* Latr. (Crustacea). Ann Med Sect Pol Acad Sci 21:7–8

Beck L, Brestowsky E (1980) Auswahl und Verwertung verschiedener Fallaubarten durch *Oniscus assellus* (Isopoda). Pedobiologia 20:428–441

Beyer R (1964) Faunistisch-ökologische Untersuchungen an Landisopoden in Mitteldeutschland. Zool Jb Syst 91:341–402

Brereton JLG (1957) The distribution of woodland isopods. Oikos 8:75–106

Cloudsley-Thompson JL (1977) The water and temperature relations of woodlice. Durham, England

Coenen-Stass D (1981) Some aspects of the water balance of two desert woodlice, *Hemilepistus aphganicus* and *Hemilepistus reaumuri* (Crustacea, Isopoda, Oniscoidea). Comp Biochem Physiol 70A:405–419

Den Boer PJ (1961) The ecological significance of activity patterns in the woodlouse *Porcellio scaber* Latr. (Isopoda). Archs Néerl Zool 14:283–408

Dunger W (1958) Über die Zersetzung der Laubstreu durch die Boden-Makrofauna im Auenwald. Zool Jb Syst 86:139–180

Dunger W (1962) Methoden zur vergleichenden Auswertung von Fütterungsversuchen in der Bodenbiologie. Abh Ber Naturk Mus Görlitz 37:143–162

Edney EB (1954) Woodlice and land habitat. Biol Rev 29:185–219

Edney EB (1960) Terrestrial adaptations. In: Waterman TH (ed) The physiology of Crustacea 1:367–393, New York

Edney EB (1977) Water balance in land arthropods. Springer, Berlin Heidelberg New York

Gruner HE (1965/66) Krebstiere oder Crustacea, V. Isopoda. Die Tierwelt Deutschlands 51:1–380

Henke G (1960) Sinnesphysiologische Untersuchungen bei Landisopoden, insbesondere bei *Porcellio scaber*. Verh Dt Zool Ges 54:167–171

Herold W (1925) Untersuchungen zur Ökologie und Morphologie einiger Landasseln. Z Morph Ökol Tiere 4:335–415

Hoese B (1981) Morphologie und Funktion des Wasserleitungssystems der terrestrischen Isopoden (Crustacea, Isopoda, Oniscoidea). Zoomorphology 98:135–167

Hoese B (1982a) Morphologie und Evolution der Lungen bei den terrestrischen Isopoden (Crustacea, Isopoda, Oniscoidea). Zool Jb Anat 107:396–422

Hoese B (1982b) Der *Ligia*-Typ des Wasserleitungssystems bei terrestrischen Isopoden und seine Entwicklung in der Familie Ligiidae (Crustacea, Isopoda, Oniscoidea). Zool Jb Anat 108:225–261

Hoese B (1983) Struktur und Entwicklung der Lungen der Tylidae (Crustacea, Isopoda, Oniscoidea). Zool Jb Anat 109:487–501

Kästner A (1967) Crustacea – Lehrbuch der Speziellen Zoologie, Bd. 1 (Wirbellose) 2. Teil. G. Fischer, Stuttgart

Kümmel G (1981) Fine structural indications of an osmoregulatory function of the "gills" in terrestrial isopods (Crustacea, Oniscoidea). Cell Tiss Res 214:663–666

Kümmel G (1984) Fine-structural investigations of the pleopodal endopods of terrestrial isopods with some remarks on their function. In: Sutton SL, Holdich DM (ed) The biology of terrestrial isopods. Symp Zool Soc Lond 53:77–93, Clarendon, Oxford

Lindqvist OV (1972) Components of water loss in terrestrial isopods. Physiol Zool 45:316–324

Linsenmair K, Linsenmair L (1971) Paarbildung und Paarzusammenhalt bei der monogamen Wüstenassel *Hemilepistus reaumuri* (Crustacea, Isopoda, Oniscoidea). Z Tierpsychol 29:134–155

Matthes D (1950) Die Kiemenfauna unserer Landasseln. Zool Jb Syst 78:573–640

Risler H (1976) Die Ultrastruktur eines Chordotonalorgans in der Geißel der Antenne von *Armadillidium nasutum* Budde-Lund (Isopoda, Crustacea). Zool Jb Anat 95:94–104

Risler H (1977) Die Sinnesorgane der Antennula von *Porcellio scaber* Latr. (Crustacea, Isopoda). Zool Jb Anat 98:29–52

Risler H (1978) Die Sinnesorgane der Antennula von *Ligidium hypnorum* (Cuvier) (Isopoda, Crustacea). Zool Jb Anat 100:514–541

Seelinger G (1977) Der Antennenendzapfen der tunesischen Wüstenassel *Hemilepistus reaumuri*, ein komplexes Sinnesorgan (Crustacea, Isopoda). J Comp Physiol 113:95–103

Schmalfuss H (1977) Morphologie und Funktion der tergalen Längsrippen bei Landisopoden (Oniscoidea, Isopoda, Crustacea). Zoomorphologie 86:155–167

Schmalfuss H (1978) Morphology und Function of Cuticular Micro-Scales and Corresponding Structures in Terrestrial Isopods (Crust., Isop., Oniscoidea). Zoomorphologie 91:263–274

Schmölzer K (1965) Ordnung Isopoda (Landasseln). Bestimmungsbücher zur Bodenfauna Europas 4/5:1–468

Schneider P (1971) Lebensweise und soziales Verhalten der Wüstenassel *Hemilepistus aphganicus* Borukky 1958. Z Tierpsychol 29:131–133

Schneider P, Jakobs B (1977) Versuche zum intra- und interspezifischen Verhalten terrestrischer Isopoden (Crustacea, Oniscoidea). Zool Anz 199:173–186

Sutton SL, Holdich DM (1984) The biology of terrestrial isopods. Symp Zool Soc Lond 53, Clarendon, Oxford

Wächtler W (1937) Isopoda (Asseln). Die Tierwelt Mitteleuropas 2:225–317

Chilopoda

Albert AM (1977) Biomasse von Chilopoden in einem Buchenaltbestand des Solling. Verh Ges Ökol Göttingen 1976:93–101

Camatini M (ed) (1979) Myriapod biology. Academic Press, London

Curry A (1974) The spiracle structure and resistence to desiccation of centipedes. Symp Zool Soc London 32:365–382

Dobroruka LJ (1961) Hundertfüßler. Die Neue Brehm-Bücherei 285, A. Ziemsen Verlag, Wittenberg-Lutherstadt

Dunger W (1983) Tiere im Boden. Die Neue Brehm-Bücherei 327 (3. Aufl.), A. Ziemsen Verlag, Wittenberg-Lutherstadt

Ernst A (1976) Die Ultrastruktur der Sinneshaare auf den Antennen von *Geophilus longicornis* Leach (Myriapoda, Chilopoda). I. Die Sensilla trichodea. Zool Jb Anat 96:586–604

Ernst A (1979) Die Ultrastruktur der Sinneshaare auf den Antennen von *Geophilus longicornis* Leach (Myriapoda, Chilopoda). II. Die Sensilla basiconica. Zool Jb Anat 102:510–532

Ernst A (1981) Die Ultrastruktur der Sinneshaare auf den Antennen von *Geophilus longicornis* Leach (Myriapoda, Chilopoda). III. Die Sensilla brachyconica. Zool Jb Anat 106:375–399

Füller H (1963) Vergleichende Untersuchungen über das Skelettmuskelsystem der Chilopoden. Abh Dtsch Akad Wiss Berlin Kl Chem Geol Biol 3:1–97

Haupt J (1979) Phylogenetic aspects of recent studies on myriapod sense organs. In: Camatini M (ed) Myriapod biology. Academic Press, London

Kästner A (1963) Lehrbuch der Speziellen Zoologie. Teil 1: Wirbellose (5. Lief). VEB Gustav Fischer Verlag, Jena

Keil T (1976) Sinnesorgane auf den Antennen von *Lithobius forficatus* L. (Myriapoda, Chilopoda). I. Die Funktions-

morphologie der „Sensilla trichodea". Zoomorphologie 84:77–102
Lewis JGE (1981) The biology of Centipedes. Cambridge Univ Press
Rilling G (1960) Zur Anatomie des braunen Steinläufers *Lithobius forficatus* L. (Chilopoda), Skelettmuskulatur, peripheres Nervensystem und Sinnesorgane des Rumpfes. Zool Jb Anat 78:39–128
Rilling G (1968) *Lithobius forficatus*. Anatomie und Biologie. G. Fischer Verlag, Stuttgart
Rosenberg J (1982) Coxal organs in geophilomorpha (Chilopoda). Organization and fine structure of the transporting epithelium. Zoomorphology 100:107–120
Rosenberg J (1983) Coxal organs of *Lithobius forficatus* (Myriapoda, Chilopoda). Fine-structural investigation with special reference to the transport epithelium. Cell Tissue Res 230:421–430
Rosenberg J, Bajorat KH (1983) Feinstruktur der Coxalorgane bei *Lithobius forficatus* und ihre Beteiligung an der Aufnahme von Wasserdampf aus der Atmosphäre. Verh Dtsch Zool Ges Bonn 1983:316
Rosenberg J, Seifert G (1977) The coxal glands of Geophilomorpha (Chilopoda): Organs of osmoregulation. Cell Tiss Res 182:247–251
Rudolph D, Knülle W (1982) Novel uptake systems for atmospheric water vapor among insects. J Exp Zool 222:321–333
Tichy H (1972) Das Tömösvárysche Sinnesorgan des Hundertfüßlers *Lithobius forficatus* – ein Hygrorezeptor. Naturwissenschaften 59:315
Tichy H (1973) Untersuchungen über die Feinstruktur des Tömösváryschen Sinnesorgans von *Lithobius forficatus* L. (Chilopoda) und zur Frage seiner Funktion. Zool Jb Anat 91:93–139
Verhoeff KW (1925) Chilopoda. In: Bronn HG: Klassen und Ordnungen des Tierreichs 5. Bd, Leipzig

Diplopoda

Bedini C, Mirolli M (1967) The fine structure of the temporal organs of a pill millipede *Glomeris romana* Verhoeff. Monit Zool Ital (NS) 1:41–63
Blower JG (1955) Millipedes and Centipedes as soil animals. In: Kevan DK McE, Soil zoology, pp 138–151, Butterworths Sci Publ, London
Dunger W (1974, 1983) Tiere im Boden. Die Neue Brehm-Bücherei 327 (2., 3. Aufl.), A. Ziemsen Verlag, Wittenberg-Lutherstadt
Edney EB (1951) The evaporation of water from woodlice and the millipede *Glomeris*. J Exp Biol 28:91–115
Edney EB (1977) Water balance in land arthropods. Springer, Berlin Heidelberg New York
Haupt J (1979) Phylogenetic aspects of recent studies on Myriapod sense organs. In: Camatini M (ed) Myriapod biology. Academic Press, London
Manton SM (1977) The Arthropoda, habits, functional morphology, and evolution. Clarendon, Oxford
Marcuzzi G (1970) Experimental observations on the role of *Glomeris* ssp. (Myriapoda, Diplopoda) in the process of humification of litter. Pedobiologia 10:401–406
Meyer E, Eisenbeis G (1985) Water relations in millipedes from some alpine habitat types (Central Alps, Tyrol) (Diplopoda). Bijdragen tot de Dierkunde 55:131–142
Nguyen Duy-Jacquemin M (1981) Ultrastructure des organes sensoriels de l'antenne de *Polyxenus lagurus* (Diplopode, Penicillate). I. Les cones sensoriels apicaux du 8e article antennaire. Ann Sci Nat Zool Paris 13e Sér 3:95–114
Nguyen Duy-Jacquemin M (1982) Ultrastructure des organes sensoriels de l'antenne de *Polyxenus lagurus* (Diplopode, Penicillate). II. Les sensilles basiconiques des 6e et 7e articles antennaires. Ann Sci Nat Zool Paris 13e Sér 4:211–229
Röper H (1978) Ergebnisse chemisch-analytischer Untersuchungen der Wehrsekrete von Spirostreptiden, Spiroboliden und Juliden (Diplopoda), von *Peripatopsis* (Onychophora) und von *Polyzonium* (Diplopoda, Colobognatha). Abh Verh Naturwiss Ver Hamburg 21/22:353–363
Schömann KH (1956) Zur Biologie von *Polyxenus lagurus* (L. 1758). Zool Jb Syst 84:195–256
Schönrock GU (1981) Zur Häutung der antennalen Sensillen bei der Bandfüßer-Art *Polydesmus coriaceus* (Diplopoda, Polydesmoidea). Ent Gen 7:157–160
Schubart O (1934) Tausendfüßler oder Myriapoda, 1: Diplopoda. Die Tierwelt Deutschlands, Jena
Seifert G (1960) Die Entwicklung von *Polyxenus lagurus* L. (Diplopoda, Pselaphognatha). Zool Jb Anat 78: 257–312
Seifert G (1961) Die Tausendfüßler (Diplopoda). Die Neue Brehm-Bücherei 273, A. Ziemsen Verlag, Wittenberg-Lutherstadt
Striganova BR (1967) Über die Zersetzung von überwinterter Laubstreu durch Tausendfüßler und Landasseln. Pedobiologia 7:125–138
Tichy H (1975) Unusual fine structure of sensory hair triad of the millipede, *Polyxenus*. – Cell Tiss Res 156, 229–238
Topp W (1981) Biologie der Bodenorganismen. UTB, Quelle und Meyer, Heidelberg
Verhoeff KW (1932) Diplopoda. In: Bronn HG: Klassen und Ordnungen des Tierreichs 5. Bd, Leipzig

Pauropoda

Hüther W (1974) Zur Bionomie mitteleuropäischer Pauropoden. Symp Zool Soc London 32:411–421
Haupt J (1973) Die Ultrastruktur des Pseudoculus von *Allopauropus* (Pauropoda) und die Homologie der Schläfenorgane. Z Morph Tiere 76:173–191
Haupt J (1976) Anpassung an einen Lebensraum – das hygrophile Edaphon. Sber Ges Naturf Freunde (Berlin) 16:89–97
Haupt J (1978) Ultrastruktur der Trichobothrien von *Allopauropus (Decapauropus)* (Pauropoda). Abh Verh Naturwiss Ver Hamburg 21/22:271–277
Verhoeff KW (1937) Progoneata: Diplopoda, Symphyla, Pauropoda, Chilopoda. Die Tierwelt Mitteleuropas, Leipzig

Symphyla

Friedel H (1928) Ökologische und physiologische Untersuchungen an *Scutigerella immaculata* (Newp.). Z Morph Ökol Tiere 10:737–797

Gill B (1981) Die Coxalblasen der Symphyla: eine elektronenmikroskopische Untersuchung des Blasenepithels. Staatsexamensarbeit, Mainz

Haupt J (1971) Beitrag zur Kenntnis der Sinnesorgane von Symphylen (Myriapoda). II. Feinstruktur des Tömösváryschen Organs von *Scutigerella immaculata* Newport. Z Zellforsch 122:172–189

Haupt J (1979) Phylogenetic Aspects of Recent Studies on Myriapod Sense Organs. In: Camatini M (ed) Myriapod biology, pp 391–406, Academic Press, London New York

Hennings C (1904) Das Tömösvárysche Organ der Myriapoden I. Z Wiss Zool 76:26–52

Hennings C (1906) Das Tömösvárysche Organ der Myriapoden II. Z Wiss Zool 80:576–641

Juberthie-Jupeau L (1956) Existence de spermatophores chez les Symphyles. CR Acad Sci Paris 243:1164–1166

Juberthie-Jupeau L (1959a) Donées sur les phénomènes externes de l'émission de spermatophores chez les Symphyles (Myriapodes). CR Acad Sci Paris 248:469–472

Juberthie-Jupeau L (1959b) Etude de la ponte chez les Symphyles (Myriapodes), avec mise en évidence d'une fécondation externe des oeufs pour la femelle. CR Acad Sci Paris 249:1821–1823

Kaestner A (1963) Lehrbuch der Speziellen Zoologie, Teil 1: Wirbellose (5. Liefg). VEB Gustav Fischer Verlag, Jena

Michelbacher AE (1938) The biology of the garden centipede *Scutigerella immaculata* Newp. Hilgardia 11:55–148

Tömösváry E (1883) Eigentümliche Sinnesorgane der Myriapoden. Math Naturw Ber Ungarn 1:324–326

Verhoeff KW (1934) Symphyla. In: Bronn HG: Klassen und Ordnungen des Tierreichs, 5. Bd, Leipzig

Diplura

Bareth C (1983) Les organes sensoriels des Diploures Campodéidés, étude ultrastructurale (Insecta apterygota). Pedobiologia 25:216

Bareth C (1986) Acquisitions récentes sur l'écologie et la biologie des Diploures Campodéidés (Insecta Apterygota). In: Dallai R (ed) 2nd Int Seminar on Apterygota, Siena 1986, pp 99–103. University of Siena, Siena

Bareth C, Juberthie-Jupeau L (1977) Ultrastructure des soies sensorielles des palpes labiaux de *Campodea sensillifera* (Conde et Mathieu) (Insecta: Diplura). Int J Insect Morph Embryol 6:191–200

Eisenbeis G (1976) Zur Feinstruktur und Histochemie des Transportepithels abdominaler Koxalblasen der Doppelschwanz-Art *Campodea staphylinus* (Diplura: Campodeidae). Ent Germ 3:185–201

Eisenbeis G (1983a) Kinetics of water exchange in soil arthropods. In: Lebrun P et al (ed): New trends in soil biology, pp 414–425, Dieu-Brichart, Louvain-La-Neuve (Belgium)

Eisenbeis G (1983b) Kinetics of transpiration in soil arthropods. In: Lebrun P et al. (ed) New trends in soil biology, pp 626–627, Dieu-Brichart, Louvain-La-Neuve (Belgium)

Endres E (1980) Untersuchung der antennalen Sinnesorgane von *Campodea* spec. (Diplura: Insecta). Staatsexamensarbeit, Univ Mainz

Francois F (1970) Squelette et musculature céphalique de *Campodea chardardi* Condé (Diplura: Campodeidae). Zool Jb Anat 87:331–376

Handschin E (1929) Urinsekten oder Apterygota. Die Tierwelt Deutschlands 16:1–150

Juberthie-Jupeau L, Bareth C (1980a) Ultrastructure des glandes dermiques à petits pores des Diploures Campodéidés (Insecta, Entognatha, Diplura). Zoomorph 95:105–113

Juberthie-Jupeau L, Bareth C (1980b) Ultrastructure des sensilles de l'organe cupuliforme de l'antenne des Campodés (Insecta: Diplura). J Insect Morph Embryol 9:255–268

Kosaroff G (1935) Beobachtungen über die Ernährung der Japygiden. Jzrest, carsz prirodononc Inst 8:181–185

Marten W (1939) Zur Kenntnis von *Campodea*. Z Morph Ökol Tiere 36:41–88

Paclt J (1956) Biologie der primär flügellosen Insekten. VEB Gustav Fischer Verlag, Jena

Pages J (1967a) La notion de territoire chez les Diploures Japygidés. Ann Soc Ent Fr (NS) 3:715–719

Pages J (1967b) Données sur la biologie de *Dipljapyx humberti* (Grassi). Rev Ecol Biol Sol 4:187–281

Pages J (1978) Les Japygoidea (Insectes, Diploures) de France. Bull Soc Zool France 103:385–394

Palissa A (1964) Apterygota – Urinsekten. Die Tierwelt Mitteleuropas IV:1–407, Quelle & Meyer, Leipzig

Schaller F (1949) *Notiophilus biguttatus* F. (Coleopt.) und *Japyx solifugus* Haliday (Diplura) als spezielle Collembolenräuber. Zool Jb Syst 78:294–296

Schaller F (1962) Die Unterwelt des Tierreiches. Springer, Berlin Göttingen Heidelberg

Simon HR (1963) Die Japygiden Deutschlands (Apterygota, Diplura). Mitt Dtsch Ent Ges 22:67–68

Simon HR (1964) Zur Ernährungsbiologie collembolenfangender Arthropoden. Biol Zbl 83:273–296

Weyda F (1976) Histology and ultrastructure of the abdominal vesicles of *Campodea franzi* (Diplura, Campodeidae). Acta ent bohem slov 73:237–242

Weyda F (1980) Diversity of cuticular types in the abdominal vesicles of *Campodea silvestri* (Diplura, Campodeidae). Acta ent bohem slov 77:297–302

Protura

Bedini C, Tongiorgi P (1971) The fine structure of the pseudoculus of Acerentomide Protura (Insecta, Apterygota). Monit Zool Ital 5:25–38

Francois J (1959) Squelette et musculature céphaliques d'*Acerentomon propinquum* (Condé) (Ins., Protoures). Trav de Lab Zool Dijon 29:1–58

Francois J (1969) Anatomie et morphologie céphalique des Protoures (Insecta, Apterygota). Mém Mus Hist Nat Paris 49:1–144

Francois J, Dallai R (1986) Les glandes abdominales des Protoures. In: Dallai R (ed) 2nd Int Seminar on Apterygota, Siena 1986, pp 273–280. University of Siena, Siena

Gunnarsson B (1980) Distribution, abundance and population structure of Protura in two woodland soils in Southwestern Sweden. Pedobiologia 20:254–262

Haupt J (1972) Ultrastruktur des Pseudoculus von *Eosentomon* (Protura, Insecta). Z Zellforsch 135:539–551

Haupt J (1979) Phylogenetic aspects of recent studies on Myriapod sense organ. In: Camatini M (ed) Myriapod biology. Academic Press, London New York, pp 391–406

Janetschek H (1970) Protura (Beintastler). Handb Zool Berlin 4 (2):1–72

Nosek J (1973) The European Protura. Mus. Hist Nat Genf

Nosek J, Ambrož Z (1964) Apterygotenbesatz und mikrobielle Aktivität in Böden der Niederen Tatra. Pedobiologia 4:222–240

Snodgrass RE (1935) Principles of insect morphology. McGraw-Hill, New York, London

Strenzke K (1942) Norddeutsche Proturen. Zool Jb Syst 75:73–102

Tuxen SL (1931) Monographie der Proturen. I, Morphologie nebst Bemerkungen über Systematik und Ökologie. Z Morph Ökol Tiere 22:671–720

Tuxen SL (1964) The Protura. Hermann, Paris

Yin W-Y, Xue L, Tang B (1986) A comparative study on pseudoculus of Protura. In: Dallai R (ed) 2nd Int Seminar on Apterygota, Siena 1986, pp 249–256. University of Siena, Siena

Collembola

Agrell I (1941) Zur Ökologie der Collembolen. Opusc Entomol Suppl 3:1–236

Aitchison CW (1979) Winter-active subnivean invertebrates in Southern Canada. I. Collembola. Pedobiologia 19:113–120

Aitchison CW (1983) Low temperature and preferred feeding by winter-active Collembola (Insecta, Apterygota). Pedobiologia 25:27–36

Altner H, Ernst KD (1974) Struktureigentümlichkeiten antennaler Sensillen bodenlebender Collembolen. Pedobiologia 14:118–122

Altner H, Ernst KD, Karuhize G (1970) Untersuchungen am Postantennalorgan der Collembolen (Apterygota). I. Feinstruktur der postantennalen Sinnesborste von *Sminthurus fuscus* (L). Z Zellforsch 111:263–285

Altner H, Thies G (1972) Reizleitende Strukturen und Ablauf der Häutung an Sensillen einer euedaphischen Collembolenart. Z Zellforsch 129:196–216

Altner H, Thies G (1976) The postantennal organ: A specialized unicellular sensory input to the protocerebrum in apterygotan insects (Collembola). Cell Tissue Res 167:97–110

Altner H, Thies G (1978) The multifunctional sensory complex in the antenne of *Allacma fusca* (Insecta). Zoomorphologie 91:119–131

Barra JA (1973) Structure et régression des photorécepteurs dans le groupe *Lepidocyrtus – Pseudosinella* (Insecta, Collembola). Ann Spéléol 28:167–175

Barra JA, Poinsot-Balaguer N (1977) Modifications ultrastructurales accompagnant l'anhydrobiose chez un Collembole: *Folsomides variabilis*. Rev Ecol Biol Sol 14:189–197

Bauer T (1979) Die Feuchtigkeit als steuernder Faktor für das Kletterverhalten von Collembolen. Pedobiologia 19:165–175

Bauer T, Christian E (1986) Flight behaviour of springtails (Collembola) with respect to their habitat. In: Dallai R (ed) 2nd Int Seminar on Apterygota, Siena 1986, pp 177–179. University of Siena, Siena

Betsch J-M, Vannier G (1977) Caractérisation des deux phases juvéniles d'*Allacma fusca* (Collembola, Symphypleona) par leur morphologie et leur écophysiologie. Z Zool Syst Evolut-forsch 15:124–141

Bleicher M (1981) Untersuchungen zur Ultrastruktur und Transportbiologie am Ventraltubus der Gattung *Tomocerus* (Collembola, Tomoceridae). Staatsexamensarbeit, Univ Mainz

Block W (1983) Low temperature tolerance of soil arthropods – Some recent advances. In: Lebrun P et al New trends in soil biology, pp 427–431, Dieu-Brichart, Louvain-La-Neuve (Belgium)

Block W, Zettel J (1980) Cold hardiness of some Alpine Collembola. Ecol Entomology 5:1–9

Blottner D, Eisenbeis G (1984) Ultrastructure of long tibiotarsal spatula-hairs in *Tomocerus flavescens* (Collembola: Tomoceridae). Ann Soc R Zool Belg 114:51–57

Bockemühl J (1956) Die Apterygoten des Spitzberges, eine faunistisch-ökologische Untersuchung. Zool Jb Syst 84:113–194

Brummer-Korvenkontio M, Brummer-Korvenkontio L (1980) Springtails (Collembola) on and in snow. Mem Soc Fauna Flora Fennica 56:91–94

Butcher JW, Snider R, Snider RJ (1971) Bioecology of edaphic Collembola and Acarina. Ann Rev Entomol 16:249–288

Cassagnau P, Lauga-Reyrel F (1985) Sur la signification adaptative de l'architecture cuticulaire chez les Collemboles Arthropléones. Rev Ecol Biol Sol 22:381–402

Christian E (1978) The jump of the Springtails. Naturwissenschaften 65:495–496

Christian E (1979) Der Sprung der Collembolen. Zool Jb Physiol 83:457–490

Christiansen K (1964) Bionomics of Collembola. Ann Rev Entomol 9:147–178

Christiansen K (1970) Experimental studies on the aggregation and dispersion of Collembola. Pedobiologia 10:180–198

Christiansen K, Bellinger P (1980) The Collembola of North America north of the Rio Grande. Grinell College, Grinell/Iowa

Dallai R (1977) Considerations on the cuticle of Collembola. Rev Ecol Biol Sol 14:117–124

Döring D (1985) Struktur, Funktion und Merkmalsabwandlung der Spermatophoren und der männlichen Fortpflanzungsorgane von Collembolen (Insecta, Entognatha). Diss Bremen

Döring D (1986) On the male reproduction biology of *Orchesella cincta* (Collembola, Entomobryidae). In: Dallai R (ed) 2nd Int Seminar on Apterygota, Siena 1986, pp 171–176. University of Siena, Siena

Dunger W (1961) Zur Kenntnis von *Tetrodontophora bielanensis* (Waga, 1842) (Collembola, Onychiuridae). Abh Ber Naturk Mus Görlitz 37:79–99

Dunger W (1983) Tiere im Boden. Die Neue Brehm-Bücherei 327 (3. Aufl), A. Ziemsen Verlag, Wittenberg-Lutherstadt

Eisenbeis G (1974) Licht- und elektronenmikroskopische Untersuchungen zur Ultrastruktur des Transportepithels am Ventraltubus arthropleoner Collembolen (Insecta). Cytobiologie 9:180–202

Eisenbeis G (1976a) Zur Feinstruktur und Funktion von Sensillen im Transport-Epithel des Ventraltubus von *Tomocerus* und *Orchesella* (Collembola: Tomoceridae/Entomobryidae). Ent Germ 2:271–295

Eisenbeis G (1976b) Zur Morphologie des Ventraltubus von *Tomocerus* ssp. (Collembola: Tomoceridae) unter besonderer Berücksichtigung der Muskulatur, der cuticularen Strukturen und der Ventralrinne. Int J Insect Morph Embryol 5:357–379

Eisenbeis G (1978) Die Thorakal- und Abdominal-Muskulatur von Arten der Springschwanz-Gattung *Tomocerus* (Collembola: Tomoceridae). Ent Germ 4:55–83

Eisenbeis G (1982) Physiological absorption of liquid water by Collembola: Absorption by the ventral tube at different salinities. J Insect Physiol 28:11–20

Eisenbeis G, Meyer E (1986) Some ultrastructural features of glacier Collembola, *Isotoma* 'sp. G' and *Isotomurus palliceps* (Uzel, 1891) from the Tyrolean Central Alps. In: Dallai R (ed) 2nd Int Seminar on Apterygota, Siena 1986, pp 257–272. University of Siena, Siena

Eisenbeis G, Ulmer S (1978) Zur Funktionsmorphologie des Sprung-Apparates der Springschwänze am Beispiel von Arten der Gattung *Tomocerus* (Collembola: Tomoceridae). Ent Germ 5:35–55

Eisenbeis G, Wichard W (1975a) Histochemischer Chloridnachweis im Transportepithel am Ventraltubus arthropleoner Collembolen. J Insect Physiol 21:231–236

Eisenbeis G, Wichard W (1975b) Feinstruktureller und histochemischer Nachweis des Transportepithels am Ventraltubus symphypleoner Collembolen (Insecta, Collembola). Z Morph Tiere 81:103–110

Eisenbeis G, Wichard W (1977) Zur feinstrukturellen Anpassung des Transportepithels am Ventraltubus von Collembolen bei unterschiedlicher Salinität. Zoomorphologie 88:175–188

Falkenhan HH (1932) Biologische Beobachtungen an *Sminthurides aquaticus* (Collembola). Z Wiss Zool 141:525–580

Ghiradella H, Radigan W (1974) Collembolan cuticle: Wax layer and antiwetting properties. J Insect Physiol 20: 301–306

Gisin H (1943) Ökologie und Lebensgemeinschaften der Collembolen im schweizerischen Exkursionsgebiet Basels. Rev Suisse Zool 50:131–224

Gisin H (1960) Collembolenfauna Europas. Mus Hist Nat Genf

Hale WG, Smith AL (1966) Scanning electron microscope studies of cuticular structures in the genus *Onychirurus* (Collembola). Rev Ecol Biol Sol 3:343–354

Handschin E (1919) Über die Collembolenfauna der Nivalstufe. Rev Suisse Zool 27:65–101

Handschin E (1926) Collembola-Springschwänze. Biologie der Tiere Deutschlands 25:7–56

Hanlon RDG, Anderson JM (1979) The effect of Collembola grazing on microbial activity in decomposing leaf litter. Oecologia 38:93–99

Jaeger G, Eisenbeis G (1984) pH-dependent absorption of solutions by the ventral tube of *Tomocerus flavescens* (Tullberg, 1871) (Insecta, Collembola). Rev Ecol Biol Sol 21:519–531

Joosse ENG (1970) The formation and biological significance of aggregation in the distribution of Collembola. Neth J Zool 20:299–314

Joosse ENG (1971) Ecological aspects of aggregation in Collembola. Rev Ecol Biol Sol 8:91–97

Joosse ENG (1981) Ecological strategies and population regulation of Collembola in heterogeneous environments. Pedobiologia 21:346–356

Joosse ENG (1983) New developments in the ecology of Apterygota. Pedobiologia 25:217–234

Joosse ENG, Verhoef HA (1974) On the aggregational habits of surface dwelling Collembola. Pedobiologia 14:245–249

Jura C, Krzystofowicz A (1977) Ultrastructural changes in embryonic midgut cells developing into larval midgut epithelium of *Tetrodontophora bielanensis* (Waga) Collembola. Rev Ecol Biol Sol 14:103–115

Karuhize GR (1971) The structure of the postantennal organ in *Onychiurus* sp. (Insecta: Collembola) and its connections to the central nervous system. Z Zellforsch 118:263–282

Koledin D, Ribarac-Stepic N, Stanković J (1981) Participation of *Tetrodontophora bielanensis* (Collembola, Insecta) in decomposition of forest litter lipid compounds. Pedobiologia 22:71–76

Konček SK (1924) Über Autohämorhoe bei *Tetrodontophora gigas* Reut. Zool Anz 61:238–242

Lan An der H (1963) Neues zur Tierwelt des Ewigschneegebietes. Zool Anz Suppl 26:673–678

Lawrence PN, Massoud Z (1973) Cuticle structures in the Collembola (Insecta). Rev Ecol Biol Sol 10:77–101

Leinaas HP (1983) Winter strategy of surface dwelling Collembola. Pedobiologia 25:235–240

Massoud Z (1969) Etude de l'ornamentation épicuticulaire du tégument des Collemboles au microscope électronique à balayage. C R Acad Sci Paris 268:1407–1409

Mayer H (1957) Zur Biologie und Ethologie einheimischer Collembolen. Zool Jb Syst 85:501–672

Mertens J, Bourgoignie R (1977) Aggregation pheromone in *Hypogastrura viatica* (Collembola). Behav Ecol Sociobiol 2:41–48

Mertens J, Blancquaert JP, Bourgoignie R (1979) Aggregation pheromone in *Orchesella cincta* (Collembola). Rev Ecol Biol Sol 16:441–447

Milne S (1962) Phenology of a natural population of soil Collembola. Pedobiologia 2:41–52

Paclt J (1956) Biologie der primär flügellosen Insekten. G. Fischer Verlag, Jena

Palissa A (1964) Apterygota – Urinsekten. Die Tierwelt Mitteleuropas 4:1–407, Quelle & Meyer, Leipzig

Paulus HF (1971) Einiges zur Cuticula-Struktur der Collembolen mit Bemerkungen zur Oberflächenskulptur der Cornea. Rev Ecol Biol Sol 8:37–44

Paulus HF (1972) Zum Feinbau der Komplexaugen einiger Collembolen. Eine vergleichend-anatomische Untersuchung (Insecta, Apterygota). Zool Jb Anat 89:1–116

Petersen H (1980) Population dynamic and metabolic characterization of Collembola species in a beech forest ecosystem. In: Dindal DL (ed), Proc VII Int Coll Soil Zool Syracuse 1979, pp 806–833

Poole TB (1964) A study of the distribution of soil Collembola in three small areas in a coniferous woodland. Pedobiologia 4:35–42

Rusek J (1975) Die bodenbildende Funktion von Collembolen und Acarina. Pedobiologia 15:299–308

Rusek J, Weyda F (1981) Morphology, ultrastructure and function of pseudocelli in *Onychiurus armatus* (Collembola: Onychiuridae). Rev Ecol Biol Sol 18:127–133

Schaller F (1960) Neues vom Gletscherfloh. Jb Dtsch Ö Alpenverein 85:160–168

Schaller F (1963) Beobachtungen am Gletscherfloh *Isotoma saltans* (Nicolet 1841). Zool Anz Suppl 26:679–682

Schaller F (1970) Collembola (Springschwänze). Handb Zool Berlin 4 (2):1–72

Schenker R (1983) Effects of temperature acclimation on cold-hardiness of alpine micro-arthropods. Rev Ecol Biol Sol 20:37–47

Sømme L (1976) Cold-hardiness in winter-active Collembola. Norweg J Ent 23:149–153

Sømme L (1976/77) Notes on the cold-hardiness of prostigmate mites from Vestfjella, Dronning Maud Land. Norwegian Ant Res Exped 9:51–55

Sømme L (1979) Overwintering ecology of alpine Collembola and oribatid mites from the Austrian Alps. Ecol Entomol 4:175–180

Sømme L (1982) Supercooling and winter survival in terrestrial arthropods. Comp Biochem Physiol 73A: 519–543

Stach J (1947–1960) The apterygotan fauna of Poland in relation to the world fauna of this group of insects. I–VIII. Polska Akad Nauk Krakow

Steinböck O (1939) Der Gletscherfloh. Z Dtsch Ö Alpenverein 70:138–147

Strebel O (1932) Beiträge zur Biologie, Ökologie und Physiologie einheimischer Collembolen. Z Morph Ökol Tiere 25:31–153

Strebel O (1963) Die Variabilität der Ommenzahl bei *Hypogastrura cavicola* Börner (Collembola), ein neuer Fall von degenerativer Evolution. Naturwissenschaften 50 (13):1–2

Takeda H (1978) Ecological studies of collembolan populations in a pine forest soil. II. Vertical distribution of Collembola. Pedobiologia 18:22–30

Usher MB (1969) Some properties of the aggregations of soil arthropods: Collembola. J Anim Ecol 38:607–622

Usher MB (1970) Seasonal and vertical distribution of a population of soil arthropods: Collembola. Pedobiologia 10:224–236

Usher MB, Balogun RA (1966) A defence mechanism in *Onychiurus* (Collembola, Onychiuridae). Entomologist, London 102:237–238

Vannier G (1974) Variation du flux d'évaporation corporelle et de la résistance cuticulaire chez *Tetrodontophora bielanensis* (Waga), Insecte Collembole, vivant dans une atmosphère à régime hygrométrique variable. Rev Ecol Biol Sol 11:201–211

Vannier G (1975) Etude de la rétention hydrique chez l'insecte Collembole *Tetrodontophora bielanensis*. Pedobiologia 15:68–80

Verhoef HA, Nagelkerke CJ (1977) Formation and ecological significance of aggregations in Collembola. Oecologia 31:215–226

Verhoef HA, Nagelkerke CJ, Joosse ENG (1977) Aggregation pheromones in Collembola. J Insect Physiol 23:1009–1013

Weissgerber J (1983) Untersuchungen am Transportepithel des Ventraltubus von *Tomocerus flavescens* (Tullberg, 1871). Diplomarbeit, Univ Mainz

Wolters V (1983) Ökologische Untersuchungen an Collembolen eines Buchenwaldes auf Kalk. Pedobiologia 25:73–85

Zettel J (1984) Cold hardiness strategies and thermal hysteresis in Collembola. Rev Ecol Biol Sol 21:189–203

Zinkler D (1966) Vergleichende Untersuchungen zur Atmungsphysiologie von Collembolen (Apterygota) und anderen Bodenkleinarthropoden. Z Vergl Physiol 52:99–144

Zinkler D, Rüssbeck R (1986) Ecophysiological adaptations of Collembola to low oxygen concentrations. In: Dallai R (ed) 2nd Int Seminar on Apterygota, Siena 1986, pp 123–127. University of Siena, Siena

Archaeognatha

Bitsch J (1974) Fonction et ultrastructure des vésicules exsertiles de l'abdomen des Machilides. Pedobiologia 14:144–145

Bitsch J, Palévody C (1973) L'épithélium absorbant des vésicules coxales des Machilides (Insecta, Thyanura). Z Zellforsch 143:169–182

Delany MJ (1959) The life histories and ecology of two species of *Petrobius* Leach, *P. brevistylis* and *P. maritimus*. Trans R Soc Edinb 63:501–533

Eisenbeis G (1983) The water balance of *Trigoniophtalmus alternatus* (Silvestri, 1904) Archaeognatha: Machilidae). Pedobiologia 25:207–215

Handschin E (1929) Urinsekten oder Apterygota. Die Tierwelt Deutschlands 16:1–150

Houlihan DF (1976) Water transport by the eversible abdominal vesicles of *Petrobius brevistylis.* J Insect Physiol 22:1683–1695

Janetschek H (1951) Über Borstenschwänze Südtirols, besonders des Schlerngebietes (Apterygota, Thysanura). Der Schlern:321–329

Joosse ENG (1976) Littoral apterygotes (Collembola and Thysanura). In: Cheng L (ed) Marine insects. North-Holland Publishing Company, Amsterdam Oxford New York, pp 151–186

Krüger G (1975) Histologische Untersuchungen an Sinnesorganen auf den Mundwerkzeugen von Machiliden (Insecta, Thysanura). Hausarbeit, Lehramt am Gymnasium, Braunschweig

Larink O (1968) Zur Biologie des küstenbewohnenden Machiliden *Petrobius brevistylis* (Thysanura, Insecta). Helgoländer Wiss Meeresunters 18:124–129

Larink O (1970) Der Felsenspringer *Petrobius* – Bau der äußeren Geschlechtsorgane. Mikrokosmos 59:67–69

Larink O (1971) Der Felsenspringer *Petrobius* – Mundwerkzeuge von ursprünglichem Bau. Mikrokosmos 60:47–49

Larink O (1972) Zur Struktur der Blastodermcuticula von *Petrobius brevistylis* und *P. maritimus* (Thysanura, Insecta). Cytobiologie 5:422–426

Larink O (1976) Entwicklung und Feinstruktur der Schuppen bei Lepismatiden und Machiliden (Insecta, Zygentoma und Archaeognatha). Zool Jb Anat 95:252–293

Larink O (1979) Struktur der Blastoderm-Cuticula bei drei Felsenspringer-Arten (Archaeognatha: Machilidae). Ent Gen 5:123–128

Palissa A (1964) Apterygota – Urinsekten. Die Tierwelt Mitteleuropas 4:1–407, Quelle & Meyer, Leipzig

Sturm H (1952) Die Paarung bei *Machilis* (Felsenspringer). Naturwissenschaften 39:308

Sturm H (1955) Beiträge zur Ethologie einiger mitteleuropäischer Machiliden. Z Tierpsychol 12:337–363

Sturm H (1960) Zur Entwicklung der in der Umgebung von Mainz vorkommenden Machilidenarten. Jb Ver Naturk Nassau 95:90–107

Sturm H (1978) Zum Paarungsverhalten von *Petrobius maritimus* Leach (Machilidae: Archaeognatha: Insecta). Zool Anz 201:5–20

Sturm H (1980) Die Machiliden (Archaeognatha, Apterygota, Insecta) Nordwestdeutschlands und die tiergeographische Bedeutung dieser Vorkommen. Drosera 80:53–62

Sturm H (1984) Zur Systematik, Biogeographie und Evolution der südamerikanischen Meinertellidae (Machiloidea, Archaeognatha, Insecta). Z Zool Syst Evolutionsforsch 22:27–44

Sturm H, Adis J (1984) Zur Entwicklung und zum Paarungsverhalten zentralamazonischer Meinertelliden (Machiloidea, Archaeognatha, Insecta). Amazonia 8:447–473

Weyda F (1974) Coxal vesicles of Machilidae. Pedobiologia 14:138–141

Wygodzinsky PW (1941) Beiträge zur Kenntnis der Dipluren und Thysanuren der Schweiz. Denkschr Schweiz Naturf Ges 74:107–227

Zygentoma

Adel T (1984) Sensilleninventar und Sensillenmuster auf den Antennen von *Thermobia domestica* und *Lepisma saccharina* (Insecta: Zygentoma). Braunschw Naturk Schr 2:191–217

Berridge MJ, Oschmann JL (1972) Transporting epithelia. Academic Press, New York London

Handschin E (1929) Urinsekten oder Apterygota. Die Tierwelt Deutschlands 16:1–150

Haupt J (1965) Zur Feinstruktur der Labialniere des Silberfischchens *Lepisma saccharina* L. (Thysanura, Insecta). Zool Beitr 15:139–170

Kränzler L, Larink O (1980) Postembryonale Veränderungen und Sensillenmuster der abdominalen Anhänge von *Thermobia domestica* (Packard) (Insecta: Zygentoma). Braunschw Naturk Schr 1:27–49

Laibach E (1952) *Lepisma saccharina* L., das Silberfischchen. Z Hyg Zool 40:1–50

Larink O (1976) Entwicklung und Feinstruktur der Schuppen bei Lepismatiden und Machiliden (Insecta, Zygentoma und Archaeognatha). Zool Jb Anat 95:252–293

Larink O (1982) Das Sensillen-Inventar der Lepismatiden (Insecta: Zygentoma). Braunschw Naturk Schr 1: 493–512

Palissa A (1964) Apterygota. Tierwelt Mitteleuropas 4: 1–407, Quelle & Meyer, Leipzig

Sahrhage D (1953) Ökologische Untersuchungen an *Thermobia domestica* (Packard) und *Lepisma saccharina* L. Z Wiss Zool 157:77–168

Sturm H (1956a) Die Paarung von *Lepisma saccharina* L. (Silberfischchen). Zool Anz Suppl 19:463–466

Sturm H (1956b) Die Paarung beim Silberfischchen *Lepisma saccharina*. Z Tierphysiol 13:1–12

Dermaptera

Beier M (1953) Dermaptera – Ohrwürmer. Biol Tiere Deutschl 26:169–321

Beier M (1959) Ohrwürmer und Tarsenspinner. Die Neue Brehm-Bücherei 251, Leipzig

Brauns A (1964) Taschenbuch der Waldinsekten. Gustav Fischer Verlag, Stuttgart

Caussanel C (1966) Etude du développement larvaire de *Labidura riparia* (Derm., Labiduridae). Ann Soc Entomol France 2:469–498

Caussanel C (1970) Principales exigences écophysiologiques du Forficule des sables, *Labidura riparia* (Derm., Labiduridae). Ann Soc Entomol France 6:589–612

Franke U (1985) Zur Biologie eines Buchenwaldbodens – 7. Der Waldohrwurm *Chelidurella acanthopygia*. Carolinea 43:105–112

Günther K, Herter K (1974) Dermaptera (Ohrwürmer). Handb Zool Berlin 4 (2) 2/11:1–158

Harz K (1957) Die Geradflügler Mitteleuropas. G. Fischer Verlag, Jena

Harz K (1960) Geradflügler oder Orthoptera. Die Tierwelt Deutschlands 46:1–232

Herter K (1963) Zur Fortpflanzungsbiologie des Sand- oder Ufer-Ohrwurms *Labidura riparia* Poll. Zoll Beitr 8:297–329

Herter K (1965) Zur Fortpflanzungsbiologie des Ohrwurms *Forficula auricularia* L. Zool Jb Syst 92:405–466

Herter K (1967) Weiteres zur Fortpflanzungsbiologie des Ohrwurmes *Forficula auricularia* L. Zool Beitr 13:213–244

Meissner B (1963) Über das Vorkommen von *Labidura riparia* (Pall.) (Dermaptera) auf den Abraumhalden der Braunkohlenertragsebene um Tröpitz und Lauchhammer. Entomol Ber Dresden 1:24–28

Popham EJ (1959) The anatomy in relation to feeding habits of *Forficula auricularia* and other Dermaptera. Proc Zool Soc London 133:251–300

Slifer EH (1967) Sense organs on the antennal flagella of earwigs (Dermaptera) with special reference to those of *Forficula auricularia*. J Morph 122:63–80

Weidner H (1941) Vorkommen und Lebensweise des Sandohrwurms *Labidura riparia* Pall. Zool Anz 133: 185–202

Blattodea

Beier M (1974) Blattariae (Schaben). Handb Zool Berlin 4 (2) 2 13:1–127

Brauns A (1964) Taschenbuch der Waldinsekten. Gustav Fischer Verlag, Stuttgart

Brown EB (1952) Observations on the life-history of the cockroach *Ectobius panzeri* Stephens. – Ent Mon Mag London 88:209–212

Chopard L (1938) La Biologie des Orthoptères. Encycl Entom Paris A 20:1–541

Eggers F (1924) Zur Kenntnis der antennalen stiftführenden Sinnesorgane der Insekten. Z Morph Ökol Tiere 2: 259–349

Harz K (1957) Die Geradflügler Mitteleuropas. G. Fischer Verlag. Jena

Harz K (1960) Geradflügler oder Orthoptera. Die Tierwelt Deutschlands 46:1–232

Harz K (1972) Der Entwicklungszyklus von *Ectobius lapponicus* L. am Polarkreis. Ber Ökol Station Messaure 16:1–8

Kupka E (1946) Über Bremsvorrichtungen an den Laufbeinen der Blattodea. Österr Zool Z Wien 1:170–175

Loftus SJR (1966) Cold receptor on the antenna of *Periplaneta americana*. Z Vergl Physiol Berlin 52:380–385

Loftus SJR (1969) Differential thermal components in the response of the antennal cold receptor of *Periplaneta americana* to slowly changing temperature. Z Vergl Physiol Berlin 63:415–433

Princis K (1965) Ordnung Blattariae (Schaben). Bestimmungsbücher zur Bodenfauna Europas 3:1–50

Roth LM, Willis ER (1952) Observations on the biology of *Ectobius pallidus* (Olivier). Trans Amer Ent Soc Philadelphia 83:31–37

Slifer EH (1968) Sense organs on the antennal flagellum of a giant cockroach, *Gromphadorhina portentosa*, and a comparison with those of several other species. J Morph Philadelphia 126:19–30

Winston PW, Green CC (1967) Humidity responses from antennae of the cockroach, *Leucophaea maderae*. Naturwissenschaften Leipzig 54:499

Ensifera

Beier M (1954) Grillen und Maulwurfsgrillen. Die Neue Brehm-Bücherei 119, A. Ziemsen Verlag, Wittenberg-Lutherstadt

Beier M (1972) Saltatoria (Grillen und Heuschrecken). Handb Zool Berlin 4 (2) 2/9:1–217

Brocksieper R (1978) Der Einfluß des Mikroklimas auf die Verbreitung der Laubheuschrecken, Grillen und Feldheuschrecken im Siebengebirge und auf dem Rodderberg bei Bonn (Orthoptera: Saltatoria). Decheniana – Beih Bonn 21:1–141

Chopard L (1951) Orthoptéroides. Faune France Paris 56:1–359

Gnatzy W, Schmidt K (1971) Die Feinstruktur der Sinneshaare auf den Cerci von *Gryllus bimaculatus* Deg. (Saltatoria, Gryllidae). I. Faden- und Keulenhaare. Z Zellforsch 122:190–209

Gnatzy W, Schmidt K (1972a) Die Feinstruktur der Sinneshaare auf den Cerci von *Gryllus bimaculatus* Deg. (Saltatoria, Gryllidae). IV. Die Häutung der kurzen Borstenhaare. Z Zellforsch 126:223–239

Gnatzy W, Schmidt K (1972b) Die Feinstruktur der Sinneshaare auf den Cerci von *Gryllus bimaculatus* Deg. (Saltatoria, Gryllidae). V. Die Häutung der langen Borstenhaare an der Cercusbasis. J Microscopie 14:75–84

Gnatzy W, Tautz J (1980) Ultrastructure and mechanical properties of an insect mechanoreceptor: Stimulus-transmitting structures and sensory apparatus of the cercal filiform hairs of *Gryllus*. Cell Tissue Res 213:441–463

Godan D (1961) Untersuchungen über die Nahrung der Maulwurfsgrille (*Gryllotalpa gryllotalpa* L.). Z Angew Zool Berlin 48:341–357

Halm E (1958) Untersuchungen über die Lebensweise und Entwicklung der Maulwurfsgrille (*Gryllotalpa vulgaris* Latr.) im Lande Brandenburg. Beitr Ent Berlin 8:334–365

Harz K (1957) Die Geradflügler Mitteleuropas. G Fischer Verlag, Jena

Harz K (1960) Geradflügler oder Orthoptera. Die Tierwelt Deutschlands 46:1–232

Nicklaus R (1969) Zur Funktion der keulenförmigen Sensillen auf den Cerci der Grillen. Zool Anz Suppl 32: 393–398

Röber H (1970) Die Saltatorienfauna montan getönter Waldgebiete Westfalens unter besonderer Berücksichtigung der Ensiferenverbreitung. Abh Landesmus Naturkd Münster 32:1–28

Schmidt K, Gnatzy W (1971) Die Feinstruktur der Sinneshaare auf den Cerci von *Gryllus bimaculatus* Deg. (Saltatoria, Gryllidae). II. Die Häutung der Faden- und Keulenhaare. Z Zellforsch 122:210–226

Schmidt K, Gnatzy W (1972) Die Feinstruktur der Sinneshaare auf den Cerci von *Gryllus bimaculatus* Deg. (Saltatoria, Gryllidae). III. Die kurzen Borstenhaare. Z Zellforsch 126:206–222

Sihler H (1924) Die Sinneshaare an den Cerci der Insekten. Zool Jb Anat Ontog Tiere 45:519–580

Hemiptera

Jordan KHC (1962) Landwanzen. Die Neue Brehm-Bücherei 294, A. Ziemsen Verlag. Wittenberg-Lutherstadt

Jordan KHC (1972) Heteroptera (Wanzen). Handb Zool Berlin 4 (2) 2/20:1–113

Remold H (1963) Über die biologische Bedeutung der Duftdrüsen bei den Landwanzen (Geocorisae). Z Vergl Physiol Berlin 45:636–694

Schorr H (1957) Zur Verhaltensbiologie und Symbiose von *Brachypelta aterrima* Först. (Cydnidae, Heteroptera). Z Morph Ökol Tiere 45:561–601

Wagner E (1966) Wanzen oder Heteroptera. I. Pentatomorpha. Tierw Deutschlands 54:1–235

Weber H (1930) Biologie der Hemipteren. Springer, Berlin

Planipennia

Aspöck H, Aspöck U, Hölzel H (1980) Die Neuropteren Europas. 2 Bde. Goecke und Evers, Krefeld

Doflein F (1916) Der Ameisenlöwe – Eine biologische, tierpsychologische und reflexbiologische Untersuchung. G. Fischer Verlag, Jena
Eglin W (1939) Zur Biologie und Morphologie der Raphidien und Myrmeleoniden (Neuropteroidea) von Basel und Umgebung. Verh Naturforsch Ges Basel 50:163–220
Geiler H (1966) Über die Wirkung der Sonneneinstrahlung auf Aktivität und Position der Larven von *Euroleon nostras* Fourcr. (= *Myrmeleon europaeus* McLachl.) in den Trichterbodenfallen. Z Morph Ökol Tiere 56:260–274
Jacobs W, Renner M (1974) Taschenlexikon zur Biologie der Insekten. Gustav Fischer Verlag, Stuttgart
Jokusch B (1967) Bau und Funktion eines larvalen Insektenauges. – Untersuchungen am Ameisenlöwen (*Euroleon nostras* Fourcroy, Planip., Myrmel.). Z Vergl Physiol 56:171–198
Matthes D (1982a) Ameisenlöwe und Ameisenjungfer 1. Gestalt und mikroskopische Anatomie des Ameisenlöwen. Mikrokosmos 71:1–8
Matthes D (1982b) Ameisenlöwe und Ameisenjungfer 2. Das Verhalten des Ameisenlöwen. Mikrokosmos 71:44–47
Neboer H (1960) Ethological observations on the ant-lion (*Euroleon nostras* Fourcroy, Neuroptera). Arch Neerl Zool 13:609–611
Plett A (1964) Einige Versuche zum Beutefangverhalten und Trichterbauen des Ameisenlöwen *Euroleon nostras* Fourcr. (Myrmeleonidae). Zool Anz 173:202–209
Principi MM (1943) Contributi allo studio dei Neurotteri italiani. II. *Mymeleon inconspicuus* Ramb. ed *Euroleon nostras* Fourcroy. Boll Ist Ent Univ Bologna 14:131–192
Steffan JR (1975) Les Larves de Fourmilions (Planipennes: Myrmeleontidae) de la faune de France. Ann Soc Ent Fr 11:383–410

Coleoptera

Bauer T (1981) Prey capture and structure of the visual space of an insect that hunts by sight on the litter layer (*Notiophilus biguttatus* F., Carabidae, Coleoptera). Behav Ecol Sociobiol 8:91–97
Bauer T (1982a) Prey-capture in a ground-beetle larva. Anim Behav 30:203–208
Bauer T (1982b) Predation by a carabid beetle specialized for catching Collembola. Pedobiologia 24:169–179
Bauer T (1985a) Different adaptation to visual hunting in three ground beetle species of the same genus. J Insect Physiol 31:593–601
Bauer T (1985b) Beetles which use a setal trap to hunt springtails: The hunting strategy and apparatus of *Leistus* (Coleoptera, Carabidae). Pedobiologia 28:275–287
Bechyně J (1954) Welcher Käfer ist das? Franckh'sche Verlagshandlung, Stuttgart
Blum P (1979) Zur Phylogenie und ökologischen Bedeutung der Elytrenreduktion und Abdomenbeweglichkeit der Staphylinidae (Coleoptera) – Vergleichend- und funktionsmorphologische Untersuchungen. Zool Jb Anat 102:533–582
Böving AG, Craighead FC (1931) An illustrated synopsis of the principal larval forms of the order Coleoptera. Ent Amer 11:1–351
Brauns A (1970) Taschenbuch der Waldinsekten. – 2 Bde. Gustav Fischer Verlag, Stuttgart
Brunne G (1976) U. Fam. Scaritinae. – In: Freude-Harde-Lohse, Die Käfer Mitteleuropas, Bd 2, pp 64–73, Goecke und Evers, Krefeld
Burmeister F (1939) Biologie, Ökologie und Verbreitung europäischer Käfer. Krefeld
Coiffait K (1958) Contribution à la connaissance des Coléopteres du sol. Vie et Milieu, Suppl. 7:1–204
Coiffait K (1959) Monographie des Leptotyphlines. Rév Franc d'Ent 26:237–438
Crowson RA (1981) The Biology of Coleoptera. London
Dunger W (1983) Tiere im Boden. Die Neue Brehm-Bücherei 327 (3. Aufl), A. Ziemsen Verlag, Wittenberg-Lutherstadt
Engel H (1961) Mitteleuropäische Insekten. – Sammlung Naturkundlicher Tafeln. Kronen-Verlag Erich Cramer, Hamburg
Evans MEG (1965) The feeding method of *Cicindela hybrida* L. (Coleoptera, Cicindelidae). Proc Roy Ent Soc London 40:61–66
Evans MEG (1975) The life of Beetles. London
Evans WG (1983) Habitat selection in the Carabidae. Col Bull 37:164–167
Faasch H (1968) Beobachtungen zur Biologie und zum Verhalten von *Cicindela hybrida* L. und *Cicindela campestris* L. und experimentelle Analyse ihres Beutefangverhaltens. Zool Jb Syst 95:477–522
Freude H, Harde KW, Lohse GA (1965ff.) Die Käfer Mitteleuropas. Goecke und Evers, Krefeld
Friebe B (1983) Zur Biologie eines Buchenwaldbodens. 3. Die Käferfauna. Carolinea 41:45–80
Funke W (1971) Food and enery turnover of leafeating insects and their influence on primary production. Ecol Studies Berlin 2:81–93
Geisthardt M (1974) Das thorakale Skelett von *Lamprohiza splendidula* (L.) unter besonderer Berücksichtigung des Geschlechtsdimorphismus (Coleoptera: Lampyridae). Zool Jb Anat 93:299–334
Geisthardt M (1977) Bemerkungen zur Frage der Mikropterie und Apterie sowie zur Biologie einiger heimischer Cantharoidea (Coleoptera). Mitt Int Ent Verein Frankfurt 3:84–91
Geisthardt M (1979) 26. Fam. Lampyridae. In: Freude-Harde-Lohse, Die Käfer Mitteleuropas, Bd 6, pp 14–17, Goecke und Evers, Krefeld
Gersdorf E (1937) Ökologisch-faunistische Untersuchungen über die Carabiden der mecklenburgischen Landschaft. Zool Jb Syst 70:17–86
Grimm R (1973) Zum Energieumsatz phytophager Insekten im Buchenwald. I. Untersuchungen an Populationen der Rüsselkäfer (Curculionidae) *Rhynchaenus fagi* L., *Strophosomus* (Schönherr) und *Otiorhynchus singularis* L. Oecologia 11:187–262
Guse GW, Honomichl K (1980) Die digitiformen Sensillen auf dem Maxillarpalpus von Coleoptera. II. Feinstruktur bei *Agabus bipustulatus* (L.) und *Hydrobius fuscipes* (L.). Protoplasma 103:55–68
Hartmann P (1979) Biologisch-ökologische Untersuchun-

gen an Staphyliniden-Populationen verschiedener Ökosysteme des Solling. Diss. Univ Göttingen

Hintzpeter U, Bauer T (1986) The antennal setal trap of the ground beetle *Loricera pilicornis*: a specialization for feeding on Collembola. J Zool (Lond) A208:615–630

Honomichl K (1980) Die digitiformen Sensillen auf dem Maxillarpalpus von Coleoptera. I. Vergleichend-topographische Untersuchung des kutikulären Apparates. Zool Anz Jena 204:1–12

Honomichl K, Guse GW (1981) Digitiform sensilla on the maxillar palp of coleoptera. III. Fine structure in *Tenebrio molitor* L. and *Dermestes maculatus* De Geer. Acta Zoologica 62:17–25

Jacobs W, Renner M (1974) Taschenlexikon zur Biologie der Insekten. Gustav Fischer Verlag, Stuttgart

Kasule FK (1966) The subfamilies of the larvae of British genera of Steninae and Proteininae. Trans R Ent Soc Lond 118:261–283

Kasule FK (1968) The larval characters of some subfamilies of British Staphylinidae (Coleoptera) with keys to the known genera. Trans R Ent Soc London 120:115–138

Kasule FK (1970) The larvae of Paederinae and Staphylininae (Coleoptera: Staphylinidae) with keys to the known British genera. Trans R Ent Soc London 122:49–80

Klausnitzer B (1978) Ordnung Coleoptera (Larven). Junk, The Hague

Klausnitzer B (1982) Wunderwelt der Käfer. Verlag Herder, Freiburg

Kosmaschewski AS (1958) On the feeding habits of the Click-Beetle larvae (Coleoptera, Elateridae). Ent Rev 37:689–697

Lampe KH (1975) Die Fortpflanzungsbiologie und Ökologie des Carabiden *Abax ovalis* Dft. und der Einfluß der Umweltfaktoren Bodentemperatur, Bodenfeuchtigkeit und Photoperiode auf die Entwicklung in Anpassung an die Jahreszeit. Zool Jb Syst 102:128–170

Larsson S (1941) Danske Billelarver, Bestemmelsesnogle til Familie. Ent Medd 22:239–259

Lengerken von H (1954) Die Brutfürsorge und Brutpflegeinstinkte der Käfer. Geest und Portig, Leipzig

Löser S (1972) Art und Ursachen der Verbreitung einiger Carabidenarten im Grenzraum Ebene–Mittelgebirge. Zool Jb Syst 99:231–262

Lohse GA (1964) 23. Familie: Staphylinidae. In: Freude-Harde-Lohse, Die Käfer Mitteleuropas. Bd 4. Goecke und Evers, Krefeld

Loreau M (1983) Trophic role of carabid beetles in a forest. Proc VIII Int Coll Soil Zoology, Louvain-La-Neuve (Belgium) 1982, pp 281–286

Lundt H (1964) Ökologische Untersuchungen über die tierische Besiedlung an Aas im Boden. Pedobiologia 4:158–180

Mann JS, Crowson RA (1984) On the digitiform sensilla of adult leaf beetles (Coleoptera: Chrysomelidae). Entomol Gener 9:121–133

Neudecker C (1974) Das Präferenzverhalten von *Agonum assimile* Payk. (Carab., Coleopt.) in Temperatur-, Feuchtigkeits- und Helligkeitsgradienten. Zool Jb Syst 101:609–627

Neudecker C, Thiele HU (1974) Die jahreszeitliche Synchronisation der Gonadenreifung bei *Agonum assimile* Payk. (Coleoptera, Carab.) durch Temperatur und Photoperiode. Oecologia 17:141–157

Paarmann W (1966) Vergleichende Untersuchungen über die Bindung zweier Carabidenarten (*P. angustatus* und *P. oblongopunctatus*) an ihre verschiedenen Lebensräume. Z Wiss Zool 174:83–176

Pukowski E (1933) Ökologische Untersuchungen an Nekrophoren. Z Morph Ökol Tiere 27:518–586

Raynaud P (1974) Stades larvaires de coléoptères Carabidae. Bull Mens Soc Linn Lyon 43:229–246

Rehage HO, Renner K (1981) Zur Käferfauna des Naturschutzgebietes Jakobsberg. Natur und Heimat 41: 124–137

Reitter E (1908–1916) Fauna Germanica, I–V. KG Lutz Verlag, Stuttgart

Roth M, Funke W, Günl W, Straub S (1983) Die Käfergesellschaften mitteleuropäischer Wälder. Verh Ges Ökologie Mainz 1981, pp 35–50

Rudolph K (1970) Zur Morphologie der Elateridenlarven. Ent Nachr 14:33–46

Rudolph K (1974) Beitrag zur Kenntnis der Elateridenlarven der Fauna der DDR und der BRD. Zool Jb Syst 101:1–151

Rudolph K (1978) Elateridae. In: Klausnitzer B: Ordnung Coleoptera (Larven). Junk, The Hague, pp 133–156

Šarova JC (1960) Die morpho-ökologischen Typen der Laufkäferlarven (Carabidae), (russisch). Zool Žurnal 39:691–708

Schaerffenberg B (1942) Die Elateridenlarven der Kiefernwaldstreu. Z Angew Ent 29:85–115

Schauermann J (1973) Zum Energieumsatz phytophager Insekten im Buchenwald. II. Die produktionsbiologische Stellung der Rüsselkäfer (Curculionidae) mit rhizophagen Larvenstadien. Oecologia 13:313–350

Schauermann J (1977) Zur Abundanz- und Biomassendynamik der Tiere in Buchenwäldern des Sollings. Verh Ges Ökol Göttingen 1976, pp 113–124

Schauermann J (1977) Energy metabolism of rhizophagous insects and their role in ecosystems. Ecol Bull 25:310–319

Scherf H (1964) Die Entwicklungsstadien der mitteleuropäischen Curculioniden (Morphologie, Bionomie, Ökologie). Abh Senckenberg Naturf Ges 506:1–335

Scherney F (1959) Unsere Laufkäfer. Die Neue Brehm-Bücherei 245, A. Ziemsen Verlag, Wittenberg-Lutherstadt

Scherney F (1961) Beiträge zur Biologie und ökonomischer Bedeutung räuberisch lebender Käferarten. Beobachtungen und Versuche zur Überwinterung, Aktivität und Ernährungsweise der Laufkäfer (Carabidae). Z Angew Entomol 48:163–175

Schildknecht H, Maschwitz E, Maschwitz U (1968) Die Explosionschemie der Bombardierkäfer (Coleoptera, Carabidae). III. Mitt.: Isolierung und Charakterisierung der Explosionskatalysatoren. Z Naturforsch 23B:1213–1218

Schnepf E, Wenneis W, Schildknecht H (1969) Über Arthropoden-Abwehrstoffe XLI. Zur Explosionschemie der Bombardierkäfer (Coleoptera, Carabidae). IV. Zur Feinstruktur der Pygidialwehrdrüsen des Bombardierkäfers (*Brachynus crepitans* L.). Z Zellforsch 96:582–599

Schuster R (1966a) Über den Beutefang des Ameisenkäfers *Cephennium austriacum* Reiter. Naturwissenschaften 53:113

Schuster R (1966b) Scydmaeniden-Larven als Milbenräuber. Naturwissenschaften 53:439–440

Schwalb HH (1961) Beiträge zur Biologie der einheimischen Lampyriden *Lampyris noctiluca* Geoffr. und *Phausis splendidula* Lec. und experimentelle Analyse ihres Beutefang- und Sexualverhaltens. Zool Jb Syst 88:399–550

Späh H (1980) Faunistisch-ökologische Untersuchungen der Carabiden- und Staphylinidenfauna verschiedener Standorte Westfalens (Coleoptera: Carabidae, Staphylinidae). Decheniana (Bonn) 133:33–56

Steel WO (1966) A revision of the staphylinid subfamily Proteininae (Coleoptera). Trans R Ent Soc London 118:285–311

Steel WO (1970) The larvae of the genera of Omaliinae (Col., Staphylinidae), with particular reference to the British fauna. Trans R Ent Soc London 122:1–47

Strey G (1972) Ökoenergetische Untersuchungen an *Athous subfuscus* Müll. und *Athous vittatus* Fbr. (Elateridae, Coleoptera) in Buchenwäldern. Diss Univ Göttingen

Sturani M (1962) Osservazioni e ricerche biologiche sul genere *Carabus* Linnaeus (Sensu lato) (Coleoptera, Carabidae). Mem Soc Entomol Ital 41:85–202

Subkew W (1934) Physiologisch-experimentelle Untersuchungen an einigen Elateriden. Z Morph Ökol Tiere 28:184–192

Thiele HU (1956) Die Tiergesellschaften der Bodenstreu in den verschiedenen Waldtypen des Niederbergischen Landes. Z. Angew Entomol 39:316–357

Thiele HU (1964) Experimentelle Untersuchungen über die Ursachen der Biotopbindung bei Carabiden. Z Morph Ökol Tiere 53:387–452

Thiele HU (1967) Ein Beitrag zur experimentellen Analyse von Euryökie und Stenökie bei Carabiden. Z Morph Ökol Tiere 58:355–372

Thiele HU (1977) Carabid beetles in their environments. – A study on habitat selection by adaptations in physiology and behaviour. Springer, Berlin Heidelberg New York

Tips W (1978a) Some aspects of the ecology of *Acrotrichis intermedia* (Col., Ptiliidae). I. Locomotory activity and gregarious behaviour. Pedobiologia 18:127–133

Tips W (1978b) Some aspects of the ecology of *Acrotrichis intermedia* (Col., Ptiliidae). II. On the basis of gregarious behaviour. Pedobiologia 18:134–137

Tips W (1978c) Some aspects of the ecology of *Acrotrichis intermedia* (Col., Ptiliidae). III. Some evidence for the presence of an aggregating substance. Pedobiologia 18:218–226

Tischler WH (1976) Untersuchungen über die tierische Besiedlung von Aas in verschiedenen Strata von Waldökosystemen. Pedobiologia 16:99–105

Topp W (1971) Zur Biologie und Larvalmorphologie von *Atheta sordida* Marsh. (Col., Staphylinidae). Ann Ent Fenn 37:85–89

Topp W (1973) Über Entwicklung, Diapause und Larvalmorphologie der Staphyliniden *Aleochara moerens* Gyll. und *Bolitochara lunulata* Payk. in Nordfinnland. Ann Ent Fenn 39:145–152

Topp W (1975a) Morphologische Variabilität, Diapause und Entwicklung von *Atheta fungi* (Grav.) (Col., Staphylinidae). Zool Jb Syst 102:101–127

Topp W (1975b) Zur Besiedlung einer neu entstandenen Insel. Untersuchungen am Hohen Knechtsand. Zool Jb Syst 102:215–240

Topp W (1978) Bestimmungstabelle für die Larven der Staphylinidae. In: Klausnitzer B: Ordnung Coleoptera (Larven). Junk, The Hague, pp 304–334

Topp W (1979) Vergleichende Dormanzuntersuchungen an Staphyliniden (Coleoptera). Zool Jb Syst 106:1–49

Topp W (1981) Biologie der Bodenorganismen. UTB Quelle u. Meyer, Heidelberg

Topp W, Hansen K, Brandl R (1982) Artengemeinschaften von Kurzflüglern an Aas (Coleoptera: Staphylinidae). Entomol Gener 7:347–364

Wasner U (1977) Die Europhilus-Arten (*Agonum*, Carabidae, Coleoptera) des Federseerieds. Diss Univ Tübingen

Weber H (1933) Lehrbuch der Entomologie. G. Fischer Verlag, Jena

Weidemann G (1972) Stellung epigäischer Raubarthropoden im Ökosystem Buchenwald. Verh Dtsch Zool Ges 65:106–116

Wichard W, Komnick H (1974) Feinstruktur und Funktion der Analpapillen aquatischer Käferlarven (Coleoptera: Helodidae). Int J Insect Morphol Embryol 3:335–341

Wilms B (1961) Untersuchungen zur Bodenkäferfauna in drei pflanzensoziologisch unterschiedenen Wäldern der Umgebung Münsters. Abh Landesmus Naturk Münster 23:1–15

Zacharuk RY, Albert PJ (1978) Ultrastructure and function of scolophorous sensilla in the mandible of an elaterid larva (Coleoptera). Can J Zool 56:246–259

Zacharuk RY, Albert PJ, Bellamy FW (1977) Ultrastructure and function of digitiform sensilla on the labial palp of a larval elaterid (Coleoptera). Can J Zool 55:569–578

Hymenoptera

Bernard F (1968) Les Fourmis (Hymenoptera Formicidae) d'Europe Occidentale et Septendrionale. Faune Europe Bass Médit 3:1–411, Masson et Cie, Paris

Bischoff H (1927) Biologie der Hymenopteren. Springer, Berlin

Brauns A (1968) Praktische Bodenbiologie. Gustav Fischer Verlag, Stuttgart

Brian MV (1977) Ants.– The New Naturalist, Collins, London Glasgow

Dircksen R, Dircksen G (1968) Tierkunde, II. Wirbellose Tiere. Bayerischer Schulbuch Verlag, München

Donisthorpe H (1927) British ants, their life history and classification. London, (2. A)

Dumpert K (1978) Das Sozialleben der Ameisen. Verlag Paul Parey, Berlin Hamburg

Dunger W (1983) Tiere im Boden. Die Neue Brehm-Bücherei 327 (3. Aufl), A. Ziemsen Verlag, Wittenberg-Lutherstadt

Escherich K (1917) Die Ameise – Schilderung ihrer Lebensweise. Vieweg und Sohn, Braunschweig

Forel A (1874) Les Fourmis de la Suisse. Zürich

Gösswald K (1951) Die rote Waldameise im Dienst der Waldhygiene. Wolf und Täuber, Lüneburg

Gösswald K (1985) Organisation und Leben der Ameisen. Wissenschaftliche Verlagsgesellschaft, Stuttgart

Goetsch W (1953) Vergleichende Biologie der Insekten-Staaten. Akadem. Verlagsgesellschaft, Leipzig

Kloft WJ (1959) Nestbautätigkeit der Roten Waldameise. Waldhygiene 3:94–98

Kloft WJ (1978) Ökologie der Tiere. UTB Ulmer, Stuttgart

Kloft WJ, Hölldobler B, Haisch A (1965) Traceruntersuchungen zur Abgrenzung von Nestarealen holzzerstörender Roßameisen (*Camponotus herculeanus* L. und *C. ligniperda* Latr.). Ent Exp Appl 8:20–26

Kneitz G (1974) Untersuchungen zur Herkunft der Nestwärme und zur Temperaturregulation im Waldameisennest. Habilitationsschrift, Würzberg

Maschwitz U, Hölldobler B (1970) Der Kartonnestbau bei *Lasius fuliginosus* Latr. (Hym., Formicidae). Z Vergl Physiol 66:176–189

Otto D (1962) Die Roten Waldameisen. Die Neue Brehm-Bücherei 293, A. Ziemsen Verlag, Wittenberg-Lutherstadt

Sudd JH (1967) An introduction to the behaviour of ants. Arnold, London

Stitz H (1939) Ameisen oder Formicidae. Tierw Deutschlands 37:1–428

Trichoptera

Anderson NH (1967) Life cycle of a terrestrial caddisfly, *Philocasca demita* (Trichoptera: Limnephilidae), in North America. Ann Ent Soc Amer 60:320–323

Brauns A (1964) Taschenbuch der Waldinsekten. Gustav Fischer Verlag, Stuttgart

Drift I, van der, Witkamp M (1958) The significance of the break-down of oak litter by *Enoicyla pusilla* Burm. Arch Néerl Zool 13:486–492

Erman NA (1981) Terrestrial feeding migration and life history of the stream-dwelling caddisfly, *Desmona betula* (Trichoptera: Limnephilidae). Can J Zool 59:1658–1665

Flint OS (1958) The larva and terrestrial pupa of *Ironoquia parvula* (Trichoptera, Limnephilidae). J New York Ent Soc 66:59–62

Hickin NE (1967) Caddis Larvae – Larvae of the British Trichoptera. Hutchinson, London

Kelner-Pillault S (1960) Biologie, écologie *d'Enoicyla pusilla* Burm. (Trichoptera, Limnophilides). Ann Biol 36:51–99

Malicky H (1973) Trichoptera (Köcherfliegen). Handb Zool Berlin 4 (2) 2/29:1–114

Malicky H (1983) Atlas of European Trichoptera. Junk, The Hague Boston London

Mey W (1983) Die terrestrischen Larven der Gattung *Enoicyla* Rambur in Mitteleuropa und ihre Verbreitung. Dt Entom Z 30:115–122

Rathjen W (1939) Experimentelle Untersuchungen zur Biologie und Ökologie von *Enoicyla pusilla* Burm. Z Morph Ökol Tiere 35:14–83

Späh H (1978) *Enoicyla pusilla* Burm. aus einem Erlenbruch Ostwestfalens (Insecta: Trichoptera). Decheniana (Bonn) 131:262–265

Wichard W (1973) Zur Morphogenese des respiratorischen Epithels der Tracheenkiemen bei Larven der Limnephilini Kol. (Insecta, Trichoptera). Z Zellforsch 144:585–592

Wichard W (1974) Zur morphologischen Anpassung der Tracheenkiemen bei Larven der Limnephilini Kol. (Insecta: Trichoptera) I–II. Oecologia 15:159–175

Wichard W (1978) Die Köcherfliegen. Die Neue Brehm-Bücherei 512, A. Ziemsen Verlag, Wittenberg-Lutherstadt

Wichard W, Komnick H (1973) Fine structure and function of the abdominal chloride epithelia in caddisfly larvae. Z Zellforsch 136:579–590

Wichard W, Schmitz M (1980) Anpassungsmechanismen der osmoregulatorischen Ionenabsorption bei Limnephilidae-Larven (Insecta, Trichoptera). Gewässer und Abwässer 66/67:102–118

Wiggins GB (1977) Larvae of the North American Caddisfly Genera (Trichoptera). Univ Toronto Press, Toronto

Lepidoptera

Davis DR (1964) Bagworm Moth of the Western Hemisphere. US Nat Mus Bull 244:1–233

Dunger W (1983) Tiere im Boden. Die Neue Brehm-Bücherei 327 (3. Aufl), A. Ziemsen Verlag, Wittenberg-Lutherstadt

Forster W, Wohlfahrt TA (1954–1982) Die Schmetterlinge Mitteleuropas, I–V. Franckh'sche Verlagshandlung, Stuttgart

Hering M (1926) Biologie der Schmetterlinge. Springer, Berlin

Kozhanchikov IV (1956) Psychidae. Fauna UdSSR, Moskow

Meyer E (1983) Struktur und Dynamik einer Population von *Epichnopterix ardua* Mann (Lep.: Psychidae) in einem Krummseggenrasen der Hohen Tauern (Kärnten, Österreich). Zool Jb Syst 110:165–177

Mecoptera

Aubrook EW (1939) A contribution to the biology and distribution in Great Britain of *Boreus hyemalis* (L.) (Mecopt., Boreidae). J Soc Brit Ent Southampton 2:13–21

Engel H (1961) Mitteleuropäische Insekten. Sammlung Naturkundlicher Tafeln. Kronen-Verlag E. Cramer, Hamburg

Kaltenbach A (1978) Mecoptera (Schnabelhafte Schnabelfliegen). Hand Zool Berlin 4 (2) 2/28:1–111

Mickoleit G, Mickoleit E (1976) Über die funktionelle Bedeutung der Tergalapophysen von *Boreus westwoodi* (Hagen) (Insecta, Mecoptera). Zoomorphologie 85:157–164

Sauer CP (1966) Ein Eskimo unter den Insekten: Der Winterhafte *Boreus westwoodi*. Mikrokosmos 55:117–120

Steiner P (1937) Beitrag zur Fortpflanzungsbiologie und Morphologie des Genitalapparates von *Boreus hiemalis* L. Z Morph Ökol Tiere 32:276–288

Stitz H (1926) Mecoptera. Biologie Tiere Deutschlands 35:1–28

Strübing H (1950) Beitrag zur Biologie von *Boreus hiemalis* L. Zool Beitr Berlin 1:51–110

Strübing H (1958) Schneeinsekten. Die Neue Brehm-Bücherei 220, A. Ziemsen Verlag, Wittenberg-Lutherstadt

Diptera

Altmüller R (1979) Untersuchungen über den Energieumsatz von Dipterenpopulationen im Buchenwald (Luzulo-Fagetum). Pedobiologia 19:245–278

Bezzi M (1917) Rinvenimento di una *Chionea* (Dipt.) neu dintorni die Torino. Bull Soc Ent Ital 49:12–49

Brauns A (1954a) Terricole Dipterenlarven. Musterschmidt, Wissenschaftl Verlag, Berlin

Brauns A (1954b) Puppen terricoler Dipterenlarven. Musterschmidt, Wissenschaftl Verlag, Berlin

Brauns A (1954c) Die Sukzession der Dipterenlarven bei der Stockhumifizierung. Z Morph Ökol Tiere 43:313–320

Brauns A (1955) Die terricolen Dipterenlarven im Verknüpfungsgefüge der Waldbiozoenose. Bonner Zool Beitr 6:223–231

Brundin L (1956) Zur Systematik der Orthocladiinae (Dipt., Chironomidae). Inst Freshwater Res Drottingholm Report 37:5–185

Bürgis H (1982) Die Schneefliege *Chionea*, ein echtes Schneeinsekt. Mikrokosmos 71:40–44

Caspers N (1980) Zur Larvalentwicklung und Produktionsökologie von *Tipula maxima* Poda (Diptera, Nematocera, Tipulidae). Arch Hydrobiol Suppl 58:273–309

Dunger W (1983) Tiere im Boden. Die Neue Brehm-Bücherei 327 (3. Aufl), A. Ziemsen Verlag, Wittenberg-Lutherstadt

Erber D (1972) Einige neue Fundorte für *Chionea lutescens* (Dipt., Tipulidae) in Hessen. Ent Z Frankfurt 82: 169–175

Feldmann R, Rehage HO (1973) Westfälische Nachweise des Winterhaftes (*Boreus westwoodi*) und der Schneefliege (*Chionea lutescens*). Natur und Heimat, Münster 33:47–50

Hågvar S (1971) Field observations on the ecology of snow insect, *Chionea araneoides* Dalm. (Dipt., Tipulidae). Norske Ent Tidskr 18:33–37

Hennig W (1948–1952) Die Larvenformen der Dipteren I–III. Berlin

Hennig W (1973) Diptera (Zweiflügler). Handb Zool Berlin 4 (2) 2/31:1–200

Hofsvang T (1973) Energy flow in *Tipula excisa* Schum. (Diptera, Tipulidae) in a high mountain area, Finse, South Norway. Norwegian J Zool 21:7–16

Hövemeyer K (1984) Die Dipterengemeinschaft eines Buchenwaldes auf Kalkgestein: Produktion an Imagines, Abundanz und räumliche Verteilung insbesondere der Larven. Pedobiologia 26:1–15

McAlpine JF, Peterson BV, Shewell GE, Teskey HJ, Vockeroth JR, Wood DM (1981) Manual of the Nearctic Diptera. Research Branch Agriculture Canada, Monograph No 27, Ottawa

Musso JJ (1983) Nutritive and ecological requirements of robber flies (Diptera: Brachycera: Asilidae). Entomol Gener 9:35–50

Oliver DR (1971) Life history of Chironomidae. Ann Rev Ent Stanford 16:211–230

Perel TS, Karpachevsky LO, Yegorova EV (1971) The role of Tipulidae (Diptera) larvae in decomposition of forest litter-fall. Pedobiologia 11:66–70

Priesner E (1961) Nahrungswahl und Nahrungsverbreitung bei der Larve von *Tipula maxima*. Pedobiologia 1:25–37

Séguy E (1950) Biologie des Diptères. Encycl Ent Paris 26:1–609

Strenzke K (1950) Systematik, Morphologie und Ökologie der terrestrischen Chironomiden. Arch Hydrobiol Stuttgart, Suppl. 18:207–414

Strenzke K (1959) Lebensformen und Phylogenese der terrestrischen Chironomiden. Proc XV Int Congr Zool, pp 351–354

Strübing H (1958) Schneeinsekten. Die Neue Brehm-Bücherei 220, A. Ziemsen Verlag, Wittenberg-Lutherstadt

Theowald B (1967) Familie Tipulidae (Diptera, Nematocera) Larven und Puppen. Bestimmungsbücher zur Bodenfauna Europas 7:1–100, Akademie-Verlag, Berlin

Thienemann A (1954) Chironomus. Leben, Verbreitung und wirtschaftliche Bedeutung der Chironomiden. Die Binnengewässer, Stuttgart 20:1–834

Volz P (1983) Zur Populationsökologie der mitteleuropäischen Walddipteren. Carolinea 41:105–126

Zinkler D (1980) Ökophysiologische Anpassung des Sauerstoffverbrauchs bodenlebender Tipulidenlarven. Verh Dtsch Zool Ges 1980, p 320

Zinkler D (1983) Ecophysiological adaptations of litter dwelling Collembola and tipulid larvae. In: Lebrun, P et al: New trends in soil biology, pp 335–343, Dieu-Brichart, Louvain-La-Neuve (Belgium)

Index of Scientific Names

Subject Index